LINEAR GROUPS

WITH AN EXPOSITION OF THE GALOIS FIELD THEORY

LEONARD EUGENE DICKSON

WITH AN INTRODUCTION BY
WILHELM MAGNUS
NEW YORK UNIVERSITY

DOVER PUBLICATIONS, INC.
Mineola, New York

Copyright

Copyright © 1958 by Dover Publications, Inc.
All rights reserved.

Bibliographical Note

This Dover edition, first published in 2003, is an unabridged republication of the 1958 Dover reprint of the first edition, originally published in 1901 by B. G. Teubner, Leipzig. A new introduction by Wilhelm Magnus of New York University was added to the 1958 Dover edition.

Library of Congress Cataloging-in-Publication Data

Dickson Leonard E. (Leonard Eugene), 1874–
  Linear groups : with an exposition of the Galois field theory / Leonard Eugene Dickson ; with an introduction by Wilhelm Magnus.
    p. cm. — (Dover phoenix editions)
  Originally published: Leipzig : B.G. Teubner, 1901.
  Includes index.
  ISBN 0-486-49548-5
  1. Group theory. 2. Galois theory. I. Title. II. Series.

QA174.2.D535 2003
512'.2—dc21

2003053258

Manufactured in the United States of America
Dover Publications, Inc., 31 East 2nd Street, Mineola, N.Y. 11501

Introduction to Dover Edition

L. E. Dickson's book on linear groups, with an exposition of the Galois Field theory, is a milestone in the development of modern algebra. We shall attempt to describe briefly its exact significance.

The natural starting point for almost any discussion of the history of algebra lies, of course, in the work of Evariste Galois [1811-1832] who established that "group theory," to quote Camille Jordan, "is the metaphysics of the theory of equations." A basic concept of Galois's theory was the invariant subgroup or normal divisor of a given group, which in turn, involved the concept of a simple group. The importance of this may be seen from the Jordan-Hoelder Theorem, which shows that the simple groups of finite order are the elements which when compounded give all groups of finite order.

It is therefore a fundamental problem in the theory of groups of finite order to find all the simple groups, especially the non-trivial simple groups whose order is not a prime number. The most important tool available for the investigation of this problem is based on another discovery of Galois's, the existence of finite fields. Galois himself used these finite fields (or as they were called later, Galois Fields) for the construction of groups which are not defined as groups of permutations.

Dr. Dickson's book is especially useful in this area since it contains the most extensive and thorough presentation of the theory of Galois Fields available in the literature. Although many of his proofs would be shortened today, and although the same subject would be expounded differently now, the first five chapters of Dickson's book are still very readable, and contain an enormous wealth of both examples and theorems. Only one important result referring to the theory of Galois Fields is missing; this is Maclagan Wedderburn's theorem[1], according to which multiplication

---

[1] "A Theorem on Finite Algebras," in *Transactions of the American Mathematical Society*, VI (1905), pp. 349-352. For a very brief proof of his result cf. E, Witt, "Ueber die Kommutativitaet endlicher Schiefkoerper," in *Abh. Math. Sem. Univ. Hamburg*, VIII (1930), p. 413.

vi INTRODUCTION TO DOVER EDITION

in a finite field is always commutative. This, however, was not available when Dickson prepared his book.

The second portion of *Linear Groups* ends with a survey of the known simple groups discovered by Mathieu, and the system of alternating groups of a degree exceeding four, all of these groups are contained in eight infinite systems which arise naturally from the preceding discussion of linear groups with coefficients in a Galois Field. Here Dickson based his work upon the discovery of predecessors, for the idea that simple groups could be discovered by studying linear groups had certainly been stimulated by the work of Lie, Cartan, Killing, and Engel, who had discovered and discussed all simple continuous groups of a finite number of parameters. In two later papers[2], indeed, Dickson refers to the work of Lie when describing the discovery of a new system of finite simple groups of composite order which are closely related to the exceptional simple continuous group of fourteen parameters.

It is a remarkable fact that after these two papers of Dickson's appeared no new simple groups of finite order were discovered for half a century. It was not until 1955 that a paper by C. Chevalley[3] appeared presenting a new infinite system of finite simple groups. In this same paper Chevalley also announced that he had been able to obtain simple linear groups in a Galois Field which are the analogue of the exceptional simple continuous groups not considered by Dickson.

It is hardly necessary to mention, of course, that not all of the results in Dickson's book are his own. Dickson mentions the most noteworthy work by other authors in his preface, placing special emphasis on the work of Camille Jordan, whose *Traité des Substitutions* [1870] established group theory as a new independent branch of mathematics. It is however noteworthy that for many years Dickson's book was the final word on the subject of linear groups in a Galois Field.

---

[2]"Theory of Linear Groups in an Arbitrary Field," in *Trans. Amer. Soc.*, II (1901), specifically, pp. 383-91; and "A New System of Simple Groups," in *Math. Ann.* LX (1905), pp. 137-50.

[3]"Sur certains groups simples," in *Tohoku Math. J.* (2), VII (1955), pp. 14-66.

INTRODUCTION TO DOVER EDITION   vii

Dickson's book, of course, was followed by a long series of papers by many authors, including Dickson himself, and several proofs were simplified and streamlined. But a new systematic review of the various systems of finite simple groups and the relations between them did not appear until J. Dieudonné's monograph *Sur les groups classiques* [1948][4]. In a later paper[5] Dieudonné settled one of the fundamental questions which Dickson had left unanswered by showing that Dickson's list of isomorphisms between the simple groups discussed in his book is complete. Finally, in 1955, E. Artin, in two astonishingly short papers[6], demonstrated that the whole table for the orders of finite simple groups and the isomorphisms between them can be derived systematically, with discussion of only a very few separate cases.

In spite of this recent work by Dieudonné and Artin, Dickson's book is still important as a source of information which cannot be obtained elsewhere without great difficulty. It contains an incredible wealth of detail about the groups he studied — groups which are interesting as individuals besides as members of an important system.

Let us mention only a few of those groups which are connected with geometrical problems. The Galois group of the equation for the twenty-seven straight lines on a cubic surface is a simple group of order 25920. It also appears in other problems, and Dickson discusses it carefully in this book and in a later paper . A second of these groups, the Galois group of the equation for the twenty-eight bitangents to a quartic without double points turns out to be the senary abelian (symplectic) groups mod. 2, and is therefore a special case of a class of groups in-

---

[4]Actualities Sci. Ind. No. 1040, Hermann, Paris, 1948.

[5]"On the Automorphisms of the Classical Groups," in *Mem. Amer. Math. Soc.* No. 2 (1951).

[6]"The Orders of the Linear Groups," and "The Orders of the Classical Simple Groups," both in *Comm. Pure Appl. Math.* VIII (1955), pp. 355-66 and 455-72.

[7]"Determination of All Subgroups of the Known Simple Group of Order 25920," *Proc. Amer. Math. Soc.*, V (1905), pp. 126-66.

vestigated in this book[8]. A year later, Dickson also dealt with certain generalizations of the problem of the bitangents[9], while the discovery of finite projective geometries by O. Veblen and W. M. Bussey[10] shows abundantly how important the linear groups in a Galois Field are for problems arising in the foundations of geometry.

The terminology in this book is still up-to-date. Such minor peculiarities as may exist are not likely to disturb the present-day reader. The most important change which has been made in the text has already been indicated; the replacement of the ambiguous term" abelian" by the more precise term"symplectic."

Dickson's book is now easily within the range of graduate students. This was not true at the time it first appeared, a fact which may be taken as evidence for the book's lasting influence upon the development of algebra.

It may be in order to mention at this point that Dickson's work on linear groups is only a fraction of his accomplishment. He did valuable work in number theory, and his monumental three valumes, *History of Theory of Numbers,* may be considered a very unselfish service rendered by him to all mathematicians working in this field. His contributions to the theory of algebras, in particular his book* *Algebras and their Arithmetics* have done much to stimulate research in this discipline. The present new edition of his first book might also serve as a reminder of the total contribution made by this eminent American mathematician.

> Wilhelm Magnus
> New York University
> New York.

---

[8]For the not purely group-theoretical aspects of the theory of the twenty-seven straight lines and the twenty-eight bitangents, see R. Fricke, *Lehrbuch der Algebra* (Vol. 2, Brunswick, 1926).

[9]"Groups of Steiner in Problems of Contact," *Proc. Am. Math. Soc.* III (1902), pp. 38-45, 377-382.

[10]In *Trans. Amer. Math. Soc.* VII (1906), pp. 241-59.

*Chicago, 1923

PREFACE.

Since the appearance in 1870 of the great work of Camille Jordan on substitutions and their applications, there have been many important additions to the theory of finite groups. The books of Netto, Weber and Burnside have brought up to date the theory of abstract and substitution groups. On the analytic side, the theory of linear groups has received much attention in view of their frequent occurrence in mathematical problems both of theory and of application. The theory of collineation groups will be treated in a forthcoming volume by Loewy. There remains the subject of linear groups in a finite field (including linear congruence groups) having immediate application in many problems of geometry and function-theory and furnishing a natural method for the investigation of extensive classes of important groups. The present volume is intended as an introduction to this subject. While the exposition is restricted to groups in a finite field (endliche Körper), the method of investigation is applicable to groups in an infinite field; corresponding theorems for continuous and collineation groups may often be enunciated without modification of the text.

The earlier chapters of the text are devoted to an elementary exposition of the theory of Galois Fields chiefly in their abstract form. The conception of an abstract field is introduced by means of the simplest example, that of the classes of residues with respect to a prime modulus. For any prime number p and positive integer n, there exists one and but one Galois Field of order p^n. In view of the theorem of Moore that every finite field may be represented as a Galois Field, our investigations acquire complete generality when we take as basis the general Galois Field. It was found to be impracticable to attempt to indicate the sources of the individual theorems and conceptions of the theory. Aside from the independent discovery of theorems by different writers and a general lack of reference to earlier papers, the later writers have given wide generalizations of the results of earlier investigators. It will suffice to give the following list of references on Galois Fields and higher irreducible congruences:

Galois, "Sur la théorie des nombres", *Bulletin des sciences mathématiques* de M. Férussac, 1830; *Journ. de mathématiques*, 1846.
Schönemann, *Crelle*, vol. 31 (1846), pp. 269—325.

Dedekind, *Crelle*, vol. 54 (1857), pp. 1—26.
Serret, *Journ. de math.*, 1873, p. 301, p. 437; Algèbre supérieure.
Jordan, Traité des substitutions, pp. 14—18, pp. 156—161.
Pellet, *Comptes Rendus*, vol. 70, p. 328, vol. 86, p. 1071, vol. 90, p. 1339, vol. 93, p. 1065; *Bull. Soc. Math. de France*, vol. 17, p. 156.
Moore, *Bull. Amer. Math. Soc.*, Dec., 1893; Congress Mathematical Papers.
Dickson, *Bull. Amer. Math. Soc.*, vol. 3, pp. 381—389; vol. 6, pp. 203—204. *Annals of Math.*, vol. 11, pp. 65—120; Chicago Univ. *Record*, 1896, p. 318.
Borel et Drach, Théorie des nombres et algèbre supérieure, 1895.

The second part of the book is intended to give an elementary exposition of the more important results concerning linear groups in a Galois Field. The linear groups investigated by Galois, Jordan and Serret were defined for the field of integers taken modulo p; the general Galois Field enters only incidentally in their investigations. The linear fractional group in a general Galois Field was partially investigated by Mathieu, and exhaustively by Moore, Burnside and Wiman. The work of Moore first emphasized the importance of employing in group problems the general Galois Field in place of the special field of integers, the results being almost as simple and the investigations no more complicated. In this way the systems of linear groups studied by Jordan have all be generalized by the author and in the investigation of new systems the Galois Field has been employed ab initio.

The method of presentation employed in the text often differs greatly from that of the original papers; the new proofs are believed to be much simpler than the old. For example, the structure of all linear homogeneous groups on six or fewer indices which are defined by a quadratic invariant is determined by setting up their isomorphism with groups of known structure. Then the structure of the corresponding groups on m indices, $m > 6$, follows without the difficult calculations of the published investigations. In view of the importance thus placed upon the isomorphisms holding between various linear groups, the theory of the compounds of a linear group has been developed at length and applied to the question of isomorphisms. Again, it was found practicable to treat together the two (generalized) hypoabelian groups. The identity from the group standpoint of the problem of the trisection of the periods of a hyperelliptic function of four periods and the problem of the determination of the 27 straight lines on a general cubic surface is developed in Chapter XIV by an analysis involving far less calculation than the proof by Jordan.

Chicago, November, 1900.

TABLE OF CONTENTS.

FIRST PART.
INTRODUCTION TO THE GALOIS FIELD THEORY.

CHAPTER I.
Definition and properties of finite fields.

Section		Page
1—3.	Classes of residues with respect to a prime modulus	3—4
4.	Fermat's theorem	4
5.	Definition of a field	5
6—7.	Definition of a Galois Field	6—8
8—10.	Order of a finite field is a power of a prime	9—10
11—17.	Period of a mark of a field; primitive roots	11—12
18.	Every finite field may be represented as a Galois Field	13—14

CHAPTER II.
Proof of the existence of the $GF[p^m]$ for every prime p and integer m.

19—22.	Decomposition of functions belonging to the $GF[p^n]$	14—15
23—25.	Irreducible factors of $x^{p^{nm}} - x$	15—16
26—27.	Expression for product of all irreducible quantics of degree m in the $GF[p^n]$. Their number	17—18
	Exercices	19

CHAPTER III.
Classification and determination of irreducible quantics.

29—30.	Exponent to which an irreducible quantic belongs	19—20
31—32.	Roots of an irreducible quantic; their exponents	21
33.	When $x^t + x^{t-1} + \cdots + x + 1$ is irreducible	21
34—38.	Determination of irreducible quantics in the $GF[p^n]$ whose degree contains no prime factor other than those of p^n-1	22—27
39—46.	Irreducible quantics of degree p^s in the $GF[p^n]$	28—32
47—49.	Miscellaneous theorems on irreducible quantics	32—34
50—58.	Primitive roots and primitive irreducible quantics	35—42
59.	Exercises	42—44
60.	Table of primitive irreducible quantics	44

xii TABLE OF CONTENTS.

CHAPTER IV.
Miscellaneous properties of Galois Fields.

Section		Page
61—62.	Squares and not-squares	44
63.	Number of mth powers in a field; extraction of roots . .	45
64—67.	Number of sets of solutions of certain quadratics . . .	46—48
68—71.	Additive-groups and their multiplier Galois Fields . . .	49—51
72.	Condition for linear independence of marks with respect to an included field	52
73.	Conjugacy of marks with respect to an included field . .	52
74.	Newton's identities for sums of powers of the roots of an equation belonging to a Galois Field	53—54

CHAPTER V.
Analytic representation of substitutions on the marks of a Galois Field.

76—78.	Definitions. Representation of a given substitution . . .	54—55
79—83.	Special functions suitable to represent substitutions . . .	56—59
84.	Necessary and sufficient conditions for a substitution quantic	59—60
85—89.	Applications of preceding theorem. Reduced form . . .	61—63
90.	Table of all substitution quantics of degree < 6	63—64
91—94.	Betti-Mathieu Group. Certain of its subgroups . . .	64—68
95.	Identity of Betti-Mathieu Group in the $GF[p^{nm}]$ with Jordan's linear homogeneous group in the $GF[p^n]$ on m indices	69—70
96.	Exercises	70—71

SECOND PART.
THEORY OF LINEAR GROUPS IN A GALOIS FIELD.

CHAPTER I.
General linear homogeneous group.

97— 98.	Two definitions of the group	75—77
99—100.	Order and generators	77—79
101—102.	Transformation of indices. Invariance of characteristic determinant	80—81
103—107.	Factors of composition of the linear homogeneous group .	81—86
108—109.	Linear fractional group. Isomorphic permutation group .	87—88

CHAPTER II.
The Abelian linear group.

110—112.	Conditions for Abelian substitutions. Inverse substitution .	89— 91
114—115.	Generators and order of Abelian group	92— 94
116—119.	Factors of composition of the Abelian linear group . . .	94—100
120—121.	Conjugacy of operators of period two of the Abelian group	100—105
122—123.	Operators of period two in the quotient-group $A(2m, p^n)$	105—109

TABLE OF CONTENTS. xiii

CHAPTER III.
A generalization of the Abelian linear group.

Section		Page
124—125.	Definition of the substitutions; their inverse	110—111
126—128.	Structure of the group	111—114

CHAPTER IV.
The hyperabelian group.

129—130.	Conditions on its substitutions; their inverse	115—116
131.	Largest subgroup containing the Abelian group self-conjugately	117—120
132—133.	Corresponding theorems for their quotient-groups	120—121
134—136.	Binary linear homogeneous subgroups of the quaternary hyperabelian group. Application to their quotient-groups	122—125
137.	Identity of binary hyperabelian and binary linear group	125

CHAPTER V.
The hyperorthogonal and related linear groups.

139—142.	Definition. Structure in the general case	126—131
143—151.	Order, generators and structure in the hyperorthogonal case	131—144

CHAPTER VI.
The compounds of a linear homogeneous group.

153.	Isomorphism of linear group with its compounds	145—146
154.	Multiplicity of isomorphism for general linear group	146—147
156.	Pfaffian invariant of the second compound	147—148
157—158.	Group induced upon certain Pfaffians by the second compound	148—150
159—162.	The second compound of the general and special Abelian groups	151—153
163—165.	The second compound of the quaternary linear group	153—155

CHAPTER VII.
Linear homogeneous group in the $GF[p^n]$, $p > 2$, defined by a quadratic invariant.

166—169.	Canonical forms of the quadratic invariant	156—158
170—171.	Orthogonal substitutions; the first and second orthogonal groups	159
172—180.	Order and generators of the orthogonal groups	160—169
178.	The ternary first orthogonal group and the linear fractional group	164
181—198.	The structure of the orthogonal groups	169—197
186—188.	Senary orthogonal groups isomorphic with quaternary linear groups	172—179
189.	Quinary orthogonal group isomorphic with quaternary Abelian group	179—182
190.	Senary orthogonal groups isomorphic with hyperabelian groups	183—186
195—198.	Quaternary orthogonal and linear fractional groups	191—196

TABLE OF CONTENTS.

CHAPTER VIII.

Linear homogeneous group in the $GF[2^n]$ defined by a quadratic invariant.

Section		Page
199.	Canonical forms of the quadratic invariant	197—199
200.	Structure of group on an odd number of indices	199—200
201—204.	Definition, order and generators of the hypoabelian groups	200—206
205.	Invariant defining the subgroup J_λ	206—208
206—208.	Isomorphism of senary group J_λ with certain quaternary groups	208—211
209.	Simplicity of J_λ on more than six indices	212—216
210.	Miscellaneous exercises on chapters I—VIII	216—218

CHAPTER IX.

Linear groups with certain invariants of degree $q > 2$.

211—213.	Definition, generators and structure of group	218—221

CHAPTER X.

Canonical form and classification of linear substitutions.

214—216.	Canonical form of linear homogeneous substitutions	221—229
217—220.	Substitutions commutative with a given linear substitution	229—236
221—223.	Distribution of the substitutions of the general ternary and quaternary linear groups into sets of conjugate substitutions	236—241

CHAPTER XI.

Operators and cyclic subgroups of the simple group $LF(3, p^n)$.

224—225.	Notations. The seven distinct canonical forms	242—244
226—237.	Conjugate operators and cyclic groups of each type	245—259
238.	$LF(3, 2^2)$ not isomorphic with the alternating group on 8 letters, each group being simple and of equal order	259—260

CHAPTER XII.

Subgroups of the linear fractional group $LF(2, p^n)$.

239.	Doubly transitive substitution group on $p^n + 1$ letters	260—261
240—244.	Commutative subgroups of order p^n; cyclic subgroups	261—265
245.	Concerning dihedron groups and their subgroups	265—266
246—248.	Subgroups of dihedron and four-group types	267—268
249—255.	Subgroups containing operators of period p	268—280
256.	Subgroups containing no operators of period p	280—282
257—259.	Subgroups of tetrahedral, octahedral and icosahedral types	282—285
260—261.	Summary of subgroups. Simplicity theorem	285—286
262.	Galois' theorem on the minimum index of a subgroup	286
263.	Lowest degree of isomorphic substitution group	287

TABLE OF CONTENTS.

CHAPTER XIII.
Auxiliary theorems on abstract groups. Abstract forms of various linear groups.

Section		Page
264—267.	Abstract groups isomorphic with the symmetric and alternating groups	287—290
268—269.	Quaternary linear group modulo 2 isomorphic with the alternating group on 8 letters	290—292
270—274.	Abstract form of quinary orthogonal group modulo 3	292—298
275—276.	Its isomorphism with a hyperabelian group	298—299
278—282.	Abstract group isomorphic with $LF(2, p^n)$	300—303

CHAPTER XIV.
Group of the equation for the 27 straight lines on a general surface of the third order.

283.	Notation for the configuration of the 27 lines	303—305
284—285.	Group of the equation. Isomorphism with linear groups	305—306
286.	Subgroups of indices 27, 36, 40, 45	306—307

CHAPTER XV.
Summary of the known systems of simple groups.

287.	The ten known infinite systems	307
288.	Isomorphisms between certain groups of the systems	308
289.	Two triply infinite systems of non-isomorphic simple groups of equal order	309
290.	Table of simple groups of orders less than a million	309—310

INDEX OF SUBJECTS . . . . 311—312

DEFINITION OF SYMBOLS.

$\Phi(t)$, 12, 19.
$A(2m, p^n)$, 100.
$FH(2m, 2^n)$, 216.
$FO(m, p^n)$, 191.
$GA(2m, p^n)$, 89.
$GLH(m, p^n)$, 76.
$G(m, q, p^n)$, 110.
$G_{m, p, s}$, 131.
$G_{M(s)}^{s+1}$, 261.
G_λ, 201.
$GF[p^n]$, 14.
$H(2m, p^{2n})$, 115.
$HA(2m, p^{2n})$, 120.

$HO(m, p^{2s})$, 138.
$IQ[m, p^n]$, 16.
J_λ, 206.
L_{s, p^n}, 191.
$LF(m, p^n)$, 87.
$NS(m, p^n)$, 191, Note 1).
$O_\mu(m, p^n)$, 159.
$PIQ[m, p^n]$, 21.
$SA(2m, p^n)$, 89.
$SH(2m, 2^n)$, 216.
$SLH(m, p^n)$, 82.
$SO(m, p^n)$, 191.
$SQ[k, p^n]$, 55.

FIRST PART.

INTRODUCTION TO THE GALOIS FIELD THEORY.

CHAPTER I.

DEFINITION AND PROPERTIES OF FINITE FIELDS.

1. If the difference of two integers t and r be divisible by a third integer p, then t and r are said to be *congruent modulo* p, or *according to the modulus* p. This property is expressed by the following notation due to Gauss:

$$t \equiv r \pmod{p}.$$

For example, $7 \equiv 1 \pmod{3}$, $7 \equiv 2 \pmod{5}$.

The totality of integers congruent modulo p with a given positive integer $r < p$ is given by the formula

$$lp + r \quad (l = 0, \pm 1, \pm 2, \ldots).$$

This totality, which will be designated C_r, is said to form *a class of residues modulo* p; it includes every integer which gives the residue r when divided by p. It follows that the p classes C_0, C_1, C_2, ..., C_{p-1} include every integer, positive or negative. They are therefore said to form *a complete system of classes of residues modulo* p.

Example. — The three classes C_0, C_1, C_2 form a complete system of classes of residues modulo 3; indeed, every integer falls under one of the three forms $3l$, $3l + 1$, $3l + 2$.

2. An instructive diagram is furnished by the regular polygon of p sides inscribed in a circle. Denote the vertices taken in positive order (counter-clockwise) by C_0, C_1, ..., C_{p-1}. Regarding C_0 to be the origin, we take as the plot of any given integer $\pm m$ that vertex which is obtained by counting off from the origin m of the divisions on the circle in the positive or the negative direction according to the sign of $\pm m$. All integers of the form $lp + r$ ($l = 0, \pm 1, \pm 2, \ldots$) are evidently plotted by the one point C_r, so that congruent integers give rise to the same point. The p classes of residues modulo p are represented unambiguously by the p vertices of the polygon.

3. From the numerical identities
$$(lp + r) \pm (tp + s) = (l \pm t)p + (r \pm s),$$
$$(lp + r)(tp + s) = (ltp + ls + rt)p + rs,$$
we obtain the following formulae for the addition, subtraction and multiplication of classes of residues:
$$C_r \pm C_s = C_{r \pm s}, \quad C_r \cdot C_s = C_{rs}.$$

If two given classes C_r and C_s, $C_s \neq C_0$, lead uniquely to a third class C_x such that $C_r = C_s C_x$, then C_x is said to be the quotient of C_r by C_s and the following notation employed
$$C_x = C_r / C_s.$$
The condition for the quotient is evidently identical with the condition that there exist a solution x of the equation

1) $$r = sx + mp.$$

In order that a solution x shall exist for r and s arbitrary integers such that s is not divisible by p, it is necessary and sufficient that p be a prime number. To prove the condition necessary, let $p = p_1 p_2$, where $p_1 > 1$, $p_2 > 1$. Then 1) can not always be satisfied; for example, when $s = p_1$ and r is not divisible by p_1. The condition that p be a prime is, moreover, a sufficient one by the corollary of § 4. Hence the division of classes of residues, the divisor being other than the class C_0, is *always* possible if, and only if, the modulus p be a prime number.

In particular, these remarks show that the classes of residues with respect to a prime modulus may be combined by the rational operations of algebra and that each result is itself one of the classes of residues. For example, let $p = 3$. Then

$$C_1 + C_2 = C_0, \ C_2 + C_2 = C_1, \ C_2 \cdot C_2 = C_1, \ C_2/C_1 = C_2, \ C_2/C_2 = C_1, \ C_1/C_2 = C_2.$$

4. Fermat's Theorem. — *If an integer a be not divisible by a prime number p, then $a^{p-1} \equiv 1 \pmod{p}$.*

Since the integers $a, 2a, 3a, \ldots, (p-1)a$ are all distinct modulo p, their residues must be identical, apart from their order, with the integers $1, 2, 3, \ldots, p-1$.

Forming the product of the integers in each set, we have
$$a^{p-1} \cdot 1 \cdot 2 \cdot 3 \ldots (p-1) \equiv 1 \cdot 2 \cdot 3 \ldots (p-1) \pmod{p}.$$

Corollary. — *If a be not divisible by the prime number p, there exists an unique solution of the congruence $ax \equiv b \pmod{p}$.*

Applying the theorem just proven, the solution is evidently
$$x \equiv a^{p-2} b \pmod{p}.$$

5. Definition of a field.

— A set of elements $u_1, u_2, \ldots, u_\sigma$, which may be combined by addition subject to the formal laws

$$u_i + u_j = u_j + u_i, \quad u_i + (u_j + u_k) = (u_i + u_j) + u_k,$$

such that the sum of any two elements is likewise an element of the set is called an *additive-field*. If two elements u_i and u_k are given, there may or may not exist a third element u_j in the set such that $u_i + u_j = u_k$. If existent, u_j is said to be determined by subtraction, $u_j \equiv u_k - u_i$. Assume[1]) that subtraction is *always* possible in the given additive-field. The set will contain the differences $u_1 - u_1, u_2 - u_2, \ldots, u_\sigma - u_\sigma$. Each has the additive property of zero, since $u_j + (u_i - u_i) = u_j$. From the latter, $u_i - u_i \equiv u_j - u_j$ follows by the definition of subtraction. Hence the above differences all have a common value u. There exists no new zero element u', since $u_j + u' = u_j$ requires $u' = u_j - u_j \equiv u$. Two elements are called equal or distinct according as their difference is or is not the zero element u. Select from the original set all the distinct elements and denote them by $u_0, u_1, u_2, \ldots, u_{s-1}$, where u_0 denotes the unique zero element.

Assume next that the s elements $u_0, u_1, \ldots, u_{s-1}$ may be combined by multiplication subject to the formal laws

$$u_i u_j = u_j u_i, \quad u_i(u_j u_k) = (u_i u_j) u_k, \quad u_i(u_j \pm u_k) = u_i u_j \pm u_i u_k,$$

such that the product of any two elements is itself an element of the set. Then the element u_0 will have the multiplicative properties of zero, viz., for any element u_j of the set,

$$u_j u_0 = u_0 u_j = u_0.$$

Indeed, since every product $u_j u_i$ is an element of the set,

$$u_j(u_i - u_i) = u_j u_i - u_j u_i \equiv u_0, \quad (u_i - u_i) u_j = u_0.$$

Given two elements u_i and u_k, $u_i \neq u_0$, there may or may not exist a third element u_j in the set such that $u_i u_j = u_k$. If existent, u_j is said to be determined by division, $u_j = u_k / u_i$. Assume[2]) lastly that division is *always* possible in the set, and in a single way, the divisor being other than the zero element. A set of s distinct elements satisfying the above four conditions is said to form *a field of order s*.

To obtain a field of *finite* order, the assumption concerning division may be replaced by the postulate that a product of two

---

[1]) In the additive-field of all positive integers, not every difference of two elements belongs to the field.

[2]) The set of all positive and negative integers satisfies the assumptions as to addition, subtraction and multiplication, but not that for division.

elements shall be the zero element u_0 only when one of the factors is u_0. Under the latter hypothesis, the series of products

$$u_0 u_i, \quad u_1 u_i, \quad u_2 u_i, \ldots, u_{s-1} u_i \quad (u_i \neq u_0)$$

are all distinct and therefore (their number s being finite) are identical in some order with the series $u_0, u_1, u_2, \ldots, u_{s-1}$. Hence if u_j be any element of the set, the equation

2) $\qquad x u_i = u_j \quad (u_i \neq u_0)$

is satisfied by one and but one element x of the given set. Hence division by any element except u_0 is always possible within the set and gives an unique result.

For a field of infinite order, the assumption that division is not possible in more than one way may be replaced by the above postulate that a product vanishes only when one factor vanishes. Indeed, if 2) be satisfied by two distinct values x_1 and x_2 of x, then $u_i(x_1 - x_2) = u_0$, whereas each factor differs from u_0.

After the above explanations, we make the formal definition:

A set of s distinct elements forms a field of order s if the elements can be combined by addition, subtraction, multiplication and division, the divisor not being the element zero (necessarily in the set), these operations being subject to the laws of elementary algebra, and if the resulting sum, difference, product or quotient be uniquely determined as an element of the set.[1])

A field may therefore be defined by the property that *the rational operations of algebra can be performed within the field*.

The results of § 3 may now be stated in the form: *The complete system of classes of residues modulo p forms a field if, and only if, p be a prime number*.

6. Definition of a *Galois Field*. — Let $P(x)$ be a rational integral function of degree n having integral coefficients not all divisible by a given integer p. If we divide an arbitrary integral function $F(x)$ having integral coefficients by the function $P(x)$, we obtain a quotient $Q(x)$ and a remainder which can be written in the form $f(x) + p \cdot q(x)$, where $f(x)$ is of the form

3) $\qquad f(x) \equiv a_0 + a_1 x + a_2 x^2 + \cdots + a_{n-1} x^{n-1}$,

each a_i belonging to the series $0, 1, 2, \ldots, p-1$. Then

4) $\qquad F(x) = f(x) + p \cdot q(x) + P(x) \cdot Q(x)$.

We say that $f(x)$ is the residue of $F(x)$ moduli p and $P(x)$ and write

4_1) $\qquad F(x) \equiv f(x) \quad [\operatorname{modd} p, P(x)]$.

---

[1]) Moore, Mathematical Papers, Chicago Congress of 1893, pp. 208—242; Bull. Amer. Math. Soc., December, 1893.

DEFINITION AND PROPERTIES OF FINITE FIELDS.

The totality of functions $F(x)$ obtained by giving to the polynomials $Q(x)$ and $q(x)$ in 4) all possible forms is said to constitute a class of residues; two functions are called congruent if, and only if, they belong to the same class of residues. From the form of 3) there are evidently p^n distinct classes.

Consider two integral functions having integral coefficients

$$F_i(x) = f_i(x) + p \cdot q_i(x) + P(x) \cdot Q_i(x) \quad [i = 1, 2].$$

It is evident that the class to which $F_1 \pm F_2$ or $F_1 F_2$ belongs depends merely upon the functions $f_1 \pm f_2$ or $f_1 f_2$ respectively, being independent of the functions q_i, Q_i. Hence classes of residues combine unambiguously under addition, subtraction and multiplication. In order that the division of an arbitrary class by any class C, not the class zero C_0, shall lead uniquely to a third class, it is necessary that the equation $C_i C = C_0$ shall require $C_i = C_0$. Evidently this will not be the case if p be composite, $p = p_1 p_2$, or if $P(x)$ be reducible modulo p, viz.,

$$P(x) = P_1(x) P_2(x) + p P_3(x)$$

where the $P_i(x)$ are integral functions having integral coefficients, the degrees of $P_1(x)$ and $P_2(x)$ being less than the degree of $P(x)$. Hence p must be prime and $P(x)$ irreducible modulo p.

Inversely, if p be prime and $P(x)$ irreducible modulo p, it follows from § 7 that to any class C_{F_1} other than the class C_0 there corresponds an unique class $C_{F'_1}$ such that $C_{F'_1} C_{F_1}$ is the class unity. Hence there exists the quotient class

$$\frac{C_{F_2}}{C_{F_1}} = \frac{C_{F_2} C_{F'_1}}{C_{F_1} C_{F'_1}} = C_{F_2} C_{F'_1} = C_{F_2 F'_1}.$$

The p^n classes of residues therefore form a field called *a Galois Field of order p^n*. Moreover, the p^n classes of residues moduli p and $P(x)$ form a field if, and only if, p be prime and $P(x)$ be irreducible modulo p.

As an example, let $p = 3$ and $P(x) = x^2 - x - 1$. The 3^2 residues are

$$0, \quad 1, \quad -1, \quad x, \quad x+1, \quad x-1, \quad -x, \quad -x+1, \quad -x-1.$$

The sum, difference or product of any two of these may evidently be reduced moduli 3 and $x^2 - x - 1$ to one of the nine residues. Moreover, the quotient of any one by any residue except 0 may be reduced to one of the set. For example,

$$\frac{1}{x} \equiv \frac{x-1}{x(x-1)} \equiv x - 1, \quad \frac{1}{x+1} \equiv \frac{x}{-x+1} \equiv \frac{-x^2}{x^2-x} \equiv -x^2 \equiv -x - 1.$$

The nine residues thus form a Galois Field of order 3^2.

CHAPTER I.

7. Theorem. — *If two integral functions $F(x)$ and $P(x)$ having integral coefficients admit of no common divisor containing x modulo p, p being prime, we can determine two integral functions $F'(x)$ and $P'(x)$ having integral coefficients such that*

$$F'(x) \cdot F(x) - P'(x) \cdot P(x) \equiv 1 \pmod{p}.$$

Applying § 4, we can set

$$F(x) \equiv a \cdot A(x), \quad P(x) \equiv b \cdot B(x) \pmod{p}$$

the coefficients of the highest power of x in $A(x)$ and $B(x)$ being unity and the remaining coefficients integers. We perform the usual process to determine the greatest common divisor of A and B, neglecting however, multiples of p. Each remainder is congruent modulo p to a product of an integer r and an integral function $R(x)$ with integral coefficients, that of the highest power of x being unity. Supposing for definiteness that the degree of A is not less than that of B, we obtain the congruences (mod p):

$$A \equiv BQ_1 + r_1 R_1$$
$$B \equiv R_1 Q_2 + r_2 R_2$$
$$R_1 \equiv R_2 Q_3 + r_3 R_3$$
$$\ldots \ldots \ldots$$
$$R_{m-2} \equiv R_{m-1} Q_m + r_m.$$

We derive at once the following congruences modulo p:

$$r_1 R_1 \equiv A - Q_1 B$$
$$r_1 r_2 R_2 \equiv - Q_2 A + (r_1 + Q_1 Q_2) B$$
$$r_1 r_2 r_3 R_3 \equiv (r_2 + Q_2 Q_3) A - (r_2 Q_1 + r_1 Q_3 + Q_1 Q_2 Q_3) B$$
$$\ldots \ldots \ldots \ldots \ldots \ldots \ldots \ldots$$
$$r_1 r_2 \ldots r_m \equiv MA - NB,$$

where M and N are integral functions of x having integral coefficients.

None of the integers $r_1 \ldots, r_m$ are divisible by p; for, A and B would then have a common divisor containing x. Hence, by § 4, there exists an integer r such that

$$r \cdot ab\, r_1 r_2 \ldots r_m \equiv 1 \pmod{p}.$$

From the last congruence in the above set, we therefore find

$$1 \equiv rab\,(MA - NB) \equiv F(x) \cdot rbM - P(x) \cdot raN \pmod{p}.$$

Corollary. — If $F(x) \not\equiv 0\,[\mathrm{modd}\ p, P(x)]$, p being prime and $P(x)$ irreducible modulo p, we can determine an integral function $F'(x)$ such that

$$F'(x) \cdot F(x) \equiv 1 \quad [\mathrm{modd}\ p, P(x)].$$

Note. — By an analogous use of the process for finding the greatest common divisor, we obtain the following theorem:

DEFINITION AND PROPERTIES OF FINITE FIELDS.

If two integers f and p be relatively prime, we can determine two integers f' and p' such that $f'f - p'p = 1$.

8. The proof of the existence of a function of degree n irreducible modulo p and hence of the existence of a Galois Field of order p^n, for every prime p and integer n, will be given in §§ 19—27. We will first prove that no other finite fields exist and that not more than one Galois Field of a given order p^n exists.

9. Consider an abstract field $F[s]$ composed of a finite number $s > 1$ of elements or *marks* $u_0, u_1, \ldots, u_{s-1}$. Having every difference $u_i - u_i$, the field contains a mark, denoted by $u_{(0)}$, which has the properties of zero viz., for every u_i,
$$u_i + u_{(0)} = u_{(0)}, \quad u_{(0)} u_i = u_i u_{(0)} = u_{(0)}.$$

Having every quotient
$$u_i / u_i \quad (u_i \neq u_{(0)}),$$
the field contains a mark $u_{(1)}$ having the properties of unity; viz., for every u_i,
$$u_i u_{(1)} = u_{(1)} u_i = u_i.$$

The field thus contains every integral mark
$$u_{(c)} = u_{(1)} + u_{(1)} + \cdots + u_{(1)} \quad (c \text{ terms}),$$
$$u_{(-c)} = u_{(0)} - u_{(c)}.$$

Since there exists only a finite number of marks in the $F[s]$, there must arise equalities in the series
$$\ldots, u_{(-2)}, u_{(-1)}, u_{(0)}, u_{(1)}, u_{(2)}, \ldots$$

If $u_{(r)} = u_{(s)}$, we have
$$u_{(0)} = u_{(r)} - u_{(s)} = u_{(r-s)}.$$

Denoting by p the least positive integer such that $u_{(p)} = u_{(0)}$, the p marks
$$u_{(0)}, u_{(1)}, u_{(2)}, \ldots, u_{(p-1)}$$
are all distinct, while
$$u_{(r)} = u_{(s)} \text{ if, and only if, } r \equiv s \pmod{p}.$$

This integer p is a prime number. For, if
$$p = p_1 p_2, \quad p > p_1,$$
we have, by hypothesis, $u_{(p_1)} \neq u_{(0)}$. Hence, from
$$u_{(p_1)} \cdot u_{(p_2)} = u_{(p)} = u_{(0)},$$
we derive $u_{(p_2)} = u_{(0)}$ and hence $p_2 \geq p$. Hence the integral marks of the $F[s]$ form a field $F[p]$ which is the abstract form of the field of the classes of residues with respect to a prime modulus p. When there is no ambiguity, we denote by c the integral mark $u_{(c)}$.

10 CHAPTER I.

10. Theorem. — *The order of $F[s]$ is a power of p.*

If u_1 be a fixed mark $\neq u_0$ of the $F[s]$, the products

$$c_1 u_1 \qquad (c_1 = 0, 1, \ldots, p-1)$$

give p distinct marks of the field. If $s > p$, there exists a mark u_2 not of the form $c_1 u_1$. Then

$$c_1 u_1 + c_2 u_2 \qquad (c_1, c_2 = 0, 1, \ldots, p-1)$$

gives p^2 distinct marks. If $s > p^2$, there exists a mark u_3 not of the form $c_1 u_1 + c_2 u_2$, so that

$$c_1 u_1 + c_2 u_2 + c_3 u_3 \qquad (c_1, c_2, c_3 = 0, 1, \ldots, p-1)$$

gives p^3 distinct marks. Proceeding similarly, we must ultimately obtain all the marks of the $F[s]$ expressed by the formula

$$c_1 u_1 + c_2 u_2 + \cdots + c_n u_n \quad (\text{every } c_i = 0, 1, \ldots, p-1),$$

not two of these p^n expressions being equal. Hence $s = p^n$.

Definition. — A set of marks $u_1, u_2, \ldots, u_k$ are said to be linearly independent with respect to the included field $F[p]$, if the equation

$$c_1 u_1 + c_2 u_2 + \cdots + c_k u_k = 0,$$

where the c's are marks of the $F[p]$, can be satisfied only when every $c_i = 0$.

Definition. — A rational integral function of any number of indeterminates $X_1, X_2, \ldots, X_k$ is said *to belong to a field* if its coefficients are marks of that field. It is *irreducible in the field* if it is not identically the product of two or more functions belonging to the field, each function involving some of the indeterminates X_i. An equation between functions belonging to a field is said to belong to the field.

11. Theorem. — *Any mark u of the $F[s = p^n]$ satisfies an equation of degree $k \leq n$,*

$$\sum_{i=0}^{k} c_i X^i = 0, \quad (c_k \neq 0)$$

belonging to and irreducible in the $F[p]$.

Indeed, a linear relation with coefficients belonging to the $F[p]$ certainly holds between any $n+1$ marks of the $F[p^n]$ and hence between

$$u^0, \quad u^1, \quad u^2, \ldots, u^n.$$

If such a relation holds between the first $k+1$ of these powers of u, u satisfies an equation of degree k.

DEFINITION AND PROPERTIES OF FINITE FIELDS.   11

12. Let u be any mark $\neq 0$ of the $F[s=p^n]$. The marks
$$u^t \qquad (t=0,\pm 1,\pm 2,\ldots)$$
belonging to our finite field are not all distinct. From $u^r = u^s$, we derive $u^{r-s} = 1$. The least positive integer e for which $u^e = 1$ is called the *period* of the mark u, while u is said *to belong to the exponent* e. The marks $1, u, u^2, \ldots, u^{e-1}$ are all distinct.

We may form a rectangular array of the marks $\neq 0$ of the field as follows:
$$\begin{array}{cccc} 1 & u & u^2 & \ldots u^{e-1} \\ u_1 & uu_1 & u^2u_1 & \ldots u^{e-1}u_1 \\ u_2 & uu_2 & u^2u_2 & \ldots u^{e-1}u_2 \\ \ldots & \ldots & \ldots & \ldots \end{array}$$
where u_1 is any mark $\neq 0$ not occurring in the first line, u_2 any mark $\neq 0$ not in the first or second lines, etc. Evidently the marks in any line are different from each other and from those in the preceding lines. Since each new mark u_i gives rise to a set of e new marks, the number $p^n - 1$ of the marks $\neq 0$ in the $F[p^n]$ is a multiple of e.

Theorem. — *The period of any mark $\neq 0$ of the $F[p^n]$ is a divisor of $p^n - 1$.*

13. Raising u^e to the power $(p^n - 1)/e$, we have
$$u^{p^n-1} = 1, \quad \text{if } u \neq 0.$$
We have thus the following generalization of Fermat's Theorem:
Every mark of the $F[p^n]$ satisfies the equation
$$X^{p^n} - X = 0.$$
We have therefore the following decomposition in the $F[p^n]$:
$$X^{p^n} - X = \prod_{i=0}^{p^n-1} (X - u_i),$$
u_i running over the p^n marks of the $F[p^n]$.

14. Theorem. — *If two marks u_1, u_2 belong respectively to exponents $e_1 e_2$ which are relatively prime, their product $u_1 u_2$ belongs to the exponent $e_1 e_2$ and the $e_1 e_2$ marks*
$$u_1^{d_1} u_2^{d_2} \qquad \begin{pmatrix} d_1 = 0, 1, \ldots, e_1 - 1 \\ d_2 = 0, 1, \ldots, e_2 - 1 \end{pmatrix}$$
are all distinct.

If $u_1 u_2$ has the period t, we have
$$(u_1 u_2)^{t e_1} = u_2{}^{t e_1} = 1,$$
whence t is divisible by e_2; similarly, t is divisible by e_1. But
$$(u_1 u_2)^{e_1 e_2} = 1.$$
Hence $t = e_1 e_2$.

15. We prove as in algebra the theorem:

An equation of degree k belonging to a field has in the field at most k roots, unless it be an identity, when every mark of the field is a root.

16. Theorem. — *For every divisor d of $s-1$, the equation*
$$X^d - 1 = 0$$
has in the $F[s = p^n]$ exactly d roots.

Setting $s - 1 = dq$, we have the identity
$$X^{s-1} - 1 \equiv (X^d - 1)(X^{d(q-1)} + X^{d(q-2)} + \cdots + X^d + 1).$$

Since the last factor belongs to the $F[s]$ and does not vanish for the mark zero, it vanishes for at most $d(q-1)$ marks of the field. But the left side of the identity vanishes for $s-1$ marks of the field. Hence the factor $X^d - 1$ must vanish for at least d marks.

17. Decompose $p^n - 1$ into its prime factors,
$$p^n - 1 = p_1^{h_1} p_2^{h_2} \cdots p_k^{h_k}.$$
For each integer i of the series $1, 2, \ldots, k$, the equation
$$X^{p_i^{h_i}} - 1 = 0$$
has by § 16 exactly $p_i^{h_i}$ roots belonging to the $F[s = p^n]$. Of these roots $p_i^{h_i - 1}$ are also roots of the equation
$$X^{p_i^{h_i - 1}} - 1 = 0$$
and thus belong to exponents less than $p_i^{h_i}$. The remaining roots u_i, in number
$$p_i^{h_i} - p_i^{h_i - 1} = p_i^{h_i}\left(1 - \frac{1}{p_i}\right),$$
belong to the exponent $p_i^{h_i}$ itself. Any product of the form
$$w = u_1 u_2 \ldots u_k$$
will by § 14 belong to the exponent $p^n - 1$. Forming in every possible way the product w, we obtain[1])

---

1) This number equals $\Phi(p^n - 1)$, where $\Phi(t)$ denotes the number of integers less than and relatively prime to the positive integer t. See Dirichlet, Vorlesungen über Zahlentheorie, § 11.

DEFINITION AND PROPERTIES OF FINITE FIELDS.

$$\prod_{i=1}^{k} p_i^{h_i}\left(1 - \frac{1}{p_i}\right)$$

such marks. Each mark w belonging to the exponent $s-1$ is called a *primitive root* of the equation

$$X^{s-1} - 1 = 0$$

and also a primitive root of the $F[s]$. Since the powers $w^1, w^2, \ldots, w^{s-1}$ are all distinct, we may state the theorem:

The $p^n - 1$ *marks* $\neq 0$ *of the* $F[s = p^n]$ *are the* $p^n - 1$ *successive powers of a primitive root of that field.*

Corollary. If d be any divisor of $p^n - 1$, the mark $w^{(p^n-1)/d}$ belongs to the exponent d.

18. We may now recognize in our $F[s]$ the abstract form of a Galois Field of order $s = p^n$. Indeed, by § 11, the primitive root w satisfies an equation of degree $k \leq n$.

$$W_k(x) = 0,$$

belonging to and irreducible in the $F[p]$. Every mark $\neq 0$ of the $F[s]$, being a power of w, can be reduced by the identity $W_k(w) \equiv 0$ to the form

$$c_1 w^{k-1} + c_2 w^{k-2} + \cdots + c_{k-1} w + c_k,$$

where the c's belong to the $F[p]$. The mark zero evidently falls under this form. Since, inversely, every one of these p^k expressions is a mark of the $F[s]$, we must have $k = n$. Hence every mark of the $F[s = p^n]$ represents a class of residues moduli p, a prime, and $W_n(x)$, a function with integral coefficients irreducible modulo p. Every existent field is therefore the abstract form of a Galois Field.

Suppose there could exist a second field $F'[p^n]$ of order equal to that of $F[p^n]$. The field $F[p^n]$ possesses a primitive root w satisfying an equation $W_n(x) = 0$, of degree n, belonging to and irreducible in the $F[p]$. The function $W_n(x)$ divides $x^{p^n} - x$ in the $F[p]$[1]). We may, indeed, apply in the $F[p]$ Euclid's process for finding the greatest common divisor of these functions. If there were no common factor, we would ultimately reach as a remainder a constant, whereas the process may be interpreted in the $GF[p^n]$, in which field the common factor $x - w$ exists. Hence W_n and $x^{p^n} - x$ have a common factor in the $F[p]$. Moreover, W_n is irreducible in that field.

Since $F[p]$ is contained in $F'[p^n]$, the division of $x^{p^n} - x$ by W_n is, à fortiori, possible in the $F'[p^n]$. It follows from § 13 that

---

1) Another proof is given in § 23.

14   CHAPTER II.

the equation $W_n(x) = 0$ completely decomposes in the $F'[p^n]$. Any one of its roots w' is a primitive root in the $F'[p^n]$. Indeed, by its definition, $W_n(x)$ does not divide $x^e - x$ in the $F[p]$ for $e < p^n$. The powers of w' therefore give all the marks of the $F'[p^n]$. Hence $F[p^n]$ and $F'[p^n]$ are abstract forms of the same Galois Field. These results, first proven by Moore (loc. cit.), may be stated as follows:

Theorem. — *Every existent field of finite order s may be represented as a Galois Field of order $s = p^n$. The $GF[p^n]$ is defined uniquely by its order; in particular, it is independent of the special irreducible congruence used in its construction.*

CHAPTER II.

PROOF OF THE EXISTENCE OF THE $GF[p^n]$ FOR EVERY PRIME p AND INTEGER m.

19. The next step is to prove the existence, for every prime number p and positive integer m, of a congruence of degree m irreducible modulo p, from which will follow the existence of the $GF[p^m]$. We will, however, make a more general investigation, taking as our basis a fixed $GF[p^n]$ (in its abstract form), whose existence is supposed known. We will prove that functions belonging to and irreducible in the $GF[p^n]$ exist for every integer m and will determine their number. Since the $GF[p]$, the field of integers taken modulo p, is known to exist, we shall have proven (taking $n = 1$) the existence, for every value of m, of functions belonging to and irreducible in the $GF[p]$, i. e., irreducible modulo p.

At the same time, we shall have deduced some important properties of the $GF[p^{nm}]$ with respect to the included field, the $GF[p^n]$.

20. Theorem. — *If two functions $F(x)$ and $P(x)$ belonging to the $GF[p^n]$ have in the field no common divisor containing x, we can determine two functions $F'(x)$ and $P'(x)$, belonging to the $GF[p^n]$ such that*
$$F'(x) \cdot F(x) - P'(x) \cdot P(x) = 1.$$

The proof is quite analogous to that of § 7.

21. Theorem. — *If, in the $GF[p^n]$, $P[x]$ has no factor involving x in common with $F(x)$ but divides the product $E(x) \cdot F(x)$, then $P(x)$ divides $E(x)$ in the $GF[p^n]$.*

Indeed, by multiplying the given equation
$$E(x) \cdot F(x) = P(x) \cdot S(x)$$
by $F'(x)$, determined as in § 20, we find
$$E(x) = P(x)[S(x) \cdot F'(x) - E(x) \cdot P'(x)].$$

22. Theorem. — *A function $E(x)$ belonging to the $GF[p^n]$ can be decomposed into factors belonging to and irreducible in the $GF[p^n]$ in a single way.*

For if
$$E(x) = f_1 f_2 \ldots f_h = F_1 F_2 \ldots F_k,$$
where $f_i(x)$ and $F_i(x)$ are irreducible, F_1 must by § 21 divide one of the factors f_i, and, since the latter are irreducible, be identical (apart from a factor independent of x) with one of them, say f_1. Proceeding similarly with the equality
$$f_2 f_3 \ldots f_h = F_2 F_3 \ldots F_k,$$
we may suppose $f_2 = F_2$, etc. In particular, $h = k$.

23. Theorem. — *Every function $F(x)$ of degree m belonging to and irreducible in the $GF[p^n]$ divides*
$$x^{p^{nm}} - x.$$
Upon dividing any function $E(x)$ belonging to the $GF[p^n]$ by $F(x)$, we obtain a residue of the form
$$a_0 + a_1 x + a_2 x^2 + \cdots + a_{m-1} x^{m-1},$$
the a's being marks of the $GF[p^n]$. We denote the p^{nm} distinct residues of the above form by

5) $\qquad\qquad X_i \qquad\qquad (i = 0, 1, \ldots, p^{nm} - 1),$

and, in particular, by X_0 the residue zero. Consider the products by a fixed residue $X \neq X_0$,

6) $\qquad\qquad X_j X_i \qquad\qquad (i = 0, 1, \ldots, p^{nm} - 1).$

By the theorem of § 21, the products 6) are all distinct and different from X_0. Hence the residues obtained on dividing them by $F(x)$ must coincide apart from their order with the residues 5). Forming the products of the residues not zero in each series,
$$\prod_{i=1}^{p^{nm}-1} (X_j X_i) \equiv \prod_{i=1}^{p^{nm}-1} X_i \quad [\text{mod } F(x)].$$
Since $\Pi X_i \not\equiv 0$, we have by § 21,
$$X_j^{p^{nm}-1} - 1 \equiv 0 \quad [\text{mod } F(x)].$$
Taking for X_j the particular residue x, the proof of the theorem follows.

24. Theorem. — *If $f(x)$ belongs to the $GF[p^n]$, we have, for every integer t, the following identity in the field:*
$$f(x^{p^{nt}}) = [f(x)]^{p^{nt}}.$$

Let $$f(x) = c_0 + c_1 x + c_2 x^2 + \cdots + c_k x^k,$$
where the c's belong to the $GF[p^n]$, so that
7) $$c_i^{p^n} = c_i \qquad (i = 0, 1, \ldots, k).$$
Raising $f(x)$ to the power p and noting that the multinomial coefficients of the product terms (viz., those not p^{th} powers) are multiples of p, we have the algebraic identity,
$$[f(x)]^p = c_0^p + c_1^p x^p + \cdots + c_k^p x^{kp} + p \cdot Q_1(x).$$
We obtain by induction the formula
$$[f(x)]^{p^s} = c_0^{p^s} + c_1^{p^s} x^{p^s} + \cdots + c_k^{p^s} x^{kp^s} + p \cdot Q_s(x).$$
Applying 7), we obtain in the $GF[p^n]$ the identity:
$$[f(x)]^{p^n} = c_0 + c_1 x^{p^n} + \cdots + c_k x^{kp^n} = f(x^{p^n}).$$
Our theorem now follows by a simple induction.

25. Theorem. — *A function $F(x)$ of degree m belonging to and irreducible in the $GF[p^n]$ divides (in the field) the function*
$$x^{p^{nt}} - x$$
only when the integer t is a multiple of m.

Let $t = sm + r$, where $0 \leqq r < m$. By the theorem of § 23, we have
$$x^{p^{nt}} - x = (x^{p^{nsm}})^{p^{nr}} - x \equiv x^{p^{nr}} - x \quad [\text{mod } F(x)].$$
Hence, if $x^{p^{nt}} - x$ be divisible by $F(x)$ in the $GF[p^n]$, we have
8) $$x^{p^{nr}} \equiv x \quad [\text{mod } F(x)].$$
Denote by $f(x)$ any one of the p^{nm} expressions
$$c_0 + c_1 x + c_2 x^2 + \cdots + c_{m-1} x^{m-1}$$
in which the c's are marks of the $GF[p^n]$. By § 24, we derive from 8)
$$[f(x)]^{p^{nr}} = f(x^{p^{nr}}) \equiv f(x) \quad [\text{mod } F(x)].$$
Hence the congruence $\xi^{p^{nr}} \equiv \xi \quad [\text{mod } F(x)]$
is satisfied by the p^{nm} expressions $f(x)$, which are distinct modulo $F(x)$, the latter being an irreducible function of degree m. Since $r < m$, it follows from § 15 that the congruence must be an identity, whence $r = 0$.

26. The number N_{m,p^n} of functions $F(x)$ of degree m belonging to and irreducible in the $GF[p^n]$ may now be readily determined. For brevity, such an irreducible quantic will be designated an $IQ[m, p^n]$.

PROOF OF THE EXISTENCE OF THE $GF[p^n]$, etc. 17

It is to be understood throughout the investigation that all our operations upon quantics are performed in the $GF[p^n]$. We may therefore state the results of §§ 23 and 25 as follows:

An $IQ[m_1, p^n]$ is a divisor of $x^{p^{nm}} - x$ if, and only if, m_1 be a divisor of m.

It follows that an irreducible factor of $x^{p^{nm}} - x$ will be of degree m if, and only if, it is a factor of none of the functions

9) $\qquad x^{p^{nm_1}} - x \quad (m_1 < m,\ m_1$ a divisor of m).

After showing that the irreducible factors of any such function are all distinct, it will follow that, if we divide $x^{p^{nm}} - x$ by the product of all the distinct irreducible factors of the expressions 9), we obtain a quotient V_{m, p^n} which equals the product of all the $IQ[m, p^n]$.

For example, if m be prime, the irreducible factors of $x^{p^{nm}} - x$ are of degree m or 1. By § 13, the product of the distinct linear factors is $x^{p^n} - x$. Hence, if m be prime,

$$V_{m, p^n} = \frac{x^{p^{nm}} - x}{x^{p^n} - x}, \quad N_{m, p^n} = \frac{p^{nm} - p^n}{m}.$$

We next prove that the irreducible factors of $x^{p^{nm}} - x$ are all distinct. If such a factor be of degree m, it can be used to define the $GF[p^{nm}]$[1]). In this field the equation

$$x^{p^{nm}} - x = 0$$

has p^{nm} distinct roots; viz., the marks of the field. Hence no factor can be a multiple factor in this field and therefore not in the included field the $GF[p^n]$. If an irreducible factor f be of degree $m_1 < m$, it cannot be a multiple factor. Indeed, m_1 must be a divisor of m, and f must divide $x^{p^{nm_1}} - x$ in the $GF[p^n]$. By the former case, f is a simple factor of the expression just given. It remains to prove that f cannot divide the quotient

$$Q \equiv (x^{p^{nm}} - x)/(x^{p^{nm_1}} - x).$$

It suffices to show that Q and $x^{p^{nm_1}} - x$ have no common factor in the $GF[p^n]$. Setting

$$y \equiv x^{p^{nm_1}} - 1, \quad r \equiv \frac{p^{nm} - 1}{p^{nm_1} - 1},$$

it suffices to prove that $y - 1$ and

$$\frac{y^r - 1}{y - 1} \equiv y^{r-1} + \cdots + y + 1$$

have no common factor. The condition for a common divisor is that r be the mark zero in the field. But $r \equiv 1 \pmod{p}$.

---

1) See § 28.

18 CHAPTER II. PROOF OF THE EXISTENCE OF THE $GF[p^n]$, etc.

27. Continuing the investigation, let
$$m \equiv q_1^{r_1} q_2^{r_2} \cdots q_s^{r_s},$$
$q_1, q_2, \ldots, q_s$ being the distinct prime factors of m. For brevity, we use the symbol
$$[t] \equiv x^{p^{nt}} - x.$$

We proceed to prove the formula, due to Dedekind for $n = 1$,

$$V_{m,p^n} = \frac{[m] \prod \left[\frac{m}{q_i q_j}\right] \prod \left[\frac{m}{q_i q_j q_k q_l}\right] \cdots}{\prod \left[\frac{m}{q_i}\right] \prod \left[\frac{m}{q_i q_j q_k}\right] \cdots}$$

In this expression, the term
$$\Pi_k \equiv \Pi \left[\frac{m}{q_{i_1} q_{i_2} \cdots q_{i_k}}\right],$$
in which the product extends over the $C_{s,k}$ combinations $q_{i_1}, \ldots, q_{i_k}$ of the integers $q_1, \ldots, q_s$ taken k together, occurs in the numerator or in the denominator according as k is even or odd. Each $IQ[m, p^n]$ occurs once as a factor in $\Pi_0 \equiv [m]$ but divides no other Π_k; it is therefore a simple factor of the fraction. If there be any factor of the fraction having the degree $m_1 < m$, we denote it by $F(x)$. By § 25, m_1 must be a divisor of m. Denote by $q_1, q_2, \ldots, q_{s_1}$ the prime factors entering in m to a higher power than in m_1. Then m_1 divides $\frac{m}{q_1 q_2 \cdots q_{s_1}}$ but not $\frac{m}{q_j}$ ($j = s_1 + 1, s_1 + 2, \ldots, s$). It follows that, if $k > s_1$, Π_k does not contain $F(x)$ of degree m_1; while, for $k \leqq s_1$, Π_k contains $F(x)$ as often as k integers can be selected from $q_1, q_2, \ldots, q_{s_1}$; viz., $C_{s_1,k}$ times. Hence $F(x)$ occurs in the numerator and denominator of our fraction to the respective degrees,
$$1 + C_{s_1,2} + C_{s_1,4} + \cdots, \quad C_{s_1,1} + C_{s_1,3} + C_{s_1,5} + \cdots$$
These numbers are equal, since their difference equals $(1-1)^{s_1} \equiv 0$. It follows that every irreducible factor of our expression is an $IQ[m, p^n]$. The number of the latter multiplied by the degree m of must equal the degree of the fraction, so that

$$N_{m,p^n} = \frac{1}{m} \left[p^{nm} - \sum p^{\frac{nm}{q_1}} + \sum p^{\frac{nm}{q_1 q_2}} - \cdots + (-1)^s p^{\frac{nm}{q_1 q_2 \cdots q_s}} \right].$$

This number cannot be zero; for, upon dividing by the last term, which is the lowest power of p entering into the expression, we would then obtain unity expressed as the algebraic sum of a series of powers of the prime number p with exponents $\geqq 1$. It follows that the number of $IQ[m, p^n]$ is $\geqq 1$. [See Ex. 2 below].

28. Let $F(x)$ be an $IQ[m, p^n]$. As in § 6, the totality of rational functions of x belonging to the $GF[p^n]$ can be separated into p^{nm}

distinct classes of residues modulo $F(x)$, each being represented by one of the p^{nm} residues

$$a_0 + a_1 x + a_2 x^2 + \cdots + a_{m-1} x^{m-1} \quad (a\text{'s in the } GF[p^n]).$$

Proceeding as in § 6, we find that these classes of residues form the $GF[p^{nm}]$. We can therefore construct the $GF[p^r]$ in as many ways as we can express r as the product of two positive integers n, m; viz., by using an $IQ[m, p^n]$. From the theorem at the beginning of § 26 it follows that the $GF[p^{nm_1}]$ is contained in the $GF[p^{nm}]$ if, and only if, m_1 divides m.

EXERCISES.

Ex. 1. Granting the existence of the $GF[p^n]$, the existence of the $GF[p^{nq}]$, q being prime, follows by § 26. By induction, the $GF[p^r]$ exists for r arbitrary.

Ex. 2. Obtain for the number of $IQ[m, p^n]$ given in § 27 the following limits:
$$\frac{p^{nm} - p^n}{m} \geqq N_{m,p^n} \geqq \frac{\Phi(m)}{m-1} \cdot \frac{p^{nm} - p^n}{m}.$$

Hint: Expand each power of p^n into a series in log p^n and apply
$$\Phi(m) \equiv m \left(1 - \frac{1}{q_1}\right)\left(1 - \frac{1}{q_2}\right)\cdots\left(1 - \frac{1}{q_s}\right).$$

Ex. 3. By decomposing modulo 2 the expression $(x^{2^4} - x)/(x^{2^2} - x)$, obtain the three $IQ[4, 2]$ given in the left members below. Defining the $GF[2^2]$ by means of the irreducible congruence
$$i^2 + i + 1 \equiv 0 \pmod{2},$$
obtain the six $IQ[2, 2^2]$ by means of the following decompositions:
$$x^4 + x + 1 \equiv (x^2 + x + i)(x^2 + x + i^2),$$
$$x^4 + x^3 + 1 \equiv (x^2 + ix + i)(x^2 + i^2 x + i^2),$$
$$x^4 + x^3 + x^2 + x + 1 \equiv (x^2 + ix + 1)(x^2 + i^2 x + 1).$$

CHAPTER III.

CLASSIFICATION AND DETERMINATION OF IRREDUCIBLE QUANTICS.

29. Definition. — An $IQ[m, p^n]$, as $F(x)$, is said *to belong to an exponent* e if e be the least positive integer for which $F(x)$ divides $x^e - 1$ in the $GF[p^n]$. [Compare § 32.]

20   CHAPTER III.

The exponent e to which $F(x)$ belongs must divide $p^{nm} - 1$.
For, if
$$p^{nm} - 1 = ke + r,$$
where $0 \leq r < e$, then $F(x)$, dividing $x^e - 1$, must divide $x^{ke} - 1$ and, by § 23, also $x^{ke+r} - 1$. It must therefore divide their difference,
$$x^{ke}(x^r - 1).$$
Hence must r be zero.

Furthermore, e must not divide $p^{nt} - 1$, for $t < m$; for, if so, $x^e - 1$ and hence also $F(x)$ would divide $x^{p^{nt}} - x$, so that the degree of $F(x)$ would be a divisor of t.

An integer which divides $a^m - 1$ but not $a^t - 1$, $t < m$, is said to be *a proper divisor* of $a^m - 1$. We may state the result:

The exponent to which an $IQ[m, p^n]$ belongs is a proper divisor of $(p^n)^m - 1$.

30. Theorem. — *The number $N_{m, p^n}^{(e)}$ of $IQ[m, p^n]$ which belong to an exponent e, a proper divisor of $(p^n)^m - 1$, is $\Phi(e)/m$.*

Let $q_1, q_2, \ldots, q_s$ be the distinct prime factors of e. Proceeding as in § 26, we rid $x^e - 1$ of those of its factors which are irreducible in the $GF[p^n]$ and belong to an exponent $< e$. We obtain the expression
$$\frac{(x^e - 1) \, \Pi \left(x^{\frac{e}{q_i q_j}} - 1 \right) \cdots}{\Pi \left(x^{\frac{e}{q_i}} - 1 \right) \Pi \left(x^{\frac{e}{q_i q_j q_k}} - 1 \right) \cdots}$$
which is therefore the product of the irreducible factors of $x^e - 1$ belonging to the exponent e. Each of them is an irreducible factor of
$$x^{p^{nm}} - x$$
and hence of degree m or a divisor of m. Since each belongs to an exponent which is a proper divisor of $(p^n)^m - 1$, the degree must be m.

The degree of the above function is clearly
$$e - \sum \frac{e}{q_i} + \sum \frac{e}{q_i q_j} - \sum \frac{e}{q_i q_j q_k} + \cdots + (-1)^s \frac{e}{q_1 q_2 \cdots q_s}$$
$$= e \left(1 - \frac{1}{q_1}\right) \left(1 - \frac{1}{q_2}\right) \cdots \left(1 - \frac{1}{q_s}\right) = \Phi(e).$$
Hence
$$m \cdot N_{m, p^n}^{(e)} = \Phi(e).$$

31. Theorem. — *If $F(x)$ and $\varphi(x)$ belong to and are irreducible in the $GF[p^n]$ and are of the respective degrees m and t, a divisor of m, the roots of the congruence*

10) $\qquad\qquad \varphi(X) \equiv 0 \pmod{F(x)}$

are
$$X_1, \; X_1^{p^n}, \; X_1^{p^{2n}}, \ldots, X_1^{p^{n(t-1)}}$$
if X_1 be one root of 10) *necessarily belonging to the $GF[p^{nm}]$.*

By § 24 we have in the $GF[p^n]$ the identity
$$\varphi(X^{p^{nr}}) = \{\varphi(X)\}^{p^{nr}}.$$
Hence, if X_1 be a root of 10), so is every $X_1^{p^{nr}}$. Since $\varphi(X)$ is an $IQ[t, p^n]$, we have (§ 23) in the $GF[p^n]$,
$$X_1^{p^{nt}} - X_1 \equiv \varphi(X_1) \cdot \psi(X_1) \equiv 0 \quad [\bmod F(x)].$$
Hence, m being a multiple of t,
$$X_1^{p^{nm}} \equiv X_1 \quad [\bmod F(x)].$$
We next prove that the above t powers of X_1 are distinct modulo $F(x)$. Indeed, if
$$X_1^{p^{na}} \equiv X_1^{p^{nb}} \quad [\bmod F(x)]$$
for $a < b < t$, we would have, upon raising it to the power $p^{n(m-a)}$,
$$X_1^{p^{nm}} \equiv X_1 \equiv X_1^{p^{n(m+b-a)}} \quad [\bmod F(x)],$$
so that, by § 25, $m + b - a$ would be divisible by m. Hence $b = a$.

Corollary. — We have in the $GF[p^{nm}]$ the decomposition
$$\varphi(X) \equiv (X - X_1)(X - X_1^{p^n}) \ldots (X - X_1^{p^{n(t-1)}}).$$
In particular, $F(x) = 0$ has in the $GF[p^{nm}]$ the distinct roots
$$x, \quad x^{p^n}, \ldots, x^{p^{n(m-1)}}.$$

32. Theorem. — *If $F(x)$ be an $IQ[m, p^n]$ belonging to the exponent e, every root of $F(x) = 0$ in the $GF[p^{nm}]$ belongs to the exponent e, and inversely.*

We may define the $GF[p^{nm}]$ by means of $F(x)$. In it, all the roots of $F(x) = 0$ satisfy the equation $x^e - 1 = 0$, but do not *all* satisfy $x^f - 1 = 0$ for $f < e$. But, p^n being relatively prime to e, a divisor of $p^{nm} - 1$, it follows from the corollary of § 31 that the roots of $F(x) = 0$ in the $GF[p^{nm}]$ all belong to the same exponent. This common exponent is therefore e.

In particular, for $e = p^{nm} - 1$, the roots of $F(x) = 0$ are primitive roots in the $GF[p^{nm}]$. Such a quantic $F(x)$ will be called a *primitive irreducible quantic* of degree m in the $GF[p^n]$ and will be referred to as a $PIQ[m, p^n]$.

33. Theorem. — *If e be a prime number, the function*
$$V \equiv \frac{x^e - 1}{x - 1} \equiv x^{e-1} + \cdots + x + 1$$
is irreducible with respect to every prime modulus p which is a primitive root of e.

By hypothesis, p belongs to the exponent $e - 1$ modulo e, so that e is a proper divisor of $p^{e-1} - 1$. Hence, by § 30 for $n = 1$, $m = e - 1$, the number of irreducible factors of V is $\frac{\Phi(e)}{e-1} = 1$.

Note. — If a be a primitive root of e, then $a + ke (k = 0, \pm 1, \pm 2, \ldots)$ are also primitive roots of e. By the theorem of Dirichlet, this arithmetical progression contains an infinity of prime numbers. With respect to any such prime p, V is irreducible modulo p. À fortiori, V is algebraically irreducible.

Determination of $IQ[m, p^n]$ whose degree m contains no prime factors other than those of $p^n - 1$, §§ 34—38.

34. Theorem. — *Let $F_1(x), F_2(x), \ldots, F_N(x)$ denote the $IQ[m, p^n]$ which belong to an exponent*

$$e \equiv (p^{nm} - 1)/d,$$

and let λ be an integer relatively prime to d and containing no prime factors other than those occurring in $p^{nm} - 1$. With the exception of the case in which λ is a multiple of 4 while p^{nm} is of the form $4l - 1$, all of the $IQ[\lambda m, p^n]$ which belong to the exponent $e\lambda$ are given by the N quantics $F_1(x^\lambda), \ldots, F_N(x^\lambda)$.

By definition, λ contains no prime factor other than those occurring in e. Hence $e\lambda$ and e contain exactly the same prime factors, so that

$$\frac{\Phi(e\lambda)}{e\lambda} = \frac{\Phi(e)}{e}.$$

By § 30, we have

$$N \equiv N_{m,\,p^n}^{(e)} = \frac{\Phi(e)}{m}.$$

If we suppose satisfied the conditions (obtained below) under which $e\lambda$ shall be a proper divisor of $(p^n)^{m\lambda} - 1$, we will have

$$N_{m\lambda,\,p^n}^{(e\lambda)} \equiv \frac{\Phi(e\lambda)}{m\lambda} = \frac{\Phi(e)}{m} = N.$$

Since e divides $p^{nm} - 1$, the irreducible factors of $x^e - 1$ are of degree $\overline{\leq} m$ (§ 25). Hence, in the notation of the theorem,

$$x^e - 1 = F_1(x) F_2(x) \ldots F_N(x) \cdot Q(x)$$

where the irreducible factors of $Q(x)$ either belong to an exponent $< e$ or else are of degree $< m$. Therefore

$$x^{e\lambda} - 1 = F_1(x^\lambda) F_2(x^\lambda) \ldots F_N(x^\lambda) \cdot Q(x^\lambda),$$

where every irreducible factor of $Q(x^\lambda)$ is of degree $< \lambda m$ or else belongs to an exponent $< e\lambda$. Since there are exactly N irreducible factors of degree $m\lambda$ which belong to the exponent $e\lambda$, they must be identical with $F_1(x^\lambda), \ldots, F_N(x^\lambda)$.

Calling ν the least integer such that $p^{n\nu} - 1$ is divisible by $e\lambda$, we seek the conditions under which $\nu = m\lambda$. Since m is by hypothesis the least integer for which $p^{nm} - 1$ is divisible by e, ν must be a multiple of m. For, if $\nu = qm + r$, $0 \overline{\leq} r < m$, then e divides $p^{n(qm+r)} - 1$ and $p^{nmq} - 1$ and hence also their difference $p^{nmq}(p^{nr} - 1)$, which requires $r = 0$. Having $v = qm$, we inquire under what conditions does $q = \lambda$? Since

11) $\quad \dfrac{p^{nv}-1}{e\lambda} = \dfrac{d}{\lambda} \cdot \dfrac{p^{nmq}-1}{p^{nm}-1},$

it follows that λ divides $(p^{nmq}-1)/(p^{nm}-1)$. Raising to the power q the identity $p^{nm} \equiv 1 + (p^{nm}-1)$, we find

12) $\dfrac{p^{nmq}-1}{p^{nm}-1} \equiv q + \dfrac{q(q-1)}{1 \cdot 2}(p^{nm}-1) + \cdots + C_{q,k}(p^{nm}-1)^{k-1} + \cdots (p^{nm}-1)^{q-1}.$

Let θ be a prime factor of λ and θ^a the highest power of θ contained in λ. Since θ divides $p^{nm}-1$ and the left member of 12), it must divide q. Further, if $\theta > 2$, θ^a divides q. Indeed, the ratio of the kth term of 12) to the first term q can be written

$$\dfrac{(q-1)(q-2)\ldots(q-k+1)}{1 \cdot 2 \ldots (k-1)} \cdot \left(\dfrac{p^{nm}-1}{\theta}\right)^{k-1} \cdot \dfrac{\theta^{k-1}}{k},$$

of which the first two factors are integers, while the third factor

$$\dfrac{\{1+(\theta-1)\}^{k-1}}{k} > \dfrac{1+(k-1)(\theta-1)}{k} \equiv 1 + \dfrac{(k-1)(\theta-2)}{k}$$

is > 1 if $k \gtreqless 2$. Hence the irreducible fraction equal to θ^{k-1}/k has the factor θ in its numerator. Hence the terms of 12) beginning with the second contain θ to a higher power than the first term q. Since θ^a divides λ, which divides the left member of 12), it follows that θ^a divides the first term q on the right. Hence, if λ be odd or the double of an odd number, q is divisible by λ. Inversely, if q be divisible by λ, λ being odd or the double of an odd number, the above argument shows that the right member of 12) will contain the factor λ and therefore that the left member of 11) will be an integer. In order that v be the least integer for which this can happen, we must have $q = \lambda$.

If λ be a multiple of 4, $p^{nm}-1$ is even by hypothesis. Then $\theta = 2$ will be a factor of q as before. The ratio of the second term of 12) to the first term will be divisible by 2 if, and only if, $p^{nm}-1$ be a multiple of 4; the ratio of the kth term to the first will, for $k \gtreqless 3$, contain the factor 2. Hence, if p^{nm} be of the form $4l+1$, we can conclude that $q = \lambda$. [The case $p^{nm} = 4l-1$ leads to the entirely different theorem of § 36.]

35. Let ϱ be a primitive root in the $GF[p^n]$. The function $x - \varrho^t$ belongs to the exponent $(p^n-1)/d$ where d is the greatest common divisor of t and p^n-1. Applying the theorem § 34 for $m = 1$, we have the result:

If λ be any integer containing no prime factor not occurring in p^n-1 and if t be an integer prime to λ, the $IQ[\lambda, p^n]$ belonging

24   CHAPTER III.

to the exponent $\lambda(p^n-1)/d$, d being the greatest common divisor of t and p^n-1, are the binomials $x^\lambda - \varrho^t$, the case $p^n = 4l-1$, $\lambda = 4\lambda_1$ being excluded.

Inversely, *we obtain by this theorem every binomial irreducible in the* $GF[p^n]$. In the first place, λ and t must have no factor in common, since otherwise $x^\lambda - \varrho^t$ would be algebraically reducible. On the other hand, if λ contains a prime factor θ, not a factor of p^n-1, we can determine (§ 7, Note) an integer θ_1, such that

$$\theta\theta_1 \equiv 1 \quad [\bmod\, p^n-1].$$

Since $\varrho^{\theta\theta_1} = \varrho$, it follows that $\varrho^{\theta_1 t} \equiv \alpha$ is a root of

$$x^\theta - \varrho^t = 0.$$

Hence $x - \alpha$ is a factor of $x^\theta - \varrho^t$, so that $x^{\lambda/\theta} - \alpha$ divides $x^\lambda - \varrho^t$.

Example. — For $p^n = 7$, we may take $\varrho = 5$. Then for $\lambda = 2$ and $t = 1, 3, 5$, we obtain the irreducible binomials $x^2 - 5$, $x^2 + 1$, $x^2 - 3$ belonging to the exponents 12, 4, 12 respectively. For $\lambda = 3$ and $t = 1, 2, 4, 5$ respectively, we obtain the binomials

$$x^3 - 5, \quad x^3 - 4, \quad x^3 - 2, \quad x^3 - 3$$

irreducible modulo 7 and belonging to the respective exponents 18, 9, 9, 18.

36. Theorem. — *Let* $p^n = 2^i t - 1$, $i \geqq 2$, t *odd*; $\lambda = 2^j s$, $j \geqq 2$, s *odd; let* k *be the smaller of the integers* i *and* j; *finally, let* m *be odd. Then if, in the* N *quantics* $IQ[m, p^n]$ *belonging to the exponent*

$$e = (p^{nm} - 1)/d,$$

we replace x *by* x^λ, *where* $\lambda \equiv 2^j s$ *is prime to* d *and contains no prime factors other than those occurring in* $p^{nm} - 1$, *we obtain* N *quantics of degree* $m\lambda$ *each decomposing into* 2^{k-1} *quantics irreducible in the* $GF[p^n]$, *so that we obtain all of the* $2^{k-1} N$ *quantics*

$$IQ\left[\frac{m\lambda}{2^{k-1}}, \; p^n\right]$$

belonging to the exponent $e\lambda$.

If ν denote the least integer such that $p^{n\nu} - 1$ is divisible by $e\lambda$, we find as in § 34 that $\nu = qm$. In the present case, q is even; for, if q be odd, ν would be odd and $p^{n\nu} - 1$ the double of an odd number, whereas λ is divisible by 4. By the restrictions on p^n and m,

13) $\qquad\qquad p^{nm} = 2^i \tau - 1 \quad (\tau \text{ odd}).$

Raising this identity to the power q, we find

14) $\dfrac{p^{nmq}-1}{p^{nm}-1} = \dfrac{2^i \tau}{2^i \tau - 2}\left[-q + \dfrac{q(q-1)}{1\cdot 2} 2^i \tau - \cdots \pm C_{q,l} 2^{i(l-1)} \tau^{l-1} \mp \ldots\right].$

CLASSIFICATION AND DETERMINATION, etc. 25

The ratio of the lth term within the parenthesis to the first term is

$$\mp \frac{(q-1)(q-2)\ldots(q-l+1)}{1\cdot 2\ldots(l-1)}\, \tau^{l-1} \cdot \frac{2^{i(l-1)}}{l},$$

where the first and second factors are integers, while the third factor, being >1 for $l \geqq 2$, equals an irreducible fraction with an *even* numerator. Hence the first term contains 2 to a lower power than the remaining terms in the above parenthesis. In order that $p^{n\nu}-1$ shall be divisible by $e\lambda$, formula 11) requires that λ shall divide the left member of 14). Hence 2^j must divide the first term of the right member and consequently also $2^{i-1}q$. Hence the even integer q must contain 2 to the power 1 or $j-i+1$ according as $j \leqq i$ or $j > i$. Furthermore, by § 34, q must contain every odd factor of λ. Hence, if ν be the least possible integer,

$$q = \frac{\lambda}{2^{j-1}} \quad \text{or} \quad q = \frac{\lambda}{2^i - 1}$$

according as $j \leqq i$ or $j > i$, i.e., according as $k = j$ or $k = i$. Hence

$$q = \frac{\lambda}{2^k - 1}, \quad \nu = \frac{m\lambda}{2^k - 1}.$$

As at the beginning of § 34, we have

$$N \equiv N_{m,p^n}^{(e)} = \frac{\Phi(e)}{m} = \frac{\Phi(e\lambda)}{m\lambda} = \frac{1}{2^k-1} \frac{\Phi(e\lambda)}{\nu},$$

so that the number of $IQ[\nu, p^n]$ belonging to the exponent $e\lambda$ is $2^{k-1}N$.

By hypothesis,

$$x^e - 1 \equiv F_1(x) F_2(x) \ldots F_N(x) \cdot Q(x),$$

where the irreducible factors of $Q(x)$ in the $GF[p^n]$ belong to exponents $< e$ or are of degree $< m$. The irreducible factors of $Q(x^\lambda)$ are therefore of degree $< \lambda m$ or else belong to exponents $< \lambda e$. Hence the irreducible factors of degree λm of the expression $x^{e\lambda}-1$ must, if they belong to the exponent $e\lambda$, be factors of $F_1(x^\lambda), \ldots, F_N(x^\lambda)$. Since the combined degree of the latter is $Nm\lambda \equiv 2^{k-1}\nu N$, and since there are exactly $2^{k-1}N$ irreducible quantics of degree ν belonging to the exponent $e\lambda$, it follows that each $F_i(x^\lambda)$ is the product of 2^{k-1} irreducible quantics of degree ν.

Corollary. — Since the distinct functions of degree $m = 1$ which belong to the exponent $e = (p^n - 1)/d$ are given by the formula

$$x - \varrho^{ad},$$

ϱ being a fixed primitive root in the $GF[p^n]$ and a being any integer prime to e, it follows that $x^\lambda - \varrho^{ad}$ decomposes in the $GF[p^n]$ into 2^{k-1} irreducible factors of degree $\lambda/2^{k-1}$ belonging to the exponent

$e\lambda$, provided p^n and λ are subject to the conditions given in the main theorem.

37. Since irreducible binomials are lacking in the case treated in the last section, we proceed to set up trinomial $IQ[\lambda, p^n]$. It is, however, not necessary to suppose that λ is a multiple of 4. We suppose merely that
$$p^n = 2^i t - 1 \quad (t \text{ odd}, i \geqq 2)$$
and that λ is an *even* integer containing no prime factor not occurring in $p^n - 1$. Set
$$v = 2^{i-1}\lambda,$$
so that v is divisible by 2^i. If ϱ be a primitive root in the $GF[p^n]$ and if s be any integer prime to λ and hence also to v, then $x - \varrho^s$ belongs to the exponent $(p^n - 1)/d$, where d is the greatest common divisor of s and $p^n - 1$, and v is prime to d. Hence (§ 36), the binomial $x^v - \varrho^s$ decomposes into 2^{i-1} irreducible quantics of degree λ. We proceed to determine them.

Since 2^{i-1} and $(p^n-1)/2$ are relatively prime, we can determine (§ 7, Note) two integers l_1 and h_1 such that
$$l_1 2^{i-1} - h_1(p^n - 1)/2 = 1.$$
Multiplying this equation by the even integer $s + (p^n-1)/2$, we obtain two integers l and h for which
$$l 2^i - h(p^n - 1) = s + (p^n - 1)/2.$$
Since the $(p^n - 1)/2$ power of the primitive root ϱ is -1, we have
$$x^v - \varrho^s \equiv x^{\lambda 2^{i-1}} + \varrho^{l 2^i}.$$

In the $GF[p^n]$ we have the decomposition

15) $$x^{\lambda 2^{i-1}} + \varrho^{l 2^i} \equiv \prod_{j=1}^{2^{i-1}} (x^\lambda - \xi_j \varrho^l x^{\lambda/2} - \varrho^{2l}),$$

where the ξ_j are marks of the $GF[p^n]$ determined as the roots of the equation
$$E(\xi) \equiv \xi^{2^{i-1}} + 2^{i-1} \sum_{k=1}^{2^{i-2}-1} \frac{(2^{i-1}-k-1)!}{k!(2^{i-1}-2k)!} \xi^{2^{i-1}-2k} + 2 = 0.$$

In fact, by Waring's formula[1]), the sum of the $(2^{i-1})^{\text{st}}$ powers of the roots u and $\dfrac{-1}{u}$ of the quadratic
$$X^2 - \xi X - 1 = 0$$
is found to be $E(\xi)$. Expressed otherwise, if $\xi \equiv u - \dfrac{1}{u}$, then
$$u^{2^{i-1}} + u^{-2^{i-1}} = E(\xi).$$

---

1) Serret, Cours d'Algèbre Supérieure, I, p. 449.

Hence, if $\xi_j \equiv u_j - \dfrac{1}{u_j}$ is a root of $E(\xi) = 0$, we have
$$u_j^{2^i} + 1 = 0.$$
Then, since $p^n + 1 = 2^i t$, t odd, we have
$$u_j^{p^n+1} + 1 = 0, \quad u_j^{p^n} = -1/u_j.$$
Applying § 24, we have modulo p,
$$\xi_j^{p^n} \equiv \left(u_j - \dfrac{1}{u_j}\right)^{p^n} \equiv u^{p^n} - 1/u_j^{p^n} = -\dfrac{1}{u_j} + u_j \equiv \xi_j,$$
so that every root[1]) of $E(\xi) = 0$ belongs to the $GF[p^n]$. Hence
$$u^{2^i-1} + u^{-2^i-1} \equiv \prod_{j=1}^{2^i-1} (\xi - \xi_j).$$
Substituting in this identity
$$u = \dfrac{x^{\lambda/2}}{\varrho^l}, \quad \xi \equiv u - \dfrac{1}{u} = \dfrac{x^\lambda - \varrho^{2l}}{\varrho^l x^{\lambda/2}},$$
and clearing the equation of fractions, we obtain formula 15).

38. As a simple example, let $p^n = 7 = 2^3 - 1$, $\lambda = 4$. The binomials
$$x^{16} - 5^s \quad (s = 1, 3, 5)$$
can be readily decomposed into irreducible quartics. The congruence
$$E(\xi) \equiv \xi^4 + 4\xi^2 + 2 \equiv 0 \pmod{7}$$
has the roots ± 1 and ± 3. Further
$$x^{16} - 5^s \equiv x^{16} + 5^{s+3} = x^{16} + 5^{2l} \quad (s + 3 = 2l = 4, 6, 8).$$
Since $5^{l \cdot 2^3} \equiv 5^{2l} \pmod 7$, equation 15) becomes
$$x^{16} + 5^{2l} \equiv \prod_{j=1}^{4} (x^4 - \xi_j 5^l x^2 - 5^{2l}) \pmod{7},$$
holding for $l = 4, 2, 3$. Taking each in turn, we have modulo 7:
$$x^{16} + 4 \equiv (x^4 - x^2 - 4)(x^4 + x^2 - 4)(x^4 - 2x^2 - 4)(x^4 + 2x^2 - 4),$$
$$x^{16} + 2 \equiv (x^4 - 2x^2 - 2)(x^4 + 2x^2 - 2)(x^4 - 4x^2 - 2)(x^4 + 4x^2 - 2),$$
$$x^{16} + 1 \equiv (x^4 - x^2 - 1)(x^4 + x^2 - 1)(x^4 - 4x^2 - 1)(x^4 + 4x^2 - 1).$$

---

1) For another proof see Serret, Cours d'Algèbre supérieure, II, pp. 160—3. Compare § 82 below.

Determination and classification[1]) *of the* $IQ[p^s, p^n]$, §§ 39—46.

39. Consider for positive integers μ the auxiliary quantics

16) $\quad X_\mu \equiv x^{p^n\mu} - \mu x^{p^n(\mu-1)} + \cdots + (-1)^k C_{\mu,k} x^{p^n(\mu-k)} + \cdots + (-1)^\mu x,$

where $C_{\mu,k}$ denotes the number of combinations of μ things k at a time. Since $C_{p^r, k}$ is a multiple of p, if $0 < k < p^r$, we have

17) $\quad\quad\quad\quad X_{p^r} \equiv x^{p^n p^r} - x \pmod{p}.$

Hence, by § 26, the product of all the $IQ[p^s, p^n]$ is given by

18) $\quad\quad\quad\quad V_{p^s, p^n} \equiv X_{p^s} / X_{p^{s-1}}.$

We derive a simple expression for the quotient 18) as follows. From

$$C_{\mu, k} + C_{\mu, k-1} = C_{\mu+1, k},$$

we deduce at once the congruence

19) $\quad\quad\quad\quad X_{\mu+1} \equiv X_\mu^{p^n} - X_\mu \pmod{p}.$

Multiplying together the congruences (for $i = 1, 2, \ldots, v$)

$$X_{u+i} \equiv X_{u+i-1}^{p^n} - X_{u+i-1} \pmod{p},$$

and dividing the resulting formula by the product

$$X_{u+1} X_{u+2} \cdots X_{u+v-1},$$

we find

20) $\quad\quad\quad\quad X_{u+v} \equiv X_u \prod_{i=u}^{u+v-1} (X_i^{p^n-1} - 1) \pmod{p}.$

Taking $u = p^{s-1}$, $u + v = p^s$, we find from 18) and 20) the result

21) $\quad\quad\quad\quad V_{p^s, p^n} \equiv \prod_{i=p^{s-1}}^{p^s-1} (X_i^{p^n-1} - 1).$

Further, if $\nu_1, \nu_2, \ldots, \nu_{p^n-1}$ denote the marks $\neq 0$ of the $GF[p^n]$, we have

$$X_i^{p^n-1} - 1 = \prod_{j=1}^{p^n-1} (X_i - \nu_j).$$

Since $X_i - \nu_j$ is of degree p^{ni} in x, it must decompose in the $GF[p^n]$ into p^{ni-s} factors each an $IQ[p^s, p^n]$.

---

1) For the case $n = 1$, Serret, *Journal de Mathématiques,* 1873, p. 301; Algèbre, II, ch. IV. For general n, Dickson, *Bull. Amer. Math. Soc.,* 1897, pp. 384—389.

40. For $s=1$, there are $p-1$ factors in the product 21), given by $i=1, 2, \ldots, p-1$. The irreducible factors of $X_i^{p^n-1}-1$ are then said to form together the ith *class* of $IQ[p, p^n]$. Consider first
$$X_1 - \nu \equiv x^{p^n} - x - \nu,$$
which is the product of p^{n-1} $IQ[p, p^n]$ of the *first class*. To decompose it, consider the equation

22) $\qquad \eta^{p^{n-1}} + \eta^{p^{n-2}} + \cdots + \eta^p + \eta = c.$

It follows at once that
$$c^p - c \equiv \eta^{p^n} - \eta \pmod{p}.$$
Hence every root η of 22) belongs to the $GF[p^n]$ if, and only if, c be an integer. Setting in 22)
$$c = \lambda \nu, \quad \eta = (\lambda x)^p - \lambda x,$$
where λ belongs to the $GF[p^n]$, we find
$$\lambda (x^{p^n} - x - \nu) \equiv 0 \pmod{p}.$$
We have therefore in the $GF[p^n]$ the decomposition[1])

23) $\qquad \lambda(x^{p^n} - x - \nu) \equiv \prod_{j=1}^{p^n-1} (\lambda^p x^p - \lambda x - \beta_j),$

where the β_j are the roots of

24) $\qquad \eta^{p^{n-1}} + \eta^{p^{n-2}} + \cdots + \eta^p + \eta = \lambda \nu,$

λ or ν being determined so that $\lambda \nu$ is an integer. We have therefore the theorem: *The quantic $\lambda^p x^p - \lambda x - \beta$ is an $IQ[p, p^n]$ if, and only if,*
$$B \equiv \beta^{p^{n-1}} + \beta^{p^{n-2}} + \cdots + \beta^p + \beta \not\equiv 0.$$

Corollary. — *If b is an integer not divisible by the prime p, $x^p - x - b$ is irreducible in the $GF[p^n]$ if, and only if, n is not divisible by p; in particular, it is always irreducible modulo p.*

In fact, the condition becomes in this case
$$B \equiv nb \not\equiv 0 \pmod{p}.$$

41. The decomposition 23) may be given a more explicit form useful below. If β be one root of 24), then is also

25) $\qquad \beta_j \equiv \alpha^p - \alpha + \beta,$

for every mark α in the $GF[p^n]$. Indeed, we have
$$\beta_j^{p^{n-1}} + \cdots + \beta_j^p + \beta_j = \alpha^{p^n} - \alpha + \beta^{p^{n-1}} + \cdots + \beta^p + \beta = \lambda \nu.$$

---

[1] For the case $\nu = 0$, this decomposition was given without proof by Mathieu, *Journal de Mathématiques*, (2) vol. 6, 1861, p. 280.

30   CHAPTER III.

Further, the formula 25) furnishes all the roots of 24). For, if
then
$$\alpha^p - \alpha + \beta = \alpha_1^p - \alpha_1 + \beta,$$
$$(\alpha - \alpha_1)^p \equiv (\alpha - \alpha_1) \quad [\text{mod } p],$$
so that $\alpha = \alpha_1 +$ an integer. Hence there are $p^n/p \equiv p^{n-1}$ distinct expressions $\alpha^p - \alpha$ and hence as many roots β_j. Hence

$23_1)$ $(\lambda x)^{p^n} - \lambda x - (\beta^{p^{n-1}} + \cdots + \beta^p + \beta) \equiv \Pi[(\lambda x - \alpha_i)^p - (\lambda x - \alpha_i) - \beta]$,

the product extending over p^{n-1} marks α_i of the $GF[p^n]$ no two of which differ by an integer.

42. Consider an irreducible factor $x^p - x - \beta$ of $x^{p^n} - x - 1$, where therefore
$$\beta^{p^{n-1}} + \beta^{p^{n-2}} + \cdots + \beta^p + \beta = 1.$$
Denote by I one root of the equation
$$x^p - x - \beta = 0.$$
Its remaining roots are $I+1$, $I+2$, ..., $I+p-1$.

Then by $23_1)$ every root of every $IQ[p, p^n]$ of the first class is a linear function of I, viz., $\lambda x - \alpha_i = I + i$, $i =$ integer:

26) $\qquad x = (I + i + \alpha_i)/\lambda,$

the coefficients $1/\lambda$ and $(i + \alpha_i)/\lambda$ being marks of the $GF[p^n]$.

Inversely, every such linear function containing I is the root of an $IQ[p, p^n]$.

43. Consider an $IQ[p, p^n]$ of class μ. Its roots belong to the $GF[p^{n p}]$ and are therefore functions of I of the form
$$f(I) \equiv \sum_{j=0}^{p-1} \alpha_j I^j,$$
where the α_j belong to the $QF[p^n]$. By § 39, $f(I)$ will be a root of

27) $\qquad X_\mu = \sigma,$

if σ be suitable chosen in the $GF[p^n]$. But, by § 42,
$$I^{p^n} = I + 1.$$
Hence, by § 24, we have for any integer m,
$$[f(I)]^{p^{nm}} = f(I^{p^{nm}}) = f(I + m).$$
Substituting $f(I)$ in equation 27), X_μ being given by 16), we find
$$f(I+\mu) - \mu f(I+\mu-1) + \cdots + (-1)^k C_{\mu,k} f(I+\mu-k) + \cdots + (-1)^\mu f(I) = \sigma.$$
The degree of this equation in I being less than p, it must be an identity. But its first member is the μth difference of the polynomial $f(I)$ with respect to the constant difference unity attributed to I.

Since it reduces to the constant $\sigma \neq 0$, the degree of $f(I)$ is exactly μ[1]). Hence

$$\alpha_\mu = \frac{\sigma}{\mu!}, \quad \alpha_{\mu+1} = \alpha_{\mu+2} = \cdots = \alpha_{p-1} = 0.$$

We have therefore proved that *the roots of every $IQ[p, p^n]$ of class μ are integral functions of I of degree μ.*

44. We can readily obtain a formula including all $IQ[p, p^n]$. In the above expression $f(I)$, let the α_j be arbitrary such, however, that $f(I)$ does not reduce to α_0. To set up the equation of which $f(I)$ is a root, consider the p equations

$$I^\lambda[f(I) - \xi] = 0 \qquad (\lambda = 0, 1, \ldots, p-1).$$

Reducing the exponents of I below p by using the identity

$$I^p = I + \beta,$$

we obtain the series of equations

$$(\alpha_0 - \xi) + \alpha_1 I + \alpha_2 I^2 + \cdots + \alpha_{p-1} I^{p-1} = 0,$$
$$\beta\alpha_{p-1} + (\alpha_0 - \xi + \alpha_{p-1})I + \alpha_1 I^2 + \cdots + \alpha_{p-2} I^{p-1} = 0,$$
$$\cdots \cdots \cdots \cdots \cdots \cdots \cdots \cdots \cdots \cdots \cdots \cdots$$
$$\beta\alpha_1 + (\beta\alpha_2 + \alpha_1)I + (\beta\alpha_3 + \alpha_2)I^2 + \cdots + (\alpha_0 - \xi + \alpha_{p-1})I^{p-1} = 0.$$

Eliminating $I^0, I^1, \ldots, I^{p-1}$ from these p equations, we reach the required irreducible quantic $F(\xi)$,

$$\begin{vmatrix} \alpha_0 - \xi & \alpha_1 & \alpha_2 & \cdots & \alpha_{p-2} & \alpha_{p-1} \\ \beta\alpha_{p-1} & \alpha_0 + \alpha_{p-1} - \xi & \alpha_1 & \cdots & \alpha_{p-3} & \alpha_{p-2} \\ \beta\alpha_{p-2} & \beta\alpha_{p-1} + \alpha_{p-2} & \alpha_0 + \alpha_{p-1} - \xi & \cdots & \alpha_{p-4} & \alpha_{p-3} \\ \cdots & \cdots & \cdots & \cdots & \cdots & \cdots \\ \beta\alpha_2 & \beta\alpha_3 + \alpha_2 & \beta\alpha_4 + \alpha_3 & \cdots & \alpha_0 + \alpha_{p-1} - \xi & \alpha_1 \\ \beta\alpha_1 & \beta\alpha_2 + \alpha_1 & \beta\alpha_3 + \alpha_2 & \cdots & \beta\alpha_{p-1} + \alpha_{p-2} & \alpha_0 + \alpha_{p-1} - \xi \end{vmatrix}$$

Setting $\alpha_{\mu+1} = \alpha_{\mu+2} = \cdots = \alpha_{p-1} = 0$ and giving to $\alpha_0, \alpha_1, \ldots, \alpha_{\mu-1}$ all possible values in the $GF[p^n]$ and to α_μ every value $\neq 0$, we obtain $p^{n\mu}(p^n - 1)$ irreducible quantics of class μ. Since $f(I + m)$ leads to the same determinant as $f(I)$, if m be an integer, the number of distinct $IQ[p, p^n]$ of class μ is $p^{n\mu-1}(p^n - 1)$, a result also following from § 39.

For $\mu = 1$, we find that

$$\frac{-F(\xi)}{\alpha_1^p} \equiv \left(\frac{\xi}{\alpha_1} - \frac{\alpha_0}{\alpha_1}\right)^p - \left(\frac{\xi}{\alpha_1} - \frac{\alpha_0}{\alpha_1}\right) - \beta,$$

so that we may derive a new proof of formula 23_1).

---

[1]) Boole, Calculus of Finite Differences, p. 5 and p. 19, formula 3).

An interesting type of $IQ[p, p^n]$ of class $p-1$ is given by setting every $\alpha_j = 0$ except α_0 and α_{p-1}; viz.,

$$F(\xi) \equiv (\xi - \alpha_0 - \alpha_{p-1})^p + \alpha_{p-1}(\xi - \alpha_0 - \alpha_{p-1})^{p-1} - \beta^{p-1}\alpha_{p-1}^p.$$

Multiplying this by $\xi - \alpha_0 - \alpha_{p-1}$ and setting $F(\xi) = 0$, we find that ξ^p is a linear fractional function of ξ. But, by § 31, the roots of $F(\xi) = 0$ may be expressed in the form

$$\xi, \ \xi^{p^n}, \ \xi^{p^{2n}}, \ \ldots, \ \xi^{p^{n(p-1)}}.$$

Hence its roots are all linear fractional functions of one of them.

This result also follows from the fact that

$$f(I) \equiv \alpha_0 + \alpha_{p-1}(I + \beta)/I, \quad I^p = I + \beta,$$

so that each root is a linear fractional function of I.

45. Formula 19) expresses the fact that X_μ becomes $X_{\mu+1}$ when x is changed into $x^{p^n} - x$. Further, if we set $X_0 \equiv x$, 19) holds true for $\mu = 0$; viz.,
$$X_1 \equiv X_0^{p^n} - X_0.$$

Hence in order to change x into $x^{p^n} - x$ in any formula involving the X_μ, we have merely to advance the subscripts of each X_μ by unity. Applying this operation to formula 21), we have the theorem:

If $F(x)$, an $IQ[p^s, p^n]$, divides $X_i^{p^{n-1}} - 1$ for $i < p^s - 1$, then $F(x^{p^n} - x)$ decomposes into p^n $IQ[p^s, p^n]$, each one being a factor of $X_{i+1}^{p^{n-1}} - 1$; but if $F(x)$ divides $X_{p^s-1}^{p^{n-1}} - 1$, then $F(x^{p^n} - x)$ decomposes into p^{n-1} factors each an $IQ[p^{s+1}, p^n]$ which divides $X_{p^s}^{p^{n-1}} - 1$.

46. As an example under the second part of the last theorem, consider the $IQ[p, p^n]$ of class $p-1$ given at the end of § 44. From it we obtain the $IQ[p^2, p^n]$,

$$F(x^p - x) \equiv (x^p - x - \alpha_0 - \alpha_{p-1})^p + \alpha_{p-1}(x^p - x - \alpha_0 - \alpha_{p-1})^{p-1} - \beta^{p-1}\alpha_{p-1}^p,$$

where $\alpha_0, \alpha_{p-1}, \beta$ are arbitrary marks of the $GF[p^n]$ such that

$$\alpha_{p-1} \neq 0, \quad \beta^{p^{n-1}} + \beta^{p^{n-2}} + \cdots + \beta^p + \beta \neq 0.$$

For an $IQ[p^3, p]$ see Serret, *Cours d'Algèbre supérieure*, II, p. 209.

Miscellaneous theorems on irreducible quantics, §§ 47—49.

47. Theorem. — *An $IQ[m, p^d]$ is irreducible in the $GF[p^{nd}]$ if n be prime to m.*

CLASSIFICATION AND DETERMINATION, etc. 33

The given quantic being $F(x)$, the roots of $F(x) = 0$ are
$$x, \ x^{p^d}, \ x^{p^{2d}}, \ldots, x^{p^{d(m-1)}}$$
all belonging to the $GF[p^{dn}]$. If $F(x)$ be reducible in the $GF[p^{dn}]$, the root x will satisfy an $IQ[t, p^{dn}]$, $t < m$, of the form
$$(X - x)(X - x^{p^{dn}})(X - x^{p^{2dn}}) \ldots (X - x^{p^{dn(t-1)}}) = 0.$$
Its constant term must be a mark of the $GF[p^{dn}]$, so that
$$[x^{1+p^{dn}+p^{2dn}+\cdots+p^{dn(t-1)}}]^{(p^{dn}-1)} \equiv x^{p^{tdn}-1} = 1,$$
in virtue of the single relation $F(x) = 0$. But this requires that tn shall be a multiple of m, and therefore that t be a multiple of m, in contradiction with $t < m$. In fact, by § 23, $F(x)$ divides in the $GF[p^d]$ the function $x^{p^{kd}} - x$ if, and only if, k be a multiple of m.

48. Theorem.[1]) — *An $IQ[\mu, p^n]$ decomposes in the $GF[p^{n\nu}]$ into δ factors each an $IQ\left[\frac{\mu}{\delta}, p^{n\nu}\right]$, δ being the greatest common divisor of μ and ν.*

The given quantic being $F(x)$, the roots of $F(x) = 0$ in the $GF[p^{n\mu}]$ are
$$x, \ x^{p^n}, \ x^{p^{2n}}, \ldots, x^{p^{n(\mu-1)}} \quad [x^{p^{n\mu}} \equiv x].$$
They may be separated into δ sets each of μ/δ roots,
$$x^{p^{ni}}, \ x^{p^{n(\delta+i)}}, \ x^{p^{n(2\delta+i)}}, \ldots, x^{p^{n\left[\left(\frac{\mu}{\delta}-1\right)\delta+i\right]}}$$
for $i = 0, 1, \ldots, \delta - 1$. A symmetric function of the roots in one set is unaltered upon being raised to the power $p^{n\delta}$ and therefore belongs to the $GF[p^{n\delta}]$. The roots of the general set therefore satisfy an equation
$$F_i(X) \equiv (X - x^{p^{ni}})(X - x^{p^{n(\delta+i)}})\cdots = 0,$$
with coefficients belonging to the $GF[p^{n\delta}]$ and à fortiori to the $GF[p^{n\nu}]$. If
$$F_0(X) \equiv X^{\mu/\delta} + A_1 X^{\mu/\delta - 1} + A_2 X^{\mu/\delta - 2} + \cdots$$
then
$$F_i(X) \equiv X^{\mu/\delta} + A_1^{p^{ni}} X^{\mu/\delta - 1} + A_2^{p^{ni}} X^{\mu/\delta - 2} + \cdots.$$

We next prove that the $F_i(X)$ are irreducible in the $GF[p^{n\delta}]$. Suppose, on the contrary, that in the latter field,
$$F_0(X) \equiv f_0(X) \cdot \varphi_0(X).$$
Then
$$F_i(X) \equiv f_i(X) \cdot \varphi_i(X),$$

---

[1]) For the case $n = 1$, this theorem and the corollary of § 49 were stated without proof by Pellet, *Comptes Rendus*, vol. 70 (1870), pp. 328—330.

the coefficients of $f_{i+1}(X)$ being the power p^n of the corresponding ones of $f_i(X)$, those of f_0 being the power p^n of those of $f_{\delta-1}$. The coefficients of the product $f_0 f_1 \ldots f_{\delta-1}$ are consequently unchanged when we replace the coefficients of f_0 by their $(p^n)^{\text{th}}$ powers and are therefore unaltered upon being raised to the power p^n. Hence that product belongs to the $GF[p^n]$, so that $F(x)$ would be reducible in that field, contrary to hypothesis.

Since the degree μ/δ of $F_i(X)$, an $IQ[\mu/\delta, p^{n\delta}]$, is relatively prime to ν/δ, $F_i(X)$ is irreducible in the $GF[p^{n\nu}]$ by § 47.

49. Theorem. — *If $F(\xi)$ be an $IQ[m, p^n]$ in which the coefficient α of ξ^{m-1} is such that in the $GF[p^n]$*

$$\alpha + \alpha^p + \alpha^{p^2} + \cdots + \alpha^{p^{n-1}} \not\equiv 0,$$

then $F(\xi^p - \xi)$ is an $IQ[mp, p^n]$.

If x be one root of $F(\xi) = 0$, its roots are

$$x, \quad x^{p^n}, \quad x^{p^{2n}}, \ldots, x^{p^{n(m-1)}}.$$

By the hypothesis concerning the coefficient

$$\alpha \equiv -x - x^{p^n} - \cdots - x^{p^{n(m-1)}},$$

we have

$$x + x^p + x^{p^2} + \cdots - + x^{p^{nm-1}} \not\equiv 0.$$

Hence, by § 40, $\xi^p - \xi - x$ is irreducible in the $GF[p^{nm}]$. The same holds for each of the quantics

$$X_i \equiv \xi^p - \xi - x^{p^{ni}} \quad (i = 0, 1, \ldots, m-1).$$

Consider the function belonging to the $GF[p^n]$,

$$F(\xi^p - \xi) \equiv \prod_{i=0}^{m-1} X_i.$$

By § 22, it has in the $GF[p^{nm}]$ no irreducible factors other than the X_i. Hence if $F(\xi^p - \xi)$ have a factor $f(\xi)$ belonging to and irreducible in the $GF[p^n]$, $f(\xi)$ must be in the $GF[p^{nm}]$ a product of the X_i,

$$f(\xi) \equiv X_r X_s X_t \ldots,$$

an identity in virtue of $F(x) = 0$. Replacing x by x^{p^n}, another root of $F(x) = 0$, and therefore X_i by X_{i+1} ($i < m$) and X_m by X_0, we obtain from the above identity,

$$f(\xi) \equiv X_{r+1} X_{s+1} X_{t+1} \cdots.$$

Hence $f(\xi)$ contains every factor X_i and therefore coincides with $F(\xi^p - \xi)$. The latter function is therefore irreducible in the $GF[p^n]$.

CLASSIFICATION AND DETERMINATION, etc. 35

Corollary. — *If $F(\xi)$ be an $IQ[m,p]$ in which the coefficient of ξ^{m-1} is not zero, $F(\xi^p - \xi)$ is an $IQ[mp,p]$.*

Examples. — The following congruences are irreducible:

$(x^2 - x)^2 + (x^2 - x) + 1 \equiv x^4 + x + 1 \equiv 0 \pmod{2}$,

$(x^3 - x)^2 + (x^3 - x) - 1 \equiv x^6 + x^4 + x^3 + x^2 - x - 1 \equiv 0 \pmod{3}$.

Primitive roots and primitive irreducible quantics, §§ 50—58.

50. Theorem. — *If R be a primitive root of the $GF[p^{nm}]$ and m_1 a divisor of m, any $IQ[m_1, p^n]$ belonging to an exponent e may be exhibited as a product*

$$28)\qquad \varphi(X) \equiv \prod_{i=0}^{m_1-1} (X - R^{tp^{ni}}),$$

where t is a multiple of $d \equiv (p^{nm}-1)/e$ such that $\dfrac{t}{d}$ is prime to e. Inversely, if e be a proper divisor of $(p^n)^{m_1}-1$ and t be a multiple of d and $\dfrac{t}{d}$ be prime to e, the above product gives an $IQ[m_1, p^n]$ belonging to the exponent e.

Suppose first that $\varphi(X)$ is an $IQ[m_1, p^n]$ belonging to the exponent e, where m_1 is a divisor of m. By § 23, $\varphi(X)$ divides $X^{p^{nm}} - X$ in the $GF[p^n]$, so that any root X_1 of $\varphi(X) = 0$ belongs to the $GF[p^{nm}]$. We may therefore set $X_1 = R^t$. Then, by § 31, we have the decomposition 28). Since $\varphi(X)$ belongs to the exponent e, $X_1 \equiv R^t$ must belong to the exponent e (§ 32). Hence t must be a multiple of $d \equiv (p^{nm}-1)/e$ and $\dfrac{t}{d}$ be prime to e.

To establish the inverse, we first prove that R^t belongs to the exponent e. Since et is assumed to be a multiple of $p^{nm}-1$, we have $R^{et} = 1$. If $R^{tj} = 1$, tj is divisible by $p^{nm}-1$. Set $t = dd'$, so that d' is prime to e. Then must

$$jd' \cdot (p^{nm}-1)/e \equiv 0 \pmod{p^{nm}-1}.$$

Hence must jd', and therefore, j be divisible by e. Hence

$$R^t,\ R^{tp^n},\ R^{tp^{2n}}, \ldots, R^{tp^{n(m_1-1)}}$$

all belong to the exponent e. Upon raising these marks to the power p^n, they are merely permuted. Hence any symmetric function of them, and consequently $\varphi(X)$ defined by 28), belongs to the $GF[p^n]$. Furthermore, $\varphi(X)$ is irreducible in the $GF[p^n]$; for, if $\varphi^1(X)$ be an irreducible factor of degree $m^1 \geqq 1$, it belongs to the exponent e. Then by § 29, e would be a proper divisor of $(p^n)^{m^1}-1$, so that $m^1 = m_1$.

36 CHAPTER III.

Corollary. — *Every $PIQ[m, p^n]$ is given by the formula*
$$F_t(x) \equiv (x - R^t)(x - R^{tp^n}) \ldots (x - R^{tp^{n(m-1)}}),$$
where t is an integer relatively prime to $p^{nm}-1$.
Evidently $F_t \equiv F_{tp^n} \equiv F_{tp^{2n}} \equiv \cdots$

51. The determination of a primitive root in the $GF[p^{nm}]$ is one of the most important as well as most difficult problems in the theory. Special methods of procedure are illustrated in §§ 54—57. We may determine simultaneously all the $PIQ[m, p^n]$ and therefore all the primitive roots of the $GF[p^{nm}]$ by the following method of undetermined coefficients.

The roots of $F_t(x) = 0$ are the tth powers of the roots of $F_1(x) = 0$. Hence the equations
$$F_t(x) = 0, \quad F_1\left(x^{\frac{1}{t}}\right) = 0$$
are equivalent in the $GF[p^{nm}]$. Since t is prime to $p^{nm}-1$, we may determine t' by the congruence
$$tt' \equiv 1 \pmod{p^{nm}-1}.$$
Hence $F_1\left(x^{\frac{1}{t}}\right) = 0$ and $F_1(x^{t'}) = 0$ are equivalent equations in virtue of $x^{p^{nm}} = x$. By § 30, the product of all the $PIQ[m, p^n]$ is given thus:

29) $$\prod_t F_t(x) \equiv \frac{(x^{p^{nm}-1}-1)\prod\left(x^{\frac{p^{nm}-1}{q_i q_j}}-1\right)\ldots}{\prod\left(x^{(p^{nm}-1)/q_j}-1\right)\ldots}$$

where $q_1, q_2, \ldots$ denote the distinct prime factors of $p^{nm}-1$.
But
$$\prod_t F_t(x) = 0, \quad \prod_{t'} F_1(x^{t'}) = 0$$
are equivalent if t and t' each run through the integers less than and relatively prime to $p^{nm}-1$, which give distinct functions $F(x)$. Giving $F_1(x)$ the undetermined form
$$F_1(x) \equiv x^m + ax^{m-1} + bx^{m-2} + \cdots,$$
and forming the product of the $\frac{1}{m}\Phi(p^{nm}-1)$ distinct quantics $F_1(x^{t'})$, the result may be identified with the above fractional expression in x, giving a series of conditions for the coefficients $a, b, \ldots$ The examples which follow will serve to make clear the method.

52. For $p^n = 3$, $m = 2$, we have $p^{nm}-1 = 2^3$. The integers less than and prime to 2^3 are 1, 3, 5, 7. But
$$F_3(x) = F_1(x), \quad F_7(x) = F_{3 \cdot 5}(x) = F_5(x).$$

Hence
$$F_1(x)F_5(x) \equiv \frac{x^8-1}{x^4-1} = x^4+1.$$
Since $5 \cdot 5 \equiv 1 \pmod{2^3}$, $F_5(x) = 0$ and $F_1(x^5) = 0$ are equivalent in the $GF[3^2]$. Let
$$F_1(x) \equiv x^2 + ax + b.$$
If x be a primitive root in the $GF[3^2]$, $x^8 = 1$, $x^4 = -1$. Hence
$$F_1(x^5) \equiv x^{10} + ax^5 + b = x^2 - ax + b.$$
Hence
$$(x^2 + ax + b)(x^2 - ax + b) \equiv x^4 + 1,$$
giving
$$a^2 \equiv 2b, \quad b^2 \equiv 1 \pmod{3}.$$
Hence $b \equiv -1$, $a \equiv \pm 1 \pmod 3$, so that the two $PIQ[2, 3]$ are $x^2 \pm x - 1$.

53. For $p^n = 5$, $m = 2$, we have
$$\prod_t F_t(x) \equiv \frac{(x^{24}-1)(x^4-1)}{(x^{12}-1)(x^8-1)} = \frac{x^{12}+1}{x^4+1} = x^8 - x^4 + 1.$$
The eight integers τ less than and prime to 24 are
$$1,\ 5;\quad 7,\ 11 \equiv 5 \cdot 7;\quad 13,\ 17 \equiv 5 \cdot 13;\quad 19,\ 23 \equiv 5 \cdot 19 \pmod{24}.$$
Each pair of integers furnishes a single $F_t(x)$. For each of the eight values of τ, we have $\tau^2 \equiv 1 \pmod{24}$. Hence $F_\tau(x) = 0$ is identical with $F_1(x^\tau) = 0$ in the $GF[5^2]$. For a primitive root x, we have $x^{12} \equiv -1$. We have therefore in the field,
$$F_1(x) \equiv x^2 + ax + b, \qquad F_1(x^{13}) \equiv x^2 - ax + b,$$
$$x^2 F_1(x^{11}) \equiv bx^2 - ax + 1, \quad x^2 F_1(x^{23}) \equiv bx^2 + ax + 1.$$
The product of these four quadratics is therefore identical modulo 5 with
$$b^2(x^8 - x^4 + 1).$$
It follows that $b^2 \equiv -1 \pmod 5$ and, by subsequent expansion,
$$2a^2 \equiv b \pmod 5.$$
Hence the four $PIQ[2, 5]$ are $x^2 + ax + 2a^2$, viz.,

30) $\qquad x^2 \pm x + 2, \quad x^2 \pm 2x - 2.$

Another method of solving this example is to require that $x^2 + ax + b$ shall divide $x^8 - x^4 + 1$ modulo 5. We reduce the latter function by means of the relation
$$x^2 = -ax - b \quad [a^4 \equiv b^4 \equiv 1 \pmod 5],$$
and find, modulo 5, that
$$x^8 - x^4 + 1 \equiv (-a^5 b - ab^3)x - a^4 b^2 - a^2 b^3 + 2.$$
Hence
$$b^2 \equiv -1, \quad 2a^2 \equiv b \pmod 5.$$

38   CHAPTER III.

54. The eight $PIQ[4,3]$ are the factors of

31) $\qquad \dfrac{(x^{80}-1)\,(x^8-1)}{(x^{40}-1)\,(x^{16}-1)} \equiv x^{32} - x^{24} + x^{16} - x^8 + 1.$

It suffices, however, in view of § 50, to determine a primitive root ϱ of the $GF[3^4]$. To get an $IQ[4,3]$, we employ the theorem of § 37 for $\lambda = 4$, $i = 2$, $\varrho = 2$, $l = 1$, giving the decomposition

$$x^8 + 1 \equiv \Pi\,(x^4 \pm x^2 - 1) \pmod{3}.$$

Hence a root i of the irreducible congruence

$$x^4 - x^2 - 1 \equiv 0 \pmod{3}$$

belongs to the exponent 16. If then we find a mark σ belonging to the exponent 5, $\varrho \equiv i\sigma$ will, by § 14, be a primitive root of $x^{80} = 1$. We readily verify that the fifth power of $i^2 \pm i$ is congruent to unity modulo 3. To find the irreducible congruence satisfied by the primitive root $\varrho = i\,(i^2 \pm i)$, we form its powers,

$$\varrho^2 = \mp i^3 \mp i - 1, \quad \varrho^3 = \mp i^2 - i \pm 1, \quad \varrho^4 = \mp i^3 - i^2 \mp i + 1.$$

Eliminating the powers of i, we have

$$\varrho^4 \pm \varrho^3 + \varrho^2 \mp \varrho - 1 \equiv 0 \pmod{3}.$$

The product of the two $PIQ[4,3]$ thus reached is $\varrho^8 + \varrho^6 + \varrho^4 + 1$. Since the expression 31) contains only exponents which are multiples of 4, we would expect the new factor $\varrho^8 - \varrho^6 + \varrho^4 + 1$. In fact, the product of these two quantics of degree 8 gives $\varrho^{16} + \varrho^{12} - \varrho^4 + 1$, which divides 31) giving the quotient

$$\varrho^{16} - \varrho^{12} + \varrho^4 + 1 \equiv (\varrho^8 + \varrho^4 + \varrho^2 + 1)\,(\varrho^8 + \varrho^4 - \varrho^2 + 1).$$

We therefore have two new $PIQ[4,3]$ given by the decomposition

$$\varrho^8 - \varrho^6 + \varrho^4 + 1 \equiv (\varrho^4 + \varrho^3 - 1)\,(\varrho^4 - \varrho^3 - 1).$$

Since $\varrho^8 + \varrho^4 \pm \varrho^2 + 1$ is derived from $\varrho^8 \pm \varrho^6 + \varrho^4 + 1$ upon replacing ϱ by $\dfrac{1}{\varrho}$ in the latter and multiplying by ϱ^8, we find

$$\varrho^8 + \varrho^4 + \varrho^2 + 1 = \Pi\,(\varrho^4 \pm \varrho^3 - \varrho^2 \mp \varrho - 1),$$
$$\varrho^8 + \varrho^4 - \varrho^2 + 1 = \Pi\,(\varrho^4 \mp \varrho - 1).$$

Hence the eight $PIQ[4,3]$ are

$$\varrho^4 \pm \varrho^3 - 1, \quad \varrho^4 \pm \varrho - 1, \quad \varrho^4 + \varepsilon\varrho^3 \pm \varrho^2 - \varepsilon\varrho - 1 \quad (\varepsilon = \pm 1).$$

55. To obtain a primitive root ϱ of the $GF[5^4]$, we define the latter by means of a root i of the irreducible congruence

$$x^4 \equiv 2 \pmod{5}.$$

Indeed, by § 35, $x^4 - 3^3$ is an $IQ[4,5]$ belonging to the exponent 16. Since $5^4 - 1 = 16 \cdot 3 \cdot 13$, we seek marks belonging to the exponents 3

CLASSIFICATION AND DETERMINATION etc.   39

and 13. We verify at once that $2i^2+2$ belongs to the exponent 3. To find the most general mark η which belongs to the exponent 13, we simplify the calculations by first determining the marks
$$\eta_1 \equiv ai^3 + bi^2 + ci + d$$
of the $GF[5^4]$ for which $\eta_1^{26} = 1$. Then either $(+\eta_1)^{13}$ or $(-\eta_1)^{13}$ equals unity. Now
$$\eta_1^{25} \equiv ai^{75} + bi^{50} + ci^{25} + d$$
$$\equiv a2^{18}i^3 + b2^{12}i^2 + c2^6 i + d$$
$$\equiv -ai^3 + bi^2 - ci + d.$$
The condition $\eta_1^{26} = 1$ thus gives
$$(bi^2 + d)^2 - (ai^3 + ci)^2 \equiv 1.$$
Reducing by $i^4 \equiv 2$, we obtain the conditions, modulo 5,
$$-2a^2 - c^2 + 2bd \equiv 0, \quad -4ac + 2b^2 + d^2 \equiv 1.$$
For $a \equiv 0$, the only solutions are seen to be
$$b^2 \equiv 1, \quad d^2 \equiv -1, \quad c^2 \equiv \pm 1; \quad b \equiv c \equiv 0, \quad d \equiv \pm 1.$$
Hence $\pm i^2 + ci \pm 2$ ($c = 1, 2, 3$ or 4), or else the negative of this expression, belongs to the exponent 13. We may verify that $i^2 + i + 3$ belongs to the exponent 13. We may therefore take
$$\varrho = i(2i^2 + 2)(i^2 + i + 3) \equiv 3i^3 + 2i^2 + 4.$$
Then
$$\varrho^2 \equiv -i^3 - i^2 - i - 1, \quad \varrho^3 \equiv i^3 - 2i^2 + i + 1, \quad \varrho^4 \equiv -i^3 + i + 2.$$
Hence we obtain the following $PIQ[4, 5]$ satisfied by the primitive root ϱ,
$$\varrho^4 - \varrho^3 - \varrho - 2 \equiv 0 \pmod{5}.$$
This quartic can be decomposed into the two $PIQ[2, 5^2]$,
$$(x - \varrho)(x - \varrho^{25}), \quad (x - \varrho^5)(x - \varrho^{125}).$$
But
$$\varrho^5 \equiv 4i^3 + 3i^2 + 4, \quad \varrho^{25} \equiv 2i^3 + 2i^2 + 4, \quad \varrho^{125} \equiv i^3 + 3i^2 + 4.$$
Hence
$$(x - \varrho)(x - \varrho^{25}) \equiv x^2 - x(-i^2 + 3) + 3i^2 + 4,$$
$$(x - \varrho^5)(x - \varrho^{125}) \equiv x^2 - x(i^2 + 3) - 3i^2 + 4.$$

56. The determination of primitive roots in the $GF[5^6]$ and in the $GF[5^3]$ may be made to depend upon the congruence

32) $\qquad x^6 + x^5 + x^4 + x^3 + x + x^2 + 1 \equiv 0 \pmod 5$,

which, by § 33, is irreducible. The root x belongs to the exponent 7. The general mark of the $GF[5^6]$ may be expressed in the form
$$\sigma \equiv \sum_{i=0}^{5} c_i x^i \quad \text{(each } c_i \text{ an integer)}.$$

It will belong to the included field $GF[5^3]$ if, and only if, $\sigma^{125} = \sigma$. Applying $x^7 \equiv 1$, we have (mod 5)
$$\sigma^{125} \equiv \sum_{i=0}^{5} c_i x^{125i} \equiv \sum_{i=1}^{5} c_i x^{7-i} + c_0.$$
Applying 32), this becomes
$$(c_0 - c_1) - c_1 x + (c_5 - c_1)x^2 + (c_4 - c_1)x^3 + (c_3 - c_1)x^4 + (c_2 - c_1)x^5.$$
The conditions that this shall be identical with σ are
$$c_1 \equiv 0, \quad c_2 \equiv c_5, \quad c_3 \equiv c_4 \pmod{5}.$$
Hence the 5^3 marks of the $GF[5^3]$ are given by
33) $\quad c_0 + c_2(x^2 + x^5) + c_3(x^3 + x^4) \quad [c_0, c_2, c_3 = 0, 1, 2, 3, 4].$
Since $\quad\quad\quad\quad\quad (x^2 + x^5)^5 \equiv x^3 + x^4,$
we infer that $\tau \equiv x^2 + x^5$ defines the $GF[5^3]$. In fact, we find
$$\tau^5 \equiv x^3 + x^4, \quad \tau^{25} \equiv x + x^6, \quad \tau^{30} \equiv x^5 + x^4 + x^3 + x^2,$$
and finally that $\tau^{31} = 1$. Hence $\lambda \equiv 2(x^2 + x^5)$ belongs to the exponent $4 \cdot 31$ and is therefore a primitive root in the $GF[5^3]$. We derive at once the $PIQ[3, 5]$ satisfied by λ, viz.,
$$2\lambda^3 = \lambda^2 + \lambda + 1 \pmod{5}.$$

We next verify that $x - 2$ belongs to the exponent $2^3 \cdot 3^2 \cdot 31$, so that $\varrho \equiv x(x - 2)$ belongs to the exponent
$$5^6 - 1 \equiv 2^3 \cdot 3^2 \cdot 7 \cdot 31,$$
so that ϱ is a primitive root in the $GF[5^6]$. We have
$$(x - 2)^{126} \equiv (x^6 - 2)(x - 2) \equiv -2(x + x^6) \equiv -2\tau^{25}.$$
But τ^{25} belongs to the exponent 31. Hence the exponent of $x - 2$ contains the factor 31 and, moreover, the factor 2^3, since
$$(x - 2)^{\frac{1}{2}(5^6 - 1)} = (x - 2)^{126 \cdot 31 \cdot 2} \equiv (-2)^{62} \equiv -1 \pmod{5}.$$
We next prove that the power $2^3 \cdot 3^2 \cdot 31$ of $x - 2$ gives unity. Indeed,
$$(x - 2)^{15} \equiv (x^5 - 2)^3 \equiv 2x^5 - x^3 + x + 2 \pmod{5},$$
and, by a slight calculation,
$$(x - 2)^{18} = 2x^5 + x^4 + x^3 + 2x^2 + 4.$$
This being of the form 33), we have
$$(x - 2)^{2^3 \cdot 3^2 \cdot 31} \equiv [(x - 2)^{18}]^{124} \equiv 1 \pmod{5}.$$
For the same reason,
$$(x - 2)^{2^3 \cdot 3 \cdot 31} \equiv [(x - 2)^6]^{124} \equiv (2x^5 - x^4 - x^3 - x^2 + 2x + 3)^{124} \not\equiv 1.$$

CLASSIFICATION AND DETERMINATION, etc. 41

To determine the $PIQ[6, 5]$ satisfied by the primitive root $\varrho \equiv x^2 - 2x$, we form the powers,
$$\varrho^3 \equiv -2x^5 + x^4 + x^3 - x^2 - x - 1; \quad \varrho^5 \equiv x^3 - 2x^5,$$
$$\varrho^4 \equiv -x^5 + 2x^4 + x^3 + x^2 + 2x + 3, \quad \varrho^6 \equiv 2x^5 - x^4 + x^3 + x^2 + x - 1.$$
We derive at once the required congruence
$$\varrho^6 - \varrho^5 + \varrho^4 - \varrho^3 + 2\varrho + 2 \equiv 0 \pmod{5}.$$

57. We can set up the $PIQ[2, 2^3]$ and $PIQ[6, 2]$ by means of the theorem:

34) $\qquad \lambda^2 x^2 + \lambda x + \beta$

is a $PIQ[2, 2^3]$ if, and only if, β is a root of

35) $\qquad j^3 \equiv j^2 + 1 \pmod{2}$

and λ is any mark except zero and β^4.

By § 40, the quadratic 34) is an $IQ[2, 2^3]$ for every mark $\lambda \neq 0$ in the $GF[2^3]$ and for every root β of the congruence
$$\beta^4 + \beta^2 + \beta + 1 \equiv (\beta + 1)(\beta^3 + \beta^2 + 1) \equiv 0 \pmod{2}.$$
Defining the $GF[2^3]$ by means of the irreducible congruence 35), we may take $\beta = 1, j, j^2$ or j^4. We first find the exponent e_β to which belongs a root ξ of the congruence
$$\xi^2 \equiv \xi + \beta \pmod{2}.$$
Since ξ belongs to the $GF[2^{2 \cdot 3}]$, e_β is a divisor of $2^6 - 1 \equiv 3^2 \cdot 7$. But
$$\xi^3 = \xi(\beta+1) + \beta, \quad \xi^7 = \xi(\beta^3+\beta+1) + \beta^3 + \beta, \quad \xi^9 = \xi(\beta+1)(\beta^3+\beta^2+1) + \beta.$$
Hence for $\beta = 1$, $e_\beta = 3$; for a root β of 35), we find
$$\xi^9 = \beta, \quad \xi^{21} = \xi^{18} \cdot \xi^3 = \xi + \beta^3, \quad \xi^7 = \xi(\beta^2 + \beta) + \beta^2 + \beta + 1,$$
so that $e_\beta = 2^6 - 1$. The theorem is therefore proven for the case $\lambda = 1$.

Setting $\xi = \lambda x$, it follows that, for $\beta \neq 1$, x belongs to the exponent $2^6 - 1$ unless $x^9 = 1$, which occurs only when $\lambda^2 = \beta$, i. e., $\lambda = \beta^4$. We therefore reach all $\frac{1}{2}\Phi(2^6 - 1) \equiv 18 \, PIQ[2, 2^3]$. Half of them are given in the left members of the identities below. To pick out a set of three whose product gives a $PIQ[6, 2]$, we select three which are like functions of respectively j, j^2, j^4, the latter being the roots of 35). We thus find

$$(x^2 + x + j)(x^2 + x + j^2)(x^2 + x + j^4) = x^6 + x^5 + x^3 + x^2 + 1,$$
$$(j^6 x^2 + j^3 x + j)(j^5 x^2 + j^6 x + j^2)(j^3 x^2 + j^5 x + j^4) = x^6 + x^5 + x^4 + x + 1,$$
$$(j^6 x^2 + j^3 x + j^4)(j^5 x^2 + j^6 x + j)(j^3 x^2 + j^5 x + j^2) = x^6 + x^5 + 1.$$

CHAPTER III.

Replacing x by $\frac{1}{x}$ and multiplying by x^6, we find
$$x^6 + x^4 + x^3 + x + 1, \quad x^6 + x^5 + x^2 + x + 1, \quad x^6 + x + 1,$$
which with the above three sextics give the six existing $PIQ[6, 2]$.

58. Theorem. — *The necessary and sufficient conditions that $x^p - x - \alpha$ shall be a $PIQ[p, p]$ are that α be a primitive root modulo p and that a root of $y^p \equiv y + 1 \pmod{p}$ belong to the exponent $(p^p - 1)/(p - 1)$.*

If α be an integer not divisible by p, the congruence
$$x^p \equiv x + \alpha \pmod{p}$$
is irreducible by § 40. The product of its roots is
$$x x^p x^{p^2} \ldots x^{p^{p-1}} = x^{\frac{p^p-1}{p-1}} = \alpha.$$
Setting $x = \alpha y$, we find that
$$y^p \equiv y + 1 \pmod{p}.$$
Hence if x belong to the exponent $p^p - 1$, then α is a primitive root modulo p and y belongs to the exponent $(p^p - 1)/(p - 1)$. The inverse is true by § 14, since $p - 1$ and $(p^p - 1)/(p - 1)$ are relatively prime.

59.   EXERCISES ON CHAPTER III.

Ex. 1. If ϱ be a root of one of the $PIQ[2, 5]$ of § 53, then $x^3 - \varrho$ is an $IQ[3, 5^2]$. Eliminate ϱ and derive the following $IQ[6, 5]$:
$$x^6 \pm x^3 + 2, \quad x^6 \pm 2x^3 - 2.$$

Ex. 2. (Moore). If x be a root of the irreducible congruence
$$x^6 - 2x^3 - 2 \equiv 0 \pmod{5},$$
a mark $c_0 + c_1 x + c_2 x^2 + c_3 x^3 + c_4 x^4 + c_5 x^5$ of the $GF[5^6]$ will belong to the included field $GF[5^3]$ if and only if
$$c_3 \equiv 0, \quad c_4 \equiv 3c_1 + 4c_2, \quad c_5 \equiv 2c_1 + 3c_2 \pmod{5}.$$
Show that $\varphi \equiv x + x^2 + 2x^4$ is a primitive root of the $GF[5^3]$ and that it satisfies the congruence $\varphi^3 \equiv 2\varphi + 3 \pmod{5}$.

Ex. 3. (Pellet). If γ belong to the $GF[p^n]$ and m be the least integer for which $\gamma^{p^m} = \gamma$, then $x^p - x - \gamma$ is irreducible in the field if neither n/m nor $\gamma + \gamma^p + \gamma^{p^2} + \cdots + \gamma^{p^{m-1}}$ be divisible by p; in the contrary case it decomposes in the field into linear factors. Prove this theorem equivalent to that of § 40 for $\lambda = 1$.

Ex. 4. (Pellet). If p be a prime number which is a primitive root of the prime number n, $\dfrac{(x^p - x)^n - 1}{x^p - x - 1}$ is irreducible modulo p.

Ex. 5. Show that the theorems of § 34 and § 36 may be combined into the theorem stated without proof by Pellet:

CLASSIFICATION AND DETERMINATION, etc.   43

If in an $IQ[\nu_1, p^\nu]$ belonging to the exponent n, we replace x by x^λ, where λ contains only the prime factors of n, the resulting quantic decomposes into $\frac{1}{n} D 2^{k-1}$ quantics $IQ\left[\frac{\lambda n \nu_1}{D 2^k - 1}, p^\nu\right]$ belonging to the exponent λn, where D is the greatest common divisor of λn and $p^{\nu \nu_1} - 1$ and where 2^{k-1} is the highest power of 2 dividing the numerators of each of the fractions $\frac{p^{\nu \nu_1} + 1}{2}$ and $\frac{\lambda n}{2D}$ when reduced to their simplest form.

Ex. 6. (Schönemann). If $F(x, \alpha)$ be an $IQ[m, p^n]$ in which the coefficient of at least one power of x satisfies the equation $c^{p^\nu - 1} \equiv 1$ if, and only if, $\nu = n$ or a multiple of n, the product

$$F(x, \alpha) \cdot F(x, \alpha^p) \ldots F(x, \alpha^{p^{n-1}})$$

gives an $IQ[mn, p]$.

Ex. 7. (Schönemann). Generalize the theorem of § 33 as follows: If p belong to the exponent t modulo e, e being prime, $(x^e - 1)/(x - 1)$ decomposes modulo p into $(e - 1)/t$ quantics irreducible modulo p.

Ex. 8.   Prove that $x^5 - x + 1$ is a $PIQ[5, 3]$.

Ex. 9.   (Pellet). If e be the exponent to which belongs

$$i^{1 + p + p^2 + \cdots + p^{\nu - 1}}$$

the product of the roots of an irreducible congruence of degree ν, $F(x) \equiv 0$ (mod p), and if λ be a prime divisor of e, then

1) $F(x^\lambda)$ is irreducible modulo p if λ does not divide $(p - 1)/e$;
2) $F(x^\lambda)$ decomposes into λ irreducible factors of degree ν if λ divides $(p - 1)/e$. According as λ divides or does not divide e, all of these factors belong or do not belong to the same exponent.

Ex. 10.   Using Jordan's irreducible congruence

$$x^9 \equiv x + 1 \pmod{2},$$

show that x belongs to the exponent 73 and $x + x^4 + x^6 + x^7 + x^8$ to the exponent 7. The product $y \equiv x(x + x^4 + x^6 + x^7 + x^8)$ belongs to the exponent $2^9 - 1$ and is therefore a primitive root of the $GF[2^9]$. Verify that it satisfies the congruence

$$y^9 + y^8 + y^4 + y^3 + y^2 + y + 1 \equiv 0 \pmod{2}.$$

Ex. 11.   If the $GF[3^2]$ be defined by $i^2 \equiv i + 1 \pmod{3}$, the 16 $PIQ[2, 3^2]$ are given by the decomposition of the $PIQ[4, 3]$ of § 54; for example,

$$x^4 \pm x - 1 \equiv \{x^2 \pm (i + 1) x - i\}\{x^2 \mp (i + 1) x + i - 1\},$$
$$x^4 \mp x^3 - 1 \equiv \{x^2 \pm (i - 1) x + i\}\{x^2 \mp ix - i + 1\}.$$

Ex. 12. (Mathieu). If H belong to the $GH[p^{nm}]$, we have the decomposition
$$H(Z^{p^{nm}} - Z) = \Pi\{(HZ)^{p^{n(m-1)}} + (HZ)^{p^{n(m-2)}} + \cdots + (HZ)^{p^n} + HZ + \mu\},$$
where μ runs through the series of marks of the $GF[p^n]$.

60. *Table of primitive irreducible quantics*[1]). When more than one $PIQ[m, p]$ is known, we choose that one $x^m \equiv \alpha x^r + \beta x^{r-1} + \cdots$ (mod p) in which the exponent r is as small as possible.

Modulo 2: $x^2 \equiv x+1$, $x^3 \equiv x+1$, $x^4 \equiv x+1$, $x^5 \equiv x^2+1$, $x^6 \equiv x+1$,
$x^7 \equiv x+1$, $x^8 \equiv x^4 + x^3 + x^2 + 1$, $x^9 \equiv x^8 + x^4 + x^3 + x^2 + x + 1$.

Modulo 3: $x^2 \equiv 2x+1$, $x^3 \equiv x+2$, $x^4 \equiv 2x^3 + 2x^2 + x + 1$, $x^5 \equiv x+2$.

Modulo 5: $x^2 \equiv 2x+2$, $x^3 \equiv 2x+3$, $x^4 \equiv x^2 + x + 2$, $x^5 \equiv x+2$,
$x^6 \equiv x^5 - x^4 + x^3 - 2x - 2$.

Modulo [2]) 7: $x^2 \equiv x-3$, $x^3 \equiv x-2$, $x^4 \equiv 2x^3 + 2x + 2$, $x^5 \equiv 6x+3$,
$x^6 \equiv -x^5 - x^4 - x^3 - x^2 - x - 3$, $x^7 \equiv x+3$.

Modulo 11: $x^2 \equiv 4x - 2$.

CHAPTER IV.

MISCELLANEOUS PROPERTIES OF GALOIS FIELDS.

Squares, not-squares, m^{th} powers in a Galois Field, §§ 61—63.

61. Every mark of the $GF[2^n]$ satisfies the equation $x^{2^n} = x$, so that x is the square of the mark $x^{2^{n-1}}$. Every mark has one and only one square root, since $-1 = +1$ in the $GF[2^n]$.

In the $GF[p^n]$, $p > 2$, a mark may or may not be the square of a mark belonging to the field, and is called *a square* or *a not-square* respectively. If ϱ be a primitive root of the $GF[p^n]$, so that

36) $\qquad \varrho^{p^n-1} = 1, \quad \varrho^{(p^n-1)/2} = -1,$

the even powers of ϱ are squares, $\varrho^{2h} = (\pm \varrho^h)^2$; while the odd powers are not-squares. In fact, $\varrho^{2h+1} = x^2$ would require

$$\varrho^{(2h+1)(p^n-1)/2} \equiv \varrho^{h(p^n-1)} \cdot \varrho^{(p^n-1)/2} = -1 = x^{p^n-1} = +1.$$

Hence there are $(p^n-1)/2$ squares and as many not-squares in the $GF[p^n]$. Furthermore, the product or quotient of two squares or of two not-squares is again a square; but the product or quotient of a square by a not-square, or *vice versa*, is a not-square.

---

1) A table of irreducible quantics (not all primitive) is given by Jordan, *Comptes Rendus*, 72 (1871), pp. 283—290. His quantic $x^5 + x^3 + x^2 + x + 1$ is divisible by $x^3 + x^2 + 1$ modulo 2, while $x^5 + x + 2$ is divisible by $x - 5$ mod 11.

2) Serret, *Cours d'Algèbre supérieure*, II, pp. 181—189.

MISCELLANEOUS PROPERTIES OF GALOIS FIELDS. 45

62. Theorem. — *The not-squares of any $GF[p^n]$, $p > 2$, are not-squares or squares in the $GF[p^{nm}]$ according as m is odd or even.*

If σ be a primitive root of the $GF[p^{nm}]$, then $\varrho \equiv \sigma^u$, where $u \equiv (p^{nm}-1)/(p^n-1)$, is a primitive root of the $GF[p^n]$. Hence the marks $\neq 0$ of the $GF[p^n]$ are given by the formula

$$\varrho^v \equiv \sigma^{uv} \quad (v = 1, 2, \ldots, p^n - 1).$$

Let ϱ^v be a not-square in the $GF[p^n]$, so that v is odd. It will be a not-square or a square in the $GF[p^{nm}]$ according as uv is odd or even, i. e., according as u is odd or even. But

$$u \equiv (p^{nm}-1)/(p^n-1) = \sum_{i=0}^{m-1} p^{ni} = \text{sum of } m \text{ odd terms}.$$

Hence u is odd or even according as m is odd or even.

63. Theorem. — *If d be the greatest common divisor of m and p^n-1, there exist exactly $(p^n-1)/d$ marks $\neq 0$ in the $GF[p^n]$ which are m^{th} powers in the field.*

If $\mu \neq 0$ be the mth power of some mark v of the field, we find, upon raising $\mu = v^m$ to the power $(p^n-1)/d$ and noting that the power p^n-1 of the mark $v^{m/d} \neq 0$ is 1, the equation

37) $$\mu^{(p^n-1)/d} = 1.$$

Inversely, there are $(p^n-1)/d$ roots of 37) in the $GF[p^n]$ by § 16 and each root is an mth power in the $GF[p^n]$. To prove the last statement, we note first that such a root μ is a dth power. In fact, the roots of 37) may be exhibited as follows:

$$\varrho^{di} \quad \left(i = 0, 1, \ldots, \frac{p^n-1}{d} - 1\right),$$

where ϱ is a primitive root of the $GF[p^n]$. That these roots are distinct is shown by supposing

$$\varrho^{di} = \varrho^{dj} \quad (j \leqq i < (p^n-1)/d).$$

Then

$$\varrho^{dj}(\varrho^{d(i-j)} - 1) = 0 \quad [d(i-j) < p^n - 1].$$

Hence $i - j = 0$. We next prove that $\mu \equiv \varrho^{di}$ is an mth power. Since m/d is relatively prime to $p^n - 1$, we can determine integers l and t satisfying the equation

$$t(p^n-1) + lm/d = 1.$$

Hence

$$\varrho = (\varrho^l)^{m/d} \cdot (\varrho^{p^n-1})^t = \varrho^{lm/d}.$$

Therefore

$$\mu = \varrho^{di} = (\varrho^{li})^m.$$

Corollary. — *Every mark of the $GF[p^n]$ will be an mth power in the field if, and only if, $d = 1$.* Extraction of the mth root of an

46   CHAPTER IV.

arbitrary mark of the $GF[p^n]$ is possible if, and only if, m be relatively prime to p^n-1. With this condition satisfied, there exists but one m^{th} root of each mark.

Number of solutions of certain quadratic equations in a Galois Field, §§ 64—67.

64. Theorem.[1]) — *If $\nu = +1$ or -1 according as $-\alpha_1\alpha_2$ is a square or a not-square in the $GF[p^n]$, $p > 2$, the equation belonging to the field,*
$$\alpha_1 \xi_1^2 + \alpha_2 \xi_2^2 = \varkappa \quad (\alpha_1 \neq 0, \ \alpha_2 \neq 0),$$
has $p^n - \nu$ or $p^n + (p^n - 1)\nu$ sets of solutions according as $\varkappa \neq 0$ or $\varkappa = 0$.

Setting $\alpha_1 \xi_1 \equiv \eta$, the equation becomes
$$\eta^2 + \alpha_1 \alpha_2 \xi_2^2 = \alpha_1 \varkappa.$$

1^0. If $-\alpha_1 \alpha_2 = \lambda^2$, a square $\neq 0$ in the $GF[p^n]$, we set
$$\eta + \lambda \xi_2 = \varrho, \quad \eta - \lambda \xi_2 = \sigma,$$
whence
$$\eta = \frac{1}{2}(\varrho + \sigma), \quad \xi_2 = \frac{1}{2\lambda}(\varrho - \sigma).$$
The equation becomes
$$\varrho \sigma = \alpha_1 \varkappa.$$

If $\varkappa \neq 0$, we can give to σ any one of the $p^n - 1$ marks $\neq 0$ in the $GF[p^n]$, when the corresponding value of ϱ is determined by the equation. There are in this case $p^n - 1$ sets of solutions ξ_1, ξ_2 in the field of the given equation.

If $\varkappa = 0$, there are evidently $1 + 2(p^n - 1)$ sets of solutions.

2^0. If $-\alpha_1 \alpha_2$ be a not-square in the $GF[p^n]$, the equation
$$\varphi^2 = -\alpha_1 \alpha_2$$
is irreducible in the field. If one root be i, the other is $i^{p^n} = -i$ by the corollary of § 31. We therefore have the identity
$$\eta^2 + \alpha_1 \alpha_2 \xi_2^2 = (\eta + i\xi_2)(\eta + i^{p^n}\xi_2) = (\eta + i\xi_2)^{p^n+1}.$$
We are thus led to determine the number of roots in the $GF[p^{2n}]$ of the equation in the unknown $Z \equiv \eta + i\xi_2$,

38) $\qquad\qquad Z^{p^n+1} = \alpha_1 \varkappa.$

If $\varkappa = 0$, we have $Z = 0$ and hence a single set of solutions $\xi_1 = 0$, $\xi_2 = 0$.

If $\varkappa \neq 0$, let R be a primitive root of the $GF[p^{2n}]$. We may set $\alpha_1 \varkappa \equiv R^k$, whence
$$R^{k(p^n-1)} = (\alpha_1 \varkappa)^{p^n-1} = 1,$$
so that $k(p^n - 1)$ is divisible by $p^{2n} - 1$, the exponent to which R belongs. We may therefore set $k = l(p^n + 1)$, l being an integer.

---

[1]) The theorems of §§ 64—67 are immediate generalizations of Nos. 197—200 of Jordan's Traité des substitutions.

MISCELLANEOUS PROPERTIES OF GALOIS FIELDS.   47

Since Z belongs to the $GF[p^{2n}]$, we may set $Z \equiv R^t$. The equation 38) becomes
$$R^{t(p^n+1)} = R^{l(p^n+1)}.$$
Hence
$$t(p^n + 1) \equiv l(p^n + 1) \quad [\text{mod } p^{2n} - 1].$$
This congruence has $p^n + 1$ distinct solutions for t, viz.,
$$t \equiv l, \quad l + (p^n - 1), \quad l + 2(p^n - 1), \ldots, l + p^n(p^n - 1).$$
The corresponding values of $R^t \equiv Z \equiv \eta + i\xi_2$ give $p^n + 1$ distinct sets of solutions ξ_1, ξ_2 of the given equation.

65. Theorem. — *The number of sets of solutions* $(\xi_1, \xi_2, \ldots, \xi_{2m})$ *in the* $GF[p^n]$, $p > 2$, *of the equation*
$$\alpha_1 \xi_1^2 + \alpha_2 \xi_2^2 + \cdots + \alpha_{2m} \xi_{2m}^2 = \varkappa,$$
where every α_j *is a mark* $\neq 0$ *in the field, is*
$$p^{n(2m-1)} - \nu p^{n(m-1)} \qquad (if \; \varkappa \neq 0)$$
$$p^{n(2m-1)} + \nu(p^{nm} - p^{n(m-1)}) \quad (if \; \varkappa = 0),$$
where ν *is* $+1$ *or* -1 *according as* $(-1)^m \alpha_1 \alpha_2 \ldots \alpha_{2m}$ *is a square or a not-square in the field.*

By § 64, the theorem is true if $m = 1$. To prove the theorem by induction, we suppose it true for equations in $2(m-1)$ variables. The proposed equation is equivalent to the system of two equations
$$\alpha_1 \xi_1^2 + \alpha_2 \xi_2^2 = \eta, \quad \alpha_3 \xi_3^2 + \cdots + \alpha_{2m} \xi_{2m}^2 = \varkappa - \eta.$$
1^0. Let $\varkappa \neq 0$. For each of the $p^n - 2$ values of η different from $\varkappa$ and 0, the first equation has $p^n - \lambda$ sets of solutions, while by hypothesis the second has $p^{n(2m-3)} - \mu p^{n(m-2)}$, where $\lambda = \pm 1$ according as $-\alpha_1 \alpha_2$ is a square or a not-square, and $\mu = \pm 1$ according as $(-1)^{m-1} \alpha_3 \alpha_4 \ldots \alpha_{2m}$ is a square or a not-square. For the value $\eta = 0$, they have respectively $p^n + (p^n - 1)\lambda$ and $p^{n(2m-3)} - \mu p^{n(m-2)}$ sets of solutions. Finally, for $\eta = \varkappa$, they have respectively $p^n - \lambda$ and $p^{n(2m-3)} + \mu(p^{n(m-1)} - p^{n(m-2)})$ sets of solutions. The total number of sets of solutions is therefore
$$(p^n - 2)(p^n - \lambda)(p^{n(2m-3)} - \mu p^{n(m-2)}) + [p^n + (p^n - 1)\lambda][p^{n(2m-3)} - \mu p^{n(m-2)}]$$
$$+ (p^n - \lambda)[p^{n(2m-3)} + \mu(p^{n(m-1)} - p^{n(m-2)})]$$
$$\equiv p^{n(m-1)}(p^{nm} - \lambda \mu).$$

By § 61, $\lambda \mu = \nu$. Hence the induction is complete.

2^0. Let $\varkappa = 0$. Separating the two cases $\eta \neq 0$ and $\eta = 0$, we find the total number of solutions to be
$$(p^n - 1)(p^n - \lambda)(p^{n(2m-3)} - \mu p^{n(m-2)})$$
$$+ [p^n + \lambda(p^n - 1)][p^{n(2m-3)} + \mu(p^{n(m-1)} - p^{n(m-2)})]$$
$$\equiv p^{n(2m-1)} + \lambda \mu(p^{nm} - p^{n(m-1)}).$$

CHAPTER IV.

66. Theorem.— *The number of sets of solutions in the $GF[p^n]$, $p>2$, of the equation*
$$\alpha_1 \xi_1^2 + \alpha_2 \xi_2^2 + \cdots + \alpha_{2m+1} \xi_{2m+1}^2 = \varkappa,$$
where each α_j is a mark $\neq 0$ in the field and $\varkappa$ belongs to the field, is $p^{2nm} + \omega p^{nm}$, where $\omega = +1, -1$ or 0 according as $(-1)^m \varkappa \alpha_1 \alpha_2 \ldots \alpha_{2m+1}$ is a square, a not-square or zero in the field.

Consider the equivalent system of equations
$$\alpha_1 \xi_1^2 = \eta, \quad \alpha_2 \xi_2^2 + \cdots + \alpha_{2m+1} \xi_{2m+1}^2 = \varkappa - \eta.$$

The first equation has one solution if $\eta = 0$. If $\eta \neq 0$, it has two or no solutions according as $\alpha_1 \eta$ is a square or a not-square. Let $\mu = 0$ if $\varkappa = 0$, and $\mu = \pm 1$ according as $\alpha_1 \varkappa$ is a square or a not-square. We may express the number of solutions of the second equation by § 65, if we set $\nu = \pm 1$ according as $(-1)^m \alpha_2 \ldots \alpha_{2m+1}$ is a square or a not-square. Evidently we have $\mu \nu = \omega$.

According as $\mu = 0, +1$, or -1, the total number of sets of solutions of the pair of equations is respectively

$$1[p^{n(2m-1)} + \nu(p^{nm} - p^{n(m-1)})] + 2\left(\frac{p^n - 1}{2}\right)[p^{n(2m-1)} - \nu p^{n(m-1)}] \equiv p^{2nm},$$

$$1[p^{n(2m-1)} - \nu p^{n(m-1)}] + 2[p^{n(2m-1)} + \nu(p^{nm} - p^{n(m-1)})]$$
$$+ 2\left(\frac{p^n - 3}{2}\right)[p^{n(2m-1)} - \nu p^{n(m-1)}] \equiv p^{2nm} + \nu p^{nm},$$

$$1[p^{n(2m-1)} - \nu p^{n(m-1)}] + 2\left(\frac{p^n - 1}{2}\right)[p^{n(2m-1)} - \nu p^{n(m-1)}] \equiv p^{2nm} - \nu p^{nm}.$$

In each of the three cases, we have enumerated separately the number of solutions arising when $\eta = 0$, when $\eta = \varkappa$ and when η is one of the values $\neq 0$ for which the first equation has solutions (viz., two).

67. Theorem. — *If S denote the number of squares[1] σ^2 in the $GF[p^n]$ for which $\sigma^2 + 1$ is a square and N the number of square τ^2 for which $\tau^2 + 1$ is a not-square, we have*

$$S = \tfrac{1}{4}(p^n - 5), \quad N = \tfrac{1}{4}(p^n - 1), \quad \text{if } -1 = \text{square};$$
$$S = \tfrac{1}{4}(p^n - 3), \quad N = \tfrac{1}{4}(p^n + 1), \quad \text{if } -1 = \text{not-square}.$$

Indeed, the number of sets of solutions ξ, η in the $GF[p^n]$ of the equation
$$\eta^2 = \xi^2 + 1$$
is always $p^n - 1$ (by § 64). These solutions are of three kinds:

1°. $\xi = 0, \quad \eta = \pm 1$;
2°. $\xi^2 = -1, \quad \eta = 0$,

occurring when -1 is a square;

3°. $\xi^2 = \alpha \neq 0, \quad \eta^2 = \alpha + 1 \neq 0$,

giving $4S$ sets of solutions ξ, η.

---

[1] The mark zero is not reckoned as a square.

Hence, if -1 be a square, we have
$$p^n - 1 = 2 + 2 + 4S, \quad N + S + 1 = \frac{1}{2}(p^n - 1).$$
If -1 be a not-square, we have
$$p^n - 1 = 2 + 4S, \quad N + S = \frac{1}{2}(p^n - 1).$$

Additive-groups in the $GF[p^n]$ and their multiplier Galois Fields[1]), §§ 68—71.

68. A set of m marks $\lambda_1, \lambda_2, \ldots, \lambda_m$ belonging to the $GF[p^n]$ and linearly independent with respect to the $GF[p]$ give rise to p^m distinct marks of the larger field,

39) $\quad c_1\lambda_1 + c_2\lambda_2 + \cdots + c_m\lambda_m \quad$ (every $c_i = 0, 1, \ldots, p-1$).

Indeed, an identity between two of the marks 39) would contradict the linear independence of $\lambda_1, \lambda_2, \ldots, \lambda_m$. Since the *sum* of any two of these p^m marks 39) may be expressed as one of the set, they are said to form *an additive-group* $[\lambda_1, \lambda_2, \ldots, \lambda_m]$ *of rank m with respect to the $GF[p]$* and the marks $\lambda_1, \lambda_2, \ldots, \lambda_m$ are said to form its *basis-system*. In particular, the $GF[p^n]$ may be exhibited as an additive-group of rank n (§ 10).

These conceptions are capable of the following direct generalization. Any m marks $\lambda_1, \lambda_2, \ldots, \lambda_m$ of the $GF[p^{nr}]$ are called linearly independent with respect to the $GF[p^r]$ if the equation
$$\gamma_1\lambda_1 + \gamma_2\lambda_2 + \cdots + \gamma_m\lambda_m = 0,$$
in which the γ_i are marks of the $GF[p^r]$, can be satisfied only in case every $\gamma_i = 0$. [See § 72]. A system of m linearly independent marks gives rise to p^{rm} distinct marks of the $GF[p^{nr}]$
$$\gamma_1\lambda_1 + \gamma_2\lambda_2 + \cdots + \gamma_m\lambda_m$$
by letting th γ_i's run independently through the series of the marks of the $GF[p^r]$. These p^{rm} marks are said to form an *additive-group* $[\lambda_1, \lambda_2, \ldots, \lambda_m]$ *of rank m with respect to the $GF[p^r]$*, the marks $\lambda_1, \ldots, \lambda_m$ forming its *basis-system*.

If λ_{m+1} be any mark of $GF[p^{nr}]$ not in the additive-group $[\lambda_1, \ldots, \lambda_m]$ of rank m with respect to the $GF[p^r]$, then the $m+1$ marks $\lambda_1, \ldots, \lambda_m, \lambda_{m+1}$ are linearly independent with respect to the $GF[p^r]$ and therefore define an additive-group $[\lambda_1, \ldots, \lambda_m, \lambda_{m+1}]$ of rank $m+1$ with respect to the $GF[p^r]$.

69. Theorem. — *Within the $GF[p^{nr}]$ the number of additive-groups $[\lambda_1, \ldots, \lambda_m]$ of rank m with respect to the $GF[p^r]$ is*
$$\frac{(p^{nr}-1)(p^{nr}-p^r)\cdots(p^{nr}-p^{(m-1)r})}{(p^{mr}-1)(p^{mr}-p^r)\cdots(p^{mr}-p^{(m-1)r})}.$$

---

1) Moore, Mathematical Papers, Congress of 1893, p. 214, p. 216; *Math. Ann.* vol. 55, § 12.

50   CHAPTER IV.

We first prove that the numerator expresses the number of sets of m marks $\lambda_1, \lambda_2, \ldots, \lambda_m$ of the $GF[p^{nr}]$ linearly independent with respect to the $GF[p^r]$. For λ_1 we may take any one of the $p^{nr}-1$ marks $\neq 0$ of the $GF[p^{nr}]$; for λ_2 any one of the $p^{nr}-p^r$ marks not of the form $\varrho_1 \lambda_1$, where ϱ_1 belongs to the $GF[p^r]$; for λ_3 any one of the $p^{nr}-p^{2r}$ marks not of the form $\varrho_1 \lambda_1 + \varrho_2 \lambda_2$, where ϱ_1 and ϱ_2 belong to the $GF[p^r]$; etc.

We next show that the denominator expresses the number of these sets of m independent marks which generate the same additive-group $[\lambda_1, \lambda_2, \ldots, \lambda_m]$. In fact, we may use as a basis-system for the latter any set of m marks $\lambda_1', \lambda_2', \ldots, \lambda_m'$ chosen as follows. λ_1' may be chosen in $p^{mr}-1$ ways:

$$\lambda_1' = \sum_{i=1}^{m} \gamma_{1i} \lambda_i,$$

each γ_{1i} being arbitrary in the $GF[p^r]$ provided not all are simultaneously zero. λ_2' may be chosen in $p^{mr}-p^r$ ways, viz.,

$$\lambda_2' = \sum_{i=1}^{m} \gamma_{2i} \lambda_i,$$

the γ_{2i} being taken arbitrarily in the $GF[p^r]$ but so as to exclude the p^r sets of values which make $\lambda_2' = \varrho \lambda_1'$, viz.,

$$\gamma_{2i} = \varrho \gamma_{1i} \quad (i = 1, 2, \ldots, m),$$

where ϱ runs through the series of marks of the $GF[p^r]$; etc.

70. If the p^m marks $c_1 \lambda_1 + \cdots + c_m \lambda_m$ of the additive-group $[\lambda_1, \ldots, \lambda_m]$ of rank m with respect to the $GF[p]$ are multiplied by any particular mark $\mu \neq 0$ of the $GF[p^n]$, the resulting p^m marks constitute the additive-group

$$\mu [\lambda_1, \ldots, \lambda_m] \equiv [\mu \lambda_1, \ldots, \mu \lambda_m]$$

likewise of rank m with respect to the $GF[p]$. We will say that $[\mu \lambda_1, \ldots, \mu \lambda_m]$ is derived from $[\lambda_1, \ldots, \lambda_m]$ by multiplication by μ. In particular, we seek those multipliers $\mu = \varkappa$ which do not alter $[\lambda_1, \ldots, \lambda_m]$, such a mark being called *a multiplier of the additive-group* $[\lambda_1, \ldots, \lambda_m]$. If $\varkappa_1$ and $\varkappa_2$ be multipliers, then will evidently the product $\varkappa_1 \varkappa_2$ be a multiplier. To prove that $\mu \equiv \varkappa_1 + \varkappa_2$ will also be a multiplier, we observe first that $[\mu \lambda_1, \ldots, \mu \lambda_m]$ is an additive-group included within $[\lambda_1, \ldots, \lambda_m]$, since $\varkappa_1$ and $\varkappa_2$ are multipliers of the latter, and further that it is of rank m if $\mu \neq 0$. Hence $\varkappa_1 + \varkappa_2$ is a multiplier unless it be zero. Hence the multipliers $\varkappa$ together with the mark zero constitute an additive, as well as a multiplicative, group and therefore constitute a Galois Field $GF[p^k]$ included within

MISCELLANEOUS PROPERTIES OF GALOIS FIELDS. 51

the fundamental $GF[p^n]$. It is called *the multiplier Galois Field* of the additive-group $[\lambda_1, \ldots, \lambda_m]$. Every $GF[p^{k'}]$ included within the $GF[p^k]$ is called *a multiplier Galois Field* of the additive-group. By § 23, k' is a divisor of k and k a divisor of n.

The additive-group $[\lambda_1, \ldots, \lambda_m]$ of rank m with respect to the $GF[p]$ may be exhibited as an additive-group $[\lambda'_1, \ldots, \lambda'_{m'}]$ of rank $m' = m/k'$ with respect to any multiplier $GF[p^{k'}]$.

In proof, let $\gamma_1, \gamma_2, \ldots, \gamma_{m'}$ run independently through the series of marks of the $GF[p^{k'}]$. Taking λ'_1 to be any particular mark $\lambda \neq 0$ in $[\lambda_1, \ldots, \lambda_m]$, the $p^{k'}$ marks $\gamma_1 \lambda'_1$ are all distinct and all belong to $[\lambda_1, \ldots, \lambda_m]$. Taking λ'_2 any mark in $[\lambda_1, \ldots, \lambda_m]$ different from the $\gamma_1 \lambda'_1$, the $p^{2k'}$ marks $\gamma_1 \lambda'_1 + \gamma_2 \lambda'_2$ are all distinct and all belong to $[\lambda_1, \ldots, \lambda_m]$. Proceeding similarly, we obtain ultimately a set of $p^{m'k'}$ distinct marks $\gamma_1 \lambda'_1 + \gamma_2 \lambda'_2 + \cdots + \gamma_{m'} \lambda'_{m'}$ giving all the marks of $[\lambda_1, \ldots, \lambda_m]$. In particular, $p^m = p^{k'm'}$, so that k' divides m.

Corollary I. Since k is a particular k', k divides m.

Corollary II. Within the $GF[p^n]$ the number $A(p, n, m, k)$ of additive-groups of rank m with respect to the $GF[p]$ which have the $GF[p^k]$ as a multiplier Galois Field equals the total number of additive-groups of rank m/k with respect to the $GF[p^k]$:

$$A(p, n, m, k) = \frac{(p^n - 1)(p^n - p^k)(p^n - p^{2k}) \cdots (p^n - p^{m-k})}{(p^m - 1)(p^m - p^k)(p^m - p^{2k}) \cdots (p^m - p^{m-k})}.$$

71. If k be a divisor of m and n and if $h_1, h_2, \ldots, h_t$ are the prime factors occurring in both m and n to a higher power than in k, there are in the $GF[p^n]$ exactly

$$A(p, n, m, k) - \sum_{i=1}^{t} A(p, n, m, kh_i) + \sum_{i,j}^{1,\ldots,t} A(p, n, m, kh_i h_j) - \cdots$$
$$+ (-1)^t A(p, n, m, kh_1 \ldots h_t)$$

additive-groups of rank m with respect to the $GF[p]$ which have the $GF[p^k]$ as *the* multiplier Galois Field.

Indeed, from the $A(p, n, m, k)$ additive-groups having the $GF[p^k]$ as *a* multiplier Galois Field, we must eliminate those having a larger multiplier Galois Field. It suffices to eliminate those having the $GF[p^{kh_i}]$, for $i = 1, 2, \ldots$, or t, as *a* multiplier Galois Field. But the $A(p, n, m, kh_1)$ additive-groups with the $GF[p^{kh_1}]$ and the $A(p, n, m, kh_2)$ additive-groups with the $GF[p^{kh_2}]$ are not distinct but have in common $A(p, n, m, kh_1 h_2)$ additive-groups each with the $GF[p^{kh_1 h_2}]$ as a multiplier Galois Field. By this principle, we readily determine the number of distinct additive-groups among the sets of $A(p, n, m, kh_i)$ with the $GF[p^{kh_i}]$ as multiplier Galois Fields. Subtracting this number from $A(p, n, m, k)$, we obtain the required number.

52   CHAPTER IV.

72. Theorem. *The marks $\lambda_1, \lambda_2, \ldots, \lambda_m$ of the $GF[p^{nm}]$ are linearly independent with respect to the $GF[p^n]$ if and only if the following determinant*[1]) *is not zero in the $GF[p^{nm}]$:*

$$\left|\lambda_j^{p^{ni}}\right| \equiv \begin{vmatrix} \lambda_1 & \lambda_2 & \ldots \ldots & \lambda_m \\ \lambda_1^{p^n} & \lambda_2^{p^n} & \ldots \ldots & \lambda_m^{p^n} \\ \lambda_1^{p^{2n}} & \lambda_2^{p^{2n}} & \ldots \ldots & \lambda_m^{p^{2n}} \\ \cdot & \cdot & \cdot \cdot \cdot \cdot \cdot & \cdot \\ \lambda_1^{p^{n(m-1)}} & \lambda_2^{p^{n(m-1)}} & \ldots & \lambda_m^{p^{n(m-1)}} \end{vmatrix}.$$

First, if $\lambda_1, \lambda_2, \ldots, \lambda_m$ be linearly dependent, i. e., if a relation

$$\gamma_1 \lambda_1 + \gamma_2 \lambda_2 + \cdots + \gamma_m \lambda_m = 0$$

holds, the coefficients γ_j being marks of the $GF[p^n]$ not all zero, then will the determinant $|\lambda|$ vanish.

Secondly, if the determinant vanish, set

$$\lambda_i \equiv \sum_{j=0}^{m-1} \mu_{ij} R^j \quad (i = 1, 2, \ldots, m),$$

where R is a primitive root of the $GF[p^{nm}]$ and therefore satisfies an equation of degree m belonging to and irreducible in the $GF[p^n]$, and where the μ_{ij} belong to the latter field. Then

$$\left|\lambda_j^{p^{ni}}\right| = \left|\mu_{ij}\right| \cdot \left|R^{jp^{ni}}\right| \quad (j = 1, \ldots, m; \; i = 0, \ldots, m-1),$$

where the determinant in R, when written in full, is[2])

$$\begin{vmatrix} 1 & 1 & \ldots 1 \\ R & R^{p^n} & \ldots R^{p^{n(m-1)}} \\ R^2 & R^{2p^n} & \ldots R^{2p^{n(m-1)}} \\ \cdot & \cdot & \cdot \cdot \cdot \cdot \cdot \\ R^{m-1} & R^{(m-1)p^n} & \ldots R^{(m-1)p^{n(m-1)}} \end{vmatrix} \equiv \prod_{\substack{s,t \\ s > t}}^{0, \ldots, m-1} \left(R^{p^{ns}} - R^{p^{nt}}\right)$$

and therefore is not zero in the $GF[p^{nm}]$. Hence, if $|\lambda| = 0$, then must $|\mu| = 0$, so that between the λ_i exists a linear relation with coefficients belonging to the $GF[p^n]$ and not all zero.

73. If λ be a mark of the $GF[p^{nm}]$, the marks

$$\lambda, \; \lambda^{p^n}, \; \lambda^{p^{2n}}, \ldots, \lambda^{p^{n(m-1)}} \quad (\lambda^{p^{nm}} = \lambda)$$

are said to be *conjugate with respect to the $GF[p^n]$*. Any symmetric function of them is unaltered upon being raised to the power p^n and

---

1) Its decomposition into linear factors is given by Moore, "A two-fold generalization of Fermat's theorem", Bull. Amer. Math. Soc., vol. 2 (1896).
2) Baltzer, Determinanten, p. 85.

MISCELLANEOUS PROPERTIES OF GALOIS FIELDS. 53

hence belongs to the $GF[p^n]$. Hence the m conjugate marks are the roots of an equation of degree m with coefficients in the $GF[p^n]$.

By § 31, the roots of an equation of degree m belonging to and irreducible in the $GF[p^n]$ are conjugate with respect to the $GF[p^n]$.

In particular, the marks A, A^{p^n} of the $GF[p^{2n}]$ are conjugate with respect to the $GF[p^n]$. The conjugate A^{p^n} of A will be designated by $\bar{A}$. Evidently $\bar{A} = A$ if, and only if, A belongs to the $GF[p^n]$. The following relations are proven at once:

$$\overline{AB} = \bar{A} \cdot \bar{B} = \bar{B}\bar{A}, \quad \overline{A^t} = (\bar{A})^t, \quad \overline{A + B} = \bar{A} + \bar{B}, \quad \overline{(A/B)} = \bar{A}/\bar{B}.$$

74. *Newton's identities.* If S_t denote the sum of the tth powers of the roots of the equation

$$f(x) \equiv x^m + a_1 x^{m-1} + a_2 x^{m-2} + \cdots + a_{m-1} x + a_m = 0,$$

in which the coefficients a_i belong to the $GF[p^n]$, then

$$0 = S_1 + a_1$$
$$0 = S_2 + a_1 S_1 + 2 a_2$$
40) $\quad 0 = S_3 + a_1 S_2 + a_2 S_1 + 3 a_3$
$$\cdots\cdots\cdots\cdots\cdots\cdots\cdots\cdots\cdots\cdots$$
$$0 = S_{m-1} + a_1 S_{m-2} + a_2 S_{m-3} + \cdots + a_{m-2} S_1 + (m-1) a_{m-1}.$$

These identities follow as in algebra upon equating the coefficients of like powers of x in the following identity, in which $\alpha_1, \ldots, \alpha_m$ are the roots of $f(x) = 0$:

41) $\quad f'(x) \equiv \sum_{i=1}^{m} \dfrac{f(x)}{x - \alpha_i} = m x^{m-1} + (m-1) a_1 x^{m-2}$
$$+ (m-2) a_2 x^{m-3} + \cdots + a_{m-1}.$$

This identity, evidently true for $m = 1$, may be proven by simple induction[1]). Supposing it true for a particular m, we have proven it true for the value $m + 1$. Let

$$F(x) \equiv f(x) \cdot (x - \alpha_{m+1}) \equiv x^{m+1} + (a_1 - \alpha_{m+1}) x^m + (a_2 - a_1 \alpha_{m+1}) x^{m-1} + \cdots$$
$$+ (a_m - a_{m-1} \alpha_{m+1}) x - a_m \alpha_{m+1}.$$

Multiplying 41) by $x - \alpha_{m+1}$ and adding $f(x)$ to the left member and $x^m + \cdots + a_m$ to te right member, we find

$$\sum_{i=1}^{m} \dfrac{F(x)}{x - \alpha_i} + \dfrac{F(x)}{x - \alpha_{m+1}} = (m+1) x^m + m (a_1 - \alpha_{m+1}) x^{m-1}$$
$$+ (m-1)(a_2 - a_1 \alpha_{m+1}) x^{m-2} + \cdots + 2 (a_{m-1} - a_{m-2} \alpha_{m+1}) x$$
$$+ (a_m - a_{m-1} \alpha_{m+1}).$$

---

1) Since equations in the $GF[p^n]$ are not algebraic identities, we avoid the consideration of derivatives. We might, however, employ Weber's definition (*Algebra*, I, § 13) of the derivatives of a polynomial in x for the derivatives up to the pth, but not for the higher derivatives on account of the denominators $\Pi(m)$.

Hence if 41) be true for $f(x) = 0$ with the roots $\alpha_1, \ldots, \alpha_m$, a like formula is true for the equation $F(x) = 0$ with the roots $\alpha_1, \ldots, \alpha_m, \alpha_{m+1}$.

Forming the sums

$$\sum_{i=1}^{m} f(\alpha_i), \quad \sum_{i=1}^{m} \alpha_i f(\alpha_i), \quad \sum_{i=1}^{m} \alpha_i^2 f(\alpha_i), \ldots$$

we derive the new identities,

42)
$$0 = S_m + a_1 S_{m-1} + a_2 S_{m-2} + \cdots + m a_m,$$
$$0 = S_{m+1} + a_1 S_m + a_2 S_{m-1} + \cdots + a_m S_1,$$
$$0 = S_{m+2} + a_1 S_{m+1} + a_2 S_m + \cdots + a_m S_2,$$
$$\cdots\cdots\cdots\cdots\cdots\cdots\cdots\cdots\cdots\cdots$$

Corollary. — If $f(x) = 0$ have a double root α, the right member of 41) must vanish for $x = \alpha$.

75. Theorem. — *If t be a positive integer and $u_0, u_1, \ldots, u_{p^n-1}$ denote the marks of the $GF[p^n]$, then*

43)
$$\sum_{i=0}^{p^n-1} u_i^t = \begin{cases} 0 & \text{for } t < p^n - 1 \\ -1 & \text{for } t = p^n - 1. \end{cases}$$

In fact, the marks u_i are the roots in the $GF[p^n]$ of the equation
$$x^{p^n} - x = 0.$$

Applying to the latter the identities 40), we find
$$S_i = 0 \quad (i = 1, 2, \ldots, p^n - 2), \quad S_{p^n-1} - (p^n - 1) = 0.$$

CHAPTER V.

ANALYTIC REPRESENTATION OF SUBSTITUTIONS ON THE MARKS OF A GALOIS FIELD.

76. Consider the problem to find every quantic $\Phi(\xi)$ belonging to the $GF[p^n]$ such that the equation $\Phi(\xi) = \beta$ has a root in the field whatever mark of the field β may be. For example,

$$\xi^3 \equiv \beta \pmod{5}$$

is solvable for every integer β, since we have

$$0^3 \equiv 0, \quad 1^3 \equiv 1, \quad 3^3 \equiv 2, \quad 2^3 \equiv 3, \quad 4^3 \equiv 4 \pmod{5}.$$

ANALYTIC REPRESENTATION OF SUBSTITUTIONS, etc.   55

If we denote the marks of the $GF[p^n]$ as follows,

44) $\qquad \mu_0, \quad \mu_1, \quad \mu_2, \ldots, \mu_{p^n-1},$

the necessary and sufficient conditions that $\Phi(\xi) = \beta$ be solvable in the field for arbitrary β are that the marks

45) $\qquad \Phi(\mu_0), \quad \Phi(\mu_1), \quad \Phi(\mu_2), \ldots, \Phi(\mu_{p^n-1})$

be identical with the series 44) apart from their order. In fact, the p^n values which $\Phi(\xi)$ takes must all be distinct, since β is to have p^n distinct values. When the conditions named are satisfied, the series 45) forms a permutation of the series 44), and the quantic $\Phi(\xi)$ is said *to represent the substitution*

$$\begin{bmatrix} \xi \\ \Phi(\xi) \end{bmatrix} \equiv \begin{bmatrix} \mu_0, & \mu_1, & \ldots, & \mu_{p^n-1} \\ \Phi(\mu_0), & \Phi(\mu_1), & \ldots, & \Phi(\mu_{p^n-1}) \end{bmatrix}$$

on the marks of the $GF[p^n]$. For example, ξ^3 represents the substitution

$$\begin{bmatrix} \xi \\ \xi^3 \end{bmatrix} \equiv \begin{bmatrix} 0, & 1, & 2, & 3, & 4 \\ 0, & 1, & 3, & 2, & 4 \end{bmatrix},$$

on the marks of the $GF[5]$, i. e., the field of integers taken modulo 5. A quantic of degree k with coefficients belonging to the $GF[p^n]$ will be called a *substitution quantic* $SQ[k, p^n]$ if it satisfy the above conditions. Its degree k will be supposed $< p^n$ in view of the equation $\xi^{p^n} = \xi$ satisfied by every mark of the field.

77. An arbitrary substitution on the marks of the $GF[p^n]$,

$$\begin{bmatrix} \mu_0, & \mu_1, & \ldots, & \mu_{p^n-1} \\ \mu_{a_0}, & \mu_{a_1}, & \ldots, & \mu_{a_{p^n-1}} \end{bmatrix}$$

can be represented by the quantic $\Phi(\xi)$ given by Lagrange's interpolation formula,

$$\Phi(\xi) \equiv \sum_{i=0}^{p^n-1} \frac{\mu_{a_i} F(\xi)}{(\xi - \mu_i) F'(\mu_i)},$$

where

$$F(\xi) \equiv \prod_{i=0}^{p^n-1} (\xi - \mu_i),$$

and $F'(\xi)$ denotes the function derived from $F(\xi)$ by formula 41). Evidently $\Phi(\xi)$ is an integral function of ξ of degree $< p^n$.

78. Theorem. *Two distinct quantics* $\Phi(\xi)$ *and* $\Psi(\xi)$ *belonging to the* $GF[p^n]$ *can not represent the same substitution on its marks.*

For, if $\qquad \Phi(\mu_i) = \Psi(\mu_i) \quad (i = 0, 1, \ldots, p^n - 1),$

the equation $\Phi(\xi) - \Psi(\xi) = 0$ would have in the field p^n distinct roots μ_i, whereas its degree is less than p^n. By § 15, it must be an identity in ξ.

56   CHAPTER V.

79. Theorem. ξ^m *is a* $SQ[m, p^n]$ *if, and only if, m be prime to $p^n - 1$.*

The theorem follows immediately from the corollary of § 63. However, to illustrate a method of proof used below, we will verify that, if m be relatively prime to $p^n - 1$, ξ^m takes p^n distinct values in the $GF[p^n]$ when ξ does. It is sufficient to prove that from

46) $$\xi_1^m = \xi_2^m$$

follows $\xi_1 = \xi_2$, provided ξ_1 and ξ_2 are marks of the field. This being evident if either be the mark zero, we suppose $\xi_1 \neq 0$, $\xi_2 \neq 0$, so that

47) $$\xi_1^{p^n-1} = \xi_2^{p^n-1} = 1.$$

Raising the members of equations 46) and 47) to the respective powers t and τ, chosen (§ 7, note) so that $tm + \tau(p^n - 1) = 1$, and forming the product of the resulting equations, we find that $\xi_1 = \xi_2$.

80. Theorem. — *For an arbitrary mark α of the $GF[p^n]$,*

$$\Phi(\xi) \equiv 5\xi^5 + 5\alpha\xi^3 + \alpha^2\xi$$

is a $SQ[5, p^n]$, if p is a prime of the form $5m \pm 2$ and n is odd.

To prove that, in the $GF[p^n]$, $\xi_1 = \xi_2$ is the only solution of

or $$\Phi(\xi_1) = \Phi(\xi_2)$$

$(\xi_1 - \xi_2)\{5(\xi_1^4 + \xi_1^3\xi_2 + \xi_1^2\xi_2^2 + \xi_1\xi_2^3 + \xi_2^4) + 5\alpha(\xi_1^2 + \xi_1\xi_2 + \xi_2^2) + \alpha^2\} = 0,$

we set
$$\xi_1 = \lambda + \mu, \quad \xi_2 = \lambda - \mu,$$

here[1]) limiting our proof to the case $p > 2$. Then 16 times the quantity within the braces becomes

$$16\{5(5\lambda^4 + 3\alpha\lambda^2 + 10\lambda^2\mu^2 + \alpha\mu^2 + \mu^4) + \alpha^2\}$$
$$\equiv \{(20\lambda^2 + \alpha) + 5(4\mu^2 + \alpha)\}^2 - 5\{2(4\mu^2 + \alpha)\}^2.$$

But[2]) $+5$ is a quadratic residue of *no* odd number of the form $5m + 2$ or $5m - 2$. Hence (§ 62) 5 is a not-square in the $GF[5^n]$, n being odd and $p = 5m \pm 2$. Hence, if the above expression vanishes, we must have

$$4\mu^2 + \alpha = 0, \quad 20\lambda^2 + \alpha = 0,$$

whence, for $p > 2$, $\mu^2 = 5\lambda^2$, so that $\lambda = \mu = 0$, $\xi_1 = \xi_2$.

---

1) An analogous proof for $p = 2$ is given in *Annals of Math.*, 1897, pp. 84—85. For an arbitrary prime p, the theorem is a special case of that of § 82.
2) Gauss, Disquisitiones Arithmeticae, Art. 121.

81. Theorem. — *The quantic belonging to the $GF[p^{nm}]$,*

$$\Psi(X) \equiv \sum_{i=1}^{m} A_i X^{p^{n(m-i)}}$$

will represent a substitution on its marks if, and only if, $X = 0$ is the only solution in the field of $\Psi(X) = 0$.

Indeed, the necessary and sufficient condition is that

$$\Psi(X_1 - X_2) \equiv \Psi(X_1) - \Psi(X_2) = 0$$

shall require $X_1 = X_2$, or $X_1 - X_2 = 0$.

Corollary. $X^{p^{nr}} - A X^{p^{ns}}$ represents a substitution on the marks of the $GF[p^{nm}]$ if, and only if, either $A = 0$ or else A is not the power $p^{nr} - p^{ns}$ of a mark of the field.

82. Theorem. — *If k be an odd integer relatively prime to $p^{2n} - 1$, and if α be an arbitrary mark of the $GF[p^n]$, the quantic*

$$\Phi_k(\xi, \alpha) \equiv \xi^k + k\alpha\xi^{k-2} + k\sum_{l=2}^{\frac{1}{2}(k-1)} \frac{(k-l-1)(k-l-2)\ldots(k-2l+1)}{2\cdot 3\ldots l}\alpha^l\xi^{k-2l}$$

represents a substitution on the marks of the $GF[p^n]$.

We are to prove that the equation

48) $\qquad\qquad \Phi_k(\xi, \alpha) = \beta$

has a solution ξ in the $GF[p^n]$, β being an arbitrary mark of the field. By Waring's formula, $\Phi_k(\xi, \alpha)$ is the sum of the kth powers of the roots of the quadratic

$$\eta^2 - \xi\eta - \alpha = 0.$$

Hence, in virtue of the equation

$$\xi = \eta - \alpha/\eta,$$

we have the identity $\quad \Phi_k(\xi, \alpha) \equiv \eta^k - (\alpha/\eta)^k.$

The equation 48) thus becomes

$$\eta^{2k} - \beta\eta^k - \alpha^k = 0.$$

Setting $Y \equiv \eta^k$, this becomes

49) $\qquad\qquad Y^2 - \beta Y - \alpha^k = 0.$

According as 49) is reducible or irreducible in the $GF[p^n]$, it is solvable in the $GF[p^n]$ or in the $GF[p^{2n}]$, and therefore always solvable in the larger field. Call its roots Y_1 and Y_2. Since k is prime to $p^{2n} - 1$, we can determine uniquely (§ 79 or § 63) two marks η_1 and η_2 belonging to the $GF[p^{2n}]$ such that

$$\eta_1^k = Y_1, \quad \eta_2^k = Y_2.$$

58   CHAPTER V.

Likewise, it follows from $Y_1 Y_2 = -\alpha^k$ that
$$\eta_1 \eta_2 = -\alpha.$$

If 49) be irreducible in the $GF[p^n]$, we have (§ 31, corollary)
$$Y_2^{p^n} = Y_1, \quad Y_1^{p^n} = Y_2,$$
and therefore
$$\eta_2^{p^n} = \eta_1, \quad \eta_1^{p^n} = \eta_2.$$
Hence
$$(\eta_1 + \eta_2)^{p^n} = \eta_2 + \eta_1.$$

It follows that 48) has the solution in the $GF[p^n]$
$$\xi = \eta_1 - \alpha/\eta_1 = \eta_2 - \alpha/\eta_2 = \eta_1 + \eta_2.$$

If 49) be reducible, Y_1 and Y_2 belong to the $GF[p^n]$. Since k is prime to $p^n - 1$, it follows that η_1 and η_2 belong to the $GF[p^n]$.

Remark 1. — We have shown in § 37 that $\Phi_{2^i-1}(\xi, 1)$ completely decomposes in the $GF[p^n]$ into linear factors, if $p^n = 2^i t - 1$, t odd and $i > 1$.

Remark 2. — If k be a prime number, $\Phi_k(\xi, \alpha)$ is the only quantic of degree k suitable to represent a substitution on the marks of every $GF[p^n]$ for which $p^{2n} - 1$ is not divisible by k (*Annals of Mathematics*, 1897, pp. 89—91).

Remark 3. — The equation 48) is algebraically solvable, having as roots
$$\varrho \varepsilon^m + \sigma \varepsilon^{k-m} \quad (m = 0, 1, \ldots, k-1)$$
where
$$\varrho, \sigma = \sqrt[k]{\beta/2 \pm \sqrt{\beta^2/4 + \alpha^k}},$$

and ε denotes a primitive k^{th} root of unity. This result is a direct generalization of Cardan's formula for the roots of the reduced cubic and of Vallès' solution of the quintic[1])
$$\xi^5 + \alpha \xi^3 + \frac{\alpha^2}{5} \xi = \beta.$$

83. Theorem.[2]) — *If d be a divisor of $p^r - 1$ and v be not a dth power in the $GF[p^n]$, the quantic*
$$\Phi(\xi) \equiv \xi(\xi^d - v)^{(p^r-1)/d}$$
is a $SQ[p^r, p^n]$.

We are to prove that $\Phi(\xi) = \beta$ has a solution ξ in the $GF[p^n]$ for β chosen arbitrarily in the field. This being evident if $\beta = 0$, we will suppose that $\beta \neq 0$. Writing
$$\eta \equiv \xi^d - v,$$

---

1) Formes imaginaires en Algèbre, 1869, vol. 1, pp. 90—92.
2) For the case $r = n$, this theorem is included in the theorem of § 85.

we are to prove that

50) $\qquad [\Phi(\xi)]^d = (\eta + \nu)\eta^{p^r-1} \equiv \eta^{p^r} + \nu\eta^{p^r-1} = \beta^d$

has a solution η in the $GF[p^n]$; for, if η be such a solution (necessarily $\neq 0$), then

$$\xi = (\eta + \nu)^{1/d} = \beta/\eta^{(p^r-1)/d}$$

will belong to the $GF[p^n]$ and will satisfy $\Phi(\xi) = \beta$.

Setting $\eta = 1/\omega$ in 50) and multiplying by ω^{p^r}, we find

$$1 + \nu\omega = \beta^d \omega^{p^r}.$$

This has a solution ω in the $GF[p^n]$ for β arbitrary in the field. Indeed, by § 81, corollary,

$$\omega^{p^r} - \left(\frac{\nu}{\beta^d}\right)\omega$$

represents a substitution on the marks of the $GF[p^n]$, since ν/β^d is not a dth power and hence not a $(p^r-1)^{st}$ power in the field.

Note. — For $p = 3, 5, 7$ and partially for $p = 11$, the author has shown[1]) that the only $SQ[p, p^n]$ which exist are reducible to the form
$$\xi(\xi^d - \nu)^{(p-1)/d},$$
where d is a divisor of $p - 1$ and ν is not a dth power in the $GF[p^n]$.

84. Theorem.[2]) — *The necessary and sufficient conditions that $\Phi(\xi)$ shall represent a substitution on the marks of the $GF[p^n]$ are:*

1^0. *Every t^{th} power of $\Phi(\xi)$, for $t \leq p^n - 2$ and prime to p, shall reduce to a degree $\leq p^n - 2$ on applying the equation $\xi^{p^n} = \xi$;*

2^0. *There shall be one and only one root in the $GF[p^n]$ of $\Phi(\xi) = 0$.*

After the exponents of ξ are reduced below p^n, let

$$[\Phi(\xi)]^t \equiv \sum_{i=0}^{p^n-1} \alpha_i^{(t)} \xi^i.$$

Put for ξ the p^n marks μ_j of the $GF[p^n]$ and add the resulting indentities. We find, on applying § 75,

$$\sum_{j=0}^{p^n-1}[\Phi(\mu_j)]^t = p^n \alpha_0^{(t)} + \alpha_1^{(t)} \sum_{j=0}^{p^n-1} \mu_j + \cdots + \alpha_{p^n-1}^{(t)} \sum_{j=0}^{p^n-1} \mu_j^{p^n-1} = -\alpha_{p^n-1}^{(t)}.$$

If $\Phi(\xi)$ represent a substitution, we must have

$$\sum_{j=0}^{p^n-1}[\Phi(\mu_j)]^t = \sum_{j=0}^{p^n-1} \mu_j^t = 0 \quad (t < p^n - 1).$$

---

1) Dissertation, *Annals of Mathematics*, 1897, pp. 101—108.
2) For the case $n = 1$, this theorem is due to Hermite, *Comptes Rendus*, vol. 57 (1863), pp. 750—757.

60   CHAPTER V.

Hence a necessary condition is that
$$\alpha^{(t)}_{p^n-1} = 0 \quad (t = 1, 2, \ldots, p^n - 2).$$
The condition 2^0 is evidently a necessary condition.

Suppose, inversely, that 1^0 and 2^0 are satisfied. Consider the equation satisfied by the marks $\Phi(\mu_j)$,
$$\prod_{j=0}^{p^n-1}[\eta - \Phi(\mu_j)] \equiv \sum_{i=0}^{p^n}\gamma_i \eta^{p^n-i} = 0,$$
the sum of the mth powers of whose roots is denoted by σ_m. Then
$$\sigma_t \equiv \sum_{j=0}^{p^n-1}[\Phi(\mu_j)]^t = -\alpha^{(t)}_{p^n-1} = 0 \quad \begin{Bmatrix} t = 1, 2, \ldots, p^n-2 \\ t \not\equiv 0 \pmod p \end{Bmatrix}$$
$$\sigma_{p^n-1} \equiv \sum_{j=0}^{p^n-1}[\Phi(\mu_j)]^{p^n-1} = p^n - 1 \equiv -1,$$
since all but one of the $\Phi(\mu_j)$ are $\neq 0$ by 2^0 and hence have unity for their $(p^n-1)^{st}$ powers. Applying Newton's identities 40), we readily find
$$\gamma_i = 0 \quad [i = 1, 2, \ldots, p^n-2;\ i \not\equiv 0 \pmod p]$$
$$\sigma_p = \sigma_{2p} = \cdots = \sigma_{p^n-p} = \sigma_{p^n} = 0, \quad \gamma_{p^n-1} = -1.$$
To determine $\gamma_p, \gamma_{2p}, \ldots$, we apply the identities 42), viz.,
$$\sigma_k + \gamma_1 \sigma_{k-1} + \gamma_2 \sigma_{k-2} + \cdots + \gamma_{p^n}\sigma_{k-p^n} = 0 \quad (k \geq p^n),$$
which here reduce to the form
$$51) \quad \sigma_k + \gamma_p \sigma_{k-p} + \gamma_{2p}\sigma_{k-2p} + \cdots + \gamma_{p^n-p}\sigma_{k-p^n+p} + \gamma_{p^n-1}\sigma_{k-p^n+1} = 0,$$
since by 2^0,
$$\gamma_{p^n} \equiv -\prod_{j=0}^{p^n-1}\Phi(\mu_j) = 0.$$
Furthermore, since any mark equals its $(p^n)^{\text{th}}$ power, we have
$$\sigma_{s+p^n-1} = \sigma_s \quad (s = 1, 2, \ldots)$$
Applying 51) for $k = p^n + p - 1$, we find
$$\gamma_p \sigma_{p^n-1} = 0.$$
More generally, for $k = p^n + lp - 1$, $l \leq p^{n-1} - 1$, we get
$$\gamma_{lp}\sigma_{p^n-1} = 0.$$
Hence $\gamma_{lp} = 0$. We have therefore the result
$$\prod_{j=0}^{p^n-1}[\eta - \Phi(\mu_j)] = \eta^{p^n} - \eta,$$
so that the marks $\Phi(\mu_j)$ form a permutation of the marks μ_j of the $GF[p^n]$.

85. Theorem.[1]) — *If r be less than and prime to p^n-1 and s be a divisor of p^n-1, and if $f(\xi^s)$ be a rational integral function of ξ^s with coefficients belonging to the $GF[p^n]$ such that $f(\xi^s) = 0$ has no root in the field, then*

$$\xi^r[f(\xi^s)]^{(p^n-1)/s}$$

represents a substitution on the p^n marks of the field.

The conditions of the theorem of § 84 for a substitution quantic are all satisfied by the given quantic. In fact, upon raising it to any power l, not divisible by s, we obtain a set of terms whose exponents are of the form $ms + lr$ and therefore not divisible by s and consequently not by p^n-1. If, however, we take the power

$$l = ts < p^n-1,$$

we get the result ξ^{lr}, since the power p^n-1 of $f(\xi^s) \neq 0$ is unity in the field. But lr is not divisible by p^n-1.

Condition 2° is satisfied by our quantic, since it vanishes in the field only when $\xi = 0$.

86. As examples under the preceding theorem, we note first ξ^r if r be prime to p^n-1 [Compare § 79]. Next, if $p > 2$,

52) $\quad \xi^r(\xi^{(p^n-1)/2} - \tau)^2 \equiv -2\tau\{\xi^{r+(p^n-1)/2} - \frac{1}{2}\left(\tau + \frac{1}{\tau}\right)\xi^r\}$

represents a substitution on the marks of the $GF[p^n]$ if τ be any mark in the field except $+1$, -1, 0. For the remaining p^n-3 marks τ, the quantics 52) coincide in pairs. We note the following special substitution quantics 52):

$n = 1, \quad p = 5: \quad \xi^3, \ \xi,$
$n = 1, \quad p = 7: \quad \xi^4 \pm 3\xi$ and $\xi^5 \pm 2\xi^2$ (Hermite),
$n = 1, \quad p = 11: \quad \xi^6 \pm 2\xi, \ \xi^6 \pm 4\xi,$
$n = 2, \quad p = 3: \quad \xi, \ \xi^3, \ \xi^5, \ \xi^5 \pm 2^{1/2}\xi.$

For $n = 1$, $p = 7$, $\nu^3 = -1$, the theorem of § 85 gives the quantics
$$\xi\,(\xi^2 - \nu)^3 \equiv -3\nu(\xi^5 - \nu\xi^3 + 3\nu^2\xi),$$
$$\xi^5(\xi^2 - \nu)^3 \equiv 2\ (\xi^5 + 2\nu\xi^3 + 3(2\nu)^2\xi),$$

which together give the following $SQ[5, 7]$ of § 80:

$$\xi^5 + \alpha\xi^3 + 3\alpha^2\xi \quad (\alpha = \text{arbitrary}).$$

---

1) For $n = 1$, this theorem is due to Rogers, *Proc. Lond. Math. Soc.*, vol. 22 (1890), pp. 210—218.

62   CHAPTER V.

87. If $\Phi(\xi) \equiv \alpha_0 \xi^k + \alpha_1 \xi^{k-1} + \cdots$ be a $SQ[k, p^n]$, then will also $\Phi_1(\xi)$, obtained by forming the compound substitution,

$$\left(\alpha\xi + \beta\atop\xi\right)\left(\Phi(\xi)\atop\xi\right)\left(\gamma\xi + \delta\atop\xi\right) \equiv \left(\Phi_1(\xi)\atop\xi\right),$$

$\Phi_1(\xi) \equiv \gamma\Phi(\alpha\xi+\beta) + \delta = \gamma\alpha_0\xi^k + \gamma(k\beta\alpha_0 + \alpha_1)\xi^{k-1} + \cdots + (\gamma\Phi(\beta) + \delta)$,

if $\alpha = 1$. We may dispose of γ, β, δ to simplify $\Phi_1(\xi)$. We take $\gamma = \alpha_0^{-1}$, and, in case k is prime to p, we choose β so that

$$k\beta\alpha_0 + \alpha_1 = 0.$$

Finally, we take $\delta = -\gamma\Phi(\beta)$. The quantic $\Phi_1(\xi)$, in which the coefficient of ξ^k is unity, the constant term zero, and, when k is not a multiple of p, the coefficient of ξ^{k-1} is zero, will be called the *reduced form* of $\Phi(\xi)$ for the $GF[p^n]$.

88. To illustrate the use of the theorem of § 84, we apply it to determine all reduced $SQ[3, p^n]$. For $p \neq 3$, the reduced cubic in the $GF[p^n]$ is $\xi^3 + \alpha\xi$. The sub-case $p^n = 3m + 1$ must be rejected, since the mth power of $\xi^3 + \alpha\xi$ contains the power $\xi^{3m} \equiv \xi^{p^n-1}$ with coefficient unity and hence $\neq 0$. For the sub-case $p^n = 3m + 2$, the condition given by the power $m + 1$ is $(m+1)\alpha = 0$. But if $m + 1$ be divisible by p, then would also $3m + 3 \equiv p^n + 1$. Hence must $\alpha = 0$. The resulting form ξ^3 is a $SQ[3, p^n \equiv 3m + 2]$ by § 79.

There remains the case $p^n = 3^n$, when the reduced cubic is

$$\Phi(\xi) \equiv \xi^3 + \alpha_1\xi^2 + \alpha_2\xi.$$

Raising it to the power $3^{n-1} + 3^{n-2} + \cdots + 3 + 1 \equiv (3^n - 1)/2$, we find (mod 3),

$\left(\xi^{3^n} + \alpha_1^{3^{n-1}}\xi^{2\cdot 3^{n-1}} + \alpha_2^{3^{n-1}}\xi^{3^{n-1}}\right)\left(\xi^{3^{n-1}} + \alpha_1^{3^{n-2}}\xi^{2\cdot 3^{n-2}} + \alpha_2^{3^{n-2}}\xi^{3^{n-2}}\right)\ldots(\xi^3 + \alpha_1\xi^2 + \alpha_2\xi)$.

The highest exponent of ξ in this product is $< 2(3^n - 1)$. The coefficient of

$$\xi^{3^n-1} \equiv \xi^{2(3^{n-1}+3^{n-2}+\cdots+3+1)}$$

is evidently $\alpha_1^{3^{n-1}+3^{n-2}+\cdots+1}$. Hence must $\alpha_1 = 0$. Applying then the corollary of § 81, the resulting form $\xi^3 + \alpha_2\xi$ is a $SQ[3, 3^n]$ if, and only if, either $\alpha_2 = 0$ or else $-\alpha_2$ is a not-square in the $GF[3^n]$.

89. To treat a more characteristic example, we seek the $SQ[5, p^n]$, when p^n is of the form $5m + 3$. The reduced quintic is

$$\xi^5 + \alpha\xi^3 + \beta\xi^2 + \gamma\xi.$$

The power $m + 1$ gives $(m+1)\beta$ as the coefficient of $\xi^{5m+2} \equiv \xi^{p^n-1}$. If $m + 1 \equiv 0 \pmod{p}$, then $5m + 5 \equiv p^n + 2 \equiv 0$ and therefore $p = 2$.

ANALYTIC REPRESENTATION OF SUBSTITUTIONS, etc.   63

Hence, for $p \neq 2$, we must have $\beta = 0$. The power $m + 2$ of 53)
$$\xi^5 + \alpha \xi^3 + \gamma \xi$$
requires
$$C_{m+2,2}\gamma^2 + C_{m+2,3} \cdot 3\alpha^2\gamma + C_{m+2,4}\alpha^4 = 0.$$
If p is neither 2 nor 7, $C_{m+2,2} \equiv 0 \pmod{p}$ and may be divided out; for, if $m + 2$ be divisible by p, then is also $5(m + 2) = p^n + 7$. Multiplying the resulting equation by 5^2 and replacing $5m$ and $5(m-1)$ by $p^n - 3$ and $p^n - 8$, respectively, we have for $p \neq 2$, $p \neq 7$,
$$25\gamma^2 - 15\alpha^2\gamma + 2\alpha^4 \equiv (5\gamma - \alpha^2)(5\gamma - 2\alpha^2) = 0.$$
The power $m + 4$ of 53) requires, if[1]) $p^n > 13$,
$$C_{m+4,5}5\alpha\gamma^4 + C_{m+4,6}20\alpha^3\gamma^3 + C_{m+4,7}21\alpha^5\gamma^2 + C_{m+4,8}8\alpha^7\gamma + C_{m+4,9}\alpha^9 = 0.$$
If p is not 2, 3, 7 or 17, we may divide out the factor
$$(m+4)(m+3)(m+2)(m+1)m.$$
Multiplying afterwards by $5^4 \cdot 7!$ and replacing $5(m-1)$ by $-8 \pmod p$, etc., we find
$$210\alpha(5\gamma)^4 - 8 \cdot 140\alpha^3(5\gamma)^3 + 8 \cdot 13 \cdot 21\alpha^5(5\gamma)^2 - 8 \cdot 13 \cdot 18\alpha^7(5\gamma) + 23 \cdot 26\alpha^9 = 0.$$
This equation is an identity for $5\gamma = \alpha^2$, but reduces to $-10\alpha^9$ for $5\gamma = 2\alpha^2$. In the latter case, $\alpha = \gamma = 0$, if $p \neq 2$. Hence, for $p^n \neq 13, 2^n, 3^n, 7^n$ or 17^n, the only possible quintic which represents a substitution on the marks of the $GF[p^n \equiv 5m + 3]$ is reducible to
$$5\xi^5 + 5\alpha\xi^3 + \alpha^2\xi.$$
We have shown in § 80 that this quintic is indeed a $SQ[5, p^n \equiv 5m + 3]$. The special cases above excluded require separate treatment.

90. The foregoing methods may be employed[2]) to show that the following table gives every reduced $SQ[k, p^n]$ for $k < 6$:

Reduced quantic	Suitable for $p^n =$
ξ	any p^n
ξ^2	2^n
ξ^3	3^n, $3m + 2$
$\xi^3 - \alpha\xi$ (α = not-square)	3^n
$\xi^4 \pm 3\xi$	7
$\xi^4 + \alpha_2\xi^2 + \alpha_3\xi$ (if it vanishes only for $\xi = 0$)	2^n
ξ^5	5^n, $5m \pm 2$, $5m + 4$
$\xi^5 - \alpha\xi$ (α not a fourth power)	5^n
$\xi^5 \pm 2^{1/2}\xi$	3^2
$\xi^5 \pm 2\xi^2$	7
$\xi^5 + \alpha\xi^3 \pm \xi^2 + 3\alpha^2\xi$ (α = not-square)	7

---

1) If $p^n = 13$, the power $m + 4 \equiv 6$ brings in terms $\xi^{24} \equiv \xi^{2(p^n-1)}$.
2) Compare the author's Dissertation, l. c. pp. 77—86 and 101—102.

CHAPTER V.

Reduced quantic		Suitable for $p^n =$
$\xi^5 + \alpha\xi^3 + \frac{\alpha^2}{5}\xi$	(α arbitrary)	$5m \pm 2$
$\xi^5 + \alpha\xi^3 + 3\alpha^2\xi$	($\alpha =$ not-square)	13
$\xi^5 - 2\alpha\xi^3 + \alpha^2\xi$	($\alpha =$ not-square)	5^n

That in fact these quantics do represent substitutions on the marks of the corresponding $GF[p^n]$ follows from §§ 79, 80, 81, 83 and 86, with the exception possibly of the eleventh and thirteenth forms. To verify[1]) that the latter two are substitution-quantics, set

$$\Phi(\xi) \equiv \xi^5 + \alpha\xi^3 + 3\alpha^2\xi, \quad \Psi(\xi) \equiv \Phi(\xi) \pm \xi^2.$$

Then
$$\mu^{-5}\Phi(\mu\xi) \equiv \xi^5 + (\mu^{-2}\alpha)\xi^3 + 3(\mu^{-2}\alpha)^2\xi.$$

Since α is a not-square, we can choose an integer μ so that $\mu^{-2}\alpha$ shall be a particular not-square ν. But

$$\mu^{-5}\Psi(\mu\xi) = \xi^5 + (\mu^{-2}\alpha)\xi^3 + 3(\mu^{-2}\alpha)^2\xi \pm \mu^{-3}\xi^2.$$

Since $\mu^3 \equiv \pm 1 \pmod{7}$, we can choose the sign of $\mu = \pm(\alpha/\nu)^{1/2}$ to make the coefficient of ξ^2 unity. It follows, therefore, from § 87 that $\Phi(\xi)$ and $\Psi(\xi)$ will be substitution-quantics modulo 13 and 7, respectively, for α an arbitrary not-square, if they be such for α a particular not-square ν and for the $+$ sign in $\Psi(\xi)$. We take $\nu = 5$, a non-residue of both 7 and 13. In the notation of § 76, these reduced forms represent the substitutions,

$$\begin{pmatrix}\xi\\ \xi^5 + 5\xi^3 + \xi^2 + 5\xi\end{pmatrix} \equiv \begin{pmatrix}0, 1, 2, 3, 4, 5, 6\\ 0, 5, 2, 3, 1, 6, 4\end{pmatrix},$$

$$\begin{pmatrix}\xi\\ \xi^5 + 5\xi^3 + 10\xi\end{pmatrix} \equiv \begin{pmatrix}0, 1, 2, 3, 4, 5, 6, -6, -5, -4, -3, -2, -1\\ 0, 3, 1, 5, 6, 4, -2, 2, -4, -6, -5, -1, -3\end{pmatrix},$$

modulo 7 and 13 respectively.

The Betti-Mathieu Group, §§ 91—94.

91. It was shown in § 81 that the quantic belonging to the $GF[p^{nm}]$,

$$\Psi_A(X) \equiv \sum_{i=1}^{m} A_i X^{p^{n(m-i)}},$$

represents a substitution upon the p^{nm} marks of the field if, and only if, $X = 0$ is the only solution in the field of the equation

$$\Psi_A(X) = 0.$$

---

[1]) For a verification by means of the theorem of § 84, see the author's paper, *American Journal*, vol. 18, pp. 210—218; in particular, §§ 7 and 9.

Suppose that this condition is satisfied by two functions $\Psi_A(X)$ and $\Psi_B(X)$ and consider the effect of applying first the substitution[1])

$$A: \qquad X' = \Psi_A(X) \equiv \sum_{i=1}^{m} A_i X^{p^{n(m-i)}}$$

and afterwards the substitution

$$B: \qquad X'' = \Psi_B(X') \equiv \sum_{j=1}^{m} B_j X'^{p^{n(m-j)}}.$$

The result is equivalent to that produced by the single substitution

$$X'' = \Psi_B[\Psi_A(X)] \equiv \sum_{i,j}^{1,\ldots,m} B_j A_i^{p^{n(m-j)}} X^{p^{n(2m-i-j)}}.$$

After reduction by means of $X^{p^{nm}} = X$, this equation may be written

$$C: \qquad X'' = \Psi_C(X) \equiv \sum_{i=1}^{m} C_i X^{p^{n(m-i)}},$$

each C_i being a definite function of the A_j's and B_j's. By hypothesis,

$$\Psi_B[\Psi_A(X)] = 0$$

requires $\Psi_A(X) = 0$, which in turn requires $X = 0$. Hence, $X = 0$ is the only solution in the field of $\Psi_C(X) = 0$. It follows that the transformation C represents a substitution upon the marks of the $GF[p^{nm}]$. C is called the compound, or product, of A and B, and the above relation is expressed in the symbolic form,

$$C = AB.$$

Giving to the coefficients A_i every possible combination of values in the $GF[p^{nm}]$ such that

$$X' = \sum_{i=1}^{m} A_i X^{p^{n(m-i)}}$$

represents a substitution on its marks, we obtain a set of substitutions having the property that the result of applying first any one of the set and afterwards any one of the set is identical with the result of applying a single substitution of the set, called the product of the two. Such a set of substitutions is said to form a *group*. In the present case, the group will be called the *Betti-Mathieu Group*.[2])

---

1) The present notation is used in place of and as equivalent to

$$A: \qquad \begin{bmatrix} X \\ \Psi_A(X) \end{bmatrix}.$$

2) For $n = 1$, this group was studied by Betti, *Annali di Scienze Mat. e Fisiche*, vol. 3 (1852), pp. 49—115, vol. 6 (1855), pp. 5—34; for general n, by Mathieu, *Journal de Math.*, (2) vol. 5 (1860), pp. 9—42, vol. 6, pp. 241—323. The theorems of §§ 92—94 are due to the author, *Annals of Math.*, (1897), pp. 94—96, 178—183.

CHAPTER V.

92. Theorem. — *The necessary and sufficient condition that the transformation*

$$54) \qquad X' = \sum_{i=1}^{m} A_i X^{p^{n(m-i)}} \qquad (X, A_i \text{ in the } GF[p^{nm}])$$

shall represent a substitution on the marks of the $GF[p^{nm}]$ is

$$A \equiv \begin{vmatrix} A_1 & A_2 & \ldots A_m \\ A_2^{p^n} & A_3^{p^n} & \ldots A_1^{p^n} \\ \cdot & \cdot & \cdot \cdot \cdot \cdot \cdot \cdot \\ A_m^{p^{n(m-1)}} & A_1^{p^{n(m-1)}} & \ldots A_{m-1}^{p^{n(m-1)}} \end{vmatrix} \neq 0.$$

We seek the condition under which 54) is solvable for X. Raising 54) to the powers $1, p^n, p^{2n}, \ldots, p^{n(m-1)}$ and reducing the powers of X by

$$X^{p^{nm}} = X,$$

we obtain the following m equations (written with detached coefficients):

$$\begin{array}{c|cccc} & X^{p^{n(m-1)}} & X^{p^{n(m-2)}} \ldots X^{p^n} & X \\ \hline X' = & A_1 & A_2 & \ldots A_{m-1} & A_m \\ X'^{p^n} = & A_2^{p^n} & A_3^{p^n} & \ldots A_m^{p^n} & A_1^{p^n} \\ \cdot & \cdot & \cdot & \cdot & \cdot \\ X'^{p^{n(m-1)}} = & A_m^{p^{n(m-1)}} & A_1^{p^{n(m-1)}} & \ldots A_{m-2}^{p^{n(m-1)}} & A_{m-1}^{p^{n(m-1)}}. \end{array}$$

The solution of this system of equations in $X, X^{p^n}, \ldots$ gives

$$55) \qquad AX = \begin{vmatrix} A_1 & A_2 & \ldots A_{m-1} & X' \\ A_2^{p^n} & A_3^{p^n} & \ldots A_m^{p^n} & X'^{p^n} \\ \cdot & \cdot & \cdot & \cdot \\ A_m^{p^{n(m-1)}} & A_1^{p^{n(m-1)}} & \ldots A_{m-2}^{p^{n(m-1)}} & X'^{p^{n(m-1)}} \end{vmatrix},$$

$$56) \; A X^{p^{n(m-i)}} = \begin{vmatrix} A_1 & \ldots A_{i-1} & X' & A_{i+1} & \ldots A_m \\ A_2^{p^n} & \ldots A_i^{p^n} & X'^{p^n} & A_{i+2}^{p^n} & \ldots A_1^{p^n} \\ \cdot & \cdot & \cdot & \cdot & \cdot \\ A_m^{p^{n(m-1)}} & \ldots A_{i-2}^{p^{n(m-1)}} & X'^{p^{n(m-1)}} & A_i^{p^{n(m-1)}} & \ldots A_{m-1}^{p^{n(m-1)}} \end{vmatrix}.$$

The condition $A \neq 0$ is necessary, since otherwise there would exist a relation between the powers of X' with exponents $< p^{nm}$. To prove that the condition $A \neq 0$ is sufficient, we need only verify that the X given by 55) satisfies the relations 56) for $i = 1, 2, \ldots, m-1$. Observing that $A^{p^n} = A$ in the field, we find the following relations upon raising 55) to the power $p^{n(m-i)}$ and moving the first i rows below the last $m - i$ rows:

ANALYTIC REPRESENTATION OF SUBSTITUTIONS, etc. 67

$$A X^{p^{n(m-i)}} = (-1)^{i(m-i)} \begin{vmatrix} A_{i+1} & A_{i+2} & \ldots & A_{i-1} & X' \\ \cdot & \cdot & \cdot & \cdot & \cdot \\ A_m^{p^{n(m-i-1)}} & A_1^{p^{n(m-i-1)}} & \ldots & A_{m-2}^{p^{n(m-i-1)}} & X'^{p^{n(m-i-1)}} \\ A_1^{p^{n(m-i)}} & A_2^{p^{n(m-i)}} & \ldots & A_{m-1}^{p^{n(m-i)}} & X'^{p^{n(m-i)}} \\ \cdot & \cdot & \cdot & \cdot & \cdot \\ A_i^{p^{n(m-1)}} & A_{i+1}^{p^{n(m-1)}} & \ldots & A_{i-2}^{p^{n(m-1)}} & X'^{p^{n(m-1)}} \end{vmatrix}.$$

Moving the last i columns before the $m-i$ preceding columns, which brings in an additional factor $(-1)^{i(m-i)}$, we obtain the determinant of formula 56).

It follows as a corollary that formula 55) gives the *reciprocal* of 54).

A second proof may be given, based on the theorem of § 72. The condition that 54) shall represent a substitution on the marks of the $GF[p^{nm}]$ is identical with the condition under which $X'_1, X'_2, \ldots, X'_m$ shall be linearly independent with respect to the $GF[p^n]$ when it is given that $X_1, X_2, \ldots, X_m$ are similarly independent. We seek the condition under which

$$\left| X_j'^{p^{ni}} \right| \equiv \begin{vmatrix} X'_1 & X'_2 & \ldots & X'_m \\ X'^{p^n}_1 & X'^{p^n}_2 & \ldots & X'^{p^n}_m \\ \cdot & \cdot & \cdot & \cdot \\ X'^{p^{n(m-1)}}_1 & X'^{p^{n(m-1)}}_2 & \ldots & X'^{p^{n(m-1)}}_m \end{vmatrix} \neq 0.$$

Substituting the values of $X'_j, X'^{p^n}_j, \ldots, X'^{p^{n(m-1)}}_j$ in terms of $X_j, X^{p^n}_j, \ldots, X^{p^{n(m-1)}}_j$ and the A_i, as given by the above table, we find that

$$\left| X_j'^{p^{ni}} \right| = \left| X_j^{p^{ni}} \right| \cdot A.$$

The required condition is therefore that $A \neq 0$.

93. To illustrate a general method[1]) of obtaining sub-groups of the Betti-Mathieu Group, we take $m = 3$ and consider the totality of substitutions in the $GF[p^{3n}]$ on a variable X of that field,

57) $\qquad X' = A_1 X^{p^{2n}} + A_2 X^{p^n} + A_3 X,$

which multiply by a factor ϱ the function

$$Z \equiv (BX)^{p^{2n}} + (BX)^{p^n} + BX \quad (B \text{ in the } GF[p^{3n}]).$$

---

1) See the author's paper in the *American Journal*, vol. 22, pp. 49—54.

68 CHAPTER V.

The conditions for the identity $Z' \equiv \varrho Z$ are readily seen to be

58) $$BA_3 + B^{p^n} A_1^{p^n} + B^{p^{2n}} A_2^{p^{2n}} = \varrho B,$$

59) $$BA_2 + B^{p^n} A_3^{p^n} + B^{p^{2n}} A_1^{p^{2n}} = \varrho B^{p^n},$$

60) $$BA_1 + B^{p^n} A_2^{p^n} + B^{p^{2n}} A_3^{p^{2n}} = \varrho B^{p^{2n}}.$$

Since the left members of 59) and 60) are the powers p^n and p^{2n}, respectively, of the left member of 58), we must have $\varrho^{p^n} = \varrho$. Hence the totality of substitutions 57) for which the expression

$$\varrho \equiv A_3 + B^{p^n-1} A_1^{p^n} + B^{p^{2n}-1} A_2^{p^{2n}}$$

is a mark of the $GF[p^n]$ form a group leaving Z relatively invariant.

94. Consider next the substitutions 57) which multiply the function
$$Y \equiv D X^{p^n+1} + D^{p^n} X^{p^{2n}+p^n} + D^{p^{2n}} X^{p^{2n}+1}$$
by a parameter ϱ, where D is a mark $\neq 0$ in the $GF[p^{3n}]$.

To form the function Y' into which 57) transforms Y, we note that
$$X'^{p^n+1} = A_3 A_1^{p^n} X^2 + A_2 A_3^{p^n} X^{2p^n} + A_1 A_2^{p^n} X^{2p^{2n}} + (A_3^{p^n+1} + A_2 A_1^{p^n}) X^{p^n+1}$$
$$+ (A_1^{p^n+1} + A_3 A_2^{p^n}) X^{p^{2n}+1} + (A_2^{p^n+1} + A_1 A_3^{p^n}) X^{p^{2n}+p^n}.$$

Denoting by W the product of the expression on the right by D and forming the sum $Y' \equiv W + W^{p^n} + W^{p^{2n}}$, we find that the conditions for the identity $Y' = \varrho Y$ are the following six relations:

$$\tau = 0, \quad \tau^{p^n} = 0, \quad \tau^{p^{2n}} = 0, \quad f = \varrho D, \quad f^{p^n} = \varrho D^{p^n}, \quad f^{p^{2n}} = \varrho D^{p^{2n}},$$

where, for brevity,

61) $$\tau \equiv D A_3 A_1^{p^n} + D^{p^n} A_1^{p^n} A_2^{p^{2n}} + D^{p^{2n}} A_3 A_2^{p^{2n}}$$

62) $$f \equiv D(A_3^{p^n+1} + A_2 A_1^{p^n}) + D^{p^n}(A_1^{p^{2n}+p^n} + A_3^{p^n} A_2^{p^{2n}})$$
$$+ D^{p^{2n}}(A_2^{p^{2n}+1} + A_3 A_1^{p^{2n}}).$$

In particular, it follows that $\varrho^{p^n} = \varrho$. Hence those substitutions 57) whose coefficients A_1, A_2, A_3 make $\tau = 0$ and give to the function f/D a value belonging to the $GF[p^n]$ form a group with the relative invariant Y.

The method may be readily extended to determine for general m the substitutions 54) which leave relatively invariant the following function
$$Y_s \equiv \sum_{j=0}^{m-1} D^{p^{nj}} X^{p^{nj}+p^{n(s+j)}} \quad (D \text{ in the } GF[p^{nm}]),$$
where s may be any integer $< m$, except perhaps $m/2$.

ANALYTIC REPRESENTATION OF SUBSTITUTIONS, etc.   69

It is found that the number of independent conditions upon the coefficients A_i, in order that 54) shall leave Y_s relatively invariant, is at most $\frac{1}{2}(m+1)$ or $\frac{1}{2}(m+2)$ according as m is odd or even. One of these conditions merely requires that a certain function of D and the A_i shall belong to the $GF[p^n]$.

95. We proceed to identify the Betti-Mathieu Group in the $GF[p^{nm}]$ with Jordan's linear homogeneous group on m indices with coefficients in the $GF[p^n]$. Let R be a primitive root of the $GF[p^{nm}]$, so that any mark of that field can be expressed in the form $\gamma_0 + \gamma_1 R + \gamma_2 R^2 + \cdots + \gamma_{m-1} R^{m-1}$, where each γ_i is a mark of the $GF[p^n]$. Consider the general substitution 54) of the Betti-Mathieu Group. We may set

$$X \equiv \sum_{i=0}^{m-1} \xi_i R^i, \quad X' \equiv \sum_{i=0}^{m-1} \xi_i' R^i, \quad A_k \equiv \sum_{i=0}^{m-1} a_i^{(k)} R^i,$$

where each ξ_i, ξ_i' and $a_i^{(k)}$ belong to the $GF[p^n]$.

Substituting these values in the identity 54) and reducing the powers of R to a degree $\leqq m-1$ by means of the equation of degree m satisfied by the primitive root R, we may equate the coefficients of like powers of R in the resulting identity. Since

$$\xi_i^{p^n} = \xi_i, \quad \xi_i'^{p^n} = \xi_i',$$

we evidently reach a set of m equations of the form

63) $\qquad \xi_i' = \sum_{i=0}^{m-1} \alpha_{ij} \xi_j \quad (i = 0, 1, \ldots, m-1),$

in which the coefficients α_{ij} belong to the $GF[p^n]$. By hypothesis, equation 54) is solvable for X in terms of X'. Starting from this solved form, our process evidently yields the ξ_j as functions of the ξ_j', so that equations 63) are solvable in the field $GF[p^n]$. Hence $|\alpha_{ij}| \neq 0$. According to the definition given in § 97, the transformation 63) belongs to Jordan's linear homogeneous group.

Inversely, every linear substitution 63), with coefficients in the $GF[p^n]$ such that the determinant $|\alpha_{ij}| \neq 0$, can be represented in the form 54). We note first that 63) transforms $X \equiv \sum_{i=0}^{m-1} \xi_i R^i$ into

$$X' \equiv \sum_{i=0}^{m-1}\left(\sum_{j=0}^{m-1} \alpha_{ij}\xi_j\right)R^i = \sum_{j=0}^{m-1} \xi_j \left(\sum_{i=0}^{m-1} \alpha_{ij} R^i\right) \equiv \sum_{j=0}^{m-1} \xi_j \tau_j,$$

where

$$\tau_j \equiv \sum_{i=0}^{m-1} \alpha_{ij} R^i.$$

Furthermore, $\tau_0, \tau_1, \ldots, \tau_{m-1}$ are linearly independent with respect to the $GF[p^n]$; for, if $\varkappa_0, \ldots, \varkappa_{m-1}$ be marks of the latter field such that
$$\varkappa_0 \tau_0 + \varkappa_1 \tau_1 + \cdots + \varkappa_{m-1} \tau_{m-1} = 0,$$
then
$$\sum_{j=0}^{m-1} \varkappa_j \alpha_{ij} = 0 \quad (i = 0, 1, \ldots, m-1),$$
and therefore, since $|\alpha_{ij}| \neq 0$, each $\varkappa_j = 0$. Hence, when each ξ_i runs independently through the series of p^n marks of the $GF[p^n]$, the expressions X and X' both run through the p^{nm} marks of the $GF[p^{nm}]$. Every substitution 63) therefore gives rise to a permutation on the marks of that field.

But we can always find a set of marks $A_1, A_2, \ldots, A_m$ of the $GF[p^{nm}]$ such that 54) will transform the set of marks $1, R, R^2, \ldots, R^{m-1}$, linearly independent with respect to the $GF[p^n]$, into an arbitrary set of m marks of the $GF[p^{nm}]$,
$$B_i \equiv \sum_{j=0}^{m-1} \beta_{ij} R^j \quad (i = 0, \ldots, m-1),$$
linearly independent with respect to the $GF[p^n]$. The conditions are
$$\sum_{i=1}^{m} A_i (R^k)^{p^{n(m-i)}} = B_k \quad (k = 0, 1, \ldots, m-1),$$
which can be solved for $A_1, A_2, \ldots, A_m$, since the determinant in R is not zero by § 72. The resulting substitution 54) will transform the p^{nm} marks $\sum_{i=0}^{m-1} \lambda_i R^i$ of the $GF[p^{nm}]$ into the marks $\sum_{i=0}^{m-1} \lambda_i B_i$ all distinct; indeed, we have the identity
$$\sum_{i=1}^{m} A_i (\lambda_0 X_0 + \lambda_1 X_1 + \cdots + \lambda_t X_t)^{p^{n(m-i)}} \equiv \sum_{j=0}^{t} \left\{ \lambda_j \sum_{i=1}^{m} A_i X_j^{p^{n(m-i)}} \right\}.$$

96. EXERCISES ON THE TEXT OF CHAPTER V.

Ex. 1. Verify that $\xi^6 + \alpha \xi^5 - \alpha^4 \xi^2$ (α arbitrary) represents a substitution on the marks of either the $GF[3^3]$ or of the $GF[2^5]$.

Ex. 2. (Hermite). A group of order 168 is generated by the substitutions
$$x' \equiv ax + b, \quad x' \equiv a\Theta(x+b) + c \pmod{7},$$
where $\Theta(x) \equiv -x^5 - 2x^2$ and a is a quadratic residue of 7.

Ex. 3. (Rogers.) In applying Hermite's conditions (§ 84) for a substitution quantic, it suffices, *when* $n=1$, to test only the first $\frac{1}{2}(p-1)$ powers. This result of Rogers does not generalize immediately

ANALYTIC REPRESENTATION OF SUBSTITUTIONS, etc. 71

to the case $n > 1$; for $SQ[6, 3^2]$ it is necessary to consider, besides the 2^d and 4th powers, also the power $5 > \frac{1}{2}(3^2-1)$.

Ex. 3. By the theorem given by Weber, Algebra, II, p. 299, every substitution on p^n letters, each affected with n indices $z_1, z_2, \ldots, z_n$ taken modulo p, may be represented by the transformation (mod p),

$$z_i' = \Phi_i(z_1, z_2, \ldots, z_n) \quad (i = 1, 2, \ldots, n)$$

where each Φ_i is a rational integral function with integral coefficients. Apply the method of § 84 and show that, on raising each Φ_i to the powers $1, 2, \ldots, p-2$ and reducing by means of $z_i^p \equiv z_i$ (mod p), the coefficient of $z_1^{p-1} z_2^{p-1} \ldots z_n^{p-1}$ in each power must be congruent to zero.

Ex. 4. The following substitution in the $GF[p^{2n}]$

a) $\quad X' = A_1 X^{p^n} + A_2 X \quad (A_1 \neq 0, \; A_1^{p^n+1} - A_2^{p^n+1} \neq 0)$

can be reduced to the form $Y' = RY$ by introducing a new index

b) $\quad Y \equiv B_1 X^{p^n} + B_2 X \quad (B_1^{p^n+1} - B_2^{p^n+1} \neq 0)$

if and only if there exists no root in the $GF[p^n]$ of the equation

$$\begin{vmatrix} A_2 - R & A_1 \\ A_1^{p^n} & A_2^{p^n} - R \end{vmatrix} \equiv 0.$$

If $A_2 + A_2^{p^n} \neq 0$, it is not possible to reduce a) to the form

$$Y' = KY^{p^n} \quad (K \text{ in the } GF[p^{2n}])$$

by a transformation of indices of the form b).
[The first result is in marked contrast to that of § 214 for $m = 2$].

Ex. 5. By the method of § 95, show that the sub-groups of the Betti-Mathieu Group defined in §§ 93—94 by means of the invariants Z and Y are identical with certain linear homogeneous groups on m indices in the $GF[p^n]$ defined by a linear and a quadratic invariant respectively.

Ex. 6. (Moore.) The multiplier-$GF[p^k]$ of the additive-group $[\lambda_1, \ldots, \lambda_m]$ is the (largest) additive-group common to the additive-groups

$$[\lambda_i^{-1} \lambda_1, \ldots, \lambda_i^{-1} \lambda_m] \quad (i = 1, \ldots, m)$$

and is contained in the $p^m - 1$ additive-groups

$$[\lambda^{-1} \lambda, \ldots, \lambda^{-1} \lambda_m] \quad (\lambda \neq 0 \text{ of } [\lambda_1, \ldots, \lambda_m]).$$

SECOND PART.

THEORY
OF LINEAR GROUPS IN A GALOIS FIELD.

CHAPTER I.

GENERAL LINEAR HOMOGENEOUS GROUP.[1]

97. First definition. — Consider the p^{nm} letters, or symbols,

$$l_{\xi_1, \xi_2, \ldots, \xi_m}$$

characterized by m indices, each running through the series of marks of the $GF[p^n]$. The general linear homogeneous substitution A on the m indices ξ_i with coefficients in the $GF[p^n]$ replaces the letter $l_{\xi_1, \xi_2, \ldots, \xi_m}$ by $l_{\xi'_1, \xi'_2, \ldots, \xi'_m}$ where

$$A: \qquad \xi'_i = \sum_{j=1}^{m} \alpha_{ij}\xi_j \quad (i = 1, \ldots, m)$$

the coefficients α_{ij} being marks of the field. But A will indeed permute the p^{nm} letters if, and only if, *the determinant of A is not zero*,

$$|A| \equiv |\alpha_{ij}| \neq 0.$$

In fact, there must be one and only one system of m indices which A replaces by a given system ξ'_i and hence an unique set of values ξ_j satisfying the equations

$$\sum_{j=1}^{m} \alpha_{ij}\xi_j = \xi'_i \quad (i = 1, 2, \ldots, m).$$

Let B denote a second substitution with coefficients in the $GF[p^u]$,

$$B: \qquad \xi'_k = \sum_{i=1}^{m} \beta_{ki}\xi_i \quad (k = 1, 2, \ldots, m)$$

where

$$|B| \equiv |\beta_{ki}| \neq 0.$$

---

[1] Jordan, Traité des substitutions, Nos. 119, 169; author's dissertation, Part II. Cf. § 95 above.

76   CHAPTER I.

The result of applying first the substitution A, which replaces $l_{\varepsilon_1}, \ldots, \varepsilon_m$ by $l_{\varepsilon'_1}, \ldots, \varepsilon'_m$ where

64) $\qquad \xi'_i = \sum_{j=1}^{m} \alpha_{ij} \xi_j \quad (i = 1, \ldots, m),$

and afterwards the substitution B, which replaces the general letter $l_{\varepsilon'_1, \ldots, \varepsilon'_m}$ by $l_{\varepsilon''_1, \ldots, \varepsilon''_m}$ where

65) $\qquad \xi''_k = \sum_{i=1}^{m} \beta_{ki} \xi'_i \quad (k = 1, 2, \ldots, m),$

is identical with the result of applying a single linear substitution, called their compound or *product* AB, which replaces $l_{\varepsilon_1, \ldots, \varepsilon_m}$ by $l_{\varepsilon''_1, \ldots, \varepsilon''_m}$, where, by eliminating the ξ'_i between 64) and 65), we have

$$\xi''_k = \sum_{j=1}^{m} \left(\sum_{i=1}^{m} \beta_{ki} \alpha_{ij} \right) \xi_j \quad (k = 1, \ldots, m).$$

Setting

$$\gamma_{kj} \equiv \sum_{i=1}^{m} \beta_{ki} \alpha_{ij} \quad (k, j = 1, \ldots, m),$$

we may write the product of A and B in the form

$AB:$ $\qquad \xi'_k = \sum_{j=1}^{m} \gamma_{kj} \xi_j \quad (k = 1, \ldots, m).$

By the theorem for the multiplication of determinants

$$|AB| \equiv |\gamma_{kj}| = |\beta_{ki}| \cdot |\alpha_{ij}| \equiv |B| \cdot |A| \neq 0.$$

Moreover, the coefficients γ_{kj} belong to the $GF[p^n]$. Hence the compound AB is indeed a substitution and has its coefficients in the same field as those of A and B. If therefore we let the coefficients of A run through all the sets of values in the $GF[p^n]$ for which the determinant $|\alpha_{ij}| \neq 0$, we obtain a set of substitutions forming a group called the *general linear homogeneous group on m indices with coefficients in the $GF[p^n]$* and denoted by the symbol $GLH(m, p^n)$.

Remark. — If the substitution A be identical with the substitution

$$\xi'_i = \sum_{j=1}^{m} \overline{\alpha}_{ij} \xi_j \quad (i = 1, \ldots, m)$$

then must $\overline{\alpha}_{ij} = \alpha_{ij}$ $(i, j = 1, \ldots, m)$. This follows by taking in turn for $j = 1, 2, \ldots, m$ the particular set of values

$$\xi_j = 1, \quad \xi_k = 0 \quad (k = 1, 2, \ldots, m; \ k \neq j).$$

GENERAL LINEAR HOMOGENEOUS GROUP. 77

98. Second definition of $GLH(m, p^n)$. — The essential thing in the substitution A is the matrix of its coefficients (α_{ij}). Taking the indices $\xi_1, \ldots, \xi_m$ to be variable marks of the $GF[p^n]$, we obtained an immediate interpretation of A as a permutation of certain p^{nm} letters, so that the linear group was recognized as a permutation-group. We may, however, let the indices $\xi_1, \ldots, \xi_m$ be arbitrary variables and consider the linear transformations

$$A: \quad \xi_i' = \sum_{j=1}^{m} \alpha_{ij} \xi_j \quad (i = 1, \ldots, m), \quad |\alpha_{ij}| \neq 0,$$

where each coefficient belongs to the $GF[p^n]$. As in § 97, the compound of two such transformations will be a linear transformation of determinant not zero and with all its coefficients in the $GF[p^n]$. Since $|\alpha_{ij}| \neq 0$, the inverse of A exists and has similar properties. Hence the totality of transformations A form a group, evidently the $GLH(m, p^n)$.

Employing this second definition, we may represent the transformation group as a group of permutations on p^{nm} letters. Consider, indeed, the p^{nm} linear functions $\lambda_1 \xi_1 + \lambda_2 \xi_2 + \cdots + \lambda_m \xi_m$ where each λ runs through the marks of the $GF[p^n]$. These functions are merely permuted by the linear transformations A.

99. Theorem. — *The order $GLH[m, p^n]$ of the group $GLH(m, p^n)$ is*

$$(p^{nm} - 1)(p^{nm} - p^n)(p^{nm} - p^{2n}) \ldots (p^{nm} - p^{n(m-1)}).$$

The number of distinct linear functions

$$f_1 \equiv \sum_{j=1}^{m} \alpha_{1j} \xi_j$$

by which the substitutions of the group can replace the index ξ_1 is $p^{nm} - 1$, since the marks α_{1j} may be chosen arbitrarily in the $GF[p^n]$ provided not all are zero. Let T be one of the substitutions which replace ξ_1 by a definite linear function f_1. If then

$$R_1 \equiv I \text{ (identity)}, \quad R_2, \quad R_3, \ldots, R_N$$

denote all the substitutions of the group which leave ξ_1 fixed, the N products,
$$T, \quad TR_2, \quad TR_3, \ldots, TR_N$$

will replace ξ_1 by f_1. No other substitution of the group has this property; for, if U replace ξ_1 by f_1, $T^{-1}U$ will leave ξ_1 fixed and hence be a certain R_i, so that $U = TR_i$. To each of the $p^{nm} - 1$ distinct functions f_1 there corresponds a set of N substitutions. Hence
$$GLH[m, p^n] = N(p^{nm} - 1).$$

The substitutions R_i are of the form
$$\xi'_1 = \xi_1, \quad \xi'_k = \sum_{j=1}^m \alpha_{kj}\xi_j \quad (k = 2, \ldots, m)$$
where the $m-1$ coefficients α_{k1} are arbitrary and the coefficients $\alpha_{kj}(k,j = 2, \ldots, m)$ are such that their determinant $\neq 0$ in the field. The latter set of coefficients can be chosen in $GLH[m-1, p^n]$ ways. Hence
$$N = p^{n(m-1)} GLH[m-1, p^n],$$
$$GLH[m, p^n] = p^{n(m-1)}(p^{nm}-1) GLH[m-1, p^n].$$
This recursion formula gives, since $GLH[1, p^n] = p^n - 1$, the result
$$GLH[m, p^n] = p^{n(m-1)}(p^{nm}-1)p^{n(m-2)}(p^{n(m-1)}-1) \ldots p^n(p^{2n}-1)(p^n-1)$$
$$= (p^{nm}-1)(p^{nm}-p^n) \ldots (p^{nm}-p^{n(m-1)}).$$

100. Theorem. — *Every linear homogeneous substitution A on m indices with coefficients in the $GF[p^n]$ can be expressed as a product BD_m, where B is derived from the totality of substitutions of the form*
$$B_{r,s\lambda}: \quad \xi'_r = \xi_r + \lambda \xi_s, \quad \xi'_i = \xi_i \quad (i = 1, \ldots, m; \ i \neq r; \ r \neq s)$$
λ being an arbitrary mark of the $GF[p^n]$, and where D_m denotes the substitution altering only the index ξ_m which it multiplies by the determinant of A.

Let the given substitution A be the following:
$$A: \quad \xi'_i = \sum_{j=1}^m \alpha_{ij}\xi_j \quad (i = 1, \ldots, m).$$
The product $AB_{1,2,\lambda}$ has the form
$$\xi'_1 = \sum_{j=1}^m (\alpha_{1j} + \lambda \alpha_{2j})\xi_j, \quad \xi'_i = \sum_{j=1}^m \alpha_{ij}\xi_j \quad (i = 2, \ldots, m),$$
the matrix of its coefficients being
$$\begin{pmatrix} \alpha_{11} + \lambda\alpha_{21} & \alpha_{12} + \lambda\alpha_{22} & \ldots & \alpha_{1m} + \lambda\alpha_{2m} \\ \alpha_{21} & \alpha_{22} & \ldots & \alpha_{2m} \\ \cdot & \cdot & & \cdot \\ \alpha_{m1} & \alpha_{m2} & \ldots & \alpha_{mm} \end{pmatrix}.$$
Similarly, the matrix for the product $B_{1,2,\lambda}A$ is
$$\begin{pmatrix} \alpha_{11} & \alpha_{12} + \lambda\alpha_{11} & \alpha_{13} & \ldots & \alpha_{1m} \\ \alpha_{21} & \alpha_{22} + \lambda\alpha_{21} & \alpha_{23} & \ldots & \alpha_{2m} \\ \cdot & \cdot & \cdot & & \cdot \\ \alpha_{m1} & \alpha_{m2} + \lambda\alpha_{m1} & \alpha_{m3} & \ldots & \alpha_{mm} \end{pmatrix}.$$

GENERAL LINEAR HOMOGENEOUS GROUP. 79

To multiply A on the right by $B_{r,s,\lambda}$, we therefore multiply the sth row of the matrix (α_{ij}) by λ and add to the rth row; to multiply A on the left by the same substitution, we multiply the rth column by λ and add to the sth column of the matrix (α_{ij}). We make use of these operations, which are recognized to be identical with the elementary operations permissible in reducing a determinant, to simplify the form of the matrix A. It is shown below that, if $m > 1$, we can set $\alpha_{11} = 1$. Then by multiplying A on the right and left by suitable generators $B_{i,j,\lambda}$, we can reach a new matrix A' having the elements of the first row and first column all zero, except α_{11} which $= 1$. After $m-1$ such steps, we would reach a matrix $A^{(m-1)}$ having every element zero except those in the main diagonal and the latter all unity except that lying in the last row. The resulting substitution would be D_m. From the identity thus established, $B_1 A B_2 \equiv D_m$, where B_1 and B_2 are products derived from the $B_{i,j,\lambda}$, we find
$$A = B_1^{-1} D_m B_2^{-1} = B_1^{-1} B_3 D_m = B D_m.$$

It remains to be shown that, if $m > 1$, a matrix can be obtained from A having $\alpha_{11} = 1$. From the given generators we derive the substitution

66) $\quad B_{i,j,\lambda} B_{j,i-\lambda^{-1}} B_{i,j,\lambda}: \quad \xi_i' = \lambda \xi_j, \quad \xi_j' = -\lambda^{-1} \xi_i,$

affecting only the indices ξ_i and ξ_j. In particular, for $\lambda = 1$, $i = 1$, we get
$$J: \quad \xi_1' = \xi_j, \quad \xi_j' = -\xi_1.$$

We determine a substitution K derived from the $B_{i,j,\lambda}$ such that the product $A' \equiv KA$ will have the coefficient $\alpha_{21}' \neq 0$. If $\alpha_{21} \neq 0$, we take $K = I$, the identity; if $\alpha_{21} = 0$, but $\alpha_{2j} \neq 0$, we take $K = J$. The product
$$A'' \equiv A' B_{1,2,\lambda}$$
has the coefficient $\alpha_{11}'' \equiv \alpha_{11}' + \lambda \alpha_{21}'$, which can be made equal to unity by choice of λ in the $GF[p^n]$.

Corollary I. *The only linear homogeneous substitutions commutative with every $B_{r,s,\lambda} (r, s = 1, \ldots, m, r \neq s)$, where λ is a fixed mark $\neq 0$ of the $GF[p^n]$, are those of the form*
$$\xi_i' = \varrho \xi_i \quad (i = 1, \ldots, m).$$

It follows by inspection of the above two matrices for $AB_{1,2,\lambda}$ and $B_{1,2,\lambda} A$ that they are identical only when
$$\alpha_{11} = \alpha_{22}, \quad \alpha_{i1} = 0 \ (i = 2, 3, \ldots, m), \quad \alpha_{2j} = 0 \ (j = 3, \ldots, m).$$

Since the indices 1, 2 can be replaced by any pair r, s of distinct integers $\leq m$, it follows that every element of the matrix (α_{ij}) must be zero except those in the main diagonal, which must all be equal.

80 CHAPTER I.

Corollary II. *The group of binary linear homogeneous substitutions of determinant unity is generated by the substitutions* $B_{1,2,\lambda}$ *and*

T: $\quad\quad\quad \xi_1' = -\xi_2, \quad \xi_2' = \xi_1.$

Indeed, T transforms $B_{1,2,-\lambda}$ into $B_{2,1,\lambda}$.

101. *Transformation of indices.* — We can introduce in place of $\xi_1, \xi_2, \ldots, \xi_m$ the m new indices

67) $\quad\quad\quad \eta_i \equiv \sum_{k=1}^{m} \beta_{ik}\xi_k \quad (i = 1, 2, \ldots, m)$

provided the determinant $|\beta_{ik}| \neq 0$. In fact, the substitution

A: $\quad\quad\quad \xi_i' = \sum_{j=1}^{m} \alpha_{ij}\xi_j \quad (i = 1, 2, \ldots, m)$

will replace η_i by $\sum_{j,k}^{1 \ldots m} \beta_{ik}\alpha_{kj}\xi_j$, which, by solving 67), can be put into the form $\sum_{j=1}^{m} \gamma_{ij}\eta_j$. The substitution A becomes

$B^{-1}AB$: $\quad\quad\quad \eta_i' = \sum_{j=1}^{m} \gamma_{ij}\eta_j \quad (i = 1, 2, \ldots, m)$

where B denotes the substitution 67) replacing the ξ_i by the η_i. In fact

$$B^{-1}AB = \begin{pmatrix}\eta_i \\ \xi_i\end{pmatrix}\begin{pmatrix}\xi_i \\ \xi_i'\end{pmatrix}\begin{pmatrix}\xi_i' \\ \eta_i'\end{pmatrix} \equiv \begin{pmatrix}\eta_i \\ \eta_i'\end{pmatrix}.$$

The determinant of the transformed substitution equals that of A,

$$|B^{-1}AB| = |B^{-1}| \cdot |A| \cdot |B| = |A|.$$

This result is, however, a special case ($\varrho = 0$) of the next theorem.

102. Theorem. — *The characteristic determinant (with parameter ϱ) of a linear homogeneous substitution A,*

$$\Delta(\varrho) \equiv \begin{vmatrix} \alpha_{11}-\varrho & \alpha_{12} & \alpha_{13} & \ldots & \alpha_{1m} \\ \alpha_{21} & \alpha_{22}-\varrho & \alpha_{23} & \ldots & \alpha_{2m} \\ \cdot & \cdot & \cdot & & \cdot \\ \alpha_{m1} & \alpha_{m2} & \alpha_{m3} & \ldots & \alpha_{mm}-\varrho \end{vmatrix}$$

is unchanged under every linear transformation of indices.

It is only necessary to prove the theorem for the following types of transformations of indices, since by § 100 every linear transformation can be derived from them:

D_1: $\quad\quad \eta_1 = D\xi_1, \quad \eta_2 = \xi_2, \ldots, \eta_m = \xi_m;$

$B_{1,2,\lambda}$: $\quad\quad \eta_1 = \xi_1 + \lambda\xi_2, \quad \eta_2 = \xi_2, \ldots, \eta_m = \xi_m.$

GENERAL LINEAR HOMOGENEOUS GROUP. 81

Under the transformation of indices D_1, A takes the form

$$\eta_1' = D\sum_{j=1}^{m} \alpha_{1j}\xi_j = \alpha_{11}\eta_1 + D\sum_{j=2}^{m} \alpha_{1j}\eta_j,$$

$$\eta_i' = \sum_{j=1}^{m} \alpha_{ij}\xi_j = \frac{1}{D}\alpha_{i1}\eta_1 + \sum_{j=2}^{m} \alpha_{ij}\eta_j \quad (i = 2, 3, \ldots, m).$$

The characteristic determinant of the transformed substitution is

$$\begin{vmatrix} \alpha_{11}-\varrho & D\alpha_{12} & \ldots & D\alpha_{1m} \\ \alpha_{21}/D & \alpha_{22}-\varrho & \ldots & \alpha_{2m} \\ \cdot & \cdot & \cdot & \cdot \\ \alpha_{m1}/D & \alpha_{m2} & \ldots & \alpha_{mm}-\varrho \end{vmatrix} = \begin{vmatrix} \alpha_{11}-\varrho & \alpha_{12} & \ldots & \alpha_{1m} \\ \alpha_{21} & \alpha_{22}-\varrho & \ldots & \alpha_{2m} \\ \cdot & \cdot & \cdot & \cdot \\ \alpha_{m1} & \alpha_{m2} & \ldots & \alpha_{mm}-\varrho \end{vmatrix} \equiv \Delta(\varrho).$$

Under the transformation of indices $B_{1,2,\lambda}$, A becomes

$$\eta_1' = \sum_{j=1}^{m}(\alpha_{1j} + \lambda\alpha_{2j})\xi_j = (\alpha_{11} + \lambda\alpha_{21})\eta_1 + (\alpha_{12} + \lambda\alpha_{22} - \lambda\alpha_{11} - \lambda^2\alpha_{21})\eta_2$$
$$+ \sum_{j=3}^{m}(\alpha_{1j} + \lambda\alpha_{2j})\eta_j,$$

$$\eta_i' = \sum_{j=1}^{m}\alpha_{ij}\xi_j = \alpha_{i1}\eta_1 + (\alpha_{i2} - \lambda\alpha_{i1})\eta_2 + \sum_{j=3}^{m}\alpha_{ij}\eta_j \quad (i = 2, \ldots, m).$$

The characteristic determinant of this substitution is

$$\begin{vmatrix} \alpha_{11}+\lambda\alpha_{21}-\varrho & \alpha_{12}+\lambda\alpha_{22}-\lambda\alpha_{11}-\lambda^2\alpha_{21} & \alpha_{13}+\lambda\alpha_{23} & \ldots & \alpha_{1m}+\lambda\alpha_{2m} \\ \alpha_{21} & \alpha_{22}-\lambda\alpha_{21}-\varrho & \alpha_{23} & \ldots & \alpha_{2m} \\ \alpha_{31} & \alpha_{32}-\lambda\alpha_{31} & \alpha_{33}-\varrho & \ldots & \alpha_{3m} \\ \cdot & \cdot & \cdot & & \cdot \\ \alpha_{m1} & \alpha_{m2}-\lambda\alpha_{m1} & \alpha_{m3} & \ldots & \alpha_{mm}-\varrho \end{vmatrix}.$$

Multiplying the second row by λ and subtracting from the first row, and afterwards adding the first column multiplied by λ to the second column, we reach the original determinant $\Delta(\varrho)$.

Corollary. — *The transformed of A by any linear substitution B has the same characteristic determinant as A.* Indeed, by § 101, A is converted into $B^{-1}AB$ by the transformation of indices indicated by the substitution B.

Factors of composition[1]) *of $GLH(m, p^n)$*, §§ 103—107.

103. Let ϱ be a primitive root of the $GF[p^n]$. If two linear substitutions have as determinants ϱ^{rl} and ϱ^{sl}, their compound has

---

1) For the case $n = 1$, Jordan, Traité, pp. 106—110; for general n, author's dissertation, Annals of Mathematics, vol. 11 (1897), pp. 168—175; also Burnside, The theory of groups, pp. 340—341.

82   CHAPTER I.

the determinant $\varrho^{(r+s)l}$. Hence the totality of substitutions in the group $G \equiv GLH(m, p^n)$ having as determinants powers of ϱ^l forms a subgroup G_l. Suppose that
$$p^n - 1 = p_1 p_2 \ldots p_k,$$
where $p_1, p_2, \ldots, p_k$ are all primes. Denote by $G_{p_1}, G_{p_1 p_2}, \ldots, G_{p^n-1} \equiv \Gamma$ the subgroups of G formed of those of its substitutions whose determinants are respectively powers of $\varrho^{p_1}, \varrho^{p_1 p_2}, \ldots, \varrho^{p^n-1} \equiv 1$. By § 63, the orders of these groups are respectively
$$\Omega/p_1, \quad \Omega/p_1 p_2, \quad \ldots, \Omega/p^n - 1 \quad (\text{where } \Omega \equiv GLH[m, p^n]).$$

In fact, by § 100, G contains substitutions of every determinant $\neq 0$ in the $GF[p^n]$ and contains the same number of one determinant as of another.

If S and T be linear substitutions, S and $T^{-1}ST$ have the same determinant (§ 101). Hence the groups $G_{p_1}, G_{p_1 p_2}, \ldots, \Gamma$ are *self-conjugate* under G, i. e., each is transformed into itself by any substitution of G. Since $p_1, \ldots, p_k$ are primes, there is no group lying between G and G_{p_1}, no one between G_{p_1} and $G_{p_1 p_2}$, etc. Hence we may descend from G to Γ by the composition-series
$$G, \quad G_{p_1}, \quad G_{p_1 p_2}, \quad \ldots, G_{p^n - 1} \equiv \Gamma.$$

The group Γ of all substitutions of determinant unity is called the *special linear homogeneous group* $SLH(m, p^n)$. It has a self-conjugate subgroup H formed of those of its substitutions which are of the form
$$M_\mu: \qquad \xi_i' = \mu \xi_i \quad [\mu^m = 1] \qquad (i = 1, 2, \ldots, m).$$

The mark μ must also satisfy the equation
$$\mu^{p^n - 1} = 1.$$

Hence, if d be the greatest common divisor of m and $p^n - 1$, we find (by the method of proof used in § 79) that

68) $\mu^d = 1.$

Inversely, each of the d distinct solutions in the $GF[p^n]$ of 68) [see § 16], leads to a substitution M_μ belonging to the group H. The order of H is therefore d.

If δ be a mark of the $GF[p^n]$ which belongs to the exponent d (§ 17, Corollary), then μ is a power of δ. Suppose that
$$d = q_1 q_2 \ldots q_l \quad (\text{each } q_i \text{ a prime}).$$
Denote by $H_{q_1}, H_{q_1 q_2}, \ldots, H_d \equiv I$ the groups formed of those substitutions of H which multiply every index by a like power of δ^{q_1}, by

a like power of $\delta^{q_1 q_2}, \ldots$, by a like power of $\delta^d \equiv 1$, respectively. Since we have, for any mark ν,
$$M_\nu^{-1} M_\mu M_\nu \equiv M_\mu,$$
a composition-series of H is given by
$$H, \ H_{q_1}, \ H_{q_1 q_2}, \ldots, \ H_{q_1 q_2 \ldots q_l} \equiv I.$$
In view of the theorem proven in §§ 104—107, we may state the complete

Theorem. — *The factors of composition of $GLH(m, p^n)$ are*
$$p_1, \ p_2, \ldots, p_k, \ \Omega/d(p^n - 1), \ q_1, \ q_2, \ldots, q_l,$$
except in the two cases $(m, p^n) = (2, 2)$ *and* $(2, 3)$, *when the factors of composition are* 2, 3 *and* 2, 3, 2, 2, 2 *respectively.*

104. Theorem. — *Excluding the above two cases, the group H is a maximal self-conjugate subgroup of Γ.*

Suppose that Γ contains a self-conjugate subgroup J which contains all the substitutions of H and still further substitutions. We will prove that, aside from the two exceptional cases mentioned, J coincides with Γ.

By hypothesis, J contains a substitution

$$S: \quad \xi_i' = \sum_{j=1}^m \alpha_{ij} \xi_j \quad (i = 1, \ldots, m)$$

which is not in H and therefore does not multiply all the indices by the same factor. Hence, by Corollary I of § 100, S is not commutative with every $B_{r,s,\lambda}$ ($r, s = 1, 2, \ldots, m$; $r \neq s$). Changing the notation if necessary, we may suppose that S is not commutative with $B_{1,2,\lambda}$, a substitution of determinant unity and therefore in the group Γ. It therefore transforms the substitution S of the self-conjugate subgroup J into a substitution belonging to J. Hence J contains the product
$$T \equiv S^{-1} \cdot B_{1,2,\lambda}^{-1} S B_{1,2,\lambda},$$
which does not reduce to the identity I. In calculating this product, let Φ be the linear function by which S^{-1} replaces ξ_2. Then T is seen to have the form, in which the values of the β_{1j} need not be determined:
$$T: \quad \xi_1' = \sum_{j=1}^m \beta_{1j} \xi_j; \quad \xi_i' = \xi_i - \lambda \alpha_{i1} \Phi \quad (i = 2, 3, \ldots, m).$$

Suppose first that the α_{i1} are not all zero, say $\alpha_{21} \neq 0$. For $m > 2$, we introduce new indices η_i defined by the substitution V of determinant unity,
$$\eta_1 \equiv \xi_1, \quad \eta_2 \equiv \xi_2, \quad \eta_i \equiv \xi_i - \frac{\alpha_{i1}}{\alpha_{21}} \xi_2 \quad (i = 3, 4, \ldots, m).$$

The resulting substitution $V^{-1}TV$ belongs to J and leaves $\eta_i (i > 2)$ unaltered:

$$\eta'_i = (\xi_i - \lambda\alpha_{i1}\Phi) - \frac{\alpha_{i1}}{\alpha_{21}}(\xi_2 - \lambda\alpha_{21}\Phi) = \xi_i - \frac{\alpha_{i1}}{\alpha_{21}}\xi_2 \equiv \eta_i.$$

If, however, every $\alpha_{i1} = 0$, T itself leaves fixed $m - 1$ indices. In either case, J contains a substitution $\neq I$ of the form[1])

$$R: \quad \eta'_1 = \sum_{j=1}^m \gamma_{1j}\eta_j, \quad \eta'_2 = \sum_{j=1}^m \gamma_{2j}\eta_j, \quad \eta'_i = \eta_i \quad (i = 3, \ldots, m).$$

Then J contains the two substitutions leaving $\eta_3, \ldots, \eta_m$ fixed:

$$T_2 \equiv R^{-1}B_{1,3,\mu}^{-1}RB_{1,3,\mu}: \begin{cases} \eta'_1 = \eta_1 - \mu(\gamma_{11} - 1)\eta_3, \\ \eta'_2 = \eta_2 - \mu\gamma_{21}\eta_3. \end{cases}$$

$$T_3 \equiv R^{-1}B_{2,3,\mu}^{-1}RB_{2,3,\mu}: \begin{cases} \eta'_1 = \eta_1 - \mu\gamma_{12}\eta_3, \\ \eta'_2 = \eta_2 - \mu(\gamma_{22} - 1)\eta_3. \end{cases}$$

These substitutions are both of the form

$$U: \quad \eta'_1 = \eta_1 + \sigma_1\eta_3, \quad \eta'_2 = \eta_2 + \sigma_2\eta_3, \quad \eta'_i = \eta_i \quad (i = 3, \ldots, m).$$

If T_2 and T_3 reduce to the identity, R itself becomes

$$R_1: \quad \eta'_1 = \eta_1 + \gamma_{13}\eta_3 + \cdots + \gamma_{1m}\eta_m, \quad \eta'_2 = \eta_2 + \gamma_{23}\eta_3 + \cdots + \gamma_{2m}\eta_m,$$

$$\eta'_i = \eta_i \quad (i > 2).$$

If $\gamma_{1j} = \gamma_{2j} = 0$ $(j = 4, \ldots, m)$, this substitution $\neq I$ is of the form U. In the contrary case, we may suppose that γ_{14} and γ_{24} are not both zero. Then

$$R_1^{-1}B_{4,3,\mu}^{-1}R_1B_{4,3,\mu}: \begin{cases} \eta'_1 = \eta_1 - \mu\gamma_{14}\eta_3, \\ \eta'_2 = \eta_2 - \mu\gamma_{24}\eta_3, \quad \eta'_i = \eta_i \quad (i > 2) \end{cases}$$

is a substitution $\neq I$ of the form U and belonging to J. Hence, in every case J contains a substitution U not the identity. For definiteness, let $\sigma_1 \neq 0$ and introduce the new indices

$$\xi_1 \equiv \eta_1, \quad \xi_2 \equiv \eta_2 - \frac{\sigma_2}{\sigma_1}\eta_1, \quad \xi_i = \eta_i \quad (i > 2).$$

Then U becomes $B_{1,3,\sigma_1}$. Transforming the latter by the substitution

$$\xi'_1 = \lambda\xi_1, \quad \xi'_2 = \lambda^{-1}\xi_2, \quad \xi'_i = \xi_i \quad (i = 3, \ldots, m),$$

where λ is an arbitrary mark $\neq 0$ of the $GF[p^n]$, we reach in J the substitution $B_{1,3,\lambda\sigma}$, and therefore every $B_{1,3,\lambda}$. The latter is transformed into $B_{k,3,\lambda} (k \neq 1, 3)$ by the following substitution of Γ:

$$\xi'_1 = -\xi_k, \quad \xi'_k = \xi_1, \quad \xi'_i = \xi_i \quad (i = 2, \ldots, k-1, k+1, \ldots, m).$$

---

[1]) From this point, the proofs by Burnside and Jordan (l. c.) are incomplete. The specific errors were made in the *Traité*, p. 108, 1° and in *The theory of groups*, p. 316, "This process may now be repeated", etc.

Finally, for $j \neq k$, $B_{k,3,\lambda}$ is transformed into $B_{k,j,\lambda}$ by the substitution
$$\xi_3' = -\xi_j, \quad \xi_j' = \xi_3, \quad \xi_i' = \xi_i \quad (i = 1, \ldots, m; \ i \neq 3, \ i \neq j).$$
It follows from § 100 that, if $m > 2$, J is identical with Γ.

105. For $m = 2$, we are given that J contains a substitution
$$S: \quad \xi_1' = \alpha\xi_1 + \beta\xi_2, \quad \xi_2' = \alpha'\xi_1 + \beta'\xi_2 \quad (\alpha\beta' - \alpha'\beta = 1),$$
which is neither the identity I nor
$$E: \quad \xi_1' = -\xi_1, \quad \xi_2' = -\xi_2.$$
We proceed to prove that, for $p^n > 3$, J contains a substitution of the form $B_{2,1,\lambda}$ in which $\lambda \neq 0$.

a) Suppose first that $\beta = 0$, so that J contains
$$S_1: \quad \xi_1' = \alpha\xi_1, \quad \xi_2' = \alpha'\xi_1 + \alpha^{-1}\xi_2,$$
where $\alpha' \neq 0$ if $\alpha = \alpha^{-1}$, since $S_1 \neq I$ or E.

a_1) If $\alpha = \alpha^{-1}$, whence $\alpha = \pm 1$, the group J contains both S_1 and $S_1 E$, one of which has the form
$$\xi_1' = \xi_1, \quad \xi_2' = \xi_2 + \lambda\xi_1 \quad (\lambda \equiv \pm \alpha' \neq 0).$$

a_2) If $\alpha \neq \alpha^{-1}$, J contains the substitution $\neq I$,
$$S_1 B_{2,1,1}^{-1} S_1^{-1} B_{2,1,1}: \quad \xi_1' = \xi_1, \quad \xi_2' = \xi_2 + (1 - \alpha^2)\xi_1.$$

b) Suppose next that $\beta \neq 0$. The following substitution
$$\Sigma: \quad \xi_1' = \varkappa\alpha\xi_1 + \varkappa\beta\xi_2, \quad \xi_2' = -\frac{1 + \varkappa^2\alpha^2}{\varkappa\beta}\xi_1 - \varkappa\alpha\xi_2$$
has determinant unity and therefore belongs to Γ. Hence J contains $S_2 \equiv \Sigma^{-1} S \Sigma S$; viz.,
$$\xi_1' = -\varkappa^{-2}\xi_1, \quad \xi_2' = \frac{-(1 + \varkappa^2)(\beta' + \alpha\varkappa^2)}{\beta\varkappa^2}\xi_1 - \varkappa^2\xi_2.$$
If $p^n = 4$ or if $p^n > 5$, $\varkappa$ can be chosen in the $GF[p^n]$ so that
$$\varkappa^4 \neq 1, \quad \varkappa^{-2} \neq \varkappa^2.$$
Proceeding with S_2 as in case a_2), we obtain in J a substitution $B_{2,1,\lambda}$, where $\lambda \neq 0$.

If $p^n = 5$, we take $\varkappa = 1$, when $S_2 E$ becomes
$$\xi_1' = \xi_1, \quad \xi_2' = \frac{2(\beta' + \alpha)}{\beta}\xi_1 + \xi_2.$$
Our result follows unless $\beta' + \alpha \equiv 0 \pmod 5$. But J contains the product $S B_{2,1,1}^{-1} S^{-1} B_{2,1,1}$, viz.,
$$\xi_1' = (1 + \alpha\beta)\xi_1 + \beta^2\xi_2, \quad \xi_2' = (1 + \alpha\beta - \alpha^2)\xi_1 + (1 - \alpha\beta + \beta^2)\xi_2,$$
for which the sum corresponding to the above $\beta' + \alpha$ is
$$(1 + \alpha\beta) + (1 - \alpha\beta + \beta^2) \equiv \beta^2 + 2 \not\equiv 0 \pmod 5.$$

86   CHAPTER I.

We have now proved that, if $p^n > 3$, J contains a substitution $B_{2,1,\lambda}(\lambda \neq 0)$. It is transformed into $B_{2,1,\lambda\varrho^2}$ by the substitution

Also
$$\xi_1' = \varrho^{-1}\xi_1, \quad \xi_2' = \varrho\,\xi_2.$$

$$B_{2,1,\lambda\varrho^2}\,B_{2,1,\lambda\sigma^2} = B_{2,1,\lambda(\varrho^2+\sigma^2)}.$$

By § 64, there exist solutions in the $GF[p^n]$ of $\varrho^2 + \sigma^2 = \varkappa/\lambda$ for $\varkappa$ arbitrary in the field. Hence J contains $B_{2,1,\varkappa}$. Transforming the latter by $(\xi_1' = \xi_2,\ \xi_2' = -\xi_1)$ we get $B_{1,2,-\varkappa}$. It follows from § 100 that $J \equiv \Gamma$. By §§ 99 and 103, the order of the group Γ of binary linear homogeneous substitutions of determinant unity is $p^n(p^{2n}-1)$.

106. For $p^n = 2$, $m = 2$, the group Γ is of order 6 and is identical with $GLH(2,2)$. It contains a subgroup of order 3 generated by the substitution

$$\xi_1' = \xi_2, \quad \xi_2' = \xi_1 + \xi_2.$$

The index of this subgroup being 2, it is self-conjugate. The factors of composition are therefore 2 and 3.

107. For $p^n = 3$, $m = 2$, the group $G \equiv GLH(m, p^n)$ is of order $48 \equiv (3^2-1)(3^2-3)$ and contains the following substitutions

A: $\quad \xi_1' = -\xi_1, \quad \xi_2' = \xi_1 + \xi_2;$
B: $\quad \xi_1' = \xi_2, \quad \xi_2' = -\xi_1 + \xi_2;$
C: $\quad \xi_1' = -\xi_2, \quad \xi_2' = \xi_1;$
D: $\quad \xi_1' = \xi_1 + \xi_2, \quad \xi_2' = \xi_1 - \xi_2;$
E: $\quad \xi_1' = -\xi_1, \quad \xi_2' = -\xi_2,$

of which A has determinant -1 and the others determinant $+1$ modulo 3. In virtue of the relations

$E^2 = 1;$
$D^2 = E, \quad DE = ED;$
$C^2 = E, \quad CE = EC, \quad CD = EDC;$
$B^3 = E, \quad BE = EB, \quad BD = CDB, \quad BC = DB;$
$A^2 = 1, \quad AE = EA, \quad AD = CA, \quad AC = DA, \quad AB = ECB^2A;$

it results that the groups generated as follows:

$$\{E\};\ \{E, D\};\ \{E, D, C\};\ \{E, D, C, B\};\ \{E, D, C, B, A\}$$

have the orders 2, 4, 8, 24, 48 respectively and that each group is self-conjugate under the following group. The last group is identical with G, whose factors of composition are therefore 2, 3, 2, 2, 2.

108. From the linear homogeneous substitution A of § 98 on the arbitrary variables $\xi_1, \xi_2, \ldots, \xi_m$, we obtain the linear fractional substitution

$$A': \quad x_i' = \frac{\alpha_{i1} x_1 + \alpha_{i2} x_2 + \cdots + \alpha_{i\,m-1} x_{m-1} + \alpha_{im}}{\alpha_{m1} x_1 + \alpha_{m2} x_2 + \cdots + \alpha_{m\,m-1} x_{m-1} + \alpha_{mm}} \quad (i = 1, \ldots, m-1)$$

upon setting $x_i \equiv \xi_i/\xi_m$ for $i = 1, \ldots, m-1$. It being only a question of the ratios of the coefficients α_{ij} in A', its determinant $|\alpha_{ij}|$ is determined only up to a factor μ^m, μ being a mark $\neq 0$. Also, A' is the identity if, and only if, A be one of the $p^n - 1$ substitutions

$$M_\mu : \quad \xi_i' = \mu \xi_i \quad (i = 1, \ldots, m).$$

The products $M_\mu A$ and no other linear homogeneous substitutions correspond to the same linear fractional substitution A'. Hence the group $G \equiv GLH(m, p^n)$ has $(p^n - 1, 1)$ isomorphism with the group L of the substitutions A'. If Ω denote the order of G, the order of L is $\Omega \div (p^n - 1)$. To the subgroup Γ formed of the substitutions of G having determinant unity there corresponds a subgroup Λ of L composed of those of its substitutions whose determinant is an mth power in the field. If d be the greatest common divisor of m and $p^n - 1$, there are exactly d substitutions of the form M_μ in Γ and they form the group H (§ 103). Hence Γ has $(d, 1)$ isomorphism with Λ. The order of Λ is therefore $\Omega \div d(p^n - 1)$. Aside from the cases $(m, p^n) = (2, 2)$ and $(2, 3)$, H was shown to be the maximal self-conjugate subgroup of Γ; hence Λ has no self-conjugate subgroup other than itself and the identity and is therefore *simple*.

The group $LF(m, p^n)$ of all linear fractional substitutions in the $GF[p^n]$ on $m - 1$ variables and having determinant unity or some m^{th} power in the field has the order

$$\frac{1}{d}(p^{nm} - 1) p^{n(m-1)} (p^{n(m-1)} - 1) p^{n(m-2)} \cdots (p^{2n} - 1) p^n$$

d being the greatest common divisor of m and $p^n - 1$. It is a simple group except in the two cases $(m, p^n) = (2, 2)$ and $(2, 3)$. The group of all linear fractional substitutions of determinants not zero has d times the order of $LF(m, p^n)$.

The notation $LF(m, p^n)$ emphasizes the point that the essential quality of the linear fractional substitution lies in the matrix (α_{ij}) of degree m and not in the $m - 1$ variables $x_1, \ldots, x_{m-1}$ which play the rôle of indeterminates. For $m = 2$, we use the suggestive notation

$$\begin{pmatrix} \alpha, \beta \\ \gamma, \delta \end{pmatrix} : \quad x' = \frac{\alpha x + \beta}{\gamma x + \delta} \quad (\Delta \equiv \alpha \delta - \beta \gamma \neq 0).$$

In virtue of the identity of the two substitutions

$$\begin{pmatrix} \alpha, \beta \\ \gamma, \delta \end{pmatrix}, \quad \begin{pmatrix} \alpha\mu, \beta\mu \\ \gamma\mu, \delta\mu \end{pmatrix} \quad (\mu \text{ any mark} \neq 0)$$

of determinants Δ and $\mu^2 \Delta$, we may choose μ so that the substitution takes its normal form, viz., of determinant unity if $p = 2$, but of

88 CHAPTER I. GENERAL LINEAR HOMOGENEOUS GROUP.

determinant unity or a particular not-square ν if $p > 2$. In fact, if Δ is a square, $\mu^2\Delta$ may be made equal to unity by choice of μ in the field; while for Δ a not-square, $\mu^2\Delta$ may be made equal to ν.

If $p^n > 3$, *the group* $LF(2, p^n)$ *of all linear fractional substitutions in the* $GF[p^n]$ *of determinant unity (when in their normal forms) is a simple*[1]) *group of order*

$$M(p^n) = \frac{p^n(p^{2n}-1)}{2;\,1} \quad (2;\,1\ according\ as\ p > 2;\ p = 2).$$

There are $p^n(p^{2n}-1)$ *linear fractional substitutions of determinant* $\neq 0$.

From the formula of composition of binary linear homogeneous substitutions (§ 97), we derive the product SS_1 of linear fractional substitutions $S = \begin{pmatrix} \alpha, \beta \\ \gamma, \delta \end{pmatrix}$, $S_1 = \begin{pmatrix} \alpha_1, \beta_1 \\ \gamma_1, \delta_1 \end{pmatrix}$:

$$z' = \frac{\alpha z + \beta}{\gamma z + \delta}, \quad z'' = \frac{\alpha_1 z' + \beta_1}{\gamma_1 z' + \delta_1}, \quad z'' = \frac{(\alpha_1 \alpha + \beta_1 \gamma) z + (\alpha_1 \beta + \beta_1 \delta)}{(\gamma_1 \alpha + \delta_1 \gamma) z + (\gamma_1 \beta + \delta_1 \delta)}.$$

Hence if S operate first and S_1 afterwards, the product SS_1 is[2])

$$\begin{pmatrix} \alpha, \beta \\ \gamma, \delta \end{pmatrix} \begin{pmatrix} \alpha_1, \beta_1 \\ \gamma_1, \delta_1 \end{pmatrix} = \begin{pmatrix} \alpha_1 \alpha + \beta_1 \gamma, & \alpha_1 \beta + \beta_1 \delta \\ \gamma_1 \alpha + \delta_1 \gamma, & \gamma_1 \beta + \delta_1 \delta \end{pmatrix}.$$

109. The quotient-group Γ/H may be readily represented as a permutation-group on $q = (p^{nm} - 1) \div (p^n - 1)$ letters[3]). Of the $p^{nm} - 1$ letters $l_{\xi_1, \xi_2, \ldots, \xi_m}$ in which $\xi_1, \xi_2, \ldots, \xi_m$ denote marks of the $GF[p^n]$ not all zero, we combine into a single system the $p^n - 1$ letters $l_{\mu\xi_1, \mu\xi_2, \ldots, \mu\xi_m}$ in which μ runs through the series of marks $\neq 0$ while $\xi_1, \xi_2, \ldots, \xi_m$ denotes a set of fixed marks not all zero. Any linear homogeneous substitution on $\xi_1, \ldots, \xi_m$ with coefficients in the field replaces the letters of any one system by letters all of some one system and therefore permutes the q systems amongst themselves. In particular, the substitutions M_μ do not displace any system. Hence the group Γ of substitutions of determinant unity corresponds to a permutation-group on the q systems, which represents concretely the quotient-group Γ/H.

---

1) Cf. Moore, Congress Mathematical Papers, pp. 208—242, *Bull. Amer. Math. Soc.*, Dec. 1893; Burnside, Proc. Lond. Math. Soc., vol. 25, pp. 113—139 (Feb., 1894); also see § 261 below.

2) For the same product of matrices, the notation $S_1 S$ is sometimes used, S operating first.

3) Compare the method of §§ 228, 224; also, for $m = 2$, that of § 239.

CHAPTER II.

THE ABELIAN LINEAR GROUP.[1])

110. A linear homogeneous substitution on $2m$ indices with coefficients belonging to the $GF[p^n]$ is called *Abelian* if, when operating simultaneously upon two sets of $2m$ indices,

$$\xi_{i1}, \quad \eta_{i1}; \quad \xi_{i2}, \quad \eta_{i2} \quad (i = 1, 2, \ldots, m),$$

it leaves formally invariant up to a factor (belonging to the field) the bilinear function

$$74) \qquad \varphi = \sum_{i=1}^{m} \begin{vmatrix} \xi_{i1} & \eta_{i1} \\ \xi_{i2} & \eta_{i2} \end{vmatrix}.$$

The totality of such substitutions constitutes a group called the *general Abelian linear group*[2]) $GA(2m, p^n)$. These of its substitutions which leave φ absolutely invariant form the *special Abelian linear group* $SA(2m, p^n)$. For other definitions of these groups see § 160 below and the author's article, *Transactions of the American Mathematical Society*, vol. 1, pp. 30—38.

The conditions that the linear substitution

$$75) \qquad S: \begin{cases} \xi_i' = \sum_{j=1}^{m}(\alpha_{ij}\xi_j + \gamma_{ij}\eta_j) \\ \eta_i' = \sum_{j=1}^{m}(\beta_{ij}\xi_j + \delta_{ij}\eta_j) \end{cases} \quad (i = 1, 2, \ldots, m)$$

shall leave φ formally[3]) invariant up to the factor μ are

$$76) \quad \sum_{i=1}^{m}\begin{vmatrix}\alpha_{ij} & \gamma_{ij}\\ \beta_{ij} & \delta_{ij}\end{vmatrix} = \mu, \quad \sum_{i=1}^{m}\begin{vmatrix}\alpha_{ij} & \gamma_{ik}\\ \beta_{ij} & \delta_{ik}\end{vmatrix} = 0,$$

$$\sum_{i=1}^{m}\begin{vmatrix}\alpha_{ij} & \alpha_{ik}\\ \beta_{ij} & \beta_{ik}\end{vmatrix} = 0, \quad \sum_{i=1}^{m}\begin{vmatrix}\gamma_{ij} & \gamma_{ik}\\ \delta_{ij} & \delta_{ik}\end{vmatrix} = 0. \quad (j, k = 1, \ldots, m;\ j \neq k)$$

For $m = 1$, the Abelian group $GA(2, p^n)$ is evidently identical with the general binary linear homogeneous group $GLH(2, p^n)$. In

---

1) Investigated by Jordan, Traité, pp. 171—186, for the case $n = 1$; by the author, *Quar. Jour. of Math.*, 1897, pp. 169—178, ibid., 1899, pp. 383—4, for general n.

2) To distinguish these groups from the ordinary Abelian, i. e. commutative, groups, we prefix the adjective *linear*. The Abelian linear group is not commutative in general.

3) The indices ξ_i and η_i are treated as arbitrary quantities. Formal invariance is used in antithesis to numerical invariance.

determining the structure of the Abelian group, we may therefore suppose $m > 1$.

111. We proceed to determine the substitution reciprocal to S,

$$S^{-1}: \begin{cases} \xi'_i = \sum_{j=1}^{m}(\alpha'_{ij}\xi_j + \gamma'_{ij}\eta_j) \\ \eta'_i = \sum_{j=1}^{m}(\beta'_{ij}\xi_j + \delta'_{ij}\eta_j) \end{cases} \quad (i = 1, 2, \ldots, m).$$

Supposing S to be Abelian, we obtain the same result upon multiplying φ by μ that we obtain upon operating the substitution S upon the two sets of indices. The identity of the two results is not destroyed by operating the substitution S^{-1} upon the indices ξ_{i1}, η_{i1} ($i = 1, \ldots, m$) of one set. The result obtained upon multiplying φ by μ and then applying the substitution S^{-1} upon the indices ξ_{i1}, η_{i1} is therefore identical with the result obtained by applying the substitution S upon the indices ξ_{i2}, η_{i2} alone. Equating the two results, we find

$$\mu \sum_{i,j}^{1,\ldots,m} \{(\alpha'_{ij}\xi_{j1} + \gamma'_{ij}\eta_{j1})\eta_{i2} - (\beta'_{ij}\xi_{j1} + \delta'_{ij}\eta_{j1})\xi_{i2}\}$$
$$\equiv \sum_{i,j}^{1,\ldots,m} \{\xi_{i1}(\beta_{ij}\xi_{j2} + \delta_{ij}\eta_{j2}) - \eta_{i1}(\alpha_{ij}\xi_{j2} + \gamma_{ij}\eta_{j2})\}.$$

From this identity in the indices ξ_{ij}, η_{ij}, we find

$$\mu\alpha'_{ij} = \delta_{ji}, \quad \mu\gamma'_{ij} = -\gamma_{ji}, \quad \mu\beta'_{ij} = -\beta_{ji}, \quad \mu\delta'_{ij} = \alpha_{ji}.$$

Hence the reciprocal of the Abelian substitution 75) is

77) $\quad S^{-1}: \begin{cases} \xi'_i = \dfrac{1}{\mu}\sum_{j=1}^{m}(\delta_{ji}\xi_j - \gamma_{ji}\eta_j) \\ \eta'_i = \dfrac{1}{\mu}\sum_{j=1}^{m}(-\beta_{ji}\xi_j + \alpha_{ji}\eta_j) \end{cases} \quad (i = 1, 2, \ldots, m).$

When S^{-1} is operated upon the two sets of indices, φ must be multiplied by $1/\mu$. Forming the relations expressing this fact, we obtain the following conditions, together entirely equivalent to the set of conditions 76):

78) $\sum_{i=1}^{m} \begin{vmatrix} \alpha_{ki} & \gamma_{ki} \\ \beta_{ki} & \delta_{ki} \end{vmatrix} = \mu, \quad \sum_{i=1}^{m} \begin{vmatrix} \alpha_{ki} & \gamma_{ki} \\ \beta_{ji} & \delta_{ji} \end{vmatrix} = 0,$
$\quad\quad\quad\quad\quad\quad\quad\quad\quad\quad\quad\quad\quad\quad\quad\quad (j, k = 1, \ldots, m; j \neq k)$
$\sum_{i=1}^{m} \begin{vmatrix} \alpha_{ki} & \gamma_{ki} \\ \alpha_{ji} & \gamma_{ji} \end{vmatrix} = 0, \quad \sum_{i=1}^{m} \begin{vmatrix} \beta_{ki} & \delta_{ki} \\ \beta_{ji} & \delta_{ji} \end{vmatrix} = 0.$

112. Since the conditions 76) and 78) will be used repeatedly in this and the succeeding chapters, it will be found to be of great assistance to apply the following scheme by which these conditions can be read off by inspection from the matrix of the coefficients of S:

$\alpha_{11}\ \gamma_{11}$ $\beta_{11}\ \delta_{11}$	$\alpha_{12}\ \gamma_{12}$ $\beta_{12}\ \delta_{12}$	. . .	$\alpha_{1m}\ \gamma_{1m}$ $\beta_{1m}\ \delta_{1m}$
$\alpha_{21}\ \gamma_{21}$ $\beta_{21}\ \delta_{21}$	$\alpha_{22}\ \gamma_{22}$ $\beta_{22}\ \delta_{22}$	. . .	$\alpha_{2m}\ \gamma_{2m}$ $\beta_{2m}\ \delta_{2m}$
. . .	. . .	. . .	. . .
$\alpha_{m1}\ \gamma_{m1}$ $\beta_{m1}\ \delta_{m1}$	$\alpha_{m2}\ \gamma_{m2}$ $\beta_{m2}\ \delta_{m2}$	. . .	$\alpha_{mm}\ \gamma_{mm}$ $\beta_{mm}\ \delta_{mm}$

The 1st and 2nd rows of this matrix will be called *complementary*, likewise the 3rd and 4th rows, . . ., finally the $2m - 1$st and the $2m$th rows. Similarly, the 1st and 2nd columns will be called complementary, also the 3rd and 4th, . . , finally, the $2m - 1$st and $2m$th columns.

The left member of each of the relations 78) is a sum of determinants built from the coefficients of two rows, the elements of each individual determinant belonging to complementary columns. If the two rows be the sth and tth, we denote this sum by R_{st}. The relations 78) may then be written (taking $s < t$)

79) $R_{2l-1,2l} = \mu, \quad R_{st} = 0 \quad$ (unless $t = s + 1 =$ even).

Similarly, if we denote by C_{st} the sum of the determinants built from the coefficients of the sth and tth columns, the elements of each individual determinant belonging to complementary rows, we may write the relations 76) in the compact form

80) $C_{2l-1\ 2l} = \mu, \quad C_{st} = 0 \quad$ (unless $t = s + 1 =$ even).

113. Theorem. — *The factors of composition of $GA(2m, p^n)$ are the prime factors of $p^n - 1$ together with the factors of composition of $SA(2m, p^n)$.*

Let ϱ be a primitive root of the $GF[p^n]$. The general Abelian group contains the substitution

$$U: \quad \xi_i' = \varrho \xi_i, \quad \eta_i' = \eta_i \quad (i = 1, 2, \ldots, m)$$

which multiplies φ by ϱ. Let S be any Abelian substitution and $\mu = \varrho^r$ the factor by which it multiplies φ. We have

92 CHAPTER II.

$$S = U^r T,$$

where T is a new Abelian substitution not altering φ and hence in the special Abelian group. Since r may be any one of the integers $1, 2, \ldots, p^n-1$, the order of $GA(2m, p^n)$ is p^n-1 times the order $SA[2m, p^n]$ of the group $SA(2m, p^n)$.

Let $\alpha, \beta, \ldots$ be the prime factors whose product gives p^n-1. Let $A, A_\alpha, A_{\alpha\beta}, \ldots, A_{p^n-1} \equiv SA(2m, p^n)$ be the groups formed by the combination of the substitutions of $SA(2m, p^n)$ with

$$U, \quad U^\alpha, \quad U^{\alpha\beta}, \ldots, U^{p^n-1} \equiv I$$

respectively. Evidently these groups have the respective orders

$$(p^n-1)SA[2m, p^n], \quad \frac{1}{\alpha}(p^n-1)SA[2m, p^n],$$

$$\frac{1}{\alpha\beta}(p^n-1)SA[2m, p^n], \ldots, SA[2m, p^n],$$

while each is self-conjugate under $A \equiv GA(2m, p^n)$.

114. Theorem. — *The group $SA(2m, p^n)$ is generated by the substitutions*[1])

$M_i:$ $\xi_i' = \eta_i, \quad\quad \eta_i' = -\xi_i;$

$L_{i,\lambda}:$ $\xi_i' = \xi_i + \lambda\eta_i;$

$N_{i,j,\lambda}:$ $\xi_i' = \xi_i + \lambda\eta_j, \quad \xi_j' = \xi_j + \lambda\eta_i,$

where $i, j = 1, 2, \ldots, m$; $i \neq j$; and where λ is an arbitrary mark of the $GF[p^n]$. Every substitution of the group has determinant unity.

From these substitutions leaving φ absolutely invariant, we obtain other simple substitutions of $SA(2m, p^n) \equiv G$ as follows:

$L_{i,\lambda}' \equiv M_i^{-1} L_{i,-\lambda} M_i:$ $\eta_i' = \eta_i + \lambda\xi_i;$

$Q_{i,j,\lambda} \equiv M_j N_{i,j,-\lambda} M_j^{-1}:$ $\xi_i' = \xi_i + \lambda\xi_j, \quad \eta_j' = \eta_j - \lambda\eta_i;$

$R_{i,j,\lambda} \equiv M_j^{-1} Q_{j,i,\lambda} M_j:$ $\eta_i' = \eta_i - \lambda\xi_j, \quad \eta_j' = \eta_j - \lambda\xi_i;$

$T_{i,\lambda} \equiv L_{i,\lambda}' L_{i,-\lambda^{-1}} L_{i,\lambda}' M_i:$ $\xi_i' = \lambda\xi_i, \quad \eta_i' = \lambda^{-1}\eta_i;$

$P_{ij} \equiv Q_{j,i,1}^{-1} Q_{i,j,1} Q_{j,i,1} = (\xi_i\xi_j)(\eta_i\eta_j).$

Let S be any substitution of G and let it replace ξ_1 by

$$\omega_1 = \sum_{j=1}^{m}(\alpha_{1j}\xi_j + \gamma_{1j}\eta_j) \quad [\alpha_{1j}, \gamma_{1j} \text{ not all zero}].$$

We can set $S = VS'$, where V is derived from the above substitutions and S' is a substitution of G in which the coefficient corresponding

---

1) In the expression for each substitution we omit the indices not altered. For example, M_i alters only the two indices η_i and ξ_i.

THE ABELIAN LINEAR GROUP. 93

to α_{11} in S is not zero. Indeed, according as $\alpha_{1j} \neq 0$ or $\gamma_{1j} \neq 0$, we may take $V = P_{1j}$ or $P_{1j}M_j$. Let S' replace ξ_1 by

$$\omega_1' = \sum_{j=1}^{m} (\alpha_{1j}'\xi_j + \gamma_{1j}'\eta_j) \quad [\alpha_{11}' \neq 0].$$

We can determine a substitution S_1 derived from the above types which shall replace ξ_1 by ω_1', viz.,

$$S_1 = L_{1,\beta} M_1 L_{1,\alpha} Q_{1,2,\alpha_{12}'} N_{1,2,\gamma_{12}'} \cdots Q_{1,m,\alpha_{1m}'} N_{1,m,\gamma_{1m}'},$$

where α and β are determined by the conditions

$$\alpha = -\alpha_{11}' + \alpha_{12}'\gamma_{12}' + \cdots + \alpha_{1m}'\gamma_{1m}', \quad 1 + \beta\alpha_{11}' = \gamma_{11}'.$$

Hence $S' = S_1 S''$, where S'' is a new substitution of G which leaves ξ_1 fixed. Let S'' replace η_1 by

$$\omega_2 = \sum_{j=1}^{m} (\beta_{1j}\xi_j + \delta_{1j}\eta_j).$$

For $\mu = 1$, $\alpha_{11} = 1$, $\gamma_{11} = \alpha_{12} = \gamma_{12} = \cdots = \alpha_{1m} = \gamma_{1m} = 0$, the relation $R_{12} = \mu$ of 79) gives $\delta_{11} = 1$ in the substitution S''. The substitution

$$S_2 = L_{1,\tau}' R_{1,2,-\beta_{12}} Q_{2,1,-\delta_{12}} \cdots R_{1,m,-\beta_{1m}} Q_{m,1,-\delta_{1m}}$$

will replace η_1 by ω_2 if we take

$$\tau = \beta_{11} - \beta_{12}\delta_{12} - \beta_{13}\delta_{13} - \cdots - \beta_{1m}\delta_{1m}.$$

Hence $S'' = S_2 S'''$, where S''' is a new substitution of G which leaves ξ_1 and η_1 unaltered and thus has the form

$$S''': \begin{cases} \xi_1' = \xi_1, \quad \xi_i' = \sum_{j=1}^{m}(\alpha_{ij}\xi_j + \gamma_{ij}\eta_j) \\ \eta_1' = \eta_1, \quad \eta_i' = \sum_{j=1}^{m}(\beta_{ij}\xi_j + \delta_{ij}\eta_j) \end{cases} \quad (i = 2, 3, \ldots, m).$$

Applying the following relations of set 79),

we find
$$R_{1t} = 0, \quad R_{2t} = 0 \quad (t = 3, 4, \ldots, 2m),$$
$$\alpha_{i1} = \beta_{i1} = \gamma_{i1} = \delta_{i1} = 0 \quad (i = 2, 3, \ldots, m).$$

The relations between the coefficients $\alpha_{ij}, \gamma_{ij}, \beta_{ij}, \delta_{ij}$ $(i,j = 2, \ldots, m)$ of S''' are seen to be precisely those holding for a special Abelian substitution on $m-1$ pairs of indices. Furthermore,

$$S = VS' = VS_1 S'' = VS_1 S_2 S''',$$

where V, S_1, S_2 were derived from the types of substitutions given in the theorem.

After m operations similar to that by which S''' was derived from S, we reach a substitution which leaves fixed all the indices

94   CHAPTER II.

and is therefore the identity. Hence S is a product of substitutions of the given types. Since the latter are all of determinant unity, so is also the general substitution S of the group.

115. Theorem. — *The order $SA[2m, p^n]$ of the special Abelian group equals*

$$(p^{n(2m)} - 1) p^{n(2m-1)} (p^{n(2m-2)} - 1) p^{n(2m-3)} \ldots (p^{2n} - 1) p^n.$$

There are $(p^n)^{2m} - 1$ sets of values of α_{1j}, γ_{1j} $(j = 1, \ldots, m)$, not all zero, which give distinct functions ω_1. In the function ω_2, $\delta_{11} = 1$ while $\beta_{11}, \beta_{1j}, \delta_{1j}$ $(j = 2, \ldots, m)$ are arbitrary in the field. Hence ω_2 may be chosen in $(p^n)^{2m-1}$ ways. We have therefore the recursion formula

$$SA[2m, p^n] = (p^{n(2m)} - 1) p^{n(2m-1)} \cdot SA[2m - 2, p^n].$$

116. Theorem. — *For $p > 2$, the factors of composition of $SA(2m, p^n)$ are $\frac{1}{2} SA[2m, p^n]$ and 2, the case $p^n = 3, m = 1$ being exceptional.*[1])

Every substitution of $G \equiv SA(2m, p^n)$ is commutative with

$$T \equiv T_{1,-1} T_{2,-1} \ldots T_{m,-1} \colon \xi_i' = -\xi_i, \quad \eta_i' = -\eta_i \quad (i = 1, \ldots, m).$$

The group $K \equiv \{I, T\}$ of order 2 is therefore self-conjugate under G. In order to show that K is the maximal self-conjugate subgroup of G, we prove that a self-conjugate subgroup J of G, which contains K without being identical with K, must coincide with G.

Let S, given by 75), be a substitution of J not in K. Then J contains the products

$$S^{-1} \cdot L_{i,\lambda}^{-1} S L_{i,\lambda}, \quad S^{-1} \cdot L'^{-1}_{i,\lambda} S L'_{i,\lambda} \quad (i = 1, \ldots, m)$$

where λ is a fixed mark $\neq 0$. Suppose first that all of these products reduce to the identity. Then, for example, S is commutative with both $L_{1,\lambda}$ and $L'_{1,\lambda}$, so that, by the proof of Corollary I of § 100, S has the form

$$\begin{pmatrix} \alpha_{11} & 0 & 0 & 0 & \ldots & 0 & 0 \\ 0 & \alpha_{11} & 0 & 0 & \ldots & 0 & 0 \\ 0 & 0 & \alpha_{22} & \gamma_{22} & \ldots & \alpha_{2m} & \gamma_{2m} \\ 0 & 0 & \beta_{22} & \delta_{22} & \ldots & \beta_{2m} & \delta_{2m} \\ \cdot & \cdot & \cdot & \cdot & & \cdot & \cdot \\ 0 & 0 & \alpha_{m2} & \gamma_{m2} & \ldots & \alpha_{mm} & \gamma_{mm} \\ 0 & 0 & \beta_{m2} & \delta_{m2} & \ldots & \beta_{mm} & \delta_{mm} \end{pmatrix}.$$

---

1) For $m = 1$, $SA(2m, p^n)$ is identical with the group of all binary linear homogeneous substitutions of determinant unity. Its factors of composition are therefore given by the theorem of § 103.

THE ABELIAN LINEAR GROUP. 95

But S is to be commutative with every pair $L_{i,\lambda}$ and $L'_{i,\lambda}$. It follows that S reduces to the form

$$\overline{S}: \quad \xi'_i = \alpha_{ii}\xi_i, \quad \eta'_i = \alpha_{ii}\eta_i \quad (i = 1, 2, \ldots, m).$$

By the first type of Abelian conditions given under 79), we have $\alpha_{ii} = \pm 1$. Since S is not in K, the α_{ii} are not all $+1$ and not all -1. Transforming $\overline{S}$ by a suitable product of the form $P_{1r}P_{2s}$, we may suppose that $\alpha_{11} = 1$, $\alpha_{22} = -1$ in $\overline{S}$. Then J contains $N^{-1}_{1,2,\mu}\overline{S}N_{1,2,\mu}$, which replaces ξ_1 by $\xi_1 - 2\mu\eta_2$ and is therefore (since $p \neq 2$) not of the form $\overline{S}$. Taking in place of our initial substitution S, we are led to the case next considered.

Suppose that not all of the above products reduce to the identity I; for example, let

$$S_1 \equiv S^{-1}L^{-1}_{1,\lambda}SL_{1,\lambda} \neq I.$$

If S^{-1} replaces η_1 by the linear function ω/λ, the product denoted by S_1 has the following form, in which the coefficients of ξ'_1 have not been calculated:

$$S_1: \begin{cases} \xi'_1 = \sum_{j=1}^{m}(\alpha_j\xi_j + \gamma_j\eta_j), & \xi'_i = \xi_i - \alpha_{i1}\omega \quad (i = 2, \ldots, m), \\ & \eta'_i = \eta_i - \beta_{i1}\omega \quad (i = 1, \ldots, m). \end{cases}$$

From S_1 we proceed to determine a substitution $\neq I$ belonging to J and leaving $2m - 3$ indices unaltered. S_1 itself is such a substitution if $\alpha_{i1} = \beta_{i1} = 0$ $(i = 2, \ldots, m)$. In the contrary case, the transformed of S by a suitable P_{2j} or $P_{2j}M_2$ will have $\alpha_{21} \neq 0$. Consider therefore S_1 when $\alpha_{21} \neq 0$, and introduce the new indices

$$\overline{\xi}_1 = \xi_1, \quad \overline{\eta}_1 = \eta_1, \quad \overline{\xi}_2 = \xi_2, \quad \overline{\eta}_2 = -\frac{1}{\alpha_{21}}\sum_{j=2}^{m}(\beta_{j1}\xi_j - \alpha_{j1}\eta_j),$$

$$\overline{\xi}_i = \xi_i - \frac{\alpha_{i1}}{\alpha_{21}}\xi_2, \quad \overline{\eta}_i = \eta_i - \frac{\beta_{i1}}{\alpha_{21}}\xi_2 \quad (i = 3, \ldots, m),$$

an operation equivalent to the transformation of S_1 by the following product T belonging to the group G:

$$Q_{3,2,-\alpha_{31}/\alpha_{21}}R_{3,2,\beta_{31}/\alpha_{21}} \cdots Q_{m,2,-\alpha_{m1}/\alpha_{21}}R_{m,2,\beta_{m1}/\alpha_{21}}L'_{2,\sigma}$$

where

$$\sigma = -\sum_{j=2}^{m}\frac{\alpha_{j1}}{\alpha_{21}}\frac{\beta_{j1}}{\alpha_{21}}.$$

We obtain the substitution $S_2 \equiv T^{-1}S_1T$, leaving fixed $2m - 3$ indices, viz.,

$$\overline{\xi}'_i = (\xi_i - \alpha_{i1}\omega) - \frac{\alpha_{i1}}{\alpha_{21}}(\xi_2 - \alpha_{21}\omega) \equiv \overline{\xi}_i, \quad \overline{\eta}'_i = \overline{\eta}_i \quad (i = 3, \ldots, m)$$

$$\overline{\eta}'_2 = -\frac{1}{\alpha_{21}}\sum_{j=2}^{m}\{\beta_{j1}(\xi_j - \alpha_{j1}\omega) - \alpha_{j1}(\eta_j - \beta_{j1}\omega)\} \equiv \overline{\eta}_2.$$

96   CHAPTER II.

Writing ξ_i, η_i for $\bar{\xi}_i$, $\bar{\eta}_i$ in S_2, and applying conditions 79), viz.,
$$R_{34} = 1, \quad R_{14} = R_{24} = 0, \quad R_{1j} = R_{2j} = R_{3j} = 0 \quad (j = 5, 6, \ldots, 2m),$$
we find that S_2 takes the form

$$S_2: \begin{cases} \xi_1' = \alpha_{11}\xi_1 + \gamma_{11}\eta_1 + \gamma_{12}\eta_2, & \eta_1' = \beta_{11}\xi_1 + \delta_{11}\eta_1 + \delta_{12}\eta_2, \\ \xi_2' = \alpha_{21}\xi_1 + \gamma_{21}\eta_1 + \xi_2 + \gamma_{22}\eta_2, & \eta_2' = \eta_2, \end{cases}$$

the indices ξ_i, η_i ($i = 3, \ldots, m$), not being altered by S_2 and the substitutions below, are not written in the formulae.

The group J contains the product

$$S_3 \equiv S_2^{-1} \cdot N_{1,2,\mu}^{-1} S_2 N_{1,2,\mu} : \begin{cases} \xi_1' = \xi_1 + \mu(1 - \alpha_{11})\eta_2, & \xi_2' = \xi_2 + \Phi, \\ \eta_1' = \eta_1 - \mu\beta_{11}\eta_2, & \eta_2' = \eta_2, \end{cases}$$

where Φ is a linear function of ξ_1, η_1, η_2.

a) Suppose first that S_3 is not the identity. If $1 - \alpha_{11} \neq 0$, we may define τ by the equation
$$(1 - \alpha_{11})\tau = \beta_{11}.$$
Then J contains $S_4 \equiv L_{1,\tau}'^{-1} S_3 L_{1,\tau}'$, which has the form

$$\begin{aligned}\xi_1' &= \xi_1 + \gamma_{12}\eta_2, & \eta_1' &= \eta_1, \\ \xi_2' &= \xi_2 + \alpha_{21}\xi_1 + \gamma_{21}\eta_1 + \gamma_{22}\eta_2, & \eta_2' &= \eta_2.\end{aligned} \quad (\gamma_{12} \neq 0)$$

Applying the conditions $R_{13} = R_{23} = 0$ of 79), we find that $\gamma_{12} = \gamma_{21}$, $\alpha_{21} = 0$, so that S_4 has the following form (with $\alpha \neq 0$):

81) $$\begin{cases} \xi_1' = \xi_1 + \alpha\eta_2, & \eta_1' = \eta_1, \\ \xi_2' = \xi_2 + \alpha\eta_1 + \beta\eta_2, & \eta_2' = \eta_2. \end{cases}$$

If, on the contrary, $1 - \alpha_{11} = 0$, J will contain $M_1^{-1} S_3 M_1$, which is not the identity and has the form 81). In either case, J contains a substitution 81) in which α and β are not both zero.

If $\alpha = 0$, $\beta \neq 0$, 81) is of the form $L_{2,\beta} \neq I$. If $\alpha \neq 0$, J contains the transformed of 81) by $Q_{2,1,\lambda}$, giving the substitution

$$\begin{aligned}\xi_1' &= \xi_1 + \alpha\eta_2, & \eta_1' &= \eta_1, \\ \xi_2' &= \xi_2 + \alpha\eta_1 + (\beta + 2\alpha\lambda)\eta_2, & \eta_2' &= \eta_2.\end{aligned}$$

Taking $\lambda = -\beta/2\alpha$, this becomes $N_{1,2,\alpha}$. Then J contains

82) $L_{1,-\alpha^2} \equiv N_{1,2,\alpha} \cdot M_2^{-1} N_{1,2,\alpha}^{-1} M_2 (M_2 L_{2,1})^{-1} N_{1,2,\alpha} (M_2 L_{2,1}).$

Transforming by P_{12}, we reach $L_{2,-\alpha^2}$. In either case, J contains a substitution of the form $L_{2,\lambda}$ ($\lambda \neq 0$).

We next prove that J contains all the generators $L_{i,\mu}$, M_l and $N_{i,j,\mu}$ of the group G. Having $L_{2,\lambda}$, J contains the product

$$T_{2,\tau}^{-1} L_{2,\lambda} T_{2,\tau} \equiv L_{2,\lambda\tau^2} \quad (\tau \text{ any mark} \neq 0).$$

The product of two such substitutions gives $L_{2,\lambda(\tau_1^2+\tau_2^2)}$. But, by § 64, marks τ_1 and τ_2 can be found in the $GF[p^n]$, $p > 2$, such that $\tau_1^2 + \tau_2^2$ has an arbitrary value μ in the field. Hence J contains $L_{2,\mu}$. Then I contains the product

$$M_2 \equiv L_{2,1} \cdot M_2^{-1} L_{2,1} M_2 \cdot L_{2,1}.$$

Hence J contains $L_{i,\mu}$ and M_i, the transformed of $L_{2,\mu}$ and M_2 respectively by P_{2i}. Finally, J contains[1])

83) $\qquad N_{i,j,\mu} \equiv Q_{i,j,1}^{-1} L_{i,\mu} L_{j,\mu} Q_{i,j,1} L_{i,\mu}^{-2} L_{j,\mu}^{-1}.$

b) Suppose, however, that $S_3 \equiv I$. Then S_2 is commutative with $N_{1,2,\mu}$, so that
$$\alpha_{11} = 1, \quad \beta_{11} = 0, \quad \delta_{11} = 1, \quad \delta_{12} = \alpha_{21}.$$

Applying the Abelian conditions $R_{13} = R_{23} = 0$, we find that $\delta_{12} = 0$, $\gamma_{12} = \gamma_{21}$, so that S_2 becomes

$S_2:\quad \begin{cases} \xi_1' = \xi_1 + \gamma_{11}\eta_1 + \gamma_{12}\eta_2, & \eta_1' = \eta_1, \\ \xi_2' = \xi_2 + \gamma_{12}\eta_1 + \gamma_{22}\eta_2, & \eta_2' = \eta_2. \end{cases}$

S_2 is not the identity since S_1 is not. If $\gamma_{11} = 0$, S_2 is of the form 81) considered under case a). If $\gamma_{11} \neq 0$, J contains S_2', the transformed of S_2 by $Q_{2,1,\lambda}$, where $\lambda \equiv -\gamma_{12}/\gamma_{11}$, viz.,

84) $\qquad S_2': \begin{cases} \xi_1' = \xi_1 + \gamma_{11}\eta_1, & \eta_1' = \eta_1, \\ \xi_2' = \xi_2 + \delta\eta_2, & \eta_2' = \eta_2. \end{cases} \quad [\delta \equiv \gamma_{22} + \lambda\gamma_{12}]$

For $\delta = 0$, $S_2' \equiv L_{1,\gamma_{11}}$. For $\delta \neq 0$, J contains the transformed of S_2' by $T_{1,\lambda} T_{2,\mu}$, λ and μ being arbitrary marks $\neq 0$, giving the substitution
$$\xi_1' = \xi_1 + \lambda^2\gamma_{11}\eta_1, \quad \eta_1' = \eta_1,$$
$$\xi_2' = \xi_2 + \mu^2\delta\eta_2, \quad \eta_2' = \eta_2.$$

Forming the product of two such substitutions and noting that, for $p > 2$, the equation $\lambda_1^2 + \lambda_2^2 = \varkappa$ has solutions in the $GF[p^n]$ for $\varkappa$ an arbitrary mark $\neq 0$ of the field, we find that J contains

$L_{1,\alpha} L_{2,\beta}: \qquad \xi_1' = \xi_1 + \alpha\eta_1, \quad \xi_2' = \xi_2 + \beta\eta_2,$

where α and β are arbitrary marks $\neq 0$. A suitable product of two such substitutions gives
$$L_{1,\alpha} L_{2,\beta} \cdot L_{1,\alpha} L_{2,-\beta} \equiv L_{1,2\alpha}.$$

In every case we reach in J a substitution $L_{1,\lambda}$, where $\lambda \neq 0$, and therefore also $L_{2,\lambda}$. It follows as in case a) that $J = G$.

117. Theorem. — *For $p = 2$, $SA(2m, p^n)$ is simple except when $m = 2$, $p^n = 2$, and when $m = 1$, $p^n = 2$.*

---

[1]) We might reach $N_{1,2,\alpha}$ by 82) and then obtain $N_{i,j,\mu}$ in the group J.

For $p = 2$, a substitution S of $G \equiv SA(2m, p^n)$ is commutative with every $L_{i,\lambda}$ and every $L'_{i,\lambda}$ only when S is the identity. Proceeding as in § 116, we find that a self-conjugate subgroup J of G, which contains a substitution $S \neq I$, will contain either a substitution of the form 81) with α and β not both zero or else a substitution S'_2 of the form 84) in which $\gamma_{11} \neq 0$.

We next prove that J contains either $L_{1,\lambda}$ ($\lambda \neq 0$) or else $N_{1,2,1} L_{2,1}$. For $\delta = 0$, $S'_2 = L_{1,\gamma_{11}}$. For $\delta \neq 0$, we transform S'_2 by a suitable $T_{1,\lambda} T_{2,\mu}$ and obtain the substitution $L'_{i,1} L_{2,1}$. Hence J contains[1])

$$Q_{1,2,1}^{-1} L_{1,1} L_{2,1} Q_{1,2,1} \equiv N_{1,2,1} L_{2,1}.$$

For $\alpha = 0$, 81) becomes $L_{2,\beta}$, so that we reach $L_{1,\beta}$ in J. If $\beta = 0$, 81) becomes $N_{1,2,\alpha}$, so that, by 82), J contains $L_{1,-\alpha^2}$. Finally, if $\alpha \neq 0$, $\beta \neq 0$, the transformed of 81) by $T_{1,\lambda} T_{2,\mu}$ gives the substitution

$$\xi'_1 = \xi_1 + \alpha\lambda\mu\eta_2, \quad \xi'_2 = \xi_2 + \alpha\lambda\mu\eta_1 + \beta\mu^2\eta_2.$$

In the $GF[2^n]$, we may take

$$\mu = \beta^{-1/2}, \quad \lambda = \alpha^{-1}\mu^{-1},$$

when the last substitution becomes $N_{1,2,1} L_{2,1}$.

Having a substitution $L_{1,\lambda}$ ($\lambda \neq 0$), J will coincide with G. Indeed, $T_{1,\tau}$ transforms $L_{1,\lambda}$ into $L_{1,\lambda\tau^2}$. Since every mark of the field is a square, we reach $L_{1,\sigma}$, σ arbitrary. Then, as at the end of case a) of § 116, J contains every $L_{i,\sigma}$, M_i, $N_{i,j,\sigma}$ and hence coincides with G.

There remains the case in which J contains $N_{1,2,1} L_{2,1}$. Then J will contain all the products, two at a time, of the substitutions

85) $\qquad L_{i,1}, \quad M_i, \quad N_{i,j,1} \quad (i,j = 1, 2, \ldots, m; \; i \neq j).$

Indeed, if i and j be any two distinct integers $\leq m$, J contains

$(P_{1i} P_{2j})^{-1} N_{1,2,1} L_{2,1} (P_{1i} P_{2j}) = N_{i,j,1} L_{i,1} = L_{i,1} N_{i,j,1},$
$L_{i,1} N_{i,j,1} \cdot L_{j,1} N_{i,j,1} = L_{i,1} L_{j,1},$
$M_i^{-1} L_{i,1} L_{j,1} M_i \cdot L_{i,1} L_{j,1} = L_{i,1} M_i, \quad (L_{i,1} M_i)^2 = M_i L_{i,1},$
$L_{i,1} L_{j,1} \cdot L_{i,1} M_i = L_{j,1} M_i = M_i L_{j,1}, \quad M_i L_{j,1} \cdot L_{j,1} M_j = M_i M_j.$

Our statement is therefore proved if $m = 2$. If $m > 2$, let i, j, k be any three distinct integers $\leq m$. Then J contains

$N_{i,j,1} L_{i,1} \cdot L_{i,1} L_{k,1} = N_{i,j,1} L_{k,1} = L_{k,1} N_{i,j,1},$
$N_{i,j,1} L_{i,1} \cdot L_{i,1} M_k = N_{i,j,1} M_k = M_k N_{i,j,1}.$

---

[1]) This relation follows from 83), if $p = 2$, by taking $i = 1$, $j = 2$, $\mu = 1$.

We next prove that, for $m > 2$, J contains $L_{1,1}$. Since, for $p = 2$,
$$L'_{i,1} = M_i L_{i,1} M_i, \quad R_{i,j,1} = M_i M_j N_{i,j,1} M_i M_j,$$
it follows that J will contain the substitution
$$D \equiv L'_{1,1} L'_{2,1} L'_{3,1} R_{1,2,1} R_{2,3,1} R_{3,1,1},$$
the latter being the product of an even number 24 of the substitutions 85). This product is seen to be
$$D: \quad \xi'_i = \xi_i, \quad \eta'_i = \eta_i + \xi_1 + \xi_2 + \xi_3 \quad (i = 1, 2, 3).$$
But D is transformed into $L_{1,1}$ by the following Abelian substitution of period two:
$$\xi'_1 = \eta_1 + \xi_2 + \xi_3, \quad \eta'_1 = \xi_1 + \xi_2 + \xi_3,$$
$$\xi'_2 = \xi_2, \quad \eta'_2 = \xi_1 + \eta_1 + \xi_2 + \eta_2 + \xi_3,$$
$$\xi'_3 = \xi_3, \quad \eta'_3 = \xi_1 + \eta_1 + \xi_2 + \xi_3 + \eta_3.$$
Hence J contains $L_{1,1}$ and therefore also
$$L_{i,1} L_{1,1} \cdot L_{1,1} = L_{i,1} \quad T_{i,\tau}^{-1} L_{i,1} T_{i,\tau} = L_{i,\tau^2}, \quad M_i L_{1,1} \cdot L_{1,1} = M_i,$$
$$N_{i,j,1} L_{1,1} \cdot L_{1,1} = N_{i,j,1}, \quad T_{i,\tau}^{-1} N_{i,j,1} T_{i,\tau} = N_{i,j,\tau}.$$
Hence, for $p = 2$, $m > 2$, J is identical with G, so that G is simple. For $p = 2$, $m = 2$, J contains $M_1 M_2$ as above, and therefore also
$$M_1 M_2 \cdot T_{1,\tau}^{-1} M_1 M_2 T_{1,\tau} \equiv T_{1,\tau^2}.$$
Hence J contains every $T_{1,\alpha}$. But $R_{1,2,\lambda}$ transforms $T_{1,\alpha}$ into $R_{1,2,\lambda(1+\alpha)} T_{1,\alpha}$. If $n > 1$, the $GF[2^n]$ contains a mark α neither zero nor unity, so that $1 + \alpha \neq 0$, $\alpha \neq 0$. Hence, for $n > 1$, the group J contains $R_{1,2,\lambda(1+\alpha)} = R_{1,2,1}$, by proper choice of λ. It therefore contains $N_{1,2,1}$. Having the products in pairs of the substitutions 85), J contains M_i and $L_{i,1}$. Thus $J \equiv G$.

The fact that the case $m = 2$, $p = 2$, $n = 1$ is exceptional is shown in the following section.

118. Theorem. — *The Abelian group $SA(4,2)$ on four indices modulo 2 is holoedrically isomorphic with the symmetric group on six letters.*[1])

By § 264 of Chapter XIII, the symmetric group on 6 letters is holoedrically isomorphic with the abstract group $G_{6!}$ generated by B_1, B_2, B_3, B_4, B_5 subject to the generational relations
$$B_1^2 = B_2^2 = B_3^2 = B_4^2 = B_5^2 = I,$$
$$(B_1 B_2)^3 = (B_2 B_3)^3 = (B_3 B_4)^3 = (B_4 B_5)^3 = I,$$
$$(B_1 B_3)^2 = (B_1 B_4)^2 = (B_1 B_5)^2 = (B_2 B_4)^2 = (B_2 B_5)^2 = (B_3 B_5)^2 = I.$$

---

1) This theorem was first proved by Jordan by means of the groups of Steiner, Traité, No. 335. The proof given in the text is due to the author, Proc. Lond. Math. Soc., vol. 31, pp. 40—41.

100   CHAPTER II.

To the operators B_i we make correspond the following substitutions of $SA(4, 2)$:

86) $B_1 \sim M_1$, $B_2 \sim L_{1,1}$, $B_3 \sim S$, $B_4 \sim L_{2,1}$, $B_5 \sim M_2$,

where S denotes the Abelian substitution of period two:

	ξ_1	η_1	ξ_2	η_2
$\xi_1' =$	0	1	1	1
$\eta_1' =$	1	0	1	1
$\xi_2' =$	1	1	0	1
$\eta_2' =$	1	1	1	0

We readily verify that the relations corresponding to the above generational relations are satisfied in virtue of the correspondences 86). Since $SA(4, 2)$ has the order

$$(2^4 - 1)2^3(2^2 - 1)2 \equiv 6!,$$

the isomorphism between $SA(4, 2)$ and $G_{6!}$ is holoedric.

119. In determining the factors of composition of the general and special Abelian groups on $2m$ indices with coefficients in the $GF[p^n]$, we have been led to a quotient-group, $SA(2m, p^n)/K$, where $K \equiv \{I, T\}$ is of order 1 or 2 according as $p = 2$ or $p > 2$. Owing to the great importance of simple groups, we will designate this quotient-group as $A(2m, p^n)$, it being a simple group except in the three cases $m = 1$, $p^n = 2$; $m = 1$, $p^n = 3$; $m = 2$, $p^n = 2$, when its factors of composition are 2, 3; 2, 2, 3; 2, $\frac{1}{2}6!$, respectively. The order $A[2m, p^n]$ of $A(2m, p^n)$ is

$$\frac{1}{a}(p^{n(2m)} - 1)p^{n(2m-1)}(p^{n(2m-2)} - 1)p^{n(2m-3)} \cdots (p^{2n} - 1)p^n,$$

where $a = 1$ or 2 according as $p = 2$ or $p > 2$.

Conjugacy of operators of period two[1] *in $SA(2m, p^n)$ and $A(2m, p^n)$.*

120. Theorem. — *Within the special Abelian group $SA(2m, p^n)$ any substitution S defined by 75) is conjugate with a substitution Σ which replaces ξ_1 and η_1 by the respective functions*

$$\alpha_{11}\xi_1 + \gamma_{11}\eta_1 + \alpha_{1m}\xi_m, \quad \beta_{11}\xi_1 + \delta_{11}\eta_1 + \delta_{1m-1}\eta_{m-1} + \delta_{1m}\eta_m,$$

where either $\delta_{1m-1} = 0$ or else $\delta_{1m} = 0$.

The theorem is evident if $\alpha_{1i} = \gamma_{1i} = \beta_{1i} = \delta_{1i} = 0$ $(i = 2, \ldots, m)$. In the contrary case, we may suppose that α_{1m}, γ_{1m}, β_{1m}, δ_{1m} are not all zero, first transforming S by P_{im} where i is a certain one

---
[1] Taken from the author's article, *Quarterly Journal*, vol. 32, pp. 42—63.

of the integers $2, 3, \ldots, m$. According as $\alpha_{1m} \neq 0$, $\gamma_{1m} \neq 0$, $\beta_{1m} \neq 0$ or $\delta_{1m} \neq 0$, we transform S by I, M_m, M_1, or $M_1 M_m$ respectively and obtain a substitution S' in which $\alpha_{1m} \neq 0$. Transforming S' by $L_{m,\lambda}$, we obtain a substitution S'' which replaces ξ_1 by

$$\alpha_{11}\xi_1 + \gamma_{11}\eta_1 + \cdots + \alpha_{1m}\xi_m + (\gamma_{1m} - \lambda\alpha_{1m})\eta_m.$$

Since $\alpha_{1m} \neq 0$, we can choose λ in the field to make the coefficient of η_m vanish. Transforming S'' (in which now $\alpha_{1m} \neq 0$, $\gamma_{1m} = 0$) by $L'_{1,\varrho}$, we reach a substitution S_1 which replaces ξ_1, η_1 by respectively

$$(\alpha_{11} - \varrho\gamma_{11})\xi_1 + \gamma_{11}\eta_1 + \cdots + \alpha_{1m}\xi_m,$$
$$()\xi_1 + ()\eta_1 + \cdots + (\beta_{1m} + \varrho\alpha_{1m})\xi_m + \delta_{1m}\eta_m.$$

We choose ϱ to make $\beta_{1m} + \varrho\alpha_{1m} = 0$. Hence S_1 has $\alpha_{1m} \neq 0$, $\gamma_{1m} = \beta_{1m} = 0$.

We next determine an Abelian substitution which affects only the indices $\xi_2, \eta_2, \xi_m, \eta_m$ and which transforms S_1 into a substitution S_2 having $\alpha_{1m} \neq 0$, $\gamma_{1m} = \beta_{1m} = \gamma_{12} = \beta_{12} = 0$.

a) Let $\alpha_{12} = \gamma_{12} = 0$. If $\delta_{12} = 0$, the transformed of S_1 by M_2 gives S_2. If β_{12} and δ_{12} are both not zero, we transform S_1 by $L'_{2,\varrho}$, where $\beta_{12} - \varrho\delta_{12} = 0$, and obtain S_2.

b) Let α_{12} and γ_{12} be not both zero. Transforming by M_2 when $\gamma_{12} \neq 0$, we may suppose that $\alpha_{12} \neq 0$ in S_1. Transforming it by $L_{2,\varrho}$, we can make $\gamma_{12} = 0$. If then $\delta_{12} \neq 0$, we transform by $L'_{2,\varrho}$ and make $\beta_{12} = 0$. Suppose, however, that $\delta_{12} = 0$. If $\delta_{1m} \neq 0$, we transform by $R_{2,m,\lambda}$, where $\beta_{12} + \lambda\delta_{1m} = 0$, and reach S_2. But if $\delta_{1m} = 0$, we have S_2 if $\beta_{12} = 0$; while for $\beta_{12} \neq 0$, we transform by $Q_{m,2,\varrho} M_2$, where $\alpha_{12} - \varrho\alpha_{1m} = 0$, and reach S_2.

In an analogous manner, we can determine an Abelian substitution which affects only $\xi_3, \eta_3, \xi_m, \eta_m$ and which transforms S_2 into a substitution S_3 having

$$\alpha_{1m} \neq 0, \quad \gamma_{12} = \beta_{12} = \gamma_{13} = \beta_{13} = \gamma_{1m} = \beta_{1m} = 0.$$

Repeating the process, we may also make

$$\gamma_{14} = \beta_{14} = \cdots = \gamma_{1m-1} = \beta_{1m-1} = 0.$$

We therefore reach a substitution $\overline{S}$ conjugate with S within the special Abelian group and replacing ξ_1, η_1 by respectively

$$\alpha_{11}\xi_1 + \gamma_{11}\eta_1 + \sum_{j=2}^{m}\alpha_{1j}\xi_j, \quad \beta_{11}\xi_1 + \sum_{j=1}^{m}\delta_{1j}\eta_j \quad [\alpha_{1m} \neq 0].$$

Transforming $\overline{S}$ by $Q_{m,2,\sigma}$, where $\alpha_{12} - \sigma\alpha_{1m} = 0$, we obtain a substitution of the form $\overline{S}$ but having $\alpha_{12} = 0$. Similarly, we may make $\alpha_{13} = \cdots = \alpha_{1m-1} = 0$. If, in the resulting substitution $\overline{S}_1$,

102   CHAPTER II.

$\delta_{12} = \cdots = \delta_{1m} = 0$, we have reached Σ. If $\delta_{1m} \neq 0$, we transform $\bar{S}_1$ by $Q_{2,m,\sigma}$, where $\delta_{12} + \sigma\delta_{1m} = 0$, and reach a substitution of the form $\bar{S}_1$ but having also $\delta_{12} = 0$. In a similar manner we make $\delta_{13} = \cdots = \delta_{1m-1} = 0$ and reach Σ. Finally, if $\delta_{1m} = 0$ but δ_{12}, $\delta_{13}, \ldots, \delta_{1m-1}$ are not all zero, we may suppose that $\delta_{1m-1} \neq 0$, first transforming by some P_{im-1}. We then transform it by $Q_{i,m-1,\varrho}$, for $i = 2, 3, \ldots, m-2$ in succession, and make

$$\delta_{12} = \delta_{13} = \cdots = \delta_{1m-2} = 0,$$

so that we reach Σ.

Corollary. — If $\alpha_{1i}, \gamma_{1i}, \beta_{1i}, \delta_{1i}$ $(i = 2, \ldots, m)$ are not all zero in S, it is conjugate within $SA(2m, p^n)$ with one of the two types of substitutions:

Σ_1: $\xi_1' = \alpha_{11}\xi_1 + \gamma_{11}\eta_1 + \xi_m,\quad \eta_1' = \beta_{11}\xi_1 + \delta_{11}\eta_1 + \delta\eta_m, \ldots$

Σ_2: $\xi_1' = \alpha_{11}\xi_1 + \gamma_{11}\eta_1 + \xi_m,\quad \eta_1' = \beta_{11}\xi_1 + \delta_{11}\eta_1 + \eta_{m-1}, \ldots$

Since the conjugate substitution Σ then has $\alpha_{1m} \neq 0$, we may transform it by T_m, α_{1m}. Then if $\delta_{1m-1} = 0$, we have Σ_1. In the contrary case, we transform also by $T_{m-1}^{-1}, \delta_{1m-1}$ and get Σ_2.

121. Theorem. — *The special Abelian group $SA(2m, p^n)$, $p > 2$, contains exactly m sets of conjugate substitutions of period 2. The rth set includes*

$$\frac{(p^{n(2m)} - 1)(p^{n(2m-2)} - 1) \cdots (p^{n(2m-2r+2)} - 1)}{(p^{n(2r)} - 1)(p^{n(2r-2)} - 1) \cdots (p^{2n} - 1)} \cdot p^{2nr(m-r)}$$

substitutions all conjugate with $T_r \equiv T_{1,-1} T_{2,-1} \ldots T_{r,-1}$.

In order that the special Abelian substitution 75) shall be identical with its reciprocal 77), for $\mu = 1$, it is necessary and sufficient that

$$\alpha_{ij} = \delta_{ji},\quad \gamma_{ij} = -\gamma_{ji},\quad \beta_{ij} = -\beta_{ji},\quad (i,j = 1, \ldots, m).$$

Every substitution of period 2 of $SA(2m, p^n)$, $p > 2$, has therefore the form

$$S = \begin{pmatrix} \alpha_{11} & 0 & \alpha_{12} & \gamma_{12} & \cdots & \alpha_{1m} & \gamma_{1m} \\ 0 & \alpha_{11} & \beta_{12} & \delta_{12} & \cdots & \beta_{1m} & \delta_{1m} \\ \delta_{12} & -\gamma_{12} & \alpha_{22} & 0 & \cdots & \alpha_{2m} & \gamma_{2m} \\ -\beta_{12} & \alpha_{12} & 0 & \alpha_{22} & \cdots & \beta_{2m} & \delta_{2m} \\ \cdots & \cdots & \cdots & \cdots & & \cdots & \cdots \\ \delta_{1m} & -\gamma_{1m} & \delta_{2m} & -\gamma_{2m} & \cdots & \alpha_{mm} & 0 \\ -\beta_{1m} & \alpha_{1m} & -\beta_{2m} & \alpha_{2m} & \cdots & 0 & \alpha_{mm} \end{pmatrix}.$$

For $m = 1$, we have $\alpha_{11}^2 = 1$, so that $T_{1,-1}$ is the only substitution S. In order to prove the first part of our theorem by induction, we assume that every special Abelian substitution in the

THE ABELIAN LINEAR GROUP. 103

$GF[p^n]$, $p > 2$, on $t < m$ pairs of indices is conjugate within the group $SA(2t, p^n)$ with one of the substitutions T_r ($r \leqq t$) and proceed to prove that a like result holds for m pairs of indices. In view of § 120, we may suppose that S has one of the three forms Σ_1, Σ_2 or S_1, the latter having $\alpha_{1i} = \gamma_{1i} = \beta_{1i} = \delta_{1i} = 0$ ($i = 2, \ldots, m$). An S of the form S_1 is evidently a product $T_{1,\pm 1} S_2$, where S_2 affects only the $m-1$ sets of indices $\xi_2, \eta_2, \ldots, \xi_m, \eta_m$. By hypothesis, S_2 is conjugate with one of the products, I, $T_{2,-1}$, $T_{2,-1} T_{3,-1}, \ldots,$ $T_{2,-1} T_{3,-1} \ldots T_{m,-1}$. Hence an S of the form S_1 is conjugate with some T_r ($r = 1, 2, \ldots, m$). We proceed to consider Σ_1 and Σ_2 in the following three cases.

Case a). $\delta \neq 0$ in Σ_1. Then S has the form

$$\Sigma_1 = \begin{pmatrix} \alpha_{11} & 0 & 0 & 0 & \ldots & 0 & 0 & 1 & 0 \\ 0 & \alpha_{11} & 0 & 0 & \ldots & 0 & 0 & 0 & \delta \\ 0 & 0 & \alpha_{22} & 0 & \ldots & \alpha_{2m-1} & \gamma_{2m-1} & \alpha_{2m} & \gamma_{2m} \\ 0 & 0 & 0 & \alpha_{22} & \ldots & \beta_{2m-1} & \delta_{2m-1} & \beta_{2m} & \delta_{2m} \\ \cdot & \cdot & \cdot & \cdot & \cdot & \cdot & \cdot & \cdot & \cdot \\ \delta & 0 & \delta_{2m} & -\gamma_{2m} & \ldots & \delta_{m-1\,m} & -\gamma_{m-1\,m} & \alpha_{mm} & 0 \\ 0 & 1 & -\beta_{2m} & \alpha_{2m} & \ldots & -\beta_{m-1\,m} & \alpha_{m-1\,m} & 0 & \alpha_{mm} \end{pmatrix}.$$

The Abelian conditions 79) give at once

$$\alpha_{im} = \beta_{im} = \gamma_{im} = \delta_{im} \quad (i = 2, \ldots, m-1), \quad \alpha_{11} + \alpha_{mm} = 0.$$

Hence $\Sigma_1 \equiv \Sigma_1' \Sigma_1''$, where Σ_1'' affects only the indices

$$\xi_i, \eta_i \quad (i = 2, \ldots, m-1),$$

while Σ_1' affects only $\xi_1, \eta_1, \xi_m, \eta_m$, viz.,

$$\Sigma_1': \begin{array}{c} \xi_1' = \\ \eta_1' = \\ \xi_m' = \\ \eta_m' = \end{array} \begin{array}{|cccc|} \xi_1 & \eta_1 & \xi_m & \eta_m \\ \hline \alpha_{11} & 0 & 1 & 0 \\ 0 & \alpha_{11} & 0 & \delta \\ \delta & 0 & -\alpha_{11} & 0 \\ 0 & 1 & 0 & -\alpha_{11} \end{array}.$$

By hypothesis Σ_1'' is conjugate with some product of the

$$T_{i,-1} \quad (i = 2, \ldots, m-1).$$

In order to make the induction from one to two pairs of indices, we must prove that Σ_1' is conjugate with a product of $T_{1,-1}$ and $T_{m,-1}$. Transforming Σ_1' by $Q_{1,m,\sigma}$, we obtain the substitution

$$\begin{pmatrix} \alpha_{11} + \sigma\delta & 0 & 1 - 2\sigma\alpha_{11} - \sigma^2\delta & 0 \\ 0 & \alpha_{11} + \sigma\delta & 0 & \delta \\ \delta & 0 & -\alpha_{11} - \sigma\delta & 0 \\ 0 & 1 - 2\sigma\alpha_{11} - \sigma^2\delta & 0 & -\alpha_{11} - \sigma\delta \end{pmatrix}.$$

104   CHAPTER II.

Taking $\alpha_{11} + \sigma\delta = 0$ and transforming the resulting substitution by $T_{m,\delta}^{-1}$, we obtain $P_{1m} \equiv (\xi_1 \xi_m)(\eta_1 \eta_m)$. The latter is transformed into $T_{1,-1}$ by the following Abelian substitutions (and by no others):

$$\begin{Bmatrix} \alpha_{11} & \gamma_{11} & -\alpha_{11} & -\gamma_{11} \\ \beta_{11} & \delta_{11} & -\beta_{11} & -\delta_{11} \\ \alpha_{m1} & \gamma_{m1} & \alpha_{m1} & \gamma_{m1} \\ \beta_{m1} & \delta_{m1} & \beta_{m1} & \delta_{m1} \end{Bmatrix} \quad \begin{bmatrix} 2(\alpha_{11}\delta_{11} - \beta_{11}\gamma_{11}) = 1 \\ 2(\alpha_{m1}\delta_{m1} - \beta_{m1}\gamma_{m1}) = 1 \end{bmatrix}.$$

It follows that, if $\delta \neq 0$, Σ_1 is conjugate with some T_r.

Case b). $\delta = 0$ in Σ_1. The Abelian conditions 79) now give

$$\gamma_{im} = \delta_{im} = 0 \quad (i = 2, \ldots, m-1), \quad \alpha_{11} + \alpha_{mm} = 0, \quad \alpha_{11}^2 = 1.$$

Transforming Σ_1 by $Q_{1,m,\lambda}$, where $1 - 2\lambda\alpha_{11} = 0$, we obtain

$$W \equiv \begin{Bmatrix} \alpha_{11} & 0 & 0 & 0 & \ldots & 0 & 0 \\ 0 & \alpha_{11} & 0 & 0 & \ldots & 0 & 0 \\ 0 & 0 & \alpha_{22} & 0 & \ldots & \alpha_{2m} & 0 \\ \multicolumn{7}{c}{\dotfill} \\ 0 & 0 & -\beta_{2m} & \alpha_{2m} & \ldots & 0 & -\alpha_{11} \end{Bmatrix}.$$

Hence $W = T_{1,\pm 1} W'$, where W' affects only ξ_i, η_i $(i = 2, \ldots, m)$ and may therefore, by hypothesis, be transformed into a product of the $T_{j,-1}$ by an Abelian substitution on the same indices. It will transform W into a product of the $T_{j,-1}$ $(j = 1, \ldots, m)$, which is conjugate with some T_r.

Case c). In virtue of the Abelian conditions, Σ_2 becomes

$$\begin{Bmatrix} \alpha_{11} & 0 & 0 & 0 & \ldots & 0 & 0 & 1 & 0 \\ 0 & \alpha_{11} & 0 & 0 & \ldots & 0 & 1 & 0 & 0 \\ 0 & 0 & \alpha_{22} & 0 & \ldots & 0 & \gamma_{2m-1} & \alpha_{2m} & 0 \\ 0 & 0 & 0 & \alpha_{22} & \ldots & 0 & \delta_{2m-1} & \beta_{2m} & 0 \\ \multicolumn{9}{c}{\dotfill} \\ 1 & 0 & \delta_{2m-1} & -\gamma_{2m-1} & \ldots & -\alpha_{11} & 0 & \alpha_{m-1m} & 0 \\ 0 & 0 & 0 & 0 & \ldots & 0 & -\alpha_{11} & 0 & 0 \\ 0 & 0 & 0 & 0 & \ldots & 0 & 0 & -\alpha_{11} & 0 \\ 0 & 1 & -\beta_{2m} & \alpha_{2m} & \ldots & 0 & \alpha_{m-1m} & 0 & -\alpha_{11} \end{Bmatrix}$$

Transforming Σ_2 by the product $Q_{1,m,\lambda} Q_{m-1,1,\sigma}$, where $1 - 2\lambda\alpha_{11} = 0$, $1 + 2\sigma\alpha_{11} = 0$, we get a similar substitution but having zeros in place of the four elements 1. Since it is of the form W, we may proceed as in case b).

To complete the proof of our theorem, we note that
$$T_1 \equiv T_{1,-1}, \quad T_2 \equiv T_{1,-1}T_{2,-1}, \ldots, T_m \equiv T_{1,-1}T_{2,-1}\cdots T_{m,-1}$$
have the respective characteristic determinants (with parameter K)
$$(1+K)^2(1-K)^{2m-2}, \quad (1+K)^4(1-K)^{2m-4}, \ldots, (1+K)^{2m}.$$
Hence no two of them are conjugate under linear transformation.

The most general substitution of $SA(2m, p^n)$ commutative with T_r is seen to be $A \equiv A_r A_{m-r}$, where A_r is an arbitrary special Abelian substitution on the indices ξ_i, η_i $(i = 1, \ldots, r)$ and A_{m-r} an arbitrary one on the indices ξ_i, η_i $(i = r+1, \ldots, m)$. By § 115 the number of substitutions A_r and A_{m-r} is respectively $SA[2r, p^n]$ and $SA[2m-2r, p^n]$. Dividing $SA[2m, p^n]$ by the product of the foregoing numbers, we obtain the number of substitutions of $SA(2m, p^n)$ conjugate with T_r within the group.

Operators of period 2 of $A(2m, p^n)$, §§ 122--123.

122. By § 119, we obtain the quotient-group $A(2m, p^n)$ by considering as identical S and $ST \equiv TS$, where S is an arbitrary substitution of $SA(2m, p^n)$ and T is the self-conjugate substitution $T_{1,-1}T_{2,-1}\cdots T_{m,-1}$. In particular, T_r and T_rT become identical in the quotient-group. But the latter is conjugate with T'_{m-r}. Furthermore, if $s = m/2$ or $(m-1)/2$ according as m is even or odd, no two of the operators $T_1, T_2, \ldots, T_s$ are conjugate within the quotient-group. The special Abelian substitutions of period 2 lead therefore to just s distinct sets of conjugate operators of $A(2m, p^n)$, $p > 2$. To complete the study of the operators of period 2 of $A(2m, p^n)$, it remains to determine the conjugacy of the special Abelian substitutions S for which $S^2 = T$. Being of period 4, such an S is not conjugate to any T_r. Moreover, no two of the corresponding operators of the quotient-group are conjugate, since that would require one of the four relations
$$A^{-1}SA = T_r \text{ or } TT_r, \quad A^{-1}(ST)A = T_r \text{ or } TT_r,$$
A being Abelian. But any of these would require that S be conjugate with some T_i within the special Abelian group, whereas their periods are different. Making use of the result of § 123, we may state the theorem:

According as m is even or odd, the group $A(2m, p^n)$, $p > 2$, has exactly $\frac{1}{2}(m+2)$ or $\frac{1}{2}(m+1)$ distinct sets of conjugate operators of period 2.

123. Theorem. — *Within the special Abelian group on $2m$ indices in the $GF[p^n]$, $p > 2$, every substitution S, such that $S^2 = T$, is conjugate with $M \equiv M_1 M_2 \ldots M_m$.*[1]

Taking as S the general substitution 75), whose reciprocal is given by 77) for $\mu = 1$, the condition $S = S^{-1}T$ is seen to require

$$\alpha_{ij} = -\delta_{ji}, \quad \gamma_{ij} = \gamma_{ji}, \quad \beta_{ij} = \beta_{ji} \quad (i, j = 1, \ldots, m).$$

The matrix of coefficients of the general S is therefore

$$S \equiv \begin{Bmatrix} \alpha_{11} & \gamma_{11} & \alpha_{12} & \gamma_{12} & \ldots & \alpha_{1m} & \gamma_{1m} \\ \beta_{11} & -\alpha_{11} & \beta_{12} & \delta_{12} & \ldots & \beta_{1m} & \delta_{1m} \\ -\delta_{12} & \gamma_{12} & \alpha_{22} & \gamma_{22} & \ldots & \alpha_{2m} & \gamma_{2m} \\ \beta_{12} & -\alpha_{12} & \beta_{22} & -\alpha_{22} & \ldots & \beta_{2m} & \delta_{2m} \\ \cdot & \cdot & \cdot & \cdot & & \cdot & \cdot \\ -\delta_{1m} & \gamma_{1m} & -\delta_{2m} & \gamma_{2m} & \ldots & \alpha_{mm} & \gamma_{mm} \\ \beta_{1m} & -\alpha_{1m} & \beta_{2m} & -\alpha_{2m} & \ldots & \beta_{mm} & -\alpha_{mm} \end{Bmatrix},$$

subject to the special Abelian conditions.

Take first $m = 1$. Then S has the form

$$S \equiv \begin{pmatrix} \alpha_{11} & \gamma_{11} \\ \beta_{11} & -\alpha_{11} \end{pmatrix} \qquad [-\alpha_{11}^2 - \beta_{11}\gamma_{11} = 1].$$

It is conjugate with a similar substitution in which $\alpha_{11} = 0$. In fact, if $\beta_{11} \neq 0$, the transformed of S by $L_{1,\lambda}$ replaces η_1 by

$$\beta_{11}\xi_1 - (\alpha_{11} + \lambda\beta_{11})\eta_1,$$

in which the coefficient of η_1 may be made zero by choice of λ. If $\beta_{11} = 0$, $\gamma_{11} \neq 0$, we first transform S by M_1 and then proceed as before. If $\beta_{11} = \gamma_{11} = 0$, we first transform S by $L'_{1,\lambda}$ and obtain a substitution which replaces η_1 by $2\lambda\alpha_{11}\xi_1 - \alpha_{11}\eta_1$, so that the new $\beta_{11} \neq 0$.

With $\alpha_{11} = 0$, S takes the form

$$\begin{pmatrix} 0 & \gamma \\ -\gamma^{-1} & 0 \end{pmatrix}$$

and is the transformed of M_1 by the special Abelian substitution

$$\begin{pmatrix} \varrho\gamma & -\sigma\gamma \\ \sigma & \varrho \end{pmatrix} \qquad [\gamma(\varrho^2 + \sigma^2) = 1].$$

Indeed, by § 64, there exist solutions in the $GF[p^n]$, $p > 2$, of

$$\varrho^2 + \sigma^2 = 1/\gamma.$$

To prove the theorem by induction for m pairs of indices, we assume it true for t pairs of indices $t < m$.

---

[1] For the number of conjugates see Ex. 8, end of Ch. VIII.

THE ABELIAN LINEAR GROUP.   107

If $\alpha_{1i} = \beta_{1i} = \gamma_{1i} = \delta_{1i} = 0$ $(i = 2, \ldots, m)$, then $S = S_1 S'$, where S_1 affects only ξ_1, η_1 and is therefore conjugate with M_1, and where S_1' affects only ξ_i, η_i $(i = 2, \ldots, m)$ and is, by assumption, conjugate with $M_2 M_3 \ldots M_m$. Hence S is conjugate with $M_1 M_2 M_3 \ldots M_m$ within $SA(2m, p^n)$.

In the contrary case, S is conjugate (by § 120) with one of the two substitutions Σ_1, Σ_2. We consider the following three cases.

Case a). If Σ_1, with $\delta \neq 0$, be of the form S above, the Abelian conditions give

$$\alpha_{im} = \beta_{im} = \gamma_{im} = \delta_{im} = 0 \quad (i = 2, \ldots, m-1),$$
$$\alpha_{mm} = -\alpha_{11}, \quad \gamma_{mm} = -\delta\gamma_{11}, \quad \delta\beta_{mm} = -\beta_{11}.$$

Hence $\Sigma_1 \equiv \Sigma_1' \Sigma_1''$, where Σ_1' has the form

$$\Sigma_1': \quad \begin{array}{c|cccc} & \xi_1 & \eta_1 & \xi_m & \eta_m \\ \hline \xi_1' = & \alpha_{11} & \gamma_{11} & 1 & 0 \\ \eta_1' = & \beta_{11} & -\alpha_{11} & 0 & \delta \\ \xi_m' = & -\delta & 0 & -\alpha_{11} & -\delta\gamma_{11} \\ \eta_m' = & 0 & -1 & -\beta_{11}/\delta & \alpha_{11} \end{array}$$

while Σ_1'' affects only ξ_i, η_i $(i = 2, \ldots, m-1)$ and is, by assumption, transformed into $M_2 M_3 \ldots M_{m-1}$ by some special Abelian substitution affecting only the same indices. We proceed to prove that Σ_1' may be transformed into $M_1 M_m$ by a special Abelian substitution on the indices $\xi_1, \eta_1, \xi_m, \eta_m$. The proposition that Σ_1 is conjugate with $M_1 M_2 \ldots M_m$ under $SA(2m, p^n)$ will then follow.

If $\beta_{11} = \gamma_{11} = 0$ in Σ_1', we transform it by $N_{1, m, \lambda}$ and get

$$\begin{Bmatrix} \alpha_{11} & -2\lambda & 1 & 0 \\ 0 & -\alpha_{11} & 0 & \delta \\ -\delta & 0 & -\alpha_{11} & 2\lambda\delta \\ 0 & -1 & 0 & \alpha_{11} \end{Bmatrix}.$$

This is of the form Σ_1', but has $\gamma_{11} \neq 0$. Next, if $\gamma_{11} = 0$, $\beta_{11} \neq 0$, we transform Σ_1' by $M_1 M_m T_{m, \delta}$ and get a substitution of the form Σ_1' in which, however, $\gamma_{11} \neq 0$, $\beta_{11} = 0$. We may therefore assume that $\gamma_{11} \neq 0$ in Σ_1'. Transforming it by $L_{1, \lambda}' L_{m, \lambda/\delta}'$, where $\lambda \equiv \alpha_{11}/\gamma_{11}$, we get a substitution of the form

$$V \equiv \begin{Bmatrix} 0 & \gamma_{11} & 1 & 0 \\ \beta & 0 & 0 & \delta \\ -\delta & 0 & 0 & -\delta\gamma_{11} \\ 0 & -1 & -\beta/\delta & 0 \end{Bmatrix}.$$

If $\beta = 0$, then $\delta = 1$ and the transformed of V by $R_{1,m}, -\gamma_{11}^{-1}$ gives

$$W \equiv \begin{pmatrix} 0 & \gamma_{11} & 0 & 0 \\ -\gamma_{11}^{-1} & 0 & 0 & 0 \\ 0 & 0 & 0 & -\delta\gamma_{11} \\ 0 & 0 & \gamma_{11}^{-1}\delta^{-1} & 0 \end{pmatrix}.$$

If $\beta \neq 0$, the transformed of V by $N_{1,m,\delta/\beta}$ gives a substitution of the form W. Since W is the product of a substitution on the indices ξ_1, η_1 and a substitution on the indices ξ_m, η_m, it is conjugate with $M_1 M_m$.

Case b). If Σ_1, with $\delta = 0$, be of the form S, the Abelian conditions give

$$\beta_{11} = \gamma_{mm} = 0, \quad \alpha_{11} + \alpha_{mm} = 0, \quad \gamma_{im} = \delta_{im} = 0 \quad (i = 2, \ldots, m-1).$$

Transforming Σ_1 by $L_{1,\sigma}$, where $\gamma_{11} - 2\sigma\alpha_{11} = 0$, we get

$$\Sigma_1' \equiv \begin{pmatrix} \alpha_{11} & 0 & 0 & 0 & \ldots & 0 & 0 & 1 & 0 \\ 0 & -\alpha_{11} & 0 & 0 & \ldots & 0 & 0 & 0 & 0 \\ 0 & 0 & \alpha_{22} & \gamma_{22} & \ldots & \alpha_{2m-1} & \gamma_{2m-1} & \alpha_{2m} & 0 \\ 0 & 0 & \beta_{22} & -\alpha_{22} & \ldots & \beta_{2m-1} & \delta_{2m-1} & \beta_{2m} & 0 \\ \cdot & \cdot & \cdot & \cdot & \cdot & \cdot & \cdot & \cdot & \cdot \\ 0 & 0 & 0 & 0 & \ldots & 0 & 0 & -\alpha_{11} & 0 \\ 0 & -1 & \beta_{2m} & -\alpha_{2m} & \ldots & \beta_{m-1m} & -\alpha_{m-1m} & \beta_{mm} & \alpha_{11} \end{pmatrix}.$$

Transforming Σ_1' by $Q_{1,m,\sigma}$, we obtain a substitution Σ_1'' which differs from Σ_1' only in having the coefficients ± 1 replaced by $\pm(1 - 2\sigma\alpha_{11})$. By choice of σ, the latter may be made zero. Hence $\Sigma_1'' \equiv S_1 S'$, where S_1 affects only ξ_1, η_1 and is therefore conjugate with M_1, while S' affects only ξ_i, η_i $(i = 2, \ldots, m)$ and is, by assumption, conjugate with $M_2 M_3 \ldots M_m$. Hence Σ_1'' is conjugate with

$$M_1 M_2 \ldots M_m.$$

Case c). If Σ_2 be an Abelian substitution of the form S, it becomes

$$\Sigma_2 \equiv \begin{pmatrix} \alpha_{11} & \gamma_{11} & 0 & 0 & \ldots & 0 & 0 & 1 & 0 \\ \beta_{11} & -\alpha_{11} & 0 & 0 & \ldots & 0 & 1 & 0 & 0 \\ 0 & 0 & \alpha_{22} & \gamma_{22} & \ldots & 0 & \gamma_{2m-1} & \alpha_{2m} & 0 \\ 0 & 0 & \beta_{22} & -\alpha_{22} & \ldots & 0 & \delta_{2m-1} & \beta_{2m} & 0 \\ \cdot & \cdot & \cdot & \cdot & \cdot & \cdot & \cdot & \cdot & \cdot \\ -1 & 0 & -\delta_{2m-1} & \gamma_{2m-1} & \ldots & -\alpha_{11} & \gamma_{m-1m-1} & \alpha_{m-1m} & -\gamma_{11} \\ 0 & 0 & 0 & 0 & \ldots & 0 & \alpha_{11} & -\beta_{11} & 0 \\ 0 & 0 & 0 & 0 & \ldots & 0 & -\gamma_{11} & -\alpha_{11} & 0 \\ 0 & -1 & \beta_{2m} & -\alpha_{2m} & \ldots & -\beta_{11} & -\alpha_{m-1m} & \beta_{mm} & \alpha_{11} \end{pmatrix}$$

THE ABELIAN LINEAR GROUP.

Suppose first that $\alpha_{11} = 0$, so that $-\beta_{11}\gamma_{11} = 1$. Transforming Σ_2 by $R_{1,m,\lambda}$, where $1 + \lambda\gamma_{11} = 0$, we reach a substitution equal to a product $S_1 S'$, where S_1 affects ξ_1, η_1 only and S' affects only.

$$\xi_i, \quad \eta_i \qquad (i = 2, \ldots, m)$$

Suppose, however, that $\alpha_{11} \neq 0$. Transforming Σ_2 by $L_{m-1,\varrho}$,

$$\gamma_{m-1\,m-1} + 2\varrho\alpha_{11} = 0,$$

we obtain a substitution Σ_2' of the form Σ_2 and having $\gamma_{m-1\,m-1} = 0$. Transforming Σ_2' by $L'_{m,\varrho}$, where $\beta_{mm} - 2\varrho\alpha_{11} = 0$, we obtain a substitution Σ_2'' of the form Σ_2, but having $\beta_{mm} = \gamma_{m-1\,m-1} = 0$.

If $\beta_{11} = \gamma_{11} = 0$, we transform Σ_2'' by $Q_{m-1,1,\lambda}$, where $1 - 2\lambda\alpha_{11} = 0$, and afterwards by $Q_{1,m,\varrho}$, where $1 - 2\varrho\alpha_{11} = 0$, and obtain a product $S_1 S'$, where S_1 affects only ξ_1, η_1 and S' affects only ξ_i, η_i $(i > 1)$.

If $\gamma_{11} = 0$, $\beta_{11} \neq 0$, we transform Σ_2'' by $P_{m-1\,m} M_1 M_{m-1} M_m$ and get

$$\begin{pmatrix} -\alpha_{11} & -\beta_{11} & 0 & 0 & \ldots & 0 & 0 & 1 & 0 \\ 0 & \alpha_{11} & 0 & 0 & \ldots & 0 & 1 & 0 & 0 \\ \cdot & \cdot & \cdot & \cdot & & \cdot & \cdot & \cdot & \cdot \\ -1 & 0 & \beta_{2m} & -\alpha_{2m} & \ldots & \alpha_{11} & 0 & -\alpha_{m-1\,m} & \beta_{11} \\ 0 & 0 & 0 & 0 & \ldots & 0 & -\alpha_{11} & 0 & 0 \\ 0 & 0 & 0 & 0 & \ldots & 0 & \beta_{11} & \alpha_{11} & 0 \\ 0 & -1 & \delta_{2m-1} & -\gamma_{2m-1} & \ldots & 0 & \alpha_{m-1\,m} & 0 & -\alpha_{11} \end{pmatrix}$$

which has the form of Σ_2'' with $\gamma_{11} \neq 0$. We therefore treat the latter case only. Transforming Σ_2'' by $L'_{m,\varrho}$, where

$$\alpha_{m-1\,m} + \varrho\gamma_{11} = 0,$$

we obtain a substitution U of the form Σ_2, but having $\alpha_{m-1\,m}$ and $\gamma_{m-1\,m-1}$ both zero. Transforming U by $R_{m-1,m,\lambda} L'_{1,-\lambda}$, where $\alpha_{11} + \lambda\gamma_{11} = 0$, we get a substitution of the form

$$\begin{pmatrix} 0 & \gamma_{11} & 0 & 0 & \ldots & 0 & 0 & 1 & 0 \\ -1/\gamma_{11} & 0 & 0 & 0 & \ldots & 0 & 1 & 0 & 0 \\ \cdot & \cdot & \cdot & \cdot & & \cdot & \cdot & \cdot & \cdot \\ -1 & 0 & -\delta_{2m-1} & \gamma_{2m-1} & \ldots & 0 & 0 & 0 & -\gamma_{11} \\ 0 & 0 & 0 & 0 & \ldots & 0 & 0 & 1/\gamma_{11} & 0 \\ 0 & 0 & 0 & 0 & \ldots & 0 & -\gamma_{11} & 0 & 0 \\ 0 & -1 & \beta_{2m} & -\alpha_{2m} & \ldots & 1/\gamma_{11} & 0 & \beta_{mm} & 0 \end{pmatrix}.$$

Transforming this by $R_{1,m,\lambda}$, where $1 + \lambda\gamma_{11} = 0$, we get a similar substitution with the elements ± 1 replaced by zeros, and therefore the product of a substitution on ξ_1, η_1 by a substitution on the indices ξ_i, η_i $(i = 2, \ldots, m)$. It is therefore conjugate with

$$M_1 M_2 \ldots M_m.$$

CHAPTER III.

A GENERALIZATION OF THE ABELIAN LINEAR GROUP.[1]

124. Those linear homogeneous substitutions in the $GF[p^n]$ on mq indices,

87) $\quad S: \quad x'_{ij} = \sum_{k=1}^{m} (\alpha_{k1}^{ij} x_{k1} + \alpha_{k2}^{ij} x_{k2} + \cdots + \alpha_{kq}^{ij} x_{kq}),$
$$(i = 1, \ldots, m; j = 1, \ldots, q)$$

which, if operating simultaneously upon q independent sets of mq variables, the jth set of which is given the notation

$$x_{i1}^{(j)}, \quad x_{i2}^{(j)}, \ldots, x_{iq}^{(j)} \qquad (i = 1, 2, \ldots, m),$$

leave formally invariant the function

$$\Phi \equiv \sum_{i=1}^{m} \begin{vmatrix} x_{i1}^{(1)} & x_{i2}^{(1)} & \cdots & x_{iq}^{(1)} \\ x_{i1}^{(2)} & x_{i2}^{(2)} & \cdots & x_{iq}^{(2)} \\ \vdots & & & \vdots \\ x_{i1}^{(q)} & x_{i2}^{(q)} & \cdots & x_{iq}^{(q)} \end{vmatrix},$$

form a group $G(m, q, p^n)$, which for $q = 2$ is the Abelian group $SA(m, p^n)$.

The conditions upon S for the absolute invariance of Φ are seen to be those given by formulae 88) and 89), viz.,

88) $\quad \sum_{i=1}^{m} \begin{vmatrix} \alpha_{j1}^{i1} & \alpha_{j2}^{i1} & \cdots & \alpha_{jq}^{i1} \\ \vdots & & & \vdots \\ \alpha_{j1}^{iq} & \alpha_{j2}^{iq} & \cdots & \alpha_{jq}^{iq} \end{vmatrix} = 1 \qquad (j = 1, \ldots, m)$

89) $\quad \sum_{i=1}^{m} \begin{vmatrix} \alpha_{j_1 k_1}^{i1} & \alpha_{j_2 k_2}^{i1} & \cdots & \alpha_{j_q k_q}^{i1} \\ \vdots & & & \vdots \\ \alpha_{j_1 k_1}^{iq} & \alpha_{j_2 k_2}^{iq} & \cdots & \alpha_{j_q k_q}^{iq} \end{vmatrix} = 0 \quad \begin{pmatrix} \text{each } k = 1, 2, \ldots, q; \\ \text{each } j = 1, 2, \ldots, m, \\ j_1, j_2, \ldots, j_q \text{ not all equal} \end{pmatrix}$

125. *The inverse of the general substitution* 87) *of* $G(m, q, p^n)$ *is*

90) $\quad S^{-1}: \quad x'_{rs} = \sum_{i=1}^{m} (A_{rs}^{i1} x_{i1} + A_{rs}^{i2} x_{i2} + \cdots + A_{rs}^{iq} x_{iq}),$
$$(r = 1, \ldots, m; s = 1, \ldots, q)$$

---

[1] Taken from the author's paper, "A class of linear groups including the Abelian group", *Quarterly Journal*, July, 1899. The group is mentioned, but not investigated, by Jordan, Traité, p. 219, No. 301.

where A_{rs}^{ij} denotes the adjoint of α_{rs}^{ij} in the determinant

$$\begin{vmatrix} \alpha_{r1}^{i1} & \alpha_{r2}^{i1} & \ldots & \alpha_{rq}^{i1} \\ \cdot & \cdot & \cdot & \cdot \\ \alpha_{r1}^{iq} & \alpha_{r2}^{iq} & \ldots & \alpha_{rq}^{iq} \end{vmatrix}.$$

In fact, the product 87) 90) replaces x_{rs} by

$$\sum_{\substack{i=1,\ldots,m \\ j=1,\ldots,q}} A_{rs}^{ij} \left(\sum_{\substack{k=1,\ldots,m \\ l=1,\ldots,q}} \alpha_{kl}^{ij} x_{kl} \right).$$

Here the coefficient of x_{kl} is

$$\sum_{\substack{i=1,\ldots,m \\ j=1,\ldots,q}} A_{rs}^{ij} \alpha_{kl}^{ij} \equiv \sum_{i=1}^{m} \begin{vmatrix} \alpha_{r1}^{i1} & \ldots & \alpha_{rs-1}^{i1} & \alpha_{kl}^{i1} & \alpha_{rs+1}^{i1} & \ldots & \alpha_{rq}^{i1} \\ \cdot & & \cdot & \cdot & \cdot & & \cdot \\ \alpha_{r1}^{iq} & \ldots & \alpha_{rs-1}^{iq} & \alpha_{kl}^{iq} & \alpha_{rs+1}^{iq} & \ldots & \alpha_{rq}^{iq} \end{vmatrix}$$

and therefore, by 88) and 89), equals unity if $(k, l) = (r, s)$, but equals zero if $(k, l) \neq (r, s)$. Hence the above product replaces x_{rs} by x_{rs}. The reciprocal of S is therefore obtained by replacing α_{kl}^{ij} by A_{ij}^{kl} for $i, k = 1, \ldots, m$; $l, j = 1, \ldots, q$.

Writing relations 88) for S^{-1} given by 90), we find

$$91) \quad \sum_{i=1}^{m} \begin{vmatrix} A_{i1}^{j1} & \ldots & A_{i1}^{jq} \\ \cdot & \cdot & \cdot \\ A_{iq}^{j1} & \ldots & A_{iq}^{jq} \end{vmatrix} = \sum_{i=1}^{m} \begin{vmatrix} \alpha_{i1}^{j1} & \ldots & \alpha_{iq}^{j1} \\ \cdot & \cdot & \cdot \\ \alpha_{i1}^{jq} & \ldots & \alpha_{iq}^{jq} \end{vmatrix}^{q-1} = 1,$$

holding for $j = 1, 2, \ldots, m$.

Note. — For substitutions 87) which multiply Φ by a constant ϱ, the reciprocal is evidently obtained by replacing α_{kl}^{ij} by $\frac{1}{\varrho} A_{ij}^{kl}$.

126. The structure of the group $G(m, q, p^n)$ is essentially different in the two cases $q = 2$ and $q > 2$. The case $q = 2$ has been investigated at length in Chapter II. In the following investigation we assume that $q > 2$, a restriction necessary for the treatment given.

Let $j_2, j_3, \ldots, j_q$ have fixed values not all equal chosen arbitrarily from $1, 2, \ldots, m$, and let $k_2, k_3, \ldots, k_q$ have fixed values chosen from $1, 2, \ldots, q$. Then for $j_1 = 1, \ldots, m$; $k_1 = 1, \ldots, q$, we obtain mq equations 89). In fact, since $q > 2$, $j_1, j_2, \ldots, j_q$ are not all equal and hence do not lead to conditions of the type 88). Expanding the determinants of 89) according to the elements in the first columns, our mq equations may be written

112   CHAPTER III.

$$\sum_{\substack{i=1,\ldots,m \\ l=1,\ldots,q}} \alpha^{il}_{j_1 k_1} B^{il}_{j_1 k_1} = 0 \qquad \begin{pmatrix} j_1 = 1, \ldots, m \\ k_1 = 1, \ldots, q \end{pmatrix}$$

where

92) $$B^{il}_{j_1 k_1} = \begin{vmatrix} \alpha^{ib_2}_{j_2 k_2} & \cdots & \alpha^{ib_2}_{j_q k_q} \\ \cdot & \cdots & \cdot \\ \alpha^{ib_q}_{j_2 k_2} & \cdots & \alpha^{ib_q}_{j_q k_q} \end{vmatrix},$$

in which $b_2, b_3, \ldots, b_q$ denote the integers $1, \ldots, l-1, l+1, \ldots, q$. Since the determinant $|\alpha^{il}_{j_1 k_1}| \neq 0$, being the determinant of S, we have

$$B^{il}_{j_1 k_1} = 0 \qquad \begin{pmatrix} i=1, \ldots, m \\ l=1, \ldots, q \end{pmatrix}.$$

Hence the determinant 92) vanishes for $i = 1, \ldots, m$ and for $b_2, b_3, \ldots, b_q$ an arbitrary combination of $q-1$ distinct integers $\leq q$.

If $q = 3$, we have reached the relations 95) below. If $q > 3$, we denote by $C^{ib_s}_{j_2 k_2}$ the adjoint of $\alpha^{ib_s}_{j_2 k_2}$ in the determinant 92) and consider the following expansions:

93) $$\sum_{s=2}^{q} \alpha^{ib_s}_{j_2 k_2} C^{ib_s}_{j_2 k_2} = 0.$$

Of these consider the mq equations in which $i, j_3, \ldots, j_q$ have fixed values chosen arbitrarily from $1, 2, \ldots, m$, but such that $j_3, j_4, \ldots, j_q$ are not all equal, and $k_3, \ldots, k_q$ fixed values chosen arbitrarily from $1, 2, \ldots, q$, while lastly j_2 takes the values $1, 2, \ldots, m$ and k_2 the values $1, 2, \ldots, q$. Since the matrix

$$\left(\alpha^{ib_s}_{j_2 k_2}\right) \quad \begin{pmatrix} j_2 = 1, \ldots, m; \ k_2 = 1, \ldots, q \\ s = 2, \ldots, q \end{pmatrix}$$

comprises $q - 1$ rows of the matrix of S, not all of its determinants of order $q - 1$ are zero. Hence the $q - 1$ determinants C, which are the same in each of the mq equations 93), must be zero, viz.,

94) $$C^{ib_s}_{j_2 k_2} = \begin{vmatrix} \alpha^{ic_3}_{j_3 k_3} & \cdots & \alpha^{ic_3}_{j_q k_q} \\ \cdot & \cdots & \cdot \\ \alpha^{ic_q}_{j_3 k_3} & \cdots & \alpha^{ic_q}_{j_q k_q} \end{vmatrix} = 0,$$

where $c_3, \ldots, c_q$ denote any $q - 2$ distinct integers $\leq q$.

If $q = 4$, we have reached the relations 95) below. If $q > 4$, we proceed as before. After $q - 2$ such steps, we reach the set of relations

95) $$\begin{vmatrix} \alpha^{ir}_{jk} & \alpha^{ir}_{j'k'} \\ \alpha^{is}_{jk} & \alpha^{is}_{j'k'} \end{vmatrix} = 0 \quad \begin{pmatrix} i, j, j' = 1, \ldots, m; \ j' \neq j \\ r, s, k, k' = 1, \ldots, q \end{pmatrix}.$$

A GENERALIZATION OF THE ABELIAN LINEAR GROUP. 113

In virtue of the relations 95), the conditions 89) all reduce to identities. In fact, in each relation 89), at least two of the j's are distinct, say $j \neq j_2$, and therefore all minors formed from the first, and second columns vanish in virtue of 95).

A substitution S belongs to the group $G(m, q, p^n)$, $q > 2$, if and only if its coefficients satisfy the conditions 88) and 95).

127. Theorem. — *Every substitution S leaving Φ invariant can be derived from the totality of linear substitutions of determinant unity on q indices*

$$x'_{1j} = \sum_{k=1}^{q} \beta_{1k}^{1j} x_{1k} \qquad (j = 1, \ldots, q),$$

together with the linear substitutions, each on $2q$ indices,

$$P_{ij} \equiv (x_{i1} x_{j1})(x_{i2} x_{j2}) \ldots (x_{iq} x_{jq}) \quad (i, j = 1, \ldots, m).$$

We can evidently derive from these generators a substitution T which belongs to $G(m, q, p^n)$ and replaces an arbitrary index x_{kl} by any particular index as x_{11}. We may therefore suppose that in the product $S' \equiv TS$, S being defined by 87), the coefficient $\alpha_{11}^{11} \neq 0$. If then we set

$$\alpha_{jk}^{11} = C_{jk} \alpha_{11}^{11} \quad (j = 2, \ldots, m; \; k = 1, \ldots, q)$$

it follows from 95), for $i = 1$, $r = 1$, $j' = 1$, $k' = 1$, $j > 1$, that

96) $\qquad \alpha_{jk}^{1s} = C_{jk} \alpha_{11}^{1s} \quad (j = 2, \ldots, m; \; k, s = 1, \ldots, q).$

Substituting these values in the relation 91) for $j = 1$, we find

$$\begin{vmatrix} \alpha_{11}^{11} & \ldots & \alpha_{1q}^{11} \\ \cdot & \cdot & \cdot \\ \alpha_{11}^{1q} & \ldots & \alpha_{1q}^{1q} \end{vmatrix}^{q-1} + \sum_{i=2}^{m} \begin{vmatrix} C_{i1} \alpha_{11}^{11} & \ldots & C_{iq} \alpha_{11}^{11} \\ \cdot & \cdot & \cdot \\ C_{i1} \alpha_{11}^{1q} & \ldots & C_{iq} \alpha_{11}^{1q} \end{vmatrix}^{q-1} = 1.$$

It follows that
$$D^{q-1} = 1, \quad D = \begin{vmatrix} \alpha_{11}^{11} & \ldots & \alpha_{1q}^{11} \\ \cdot & \cdot & \cdot \\ \alpha_{11}^{1q} & \ldots & \alpha_{1q}^{1q} \end{vmatrix}.$$

Hence the following substitution is of determinant $D \neq 0$.

$$R: \quad \begin{cases} x'_{11} = \alpha_{11}^{11} x_{11} + \cdots + \alpha_{1q}^{11} x_{1q}, \\ \cdot \cdot \cdot \cdot \cdot \cdot \cdot \cdot \cdot \cdot \\ x'_{1q} = \alpha_{11}^{1q} x_{11} + \cdots + \alpha_{1q}^{1q} x_{1q}. \end{cases}$$

114 CHAPTER III. A GENERALIZATION OF THE ABELIAN LINEAR GROUP.

If we denote the determinants of Φ by D_i so that $\Phi \equiv \sum_{i=1}^{m} D_i$, we readily see that R multiplies D_1 by the factor D but leaves unaltered D_i ($i = 2, \ldots, m$). Hence, if W denote the substitution

$W:$ $\qquad x'_{i1} = D x_{i1} \qquad (i = 2, \ldots, m)$,

the product WR multiplies Φ by the factor D. The product $S_1 \equiv (WR)^{-1} S'$ multiplies Φ by D^{-1} and therefore satisfies the relations 89) and consequently also relations 95), derived from them. But S_1 affects the indices $x_{11}, x_{12}, \ldots, x_{1q}$ as follows:

$$x'_{1j} = x_{1j} + \sum_{k=2}^{m} (\alpha_{k1}^{1j} x_{k1} + \cdots + \alpha_{kq}^{1j} x_{kq}),$$

where α_{k1}^{1j} denotes D^{-1} times the earlier α_{k1}^{1j}, for $k = 2, \ldots, m$. For the substitution S_1 we have $\alpha_{11}^{1s} = 0$ ($s = 2, \ldots, q$). Hence by 96),

$$\alpha_{jk}^{1s} = 0 \quad (j = 2, \ldots, m; \ k = 1, \ldots, q; \ s = 2, \ldots, q).$$

Also $\alpha_{1s}^{11} = 0$ ($s = 2, \ldots, q$), $\alpha_{1s}^{1s} = 1$. Hence, by the following cases of 95),

$$\begin{vmatrix} \alpha_{jk}^{11} & \alpha_{1s}^{11} \\ \alpha_{jk}^{1s} & \alpha_{1s}^{1s} \end{vmatrix} = 0, \quad \begin{pmatrix} j = 2, \ldots, m; \ k = 1, \ldots, q \\ s = 2, \ldots, q \end{pmatrix}$$

we find $\alpha_{jk}^{11} = 0$. Hence every $\alpha_{jk}^{1s} = 0$, for $j > 1$, so that S_1 leaves fixed $x_{11}, x_{12}, \ldots, x_{1q}$.

Applying the Note of § 125 to form the reciprocal of S_1, we find that the matrix of S_1^{-1} has zeros throughout the first q columns, except the diagonal terms D in the first q rows. By the above argument, the remaining elements of the first q rows must be zeros. Reciprocating this matrix by the same rule, we find that $D = 1$ and that S_1 reduces to a substitution on the indices

$$x_{j2}, \ x_{j3}, \ldots, \ x_{jq} \qquad (j = 2, \ldots, m).$$

Since W is the identity, $S = T^{-1} S' = T^{-1} R S_1$, where T and R are derived from the generators given in the theorem. Proceeding with S_1 as we did with S, we reach a substitution S_2 on the indices $x_{j3}, \ldots, x_{jq}$. Finally, we reach the identity.

128. It follows from § 127 that the group $G(m, q, p^n)$, $q > 2$, has an invariant subgroup Γ composed of the substitutions

$$x'_{ij} = \sum_{k=1}^{q} \beta_{ik}^{ij} x_{ik} \quad (i = 1, \ldots, m; \ j = 1, \ldots, q)$$

where, for $i = 1, 2, \ldots, m$, the determinant

$$|\beta_{ik}^{ij}| = 1 \qquad (j, k = 1, \ldots, q).$$

The quotient-group is generated by the substitutions P_{ij} and is thus holoedrically isomorphic with the symmetric group on m letters. The group Γ is the direct product of m groups each the special linear homogeneous group in the $GF[p^n]$ on q indices (§ 103). The substitutions of the ith group are given as follows

$$x'_{ij} = \sum_{k=1}^{q} \beta_{ik}^{ij} x_{ik}, \quad x'_{sj} = x_{sj} \quad (s = 1, \ldots, m; \; s \neq i; \; j = 1, \ldots, q).$$

The structure of the group $G(m, q, p^n)$ is therefore completely determined.

CHAPTER IV.

THE HYPERABELIAN GROUP.

129. The totality of linear homogeneous substitutions in the $GF[p^{2n}]$

$$S: \quad \xi'_i = \sum_{j=1}^{2m} \alpha_{ij} \xi_j \quad (i = 1, \ldots, 2m)$$

which leave absolutely invariant the function

$$\Psi \equiv \sum_{l=1}^{m} \begin{vmatrix} \xi_{2l-1} & \xi_{2l} \\ \xi_{2l-1}^{p^n} & \xi_{2l}^{p^n} \end{vmatrix}$$

forms the *hyperabelian group*[1]) $H(2m, p^{2n})$. Its name is derived from the fact that the totality of its substitutions whose coefficients belong to the included field $GF[p^n]$ constitutes the Abelian group $SA(2m, p^n)$, which is therefore a subgroup of the hyperabelian group.

A general substitution S transforms Ψ into

$$\sum_{l=1}^{m} \left\{ \sum_{j=1}^{2m} \begin{vmatrix} \alpha_{2l-1\,j} & \alpha_{2l\,j} \\ \alpha_{2l-1\,j}^{p^n} & \alpha_{2l\,j}^{p^n} \end{vmatrix} \xi_j^{p^n+1} + \sum_{j=k}^{j,k=1,\ldots,2m} \begin{vmatrix} \alpha_{2l-1\,j} & \alpha_{2l\,j} \\ \alpha_{2l-1\,k}^{p^n} & \alpha_{2l\,k}^{p^n} \end{vmatrix} \xi_j \xi_k^{p^n} \right\}.$$

The conditions upon S for the absolute invariance of Ψ are thus

$$97) \quad \sum_{l=1}^{m} \begin{vmatrix} \alpha_{2l-1\,j} & \alpha_{2l\,j} \\ \alpha_{2l-1\,k}^{p^n} & \alpha_{2l\,k}^{p^n} \end{vmatrix} = \varepsilon_{jk} \quad (j, k = 1, \ldots, 2m)$$

where $\varepsilon_{jk} = 0$, unless j and k differ by unity, when

$$\varepsilon_{2i-1\,2i} = 1, \quad \varepsilon_{2i\,2i-1} = -1 \quad (i = 1, \ldots, m).$$

---

1) Introduced by the author, *Proc. Lond. Math. Soc.*, vol. 31, pp. 30—68. It will hardly be confused with Picard's hyperabelian group of infinite order.

116   CHAPTER IV.

The reciprocal of the hyperabelian substitution S is

$$S^{-1}: \begin{cases} \xi'_{2l-1} = \sum_{j=1}^{m} \left(\alpha^{p^n}_{2j\,2l} \xi_{2j-1} - \alpha^{p^n}_{2j-1\,2l} \xi_{2j} \right) \\ \xi'_{2l} = \sum_{j=1}^{m} \left(-\alpha^{p^n}_{2j\,2l-1} \xi_{2j-1} + \alpha^{p^n}_{2j-1\,2l-1} \xi_{2j} \right) \end{cases}$$
$$(l = 1, \ldots, m).$$

Indeed, the product SS^{-1} replaces ξ_{2l-1} by

$$\sum_{j,k}^{1,\ldots,m} \begin{vmatrix} \alpha_{2j-1\,2k-1} & \alpha_{2j\,2k-1} \\ \alpha^{p^n}_{2j-1\,2l} & \alpha^{p^n}_{2j\,2l} \end{vmatrix} \xi_{2k-1} + \sum_{j,k}^{1,\ldots,m} \begin{vmatrix} \alpha_{2j-1\,2k} & \alpha_{2j\,2k} \\ \alpha^{p^n}_{2j-1\,2l} & \alpha^{p^n}_{2j\,2l} \end{vmatrix} \xi_{2k}$$

$$= \sum_{k=1}^{m} \left(\varepsilon_{2k-1\,2l} \xi_{2k-1} + \varepsilon_{2k\,2l} \xi_{2k} \right) \equiv \xi_{2l-1}.$$

Similarly, SS^{-1} replaces ξ_{2l} by

$$-\sum_{k=1}^{m} \left(\varepsilon_{2k-1\,2l-1} \xi_{2k-1} + \varepsilon_{2k\,2l-1} \xi_{2k} \right) \equiv \xi_{2l}.$$

The relations 97) in which $j > k$ are derived from those in which $j < k$ by raising the latter to the power p^n. We may therefore express the hyperabelian conditions in the convenient form

98) $\quad \sum_{l=1}^{m} \begin{vmatrix} \alpha_{2l-1\,j} & \alpha_{2l\,j} \\ \alpha^{p^n}_{2l-1\,k} & \alpha^{p^n}_{2l\,k} \end{vmatrix} = \begin{cases} 1 & (\text{if } k = j+1 = \text{even}) \\ 0 & (\text{unless } k = j+1 = \text{even}) \end{cases}$

$$(j, k = 1, \ldots, 2m;\ j \lessgtr k).$$

The corresponding relations for S^{-1} are found by replacing by respectively $\quad \alpha_{2l-1\,2j-1}, \quad \alpha_{2l-1\,2j}, \quad \alpha_{2l\,2j-1}, \quad \alpha_{2l\,2j}$

$$\alpha^{p^n}_{2j\,2l}, \quad -\alpha^{p^n}_{2j-1\,2l}, \quad -\alpha^{p^n}_{2j\,2l-1}, \quad \alpha^{p^n}_{2j-1\,2l-1}.$$

Writing out the four sets of relations 98) according to the evenness or oddness of j and k, and making the replacement just indicated, we obtain four sets of relations for the invariance of Ψ by the substitution S^{-1} and therefore together equivalent to the relations 98). We may combine the four sets into the single formula

99) $\quad \sum_{l=1}^{m} \begin{vmatrix} \alpha^{p^n}_{j\,2l-1} & \alpha^{p^n}_{j\,2l} \\ \alpha_{k\,2l-1} & \alpha_{k\,2l} \end{vmatrix} = \begin{cases} 1 & (\text{if } k = j+1 = \text{even}) \\ 0 & (\text{unless } k = j+1 = \text{even}) \end{cases}$

$$(j, k = 1, \ldots, 2m;\ j \lessgtr k).$$

130. *The determinant Δ of the hyperabelian substitution S must satisfy the relation*

100) $\qquad\qquad\qquad \Delta^{p^n+1} = 1.$

THE HYPERABELIAN GROUP.   117

For proof, we reflect on its main diagonal the determinant of S^{-1}, then change the signs of the $2l - 1^{st}$ row and column for $l = 1, \ldots, m$, and finally interchange the $2l - 1^{st}$ row with the $2l^{th}$ row for $l = 1, \ldots, m$, and likewise interchange the corresponding columns. We obtain the determinant

$$\left| \alpha_{ij}^{p^n} \right| = \left| \alpha_{ij} \right|^{p^n} = \Delta^{p^n}.$$

Hence $\Delta \Delta^{p^n} = 1$, being the determinant of the product SS^{-1}.

131. Theorem. — *The maximal subgroup M of the hyperabelian group $H(2m, p^{2n})$ which transforms into itself the Abelian group $SA(2m, p^n)$ is given by the extension of the latter by the substitution*

$$V_\varrho: \quad \xi'_{2l-1} = \varrho \xi_{2l-1}, \quad \xi'_{2l} = \varrho^{-p^n} \xi_{2l} \quad (l = 1, \ldots, m),$$

where ϱ is a primitive root in the $GF[p^{2n}]$. The index of $SA(2m, p^n)$ under M is $p^n + 1$.

We determine all hyperabelian substitutions

$$S: \qquad \xi'_i = \sum_{j=1}^{2m} \alpha_{ij} \xi_j \qquad (i = 1, \ldots, 2m)$$

which transform the Abelian group into itself. Now S transforms the Abelian substitution, affecting a single index,

$$\xi'_{2r-1} = \xi_{2r-1} + \xi_{2r}$$

into the substitution

$$\xi'_i = \xi_i + \alpha_{i\,2r-1} \sum_{j=1}^{m} \left(-\alpha_{2j\,2r-1}^{p^n} \xi_{2j-1} + \alpha_{2j-1\,2r-1}^{p^n} \xi_{2j} \right)$$
$$(i = 1, \ldots, 2m),$$

whose coefficients must therefore belong to the $GF[p^n]$, viz.,

$$\alpha_{i\,2r-1} \alpha_{j\,2r-1}^{p^n} \quad (i, j = 1, \ldots, 2m; \; r = 1, \ldots, m).$$

Likewise, S must transform the Abelian substitution

$$\xi'_{2r} = \xi_{2r} + \xi_{2r-1}$$

into a substitution belonging to the $GF[p^n]$. Hence the products

$$\alpha_{i\,2r} \alpha_{j\,2r}^{p^n} \quad (i, j = 1, \ldots, 2m; \; r = 1, \ldots, m)$$

must belong to the $GF[p^n]$. The reciprocal S^{-1} must transform the Abelian group into itself. From the above results, it follows therefore that the products

$$\alpha_{2rs} \alpha_{2rt}^{p^n}, \quad \alpha_{2r-1\,s} \alpha_{2r-1\,t}^{p^n} \quad (s, t = 1, \ldots, 2m; \; r = 1, \ldots, m)$$

must belong to the $GF[p^n]$. Combining our results, every product

101) $\qquad \alpha_{ir}\alpha_{jr}^{p^n}, \quad \alpha_{ri}\alpha_{rj}^{p^n} \qquad (i, j, r = 1, \ldots, 2m)$

must belong to the $GF[p^n]$.

But, if β, γ be marks of the $GF[p^{2n}]$ such that
$$\beta\gamma^{p^n} = \mu = \text{mark of } GF[p^n],$$
then, if $\gamma \neq 0$, $\beta/\gamma = \mu\gamma^{-p^n-1}$ is a mark of the $GF[p^n]$. Hence by 101), the ratios of the non-vanishing coefficients in any row or any column of the matrix of S must all belong to the $GF[p^n]$.

Suppose first that $m = 1$. If $\alpha_{11} \neq 0$, we have
$$\alpha_{21} = \lambda\alpha_{11}, \quad \alpha_{12} = \mu\alpha_{11} \quad (\lambda, \mu \text{ in the } GF[p^n]).$$
Then if λ and μ be not both zero, $\alpha_{22} = \nu\alpha_{11}$, ν being in the $GF[p^n]$. For $\lambda = \mu = 0$, the hyperabelian condition gives $\alpha_{11}\alpha_{22}^{p^n} = 1$, whence $\alpha_{22} = \nu\alpha_{11}$. If, however, $\alpha_{11} = 0$, both α_{12} and α_{21} are not zero. Hence $\alpha_{22} = \varrho\alpha_{12}$, ϱ in the $GF[p^n]$. By the hyperabelian condition, $-\alpha_{12}\alpha_{21}^{p^n} = 1$, whence $\alpha_{21} = \sigma\alpha_{12}$, σ in the $GF[p^n]$. In either case, we have reached a substitution of the form 103) below.

For $m > 1$, S transforms the Abelian substitution
$$\xi'_{2r-1} = \xi_{2r-1} + \xi_{2s}, \quad \xi'_{2s-1} = \xi_{2s-1} + \xi_{2r} \quad (r \neq s)$$
into
$$\xi'_i = \xi_i + \alpha_{i\,2r-1}\sum_{j=1}^{m}\left(-\alpha_{2j\,2s-1}^{p^n}\xi_{2j-1} + \alpha_{2j-1\,2s-1}^{p^n}\xi_{2j}\right)$$
$$+ \alpha_{i\,2s-1}\sum_{j=1}^{m}\left(-\alpha_{2j\,2r-1}^{p^n}\xi_{2j-1} + \alpha_{2j-1\,2r-1}^{p^n}\xi_{2j}\right).$$

Hence the sums
$$\alpha_{i\,2r-1}\alpha_{j\,2s-1}^{p^n} + \alpha_{i\,2s-1}\alpha_{j\,2r-1}^{p^n} \quad (i, j = 1, \ldots, 2m;\ r, s = 1, \ldots, m)$$
must all belong to the $GF[p^n]$. In like manner, if S transform each of the following three Abelian substitutions (in which $r \neq s$),
$$\begin{aligned}\xi'_{2r-1} &= \xi_{2r-1} + \xi_{2s-1}, & \xi'_{2s} &= \xi_{2s} - \xi_{2r};\\ \xi'_{2r} &= \xi_{2r} + \xi_{2s}, & \xi'_{2s-1} &= \xi_{2s-1} - \xi_{2r-1};\\ \xi'_{2r} &= \xi_{2r} + \xi_{2s-1}, & \xi'_{2s} &= \xi_{2s} + \xi_{2r-1}\end{aligned}$$
into substitutions belonging to the $GF[p^n]$, then must the respective sums
$$\begin{cases}\alpha_{i\,2r-1}\alpha_{j\,2s}^{p^n} + \alpha_{i\,2s}\alpha_{j\,2r-1}^{p^n}\\ \alpha_{i\,2r}\alpha_{j\,2s-1}^{p^n} + \alpha_{i\,2s-1}\alpha_{j\,2r}^{p^n}\\ \alpha_{i\,2r}\alpha_{j\,2s}^{p^n} + \alpha_{i\,2s}\alpha_{j\,2r}^{p^n}\end{cases} \quad \begin{pmatrix}i, j = 1, \ldots, 2m\\ r, s = 1, \ldots, m\end{pmatrix}$$
belong to the $GF[p^n]$. Combining our results, every sum

102) $\qquad \alpha_{ir}\alpha_{js}^{p^n} + \alpha_{is}\alpha_{jr}^{p^n} \quad (i, j, r, s = 1, \ldots, 2m;\ r \neq s)$

belongs to the $GF[p^n]$.

Of the coefficients in the ith row of the matrix of S, we may suppose that $\alpha_{ir} \neq 0$, for example. If, then, $\alpha_{jr} \neq 0$, the ratios of the coefficients in the ith and j^{th} rows must all belong to the $GF[p^n]$ [by the result following from 101)]. If, however, $\alpha_{jr} = 0$, we may suppose that, for example, $\alpha_{js} \neq 0$ $(s \neq r)$. Then, by 102), the products $\alpha_{ir}\alpha_{js}^{p^n}$ belong to the $GF[p^n]$. We have in either case the result that the ratios of the coefficients in the ith and j^{th} rows belong to the $GF[p^n]$. Hence the ratios of all the coefficients in S to any one non-vanishing coefficient belong to the $GF[p^n]$, so that S may be written

103) $$\xi_i' = \alpha \sum_{j=1}^{2m} \lambda_{ij} \xi_j \qquad (i = 1, \ldots, 2m),$$

where the λ_{ij} belong to the $GF[p^n]$.

Inversely, every hyperabelian substitution of the form 103) transforms into itself the Abelian group defined for the $GF[p^n]$.

The conditions that 103) shall be hyperabelian are

104) $$\sum_{l=1}^{m} \begin{vmatrix} \lambda_{2l-1,j} & \lambda_{2lj} \\ \lambda_{2l-1,k} & \lambda_{2lk} \end{vmatrix} = \begin{cases} \alpha^{-p^n-1} & (\text{if } k = j+1 = \text{even}) \\ 0 & (\text{unless } k = j+1 = \text{even}) \end{cases}$$
$$(i, j = 1, \ldots, 2m; i \leq j).$$

The substitution (λ_{ij}), or 103) with the factor α deleted, therefore belongs to the general Abelian group $GA(2m, p^n)$ and multiplies Ψ by the mark α^{-p^n-1} of the $GF[p^n]$. If then we set

105) $$\mu_{2l-1,j} = \lambda_{2l-1,j}, \quad \mu_{2lj} = \alpha^{p^n+1}\lambda_{2lj} \qquad \binom{l=1,\ldots,m}{j=1,\ldots,2m}$$

we find that $S = UV_\alpha$, where

$U:$ $$\xi_i' = \sum_{j=1}^{2m} \mu_{ij}\xi_j \qquad (i = 1, \ldots, 2m),$$

$V_\alpha:$ $$\xi_{2l-1}' = \alpha\xi_{2l-1}, \quad \xi_{2l}' = \alpha^{-p^n}\xi_{2l} \qquad (l = 1, \ldots, m),$$

so that V_α, and therefore also U, is a hyperabelian substitution. Moreover, in virtue of the relations 104) and 105), U belongs to the special Abelian group $SA(2m, p^n)$ and is therefore of determinant unity. The first part of our theorem is therefore proven.

If we form a rectangular array of the marks $\neq 0$ of the $GF[p^{2n}]$ with those belonging to the $GF[p^n]$ as first row, the

$$p^n + 1 = (p^{2n} - 1)/(p^n - 1)$$

"multipliers" form a set of marks $\alpha_1, \alpha_2, \ldots, \alpha_{p^n+1}$ such that none of their ratios belong to the $GF[p^n]$, while every mark of the $GF[p^{2n}]$ not of this set has with some mark of the set a ratio belonging to the $GF[p^n]$. Furthermore, the product

120   CHAPTER IV.

$V_{\alpha'} V_\alpha^{-1}$: $\xi'_{2l-1} = \alpha^{-1} \alpha' \xi_{2l-1}$, $\xi'_{2l} = \alpha^{p^n} \alpha'^{-p^n} \xi_{2l}$ $(l = 1, \ldots, m)$

belongs to $SA(2m, p^n)$ if and only if $\alpha^{-1} \alpha'$ belongs to the $GF[p^n]$. It follows that the substitutions V_{α_i} $(i = 1, 2, \ldots, p^n + 1)$ give the totality of substitutions V_α such that $V_{\alpha'} V_\alpha^{-1}$ does not belong to $SA(2m, p^n)$. Hence an identity of the form

$$U V_{\alpha_i} \equiv U' V_{\alpha_j} \quad (i \text{ and } j \leqq p^n + 1;\ i \neq j)$$

is impossible when U and U' both belong to $SA(2m, p^n)$. Every hyperabelian substitution 103) is therefore of the form $U V_{\alpha_i}$, i being chosen from the series $1, 2, \ldots, p^n + 1$, while an identity $U V_{\alpha_i} = U' V_{\alpha_j}$ requires $i = j$, $U = U'$. Hence the number of distinct substitutions 103) is $(p^n + 1) SA[2m, p^n]$. The second part of our theorem is therefore proven.

132. Those substitutions of the hyperabelian group $H(2m, p^{2n})$ which have determinant unity form a self-conjugate subgroup H' of index $p^n + 1$. In fact, for σ any mark $\neq 0$ of the $GF[p^{2n}]$, the substitution

$$\xi'_1 = \sigma \xi_1, \quad \xi'_2 = \sigma^{-p^n} \xi_2, \quad \xi'_i = \xi_i \quad (i = 3, \ldots, 2m)$$

belongs to $H(2m, p^{2n})$. Its determinant $\sigma^{-(p^n-1)}$ can, by choice of σ, be made equal to any one of the $p^n + 1$ roots of $\Delta^{p^n+1} = 1$. Hence there exist hyperabelian substitutions whose determinant Δ is any root of this equation. By § 130, there are no other values of Δ.

The group H' contains a self-conjugate subgroup formed by the substitutions

106) $T_\varkappa$: $\xi'_i = \varkappa \xi_i$ $(i = 1, \ldots, 2m)$ $[\varkappa^{2m} = 1,\ \varkappa^{p^n+1} = 1]$.

The quotient-group will be denoted by the symbol $HA(2m, p^{2n})$. It will be proven *simple* except in the special cases $m = 1$, $p^n = 2$ or 3 (§§ 138, 145, 148). By the same references its order $HA[2m, p^{2n}]$ is

$$\frac{1}{q}(p^{2mn} - 1) p^{n(2m-1)} (p^{n(2m-1)} + 1) p^{n(2m-2)} \cdots (p^{2n} - 1) p^n,$$

where q denotes the greatest common divisor of $2m$ and $p^n + 1$. The order of $H(2m, p^{2n})$ is

$$(p^{2mn} - 1) p^{n(2m-1)} (p^{n(2m-1)} + 1) p^{n(2m-2)} \cdots (p^{2n} - 1) p^n (p^n + 1).$$

The Abelian group $SA(2m, p^n)$ has an invariant subgroup formed by the identity and T_{-1}. The quotient-group $A(2m, p^n)$ is simple except in the three cases $m = 1$, $p^n = 2$; $m = 1$, $p^n = 3$; $m = 2$, $p^n = 2$ (§ 119). But $H(2m, p^{2n})$ contains $SA(2m, p^n)$ as a subgroup. In order that $T_\varkappa$ shall belong to the latter, the coefficient $\varkappa$ must belong to the $GF[p^n]$. But $\varkappa^{p^n} = \varkappa$ and $\varkappa^{p^n+1} = 1$ require $\varkappa^2 = 1$. Hence would $T_\varkappa$ be the identity or T_{-1}. It follows that $A(2m, p^n)$ is a subgroup of $HA(2m, p^n)$. We proceed to determine the number of conjugates to the former group within the latter group, using the result of § 131.

133. Theorem. — *The largest subgroup M' of $HA(2m, p^{2n})$ which transforms $A(2m, p^n)$ into itself is identical with $A(2m, p^n)$ if $p = 2$ or if $p > 2$ and $p^n + 1$ contains a higher power of 2 than m contains; in the remaining case, the order of M' is double the order of $A(2m, p^n)$.*

The determinant of $S \equiv UV_\alpha$ being supposed to be unity and that of U being unity, it follows that V_α has determinant

107) $$\alpha^{-m(p^n-1)} = 1.$$

Now V_α and $T_\varkappa V_\alpha$ correspond in the quotient-group $HA(2m, p^{2n})$ to the same operator. We investigate the conditions under which $T_\varkappa V_\alpha$ has its coefficients in the $GF[p^n]$. The necessary and sufficient condition is seen to be
$$(\varkappa\alpha)^{p^n-1} = 1.$$

Hence must $\varkappa^2 = \alpha^{p^n-1}$ and therefore

$$\alpha^{\frac{1}{2}(p^{2n}-1)} = \varkappa^{p^n+1} = 1,$$

or α must be a square in the $GF[p^{2n}]$. The remaining condition $\varkappa^{2m} = 1$ becomes an identity in virtue of 107). Hence, if the solutions of 107) are all squares in the $GF[p^{2n}]$, the substitution $S = UV_\alpha$ will correspond in the quotient-group to an operator belonging to $A(2m, p^n)$. But, if there occur not-squares as solutions of 107), the resulting substitutions V_α may be expressed as products $V_\nu V_{\beta^2}$, ν being a particular not-square. Then V_{β^2} corresponds in the quotient-group to an operator of $A(2m, p^n)$, while V_ν does not. In this case the group $A(2m, p^n)$ is transformed into itself by a subgroup of $HA(2m, p^{2n})$ of double the order of $A(2m, p^n)$.

For $p = 2$, the theorem follows at once since every mark of the $GF[2^{2n}]$ is a square. For $p > 2$, we are to determine in what cases 107) has as its solutions in the $GF[p^{2n}]$ only squares. A common solution of the pair of equations

108) $$\alpha^{m(p^n-1)} = 1, \quad \alpha^{p^{2n}-1} = 1$$

is required to be a solution of $\alpha^{\frac{1}{2}(p^{2n}-1)} = 1$. A common solution of 108) satisfies $\alpha^{d(p^n-1)} = 1$, where d is the greatest common divisor of m and $p^n + 1$. The condition is therefore that d shall divide $\frac{1}{2}(p^n + 1)$. It is satisfied if, and only if, $p^n + 1$ contains 2 to a higher power than m does.

122   CHAPTER IV.

Corollary. — If $g = 1$ or 2 according as the order of M' is equal or is double the order of $A(2m, p^n)$, the number of subgroups of $HA(2m, p^{2n})$ conjugate with $A(2m, p^n)$ is

$$HA[2m, p^{2n}] \div g \cdot A[2m, p^n] \equiv$$
$$\frac{a}{gq}(p^{n(2m-1)} + 1)p^{n(2m-2)}(p^{n(2m-3)} + 1)p^{n(2m-4)} \ldots (p^{3n} + 1)p^{2n},$$

where $a = 1$ if $p = 2$, $a = 2$ if $p > 2$, and q denotes the greatest common divisor of $2m$ and $p^n + 1$.

134. The conditions that the quaternary substitution in the $GF[p^{2n}]$

109) $\quad \xi_1' = \alpha_{11}\xi_1 + \alpha_{13}\xi_3, \quad \xi_2' = \alpha_{22}\xi_2 + \alpha_{24}\xi_4,$
$\quad\quad \xi_3' = \alpha_{31}\xi_1 + \alpha_{33}\xi_3, \quad \xi_4' = \alpha_{42}\xi_2 + \alpha_{44}\xi_4,$

shall be hyperabelian include the following:

$$\alpha_{11}\alpha_{22}^{p^n} + \alpha_{31}\alpha_{42}^{p^n} = 1, \quad \alpha_{11}\alpha_{24}^{p^n} + \alpha_{31}\alpha_{44}^{p^n} = 0,$$
$$\alpha_{13}\alpha_{22}^{p^n} + \alpha_{33}\alpha_{42}^{p^n} = 0, \quad \alpha_{13}\alpha_{24}^{p^n} + \alpha_{33}\alpha_{44}^{p^n} = 1.$$

Setting $\Delta \equiv \alpha_{22}\alpha_{44} - \alpha_{24}\alpha_{42}$, we find from these conditions that

$$\alpha_{11}\Delta^{p^n} = \alpha_{44}^{p^n}, \quad \alpha_{31}\Delta^{p^n} = -\alpha_{24}^{p^n}, \quad \alpha_{13}\Delta^{p^n} = -\alpha_{42}^{p^n}, \quad \alpha_{33}\Delta^{p^n} = \alpha_{22}^{p^n}.$$

The above substitution then takes the form

$$T: \begin{cases} \xi_1' = \left(\frac{\alpha_{44}}{\Delta}\right)^{p^n}\xi_1 - \left(\frac{\alpha_{42}}{\Delta}\right)^{p^n}\xi_3, & \xi_2' = \alpha_{22}\xi_2 + \alpha_{24}\xi_4, \\ \xi_3' = -\left(\frac{\alpha_{24}}{\Delta}\right)^{p^n}\xi_1 + \left(\frac{\alpha_{22}}{\Delta}\right)^{p^n}\xi_3, & \xi_4' = \alpha_{42}\xi_2 + \alpha_{44}\xi_4. \end{cases}$$

Inversely, the substitution T is seen to leave absolutely invariant

$$\xi_1\xi_2^{p^n} - \xi_2\xi_1^{p^n} + \xi_3\xi_4^{p^n} - \xi_4\xi_3^{p^n}$$

if $\alpha_{22}, \alpha_{24}, \alpha_{42}, \alpha_{44}$ belong to the $GF[p^{2n}]$, so that T belongs to $H(4, p^{2n})$. The totality of the substitutions T forms a group G holoedrically isomorphic with the general binary linear group $GLH(2, p^{2n})$. Among the substitutions T occur the simple ones of the form

$$T_1: \begin{array}{l} \xi_1' = A^{-p^n}\xi_1, \quad \xi_3' = -B^{p^n}\xi_1 + A^{p^n}\xi_3, \\ \xi_2' = A\xi_2 + B\xi_4, \quad \xi_4' = A^{-1}\xi_4, \end{array}$$

where A and B are arbitrary marks of the $GF[p^{2n}]$ such that $A \neq 0$.

We proceed to determine every hyperabelian substitution

$$S: \quad \xi_i' = \sum_{j=1}^{4} \alpha_{ij}\xi_j \quad\quad (i = 1, \ldots, 4)$$

which transforms the subgroup G into itself. The product $S^{-1}T_1S$,

$$\xi'_i = (\alpha_{i1} A^{-p^n} - \alpha_{i3} B^{p^n}) \sum_{j=1,2} (\alpha^{p^n}_{2j2} \xi_{2j-1} - \alpha^{p^n}_{2j-1\,2} \xi_{2j})$$

$$+ \alpha_{i2} A \sum_j (-\alpha^{p^n}_{2j1} \xi_{2j-1} + \alpha^{p^n}_{2j-1\,1} \xi_{2j}) + \alpha_{i3} A^{p^n} \sum_j (\alpha^{p^n}_{2j4} \xi_{2j-1} - \alpha^{p^n}_{2j-1\,4} \xi_{2j})$$

$$+ (\alpha_{i2} B + \alpha_{i4} A^{-1}) \sum_j (-\alpha^{p^n}_{2j3} \xi_{2j-1} + \alpha^{p^n}_{2j-1\,3} \xi_{2j}),$$

must belong to G. Hence the coefficient of ξ_{2j-1} must vanish if i be even and that of ξ_{2j} if i be odd. Taking first $B = 0$, we find, after dropping the common factor $(-1)^k$,

$$\alpha_{i1} \alpha^{p^n}_{k2} A^{-p^n} - \alpha_{i2} \alpha^{p^n}_{k1} A + \alpha_{i3} \alpha^{p^n}_{k4} A^{p^n} - \alpha_{i4} \alpha^{p^n}_{k3} A^{-1} = 0,$$

where i and k are both even or both odd.

If $p^n > 2$, this leads to an equation in A of degree $2p^n < p^{2n} - 1$. Being true for every $A \neq 0$, it is therefore an identity, so that

110) $\alpha_{i1} \alpha_{k2} = 0$, $\alpha_{i3} \alpha_{k4} = 0$ (i, k both even or both odd).

Taking next the terms in B, which can have two values $\neq 0$, we find

111) $\alpha_{i3} \alpha_{k2} = 0$ (i, k both even or both odd).

Similarly, if S transform the following substitution of the form T,

$$T_2: \quad \begin{aligned} \xi'_1 &= D^{p^n} \xi_1 - C^{p^n} \xi_3, & \xi'_3 &= D^{-p^n} \xi_3, \\ \xi'_2 &= D^{-1} \xi_2, & \xi'_4 &= C \xi_2 + D \xi_4, \end{aligned}$$

into a substitution of G, we find from the terms in C that

112) $\alpha_{i1} \alpha_{k4} = 0$ (i, k both even or both odd).

If any $\alpha_{ij} \neq 0$, i and j being both even or both odd, the substitution S reduces to the form 109) and must therefore belong to G. In fact, the relations 110), 111) and 112), holding if $p^n > 2$, may be combined as follows:

113) $\alpha_{2i-1\,2l-1} \alpha_{2k-1\,2\lambda} = 0$, $\alpha_{2i\,2l-1} \alpha_{2k\,2\lambda} = 0$ (i, l, k, $\lambda = 1, 2$).

Hence, if $\alpha_{2i_1-1\,2l_1-1} \neq 0$, we get $\alpha_{2k-1\,2\lambda} = 0$ (k, $\lambda = 1, 2$). Then, for fixed λ, $\alpha_{2k\,2\lambda}$ is not zero for both $k = 1$ and $k = 2$, since otherwise all the coefficients in the $2\lambda^{\text{th}}$ column would be zero and therefore the determinant of S would vanish. It follows therefore from the second set of relations 113) that $\alpha_{2i\,2l-1} = 0$ (i, $l = 1, 2$). Hence S has the form 109). Similarly, the hypothesis $\alpha_{2k_1\,2\lambda_1} \neq 0$ requires, successively,

$$\alpha_{2i\,2l-1} = 0 \quad (i, l = 1, 2); \quad \alpha_{2k-1\,2\lambda} = 0 \quad (k, \lambda = 1, 2).$$

124 CHAPTER IV.

If every $a_{ij} = 0$, when i and j are both even or both odd, for $p^n > 2$, S reduces at once to the form

$$\xi'_1 = a_{12}\xi_2 + a_{14}\xi_4, \quad \xi'_2 = a_{21}\xi_1 + a_{23}\xi_3,$$
$$\xi'_3 = a_{32}\xi_2 + a_{34}\xi_4, \quad \xi'_4 = a_{41}\xi_1 + a_{43}\xi_3.$$

This is of the form Vg, where g is of the form 109) and V denotes the hyperabelian substitution not in G,

$$V: \quad \xi'_1 = -\xi_2, \quad \xi'_2 = \xi_1, \quad \xi'_3 = -\xi_4, \quad \xi'_4 = \xi_3.$$

The theorem stated below has thus been proven for $p^n > 2$.

For $p^n \gtreqless 2$, we consider the reciprocal of S and find the conditions corresponding to 111) and 112) that S^{-1} shall transform T_1 and T_2 into substitutions belonging to G, viz.,

114) $\qquad a_{1i}a_{4k} = 0, \quad a_{2i}a_{3k} = 0 \quad (i, k \text{ both even or both odd}).$

By 111), 112), 114), S must be of the form g or Vg, g being of the type 109). To illustrate the method of proof, let $a_{13} \neq 0$. Then $a_{41} = a_{43} = 0$ by 114). Since a_{42} and a_{44} can therefore not both vanish, $a_{12} = a_{14} = 0$ by 114). Likewise from 111) $a_{12} = a_{32} = 0$, $a_{23} = a_{43} = 0$. The hyperabelian condition involving the coefficients of the first and third rows then gives $a_{13}a_{34}^{p^n} = 0$, whence $a_{34} = 0$. Then a_{31} and a_{33} can not both vanish, so that $a_{21} = 0$ by 114). Hence S has the form 109).

The order of G is $(p^{4n} - 1)(p^{4n} - p^{2n})$ by § 99. The order of $H(4, p^{2n})$ is $(p^{4n} - 1)p^{3n}(p^{3n} + 1)p^{2n}(p^{2n} - 1)p^n(p^n + 1)$ by § 132.

Theorem. — *The quaternary hyperabelian substitutions T with coefficients in the $GF[p^{2n}]$ form a group G holoedrically isomorphic with $GLH(2, p^{2n})$. The only substitutions of $H(4, p^{2n})$ which transform the subgroup G into itself are of the form T or VT. $H(4, p^n)$ contains exactly $N = \frac{1}{2}(p^{3n} + 1)p^{3n}(p^n + 1)p^n$ subgroups conjugate with G.*

135. Consider the subgroup H' formed of the substitutions of $H(4, p^{2n})$ of determinant unity. By § 132, its index is $p^n + 1$. The determinant of the substitution T is seen to equal Δ^{-p^n+1}. Those substitutions T in the $GF[p^{2n}]$ whose determinant is unity form a group G' of order $(p^{4n} - 1)p^{2n}(p^n - 1)$. Since T_1 and T_2 are of determinant unity, the proof in § 134 leads to the following theorem:

Within the group H' of quaternary hyperabelian substitutions in the $GF[p^{2n}]$ of determinant unity, the subgroup G' of the substitutions T of determinant unity forms one of a complete set of N conjugate subgroups, each being holoedrically isomorphic with the group of binary linear substitutions in the $GF[p^{2n}]$ with determinant in the $GF]p^n]$. The only substitutions of H' which transform G' into itself are the substitutions g' of G' and the products Vg'.

136. The substitutions T for which $\Delta = 1$ form a group G_1 holoedrically isomorphic with the group of binary linear homogeneous substitutions of determinant unity in the $GF[p^{2n}]$. Since G_1 contains T_1 and T_2, it follows from § 134 that g' and Vg' (g' in G') are the only substitutions of H' which transform G_1 into itself. Hence G_1 is one of a complete set of N conjugate subgroups of H'.

For $p = 2$, H' is the simple group $HA(4, 2^{2n})$ and G_1 is the simple group $LF(2, 2^{2n})$. For $p > 2$, we pass from H' to the simple quotient-group $HA(4, p^{2n})$ by making the substitutions $T_\varkappa$ 106) correspond to the identity. In particular, T_{-1} corresponds to the identity, so that G_1 becomes $LF(2, p^{2n})$. The only $T_\varkappa$ belonging to G_1 are T_{-1} and the identity. We have therefore proven the

Theorem. — *The simple group $HA(4, p^{2n})$ contains a complete set of $\frac{1}{2}(p^{3n}+1)p^{3n}(p^n+1)p^n$ simple conjugate subgroups $LF(2, p^{2n})$.*

137. Theorem. — *The group of hyperabelian substitutions S of determinant unity on 2 indices with coefficients in the $GF[p^{2n}]$ is identical with the group of binary linear substitutions of determinant unity with coefficients in the $GF[p^n]$.*

For $m = 1$, the conditions 98) and 99) that S shall be hyperabelian are

$$\alpha_{11}\alpha_{22}^{p^n} - \alpha_{21}\alpha_{12}^{p^n} = 1, \quad \alpha_{11}\alpha_{21}^{p^n} - \alpha_{21}\alpha_{11}^{p^n} = 0, \quad \alpha_{12}\alpha_{22}^{p^n} - \alpha_{22}\alpha_{12}^{p^n} = 0,$$
$$\alpha_{11}^{p^n}\alpha_{22} - \alpha_{12}^{p^n}\alpha_{21} = 1, \quad \alpha_{11}^{p^n}\alpha_{12} - \alpha_{12}^{p^n}\alpha_{11} = 0, \quad \alpha_{21}^{p^n}\alpha_{22} - \alpha_{22}^{p^n}\alpha_{21} = 0.$$

Hence the products $\alpha_{11}\alpha_{22}^{p^n}$, $\alpha_{11}\alpha_{21}^{p^n}$, $\alpha_{11}\alpha_{12}^{p^n}$ belong to the $GF[p^n]$, being equal to their own $(p^n)^{\text{th}}$ powers. Hence if $\alpha_{11} \neq 0$, the ratios of $\alpha_{22}, \alpha_{21}, \alpha_{12}$ to α_{11} all belong to the $GF[p^n]$. Similarly, the products $\alpha_{22}\alpha_{11}^{p^n}$, $\alpha_{22}\alpha_{12}^{p^n}$, $\alpha_{22}\alpha_{21}^{p^n}$ all belong to the $GF[p^n]$ and therefore, if $\alpha_{22} \neq 0$, the ratios of $\alpha_{21}, \alpha_{12}, \alpha_{11}$ to α_{22} all belong to the $GF[p^n]$. Finally, if $\alpha_{11} = \alpha_{22} = 0$, we have $\alpha_{21}\alpha_{12}^{p^n} = -1$, so that the ratio of α_{21} to α_{12} belongs to the $GF[p^n]$. In every case, S has the form

$$\xi_1' = \alpha(a_{11}\xi_1 + a_{12}\xi_2), \quad \xi_2' = \alpha(a_{21}\xi_1 + a_{22}\xi_2)$$

where the a_{ij} belong to the $GF[p^n]$. Since it is to be hyperabelian and since it is to have determinant unity, we have the respective relations

$$\alpha^{p^n+1}(a_{11}a_{22} - a_{12}a_{21}) = 1, \quad \alpha^2(a_{11}a_{22} - a_{12}a_{21}) = 1.$$

Hence, by division, $\alpha^{p^n-1} = 1$, or α belongs to the $GF[p^n]$.

126   CHAPTER V.

Corollary I. — $HA(2, p^{2n}) \equiv A(2, p^n) \equiv LF(2, p^n)$.

Corollary II. — The group of all binary hyperabelian substitutions in the $GF[p^{2n}]$ taken fractionally is the group of all linear fractional substitutions in the $GF[p^n]$.

138. In virtue of the transformation of indices,
$$\eta_1 \equiv J\xi_1 + \xi_2, \quad \eta_2 \equiv \varrho J^{p^n}\xi_1 + \varrho\xi_2,$$
where J and ϱ are primitive roots of the respective equations
$$J^{p^n+1} = 1, \quad \varrho^{p^n+1} = -1,$$
we have the following identity
$$\eta_1^{p^n+1} + \eta_2^{p^n+1} \equiv (J - J^{p^n})(\xi_1 \xi_2^{p^n} - \xi_2 \xi_1^{p^n}).$$

Hence the hyperabelian group on $2m$ indices with coefficients in the $GF[p^{2n}]$ is holoedrically isomorphic with the group on $2m$ indices in the $GF[p^{2n}]$ defined by the invariant
$$\eta_1^{p^n+1} + \eta_2^{p^n+1} + \cdots + \eta_{2m}^{p^n+1}.$$

CHAPTER V.

THE HYPERORTHOGONAL AND RELATED LINEAR GROUPS.[1]

139. We first investigate the linear homogeneous group in the $GF[p^n]$ defined by an absolute invariant of the general type
$$\Phi_r \equiv \lambda_1 \xi_1^r + \lambda_2 \xi_2^r + \cdots + \lambda_m \xi_m^r,$$
where each λ is a mark $\neq 0$ of the $GF[p^n]$.

If $r = p^\varrho r_1$, we have in the $GF[p^n]$ the identity
$$\Phi_r \equiv \sum_{i=1}^m \lambda_i \xi_i^r = \left(\sum_{i=1}^m \lambda_i' \xi_i^{r_1}\right)^{p^\varrho} \qquad [\lambda_i' \equiv \lambda_i^{p^{-\varrho}}].$$

Hence a substitution which leaves Φ_r absolutely invariant will at most multiply the function
$$\Phi'_{r_1} \equiv \sum_{i=1}^m \lambda_i' \xi_i^{r_1}$$
by a mark η which satisfies the equations

---
1) Dickson, *Mathematische Annalen*, vol. 52, pp. 561—581.

$$\eta^{p^q}=1, \quad \eta^{p^n-1}=1,$$

from which $\eta=1$. We may therefore limit our discussion to the case in which r is prime to p.

In order that the linear substitution on $m>1$ indices

$$S: \qquad \xi_i' = \sum_{j=1}^{m} \alpha_{ij}\xi_j \qquad (i=1,\ldots,m)$$

shall leave Φ_r formally invariant, the following conditions upon its coefficients must be satisfied[1]):

115) $$\sum_{i=1}^{m} \lambda_i \alpha_{ij}^r = \lambda_j \qquad (j=1,\ldots,m)$$

116) $$\frac{r!}{r_1!\,r_2!\ldots r_s!}\sum_{i=1}^{m} \lambda_i \alpha_{ij_1}^{r_1} \alpha_{ij_2}^{r_2}\ldots \alpha_{ij_s}^{r_s} = 0,$$

holding for every partition of r into s integral parts

$$r = r_1 + r_2 + \cdots + r_s, \qquad m \geqq s > 1,$$

while for each partition $j_1, j_2 \ldots, j_s$ may take every combination of s distinct integers chosen from $1, 2, \ldots, m$.

If r be not divisible by p, the inverse of S is

$$S^{-1}: \qquad \xi_k' = \frac{1}{\lambda_k}\sum_{i=1}^{m} \lambda_i \alpha_{ik}^{r-1}\xi_i \qquad (k=1,\ldots,m).$$

Indeed, the product SS^{-1} replaces ξ_k by

$$\frac{1}{\lambda_k}\sum_{j=1}^{m}\left(\sum_{i=1}^{m}\lambda_i\alpha_{ik}^{r-1}\alpha_{ij}\right)\xi_j = \xi_k,$$

upon applying 115) and 116) for $r_1 = r-1$, $r_2 = 1$.

140. Theorem. — *If $r>2$, if r be not a multiple of p, and if $r-1$ be not a power of p, the only linear homogeneous substitutions in the $GF[p^n]$ which leave Φ_r invariant are those which merely permute the terms $\lambda_1\xi_1^r, \ldots, \lambda_m\xi_m^r$ amongst themselves.*

Consider for $r>2$ the following equations of the set 116), in which j_1 and j_2 denote two arbitrarily fixed distinct integers $\leqq m$:

---

1) If, as in § 97, the indices are to belong to the $GF[p^n]$ so that the invariance of Φ_r is numerical and not formal, we must take $r < p^n$ in order that our results shall still hold true. Cf. § 152.

128   CHAPTER V.

$$r\sum_{i=1}^{m}(\lambda_i\alpha_{ij_1}^{r-2}\alpha_{ij_2})\alpha_{ij_1}=0,$$

$$\frac{1}{2}r(r-1)\sum_{i=1}^{m}(\lambda_i\alpha_{ij_1}^{r-2}\alpha_{ij_2})\alpha_{ij_2}=0,$$

$$r(r-1)\sum_{i=1}^{m}(\lambda_i\alpha_{ij_1}^{r-2}\alpha_{ij_2})\alpha_{ij_3}=0 \quad (j_3=1,\ldots,m;\ j_3\neq j_1,j_2).$$

If neither r nor $r-1$ is divisible by p, we may drop the numerical factors from these m equations.[1]) But

$$|\alpha_{ij}|\neq 0 \qquad (i,j=1,\ldots,m)$$

being the determinant of S. Hence we have

$$\lambda_i\alpha_{ij_1}^{r-2}\alpha_{ij_2}=0 \qquad (i,j_1,j_2=1,\ldots,m;\ j_1\neq j_2).$$

Hence only one element of each row of the matrix for S is not zero. The determinant of S being not zero, the non-vanishing coefficients lie in different columns as well as in different rows. Hence S merely permutes the terms of the sum Φ_r.

Suppose next that $r-1$ is divisible by p and set

$$r-1=gp^s \qquad\qquad (s\geqq 1),$$

where g is not divisible by p. We now consider the case $g>1$. We make use of the following equations of the set 116):

$$\frac{(gp^s+1)!}{[(g-1)p^s+1]!\,p^s!}\sum_{i=1}^{m}\left(\lambda_i\alpha_{ij_1}^{(g-1)p^s}\alpha_{ij_2}^{p^s}\right)\alpha_{ij_1}=0,$$

$$\frac{(gp^s+1)!}{[(g-1)p^s]!\,(p^s+1)!}\sum_{i=1}^{m}\left(\lambda_i\alpha_{ij_1}^{(g-1)p^s}\alpha_{ij_2}^{p^s}\right)\alpha_{ij_2}=0,$$

$$\frac{(gp^s+1)!}{[(g-1)p^s]!\,p^s!\,1!}\sum_{i=1}^{m}\left(\lambda_i\alpha_{ij_1}^{(g-1)p^s}\alpha_{ij_2}^{p^s}\right)\alpha_{ij_3}=0 \quad (j_3=1,\ldots,m;\ j_3\neq j_1,j_2)$$

of which the first two alone occur when $m=2$. We may verify that the numerical factors are not divisible by p.[2]) Then, since $|\alpha_{ij}|\neq 0$

$$\lambda_i\alpha_{ij_1}^{(g-1)p^s}\alpha_{ij_2}^{p^s}=0 \qquad (i,j_1,j_2=1,\ldots,m;\ j_1\neq j_2).$$

It follows as before that S at most permutes the terms of Φ_r.

---

1) If $m=2$, only the first two equations occur. The same conclusion follows in this case that was derived for $m>2$.

2) This result follows by inspection from a general theorem on the residue of a multinomial coefficient taken modulo p given in the author's Dissertation, *Annals of Mathematics*, 1897, § 14, p. 75.

THE HYPERORTHOGONAL AND RELATED LINEAR GROUPS. 129

141. If r is not divisible by p and if $r \not\equiv p^s + 1$, the structure of the largest linear homogeneous group leaving Φ_r ($r > 2$) invariant is now evident. Indeed, the group has as a self-conjugate subgroup the commutative group of the substitutions

$$\xi_i' = \alpha_{ii} \xi_i \qquad (i = 1, \ldots, m) \qquad [\alpha_{ii}^r = 1],$$

the quotient-group being the symmetric group on the m letters ξ_i.

142. Theorem. — *The structure of the linear group in the $GF[p^n]$ which is defined by the absolute invariant Φ_r, $r \equiv p^s + 1 > 2$, results immediately from the structures of the groups in the $GF[p^{2s}]$ defined by absolute invariants of the type*

$$\Phi = \xi_1^{p^s+1} + \xi_2^{p^s+1} + \cdots + \xi_m^{p^s+1}.$$

For the case $r = p^s + 1$, the conditions that S shall leave Φ_r invariant may be derived as special cases of 115) and 116), but are given by inspection from the identity,

$$\sum_{i=1}^{m} \lambda_i \left(\sum_{j=1}^{m} \alpha_{ij} \xi_j \right)^{p^s+1} \equiv \sum_{i=1}^{m} \lambda_i \left(\sum_{i=1}^{m} \alpha_{ij}^{p^s} \xi_j^{p^s} \right) \left(\sum_{k=1}^{m} \alpha_{ik} \xi_k \right) = \sum_{i=1}^{m} \lambda_i \xi_i^{p^s+1}.$$

By either method, the conditions in question are seen to be:

117) $$\sum_{i=1}^{m} \lambda_i \alpha_{ij}^{p^s+1} = \lambda_j \qquad (j = 1, \ldots, m),$$

118) $$\sum_{i=1}^{m} \lambda_i \alpha_{ij}^{p^s} \alpha_{ik} = 0 \qquad (j, k = 1, \ldots, m; j \neq k).$$

By § 139, the inverse of S has the form

$$S^{-1}: \qquad \xi_i' = \sum_{j=1}^{m} \frac{\lambda_j}{\lambda_i} \alpha_{ji}^{p^s} \xi_j \qquad (i = 1, \ldots, m).$$

By the same rule, the inverse of the latter substitution is

$$\xi_i' = \sum_{j=1}^{m} \frac{\lambda_j}{\lambda_i} \left(\frac{\lambda_i}{\lambda_j} \alpha_{ij}^{p^s} \right)^{p^s} \xi_j \qquad (i = 1, \ldots, m).$$

Hence this substitution must be identical with S. Hence

119) $$\left(\frac{\lambda_i}{\lambda_j} \right)^{p^s - 1} \alpha_{ij}^{p^{2s}} = \alpha_{ij} \qquad (i, j = 1, \ldots, m).$$

The determinant of S^{-1} is

$$\left| \frac{\lambda_j}{\lambda_i} \alpha_{ji}^{p^s} \right| = \left| \alpha_{ji}^{p^s} \right| = \left| \alpha_{ji} \right|^{p^s} \qquad (i, j = 1, \ldots, m).$$

130   CHAPTER V.

Hence, since the product $SS^{-1}=1$ has the determinant unity, we have

120) $\qquad |\alpha_{ij}|^{p^s+1}=1.$

From the form of the reciprocal S^{-1}, it follows that

121) $\qquad \dfrac{\lambda_j}{\lambda_i}\alpha_{ji}^{p^s}=\dfrac{A_{ji}}{D} \qquad (i,j=1,\ldots,m)$

where A_{ji} denotes the adjoint of α_{ji} in the determinant

$$D \equiv |\alpha_{ij}| \qquad (i,j=1,\ldots,m).$$

The value of n, defining the $GF[p^n]$ to which the coefficients of our substitution S and the quantities λ_i were assumed to belong, has played no part in the above formulae. We proceed to prove that our problem can be reduced to a series of similar ones in which $n=2s$. Consider the $GF[p^{2ns}]$, which includes the $GF[p^n]$ and the $GF[p^{2s}]$. Raising 119) to the power $\dfrac{p^{2ns}-1}{p^{2s}-1}$, we have

$$\left(\dfrac{\lambda_i}{\lambda_j}\right)^{\tfrac{p^{2ns}-1}{p^s+1}}=1,$$

if $\alpha_{ij} \neq 0$. Hence $\dfrac{\lambda_i}{\lambda_j}$ would be the power p^s+1 of some quantity μ in the $GF[p^{2ns}]$. The substitution T_j

$$\xi'_k=\xi_k \quad (k=1,\ldots,m;\ k \neq j), \qquad \xi'_j=\mu\xi_j$$

transforms φ_r into

$$\sum_{k=1,\ k\neq j}^{k=1\ldots m} \lambda_k \xi_k^{p^s+1}+\lambda_j(\mu\xi_j)^{p^s+1} \equiv \sum_{k=1}^{m} \lambda'_k \xi_k^{p^s+1}$$

in which the coefficients λ'_i and λ'_j are equal. Evidently T_j transforms S into a substitution with coefficients in the $GF[p^{2ns}]$.

Suppose that the coefficients $\alpha_{12}, \alpha_{13}, \ldots, \alpha_{1m_1}$ do not vanish, while $\alpha_{1j}=0$ for $j>m_1$, in all of the substitutions leaving φ_r invariant. Then the group is isomorphic with a group of substitutions in the $GF[p^{2ns}]$ leaving invariant

$$\varphi'_r=\sum_{k=1}^{m}\lambda'_k\xi_k^{p^s+1} \qquad (\lambda'_1=\lambda'_2=\cdots=\lambda'_{m_1}).$$

In the latter substitutions the coefficients $\alpha_{1j}\ (j>m_1)$ are all zero. If, among the coefficients $\alpha_{2j}\ (j>m_1)$, any one as $\alpha_{2j_1} \neq 0$, we transform the invariant φ'_r by T_{j_1}, giving the function

$$\sum_{k=1}^{m}\lambda''_k\xi_k^{p^s+1} \qquad (\lambda''_1=\lambda''_2=\cdots=\lambda''_{m_1}=\lambda''_{j_1}).$$

But this function is invariant under the transposition $(\xi_1 \xi_{j_1})$ and hence φ_r must have been invariant under a substitution in which $\alpha_{1j_1} \neq 0$. It follows that
$$\alpha_{ij} = 0 \quad (i = 1, \ldots, m_1;\ j = m_1 + 1, \ldots, m)$$
in every substitution leaving φ_r invariant. Considering the form of the reciprocal, we have
$$\alpha_{ji} = 0 \quad (i = 1, \ldots, m_1;\ j = m_1 + 1, \ldots, m).$$
Hence every substitution leaving φ_r invariant is the product of two commutative substitutions, the one affecting the indices $\xi_1, \ldots, \xi_{m_1}$ only and leaving invariant
$$\xi_1^{p^s+1} + \cdots + \xi_{m_1}^{p^s+1}$$
and the other affecting only $\xi_{m_1+1}, \ldots, \xi_m$ and leaving invariant
$$\sum_{i=m_1+1}^{m} \lambda_i' \xi_i^{p^s+1}.$$

Proceeding with the latter substitutions in the same manner, it follows that the structure of the group in the $GF[p^n]$ leaving Φ_r invariant results immediately from the structures of various linear groups in the $GF[p^{2ns}]$ defined by invariants of the type Φ. But the relations 119) for substitutions of the latter groups become
$$\alpha_{ij}^{p^{2s}} = \alpha_{ij} \qquad (i, j = 1, \ldots, m).$$
Hence there is no limitation imposed in assuming that the field to which the substitutions belong is the $GF[p^{2s}]$.

143. We designate by $G_{m,p,s}$ the group of all linear homogeneous m-ary substitutions in the $GF[p^{2s}]$ which leave Φ invariant. For $p > 2$, those of its substitutions whose coefficients belong to the $GF[p^s]$ constitute the first orthogonal group[1]) in the $GF[p^s]$ on m indices. Indeed, relations 117) and 118), for $\lambda_i = 1$, then become
$$\sum_{i=1}^{m} \alpha_{ij}^2 = 1, \quad \sum_{i=1}^{m} \alpha_{ij} \alpha_{ik} = 0 \quad (j, k = 1, \ldots, m;\ j \neq k).$$

The group $G_{m,p,s}$, having the orthogonal group as a subgroup, will be called the *hyperorthogonal group* in the $GF[p^{2s}]$ on m indices. We proceed to determine its structure, treating first the case $m = 2$.

144. Theorem. — *If $p^s > 3$, the group of the substitutions of $G_{2,p,s}$ of determinant unity has a maximal invariant subgroup of order 1 or 2 according as $p = 2$ or $p > 2$; the quotient-group is $LF(2, p^s)$.*

---

1) See Chapter VII, § 171. For $p = 2$, see Ex. 4 of § 210.

132   CHAPTER V.

For $m=2$, we have by 117) and 121), when $\lambda_1 = \lambda_2 = 1$,
$$\alpha_{22} = D\alpha_{11}^{p^s}, \quad \alpha_{21} = -D\alpha_{12}^{p^s}, \quad \alpha_{11}^{p^s+1} + \alpha_{12}^{p^s+1} = 1.$$

Inversely, every substitution satisfying these relations is seen to leave $\xi_1^{p^s+1} + \xi_2^{p^s+1}$ absolutely invariant. Every such substitution is the product of a substitution

122) $\quad \begin{cases} \xi_1' = \alpha_{11}\xi_1 + \alpha_{12}\xi_2 \\ \xi_2' = -\alpha_{12}^{p^s}\xi_1 + \alpha_{11}^{p^s}\xi_2 \end{cases} \quad \left(\alpha_{11}^{p^s+1} + \alpha_{12}^{p^s+1} = 1\right)$

by one of the p^s+1 distinct substitutions
$$\xi_1' = \xi_1, \quad \xi_2' = D\xi_2 \qquad (D^{p^s+1} = 1).$$

The number of distinct substitutions 122) is $(p^{2s}-1)p^s$. Indeed, for the p^s+1 values of α_{12} for which $\alpha_{12}^{p^s+1} = 1$, we must have $\alpha_{11} = 0$; while for each of the remaining $(p^{2s} - p^s - 1)$ values of α_{12} in the $GF[p^{2s}]$, there exist p^s+1 solutions in the field of
$$\alpha_{11}^{p^s+1} = 1 - \alpha_{12}^{p^s+1};$$
for, the second member belongs to the $GF[p^s]$ and is therefore the p^s+1 power of some mark in the $GF[p^{2s}]$. But
$$(p^s+1) + (p^s+1)(p^{2s}-p^s-1) = (p^s+1)(p^s-1)p^s.$$

The group of the substitutions 122) has an invariant subgroup of order 1 or 2, according as $p=2$ or $p>2$, generated by the substitution
$$C_1C_2: \qquad \xi_1' = -\xi_1, \quad \xi_2' = -\xi_2.$$

The quotient group (obtained concretely by taking the substitutions 122) fractionally) is, by §§ 137—138, simply isomorphic with the group of linear fractional substitutions of determinant unity in the $GF[p^s]$. By § 109, it is a simple group when $p^s > 3$.

Corollary. — Every binary hyperorthogonal substitution in the $GF[p^{2s}]$ taken fractionally may be given the form
$$\begin{pmatrix} A, & B \\ -B^{p^s}, & A^{p^s} \end{pmatrix}$$
of determinant a mark of the $GF[p^s]$, where A, B belong to the $GF[p^{2s}]$.

Indeed, since $D^{p^s+1} = 1$, we may set $D = R^{p^s-1}$, R belonging to the $GF[p^{2s}]$. The fractional binary hyperabelian substitution becomes
$$\begin{pmatrix} \alpha_{11}, & \alpha_{12} \\ -D\alpha_{12}^{p^s}, & D\alpha_{11}^{p^s} \end{pmatrix} \equiv \begin{pmatrix} R\alpha_{11}, & R\alpha_{12} \\ -R^{p^s}\alpha_{12}^{p^s}, & R^{p^s}\alpha_{11}^{p^s} \end{pmatrix}.$$

THE HYPERORTHOGONAL AND RELATED LINEAR GROUPS. 133

The group may be transformed into the group of all linear fractional substitutions in the $GF[p^s]$ (see § 138, § 137, corollary II).

145. For m general, let S be an arbitrary substitution of $G_{m,p,s}$,

$$S: \qquad \xi'_i = \sum_{j=1}^{m} \alpha_{ij} \xi_j \qquad (i = 1, \ldots, m).$$

By § 139, its inverse is obtained by replacing α_{ij} by $\alpha_{ji}^{p^s}$. Hence the relations 117) and 118), for $\lambda_i = 1$, when written for the inverse S^{-1}, give the equivalent set of conditions for the invariance of $\sum_{i=1}^{m} \xi_i^{p^s+1}$:

123) $$\sum_{i=1}^{m} \alpha_{ji}^{p^s+1} = 1 \qquad (j = 1, \ldots, m),$$

124) $$\sum_{i=1}^{m} \alpha_{ji} \alpha_{ki}^{p^s} = 0 \qquad (j, k = 1, \ldots, m; j \neq k).$$

By § 146, the number of distinct linear functions

$$f_1 \equiv \sum_{j=1}^{m} \alpha_{1j} \xi_j$$

by which the substitutions of $G_{m,p,s}$ can replace ξ_1 is the number $\mathrm{P}_{m,p,s}$ of distinct sets of solutions in the $GF[p^{2s}]$ of the equation

125) $$\sum_{j=1}^{m} \alpha_{1j}^{p^s+1} = 1.$$

Let T be a substitution of the group which replaces ξ_1 by a definite function f_1. Then, if $\Sigma, \Sigma', \ldots$ denote all of the $Q_{m,p,s}$ substitutions of the group which leave ξ_1 fixed, the products $T\Sigma, T\Sigma', \ldots$ and no other substitutions of the group will replace ξ_1 by f_1. Hence the order $\Omega_{m,p,s}$ of the group $G_{m,p,s}$ is

$$\Omega_{m,p,s} = Q_{m,p,s} \mathrm{P}_{m,p,s}.$$

But the substitutions $\Sigma, \Sigma', \ldots$, have

$$\alpha_{11} = 1, \quad \alpha_{1i} = 0 \qquad (i = 2, \ldots, m).$$

Hence by 124), for $j = 1$, we have

$$\alpha_{k1}^{p^s} = 0 \qquad (k = 2, \ldots, m)$$

Hence $\Sigma, \Sigma', \ldots$ are substitutions of the group $G_{m-1,p,s}$ on the indices $\xi_2, \ldots, \xi_m$, so that $Q_{m,p,s} = \Omega_{m-1,p,s}$. Hence, since

134   CHAPTER V.

$$\Omega_{1,p,s} = p^s + 1 = \mathsf{P}_{1,p,s}$$

is the number of substitutions affecting one index only, we have

$$\Omega_{m,p,s} = \mathsf{P}_{m,p,s}\mathsf{P}_{m-1,p,s}\ldots\mathsf{P}_{1,p,s}.$$

To evaluate $\mathsf{P}_{n,p,s}$, the number of sets of solutions of

$$\sum_{i=1}^{n}\eta_i^{p^s+1} = 1,$$

we note that, for the $\mathsf{P}_{n-1,p,s}$ sets of values of $\eta_2, \ldots, \eta_n$ which make $\sum_{i=2}^{n}\eta_i^{p^s+1} = 1$, the corresponding value of η_1 is zero; while, for each of the $p^{2s(n-1)} - \mathsf{P}_{n-1,p,s}$ sets of values in the $GF[p^{2s}]$ for which that sum $\neq 1$, there exist $p^s + 1$ values in the $GF[p^{2s}]$ for η_1. Indeed,

$$1 - \sum_{i=2}^{n}\eta_i^{p^s+1}$$

belongs to the $GF[p^s]$ and is therefore the power $p^s + 1$ of a mark in the $GF[p^{2s}]$. Hence we have

$$\mathsf{P}_{n,p,s} = \mathsf{P}_{n-1,p,s} + (p^s + 1)(p^{2s(n-1)} - \mathsf{P}_{n-1,p,s}).$$

Since $\mathsf{P}_{1,p,s} = p^s + 1$, we find by mathematical induction that

$$\mathsf{P}_{n,p,s} = p^{s(2n-1)} - (-1)^n p^{s(n-1)}.$$

For another proof of this result, we consider only the case $p > 2$. Then if ν be a not-square in the $GF[p^s]$, the $GF[p^{2s}]$ may be defined by means of the irreducible equation

$$I^2 = \nu \qquad\qquad (I^{p^s} = -I).$$

Setting $\qquad\qquad \eta_i \equiv \alpha_i + \beta_i I \qquad\qquad (i = 1, \ldots, n)$

we have $\qquad\qquad \eta_i^{p^s} \equiv \alpha_i - \beta_i I.$

Hence

$$\sum_{i=1}^{n}\eta_i^{p^s+1} \equiv \sum_{i=1}^{n}(\alpha_i^2 - \nu\beta_i^2) = 1.$$

By § 65, this quadratic equation has $p^{s(2n-1)} - (-1)^n p^{s(n-1)}$ sets of solutions $\alpha_1, \ldots, \alpha_n, \beta_1, \ldots, \beta_n$ in the $GF[p^s]$. Hence $\Omega_{m,p,s}$ equals

$$[p^{sm} - (-1)^m]p^{s(m-1)}[p^{s(m-1)} - (-1)^{m-1}]p^{s(m-2)}\ldots[p^{2s}-1]p^s[p^s+1].$$

146. Theorem. — *If $\alpha_{11}, \alpha_{12}, \ldots, \alpha_{1m}$ be any system of solutions in the $GF[p^{2s}]$ of the equation* 125), *there exists a substitution S in the group $G_{m,p,s}$ which replaces ξ_1 by*

THE HYPERORTHOGONAL AND RELATED LINEAR GROUPS. 135

$$f_1 \equiv \sum_{j=1}^{m} \alpha_{1j} \xi_j$$

and which is generated by the following substitutions [in which only the indices altered are written]:

$T_{i,\tau}:$ $\qquad \xi'_i = \tau \xi_i \qquad (\tau^{p^s+1} = 1),$

$O^{\alpha,\beta}_{i,j}:$ $\begin{cases} \xi'_i = \alpha \xi_i + \beta \xi_j \\ \xi'_j = - \beta^{p^s} \xi_i + \alpha^{p^s} \xi_j \end{cases} \qquad (\alpha^{p^s+1} + \beta^{p^s+1} = 1),$

an additional generator being necessary if $p^s = 2$, $m \geq 3$, viz.,

$W:$ $\begin{cases} \xi'_1 = \xi_1 + \xi_2 + \xi_3 \\ \xi'_2 = \xi_1 + I\,\xi_2 + I^2 \xi_3 \\ \xi'_3 = \xi_1 + I^2 \xi_2 + I\,\xi_3 \end{cases} \qquad [I^2 \equiv I+1 \pmod 2].$

If $m = 1$, we may take $S = T_{1,\alpha_{11}}$. If $m = 2$, we take
$$S = O^{\alpha_{11}, \alpha_{12}}_{1,2}.$$
If $m > 2$, we prove the proposition by induction. Suppose first that the $\alpha_{1i}^{p^s+1}$ ($i = 1, \ldots, m$) are not all unity, for example,
$$1 - \alpha_{12}^{p^s+1} \neq 0.$$
The left member belongs to the $GF[p^s]$. Hence we may write

126) $\qquad \alpha_{12}^{p^s+1} + \mu^{p^s+1} = 1,$

μ being a mark $\neq 0$ in the $GF[p^{2s}]$. The group therefore contains a substitution of the form $O^{\mu,\alpha_{12}}_{1,2}$. By 125) and 126), we have
$$\left(\frac{\alpha_{11}}{\mu}\right)^{p^s+1} + \left(\frac{\alpha_{13}}{\mu}\right)^{p^s+1} + \cdots + \left(\frac{\alpha_{1m}}{\mu}\right)^{p^s+1} = 1.$$
Assuming our theorem to be true for $m-1$ indices, the group contains a substitution S' replacing ξ_1 by
$$\frac{\alpha_{11}}{\mu}\xi_1 + \frac{\alpha_{13}}{\mu}\xi_3 + \cdots + \frac{\alpha_{1m}}{\mu}\xi_m.$$
Hence the product $S \equiv S'\, O^{\mu,\alpha_{12}}_{1,2}$ will replace ξ_1 by f_1.

Suppose on the contrary that
$$\alpha_{1i}^{p^s+1} = 1 \qquad (i = 1, 2, \ldots, m).$$
If the group contains a substitution S_1 replacing ξ_1 by $\xi_1 + \xi_2 + \cdots + \xi_m$, the product
$$S \equiv T_{1,\alpha_{11}} T_{2,\alpha_{22}} \ldots T_{m,\alpha_{1m}} S_1$$
will replace ξ_1 by f_1. But the group will contain a substitution of the form S_1 if it contains $S_2 \equiv O^{\alpha,\beta}_{1,2} S_1$, which replaces ξ_1 by
$$(\alpha - \beta^{p^s})\xi_1 + (\beta + \alpha^{p^s})\xi_2 + \xi_3 + \cdots + \xi_m.$$

136   CHAPTER V.

If $p \neq 2$, we can take $\alpha = \beta^{p^s}$, since the condition

$$\alpha^{p^s+1} + \beta^{p^s+1} \equiv 2\beta^{p^s+1} = 1$$

can be satisfied by a mark β in the $GF[p^{2s}]$. In this case, $S_2 O_{2,1}^{0,1}$ replaces ξ_2 by the function

$$2\beta\xi_2 + \xi_3 + \cdots + \xi_m,$$

and therefore belongs to the group by our assumption on $m-1$ indices. If $p = 2$, $s > 1$, we can choose α and β among the sets of solutions in the $GF[p^{2s}]$ of

127) $\qquad \alpha^{p^s+1} + \beta^{p^s+1} = 1$

in such a manner that

$$(\alpha - \beta^{p^s})^{p^s+1} \equiv \alpha^{p^s+1} + \beta^{p^s+1} - \alpha\beta - \alpha^{p^s}\beta^{p^s} \neq 1.$$

Indeed, the condition is (since $p = 2$),

$$\alpha^{p^s}\beta^{p^s} \neq \alpha\beta.$$

Since $p^s > 2$, we may take for α a mark neither zero nor unity in the $GF[p^s]$ and then determine a solution β of 127) such that $\beta \neq \beta^{p^s}$. Then will $\alpha^{p^s}\beta^{p^s} \neq \alpha\beta$. To prove that such a choice for β is possible, we note first that

$$\alpha^{p^s} = \alpha, \quad \alpha^2 \neq \alpha; \quad \text{hence } \alpha^{p^s+1} \neq 1, \quad \beta \neq 0.$$

Further, if α', β' be one set of solutions of 127), then is also α', $\tau\beta'$, where τ is any root of

$$\tau^{p^s+1} = 1.$$

Not every root τ belongs to the $GF[p^s]$, and therefore not every solution β corresponding to a given α belongs to the $GF[p^s]$. Hence, if $p = 2$, $p^s > 2$, we may suppose that in the substitution S_2 the coefficient α_{11} is such that $\alpha_{11}^{p^s+1} \neq 1$, when the proposition follows as above.

For $p^s = 2$, an additional generator W, for example, is necessary since the only substitutions of the form $O_{1,2}^{\alpha,\beta}$ are the products

$$T_{1,\varrho}T_{2,\varrho^{-1}} \text{ and } (\xi_1\xi_2)T_{1,\varrho}T_{2,\varrho^{-1}} \qquad (\varrho^3 = 1).$$

Indeed, there exists in the $GF[2^2]$ only six sets of solutions of

$$\alpha^3 + \beta^3 = 1,$$

viz., $\alpha = \varrho$, $\beta = 0$ and $\alpha = 0$, $\beta = \varrho$, where $\varrho^3 = 1$. Hence the substitutions $T_{i,\tau}$ and $O_{i,j}^{\alpha,\beta}$ can not combine to give a substitution replacing ξ_1 by $\xi_1 + \xi_2 + \xi_3$, for example. It follows readily that the additional generator W is sufficient, together with the substitutions T and O, to generate the group $G_{m,2,1}$.

147. Lemma. — *If a substitution S of the group $G_{m,p,s}$ be commutative with $O_{r,t}^{\alpha,\beta}$, for certain values of α, then the following coefficients of S must be zero,*
$$\alpha_{rj}, \quad \alpha_{tj}, \quad \alpha_{jr}, \quad \alpha_{jt} \qquad (j=1,\ldots,m;\ j \neq r,t).$$

Among the conditions for the identity $S O_{r,t}^{\alpha,\beta} = O_{r,t}^{\alpha,\beta} S$ occur

$$\begin{cases} (\alpha-1)\alpha_{rj} + \beta\alpha_{tj} = 0, \\ -\beta^{p^s}\alpha_{rj} + (\alpha^{p^s}-1)\alpha_{tj} = 0, \end{cases} \qquad \begin{cases} (\alpha-1)\alpha_{jr} - \beta^{p^s}\alpha_{jt} = 0, \\ \beta\alpha_{jr} + (\alpha^{p^s}-1)\alpha_{jt} = 0 \end{cases}$$
$$(j=1,\ldots,m;\ j \neq r,t).$$

Hence the theorem follows if the determinant

$$(\alpha-1)(\alpha^{p^s}-1) + \beta^{p^s+1} \equiv 2 - \alpha - \alpha^{p^s} \neq 0.$$

The equation $2 - \alpha - \alpha^{p^s} = 0$ has p^s solutions in the $GF[p^{2s}]$; indeed,
$$\alpha^{p^{2s}} = (2-\alpha)^{p^s} = 2 - \alpha^{p^s} = \alpha.$$

But for α arbitrary there exists a mark β in the $GF[p^{2s}]$ such that
$$\alpha^{p^s+1} + \beta^{p^s+1} = 1.$$

Hence there are sets of solutions α, β for which the above determinant does not vanish, as well as sets for which it vanishes.

Note. Another statement of our result is that S breaks up into the product of a substitution affecting only ξ_r and ξ_t by a substitution affecting only ξ_j $(j=1,\ldots,m;\ j \neq r,t)$.

148. We proceed to determine the structure of the group $G_{m,p,s}$ of order $\Omega_{m,p,s}$. For $m=1$, the group is a commutative (cyclic) group of order p^s+1. For $m=2$, its structure was determined in § 144.

The substitutions of $G_{m,p,s}$ of determinant $D=1$ form an invariant subgroup $H_{m,p,s}$ of order $\Omega_{m,p,s}/(p^s+1)$. Indeed, we have shown that D must be a root of

120) $\qquad\qquad\qquad D^{p^s+1} = 1.$

Inversely, substitutions do exist in the group $G_{m,p,s}$ having as determinants every root of 120); for example, $T_{1,\tau}$ and its powers, where τ is a primitive root of 120). Hence the factors of composition of $G_{m,p,s}$ are those of $H_{m,p,s}$ together with the prime factors of p^s+1.

Supposing $m \geq 3$, let I be an invariant subgroup of $H_{m,p,s}$ containing a substitution

$S: \qquad \xi_i' = \sum_{j=1}^{m} \alpha_{ij}\xi_j \qquad (i=1,\ldots,m)$

not of the form

$T: \qquad \xi_i' = \tau\xi_i \quad (i=1,\ldots,m) \qquad [\tau^{p^s+1}=1,\ \tau^m=1].$

138   CHAPTER V.

With the single exception $m = 3$, $p^s = 2$, when $\mathsf{H}_{3,2,1}$ is of order 72, we shall prove that I coincides with H. Therefore the substitutions T form a cyclic group of order d, the greatest common divisor of m and $p^s + 1$, which is the maximal invariant subgroup of $\mathsf{H}_{m,p,s}$. Hence the quotient-group gives a simple group of order $\frac{\Omega_{m,p,s}}{d(p^s+1)}$. We shall designate it by the symbol $HO(m, p^{2s})$.

149. Theorem. — *There exists in the group I a substitution replacing ξ_1 by $\varkappa \xi_1 + \sigma \xi_2$ and not reducing to the identity.*

Suppose that $\alpha_{13} \neq 0$, for example. Transforming S by $O_{2,3}^{\lambda,\mu}$, we obtain a substitution S' replacing ξ_1 by

$$\alpha_{11}\xi_1 + (\alpha_{12}\lambda^{p^s} + \alpha_{13}\mu^{p^s})\xi_2 + (-\mu\alpha_{12} + \lambda\alpha_{13})\xi_3 + \sum_{j=4}^{m}\alpha_{1j}\xi_j.$$

To make the coefficient of ξ_3 zero, we have the conditions

$$\lambda = \frac{\alpha_{12}}{\alpha_{13}}\mu, \quad \lambda^{p^s+1} + \mu^{p^s+1} = 1.$$

The condition for μ is therefore

$$\mu^{p^s+1}\left(\alpha_{12}^{p^s+1} + \alpha_{13}^{p^s+1}\right) = \alpha_{13}^{p^s+1}.$$

Unless $\alpha_{1}^{p^s+1} + \alpha_{13}^{p^s+1} = 0$, there exists a solution μ in the $GF[p^{2s}]$ of this relation; indeed, the value of μ^{p^s+1} belongs to the $GF[p^s]$ and is therefore the $(p^s+1)^{st}$ power of a quantity μ in the $GF[p^{2s}]$.

It follows that we can assume that the only coefficients $\alpha_{1j}(j>1)$ which do not vanish are $\alpha_{12}, \ldots, \alpha_{1m_1}$ and that, if $m_1 > 2$, they have the property that

128) $\alpha_{1j}^{p^s+1} + \alpha_{1k}^{p^s+1} = 0 \quad (j, k = 2, \ldots, m_1; j \neq k).$

If $m_1 = 2$, the theorem is proven. If $m_1 > 3$, the terms in 128) must all be equal and therefore zero unless $p = 2$. Supposing first that $p \neq 2$, our theorem is proven unless $m_1 = 3$, when we have

129) $\alpha_{11}^{p^s+1} = 1, \quad \alpha_{12}^{p^s+1} + \alpha_{13}^{p^s+1} = 0, \quad \alpha_{1j} = 0 \quad (j = 4, \ldots, m).$

In the latter case we may assume that not both

$$\alpha_{11}^{p^s+1} + \alpha_{1i}^{p^s+1} = 0 \qquad (i = 2, 3);$$

for, if so, $\alpha_{12}^{p^s+1} = \alpha_{13}^{p^s+1}$ and hence each is zero by 129), since $p \neq 2$. For definiteness, let

$$\alpha_{11}^{p^s+1} + \alpha_{12}^{p^s+1} \neq 0.$$

If the left member be unity, then $\alpha_{12} = 0$ by 129) and the theorem is proven. Suppose therefore that the left member is neither zero nor unity and consider the substitution

THE HYPERORTHOGONAL AND RELATED LINEAR GROUPS. 139

$$\bar{S} \equiv S^{-1} C_1 C_2 S C_1 C_2 \equiv S_\alpha C_1 C_2,$$

where $S_\alpha \equiv S^{-1} C_1 C_2 S$ is seen to be the substitution

$$\xi'_i = \xi_i - 2\alpha_{i1} \sum_{j=1}^{m'} \alpha_{j1}^{p^s} \xi_j - 2\alpha_{i2} \sum_{j=1}^{m'} \alpha_{j2}^{p^s} \xi_j \quad (i = 1, \ldots, m).$$

The coefficient $\bar{\alpha}_{11}$ in $\bar{S}$ is therefore

$$\bar{\alpha}_{11} \equiv - \left(1 - 2\alpha_{11}^{p^s+1} - 2\alpha_{12}^{p^s+1}\right).$$

Hence

$$\bar{\alpha}_{11}^{p^s+1} \equiv \bar{\alpha}_{11}^{p^s} \bar{\alpha}_{11} = \left(1 - 2\alpha_{11}^{p^s+1} - 2\alpha_{12}^{p^s+1}\right)^2,$$

which $\neq 1$ since $\alpha_{11}^{p^s+1} + \alpha_{12}^{p^s+1}$ is neither zero nor 1. Applying the above process to $\bar{S}$ in which $\bar{\alpha}_{11}^{p^s+1} \neq 1$, we reach a substitution in the group I in which all but one of the α_{1j} $(j = 2, \ldots, m)$ are zero.

Suppose next that $p = 2$. We have by 128)

$$\alpha_{12}^{p^s+1} = \alpha_{13}^{p^s+1} = \cdots = \alpha_{1m_1}^{p^s+1} \neq 0, \quad \alpha_{1j} = 0 \quad (j = m_1 + 1, \ldots, m).$$

The ratios of $\alpha_{12}, \alpha_{13}, \ldots, \alpha_{1m_1}$ therefore satisfy the equation

130) $$\tau^{p^s+1} = 1.$$

Hence by transforming S by suitable products of the form

$$T_{1,\tau_i^{-1}} T_{i,\tau_i} \quad (i = 3, \ldots, m),$$

where the τ_i are roots of 130), we reach a substitution S' belonging to I in which $\alpha_{12} = \alpha_{13} = \cdots = \alpha_{1m_1}$. Transforming S' by the reciprocal of $O_{2,3}^{\lambda,\mu}$, we obtain in I a substitution S'' which replaces ξ_1 by

$$\alpha_{11}\xi_1 + \alpha_{12}\{(\lambda - \mu^{p^s})\xi_2 + (\mu + \lambda^{p^s})\xi_3 + \xi_4 + \cdots + \xi_{m_1}\}.$$

If $p^s \neq 2$, we can choose λ and μ [see § 146] such that

$$\lambda^{p^s+1} + \mu^{p^s+1} = 1, \quad (\lambda - \mu^{p^s})^{p^s+1} \neq 1.$$

Hence in S'' the sum of the $(p^s + 1)^{st}$ powers of the coefficients α''_{12} and α''_{14} is not zero in the $GF[2^{2s}]$. As above we can therefore make $\alpha''_{14} = 0$. If $p^s = 2$, we reach at once the same result by transforming S' by $(\xi_1 \xi_4) W (\xi_1 \xi_4)$, W being defined at the beginning of § 146.

Repeating the process, we reach finally a substitution in I, not the identity, in which either

$$\alpha_{1j} = 0 \qquad (j = 3, \ldots, m)$$

or else

$$\alpha_{12}^{p^s+1} = \alpha_{13}^{p^s+1} \neq 0, \quad \alpha_{1j} = 0 \qquad (j = 4, \ldots, m).$$

140   CHAPTER V.

In the latter case, the substitution S thus obtained has (since $p=2$)
$$\alpha_{11}^{p^s+1}=1.$$
Transforming it by $T \equiv T_{1,\tau}T_{2,\tau^{-1}}$, we obtain in I the substitution
$$S' \equiv S^{-1} \cdot T^{-1}ST \equiv S_1 T,$$
where S_1 denotes the substitution
$$\xi_i' = \xi_i + \alpha_{i1}(\tau^{-1}-1)\sum_{j=1}^m \alpha_{j1}^{p^s}\xi_j + \alpha_{i2}(\tau-1)\sum_{j=1}^m \alpha_{j2}^{p^s}\xi_j.$$
Hence for $S' \equiv S_1 T$ the coefficient of ξ_1 in ξ_1' is
$$\bar{\alpha}_{11} \equiv \tau + (1-\tau)\alpha_{11}^{p^s+1} + \tau(\tau-1)\alpha_{12}^{p^s+1} = 1 + \tau(\tau-1)\alpha_{12}^{p^s+1}.$$
Setting for brevity $\alpha_{12}^{p^s+1} \equiv \alpha$, a mark $\neq 0$ in the $GF[p^s]$, we find, since $\tau^{p^s+1}=1$, that
$$\bar{\alpha}_{11}^{p^s+1} = 1 + \alpha(\tau-1)(\tau^{p^s}-1)(\alpha - \tau^{p^s} - \tau - 1).$$
Since the theorem follows as above if $\bar{\alpha}_{11}^{p^s+1} \neq 1$, we seek to prove that a value τ can be found for which
$$\tau^{p^s+1}=1, \quad \tau \neq 1, \quad \tau^{p^s}-\tau-1 \neq \alpha.$$
But a root of $\tau^{p^s+1}=1$ will satisfy
$$\tau^{p^s}-\tau-1 = \alpha$$
only when
131)
$$1-\tau^2-\tau = \alpha\tau.$$

The desired value of τ certainly exists if $p^s+1>3$. But if $p^s=2$, we have $\alpha=1$, whence the equation 131) has the single root $\tau=1$ in the $GF[2^2]$. The theorem has therefore been proven for all cases.

150. Theorem. — *Excluding the case $m=3$, $p^s=2$, the group I contains a substitution leaving one index fixed and not reducing to the identity.*

By § 149, I contains a substitution $S \neq 1$ which replaces ξ_1 by a function of the form $\varkappa\xi_1 + \sigma\xi_2$. Hence
$$S = O_{1,2}^{\varkappa,\sigma}S_1,$$
where S_1 is a substitution of $\mathsf{H}_{m,p,s}$ of the form
$$\xi_1' = \xi_1, \quad \xi_i' = \sum_{j=2}^m \alpha_{ij}'\xi_j \qquad (i=2,\ldots,m).$$

Consider the substitution belonging to H,
$$T \equiv T_{1,\tau} T_{2,\tau} T_{i,\tau^{-2}} \qquad (\tau^{p^s+1} = 1),$$
where $i > 2$. The group I will contain the product
$$S' \equiv S^{-1} T^{-1} S T \equiv S_1^{-1} T^{-1} S_1 T,$$
since T and $O_{1,2}^{\varkappa,\sigma}$ are commutative. Since S' leaves ξ_1 fixed, our theorem is proven unless S' reduces to the identity. In the latter case, we find by comparing the values by which $S_1 T$ und TS_1 replace ξ_2 that
$$\alpha'_{2j} = 0 \quad (j = 3, \ldots, m; \, j \neq i), \quad \tau \alpha'_{2i} = \tau^{-2} \alpha'_{2i}.$$
If $m > 3$, i has at least two values and therefore
$$\alpha'_{2j} = 0 \qquad (j = 3, \ldots, m).$$
If $m = 3$, the same result holds if $p^s > 2$. For then a value of τ exists satisfying $\tau^{p^s+1} = 1$ but not $\tau^3 = 1$. Hence must $\alpha'_{2i} = 0$. Excluding the case $m = 3$, $p^s = 2$, it follows that S_1 (which was seen to leave ξ_1 fixed) alters ξ_2 at most by a constant factor λ. Hence
$$S = O_{1,2}^{\varkappa,\sigma} T'_{2,\lambda} \Sigma,$$
where Σ leaves ξ_1 and ξ_2 fixed. Hence I contains
$$S' \equiv S^{-1} (T_{1\tau} T_{2\tau^{-1}})^{-1} S (T_{1\tau} T_{2\tau^{-1}}) \qquad [\tau^{p^s+1} = 1]$$
which leaves $\xi_3, \ldots, \xi_m$ fixed. If $S' \neq 1$, the theorem is proven. If $S' = 1$, we find by comparing the values by which $ST_{1\tau} T_{2\tau^{-1}}$ and $T_{1\tau} T_{2\tau^{-1}} S$ replace ξ_1 that
$$\tau \sigma = \tau^{-1} \sigma.$$
Hence, taking for τ a value for which $\tau^2 \neq 1$, we have $\sigma = 0$. The only case left for consideration is therefore that in which
$$S = T_{1,\varkappa} T_{2,\varkappa^{-1}} T_{2,\lambda} \Sigma.$$
If S be not commutative with every $O_{1,2}^{\alpha,\beta}$, we obtain at once a substitution $\neq 1$ in I which leaves $\xi_3, \ldots, \xi_m$ fixed. In the contrary case, $\lambda = \varkappa^2$, and therefore
$$S = T_{1,\varkappa} T_{2,\varkappa} \Sigma.$$
If $m = 3$, $\Sigma = T_{3,\varkappa^{-2}}$, the determinant of S being unity. Transforming S by $(\xi_1 \xi_3) C_1$, we obtain the substitution
$$S_2 = T_{3,\varkappa} T_{2,\varkappa} T_{1,\varkappa^{-2}},$$
belonging to I. Then I contains
$$SS_2^{-1} \equiv T_{1,\varkappa^3} T_{3,\varkappa^{-3}}$$
leaving ξ_2 fixed and not reducing to the identity. For that requires $\varkappa^3 = 1$, when we should have
$$S = T_{1,\varkappa} T_{2,\varkappa} T_{3,\varkappa}$$
contrary to the hypothesis made in § 148.

142   CHAPTER V.

Let $m \geqq 4$. If Σ be not commutative with every

$$O \equiv O_{i,j}^{\alpha,\beta} \quad (i,j = 3, \ldots, m; \; \alpha^{p^s+1} + \beta^{p^s+1} = 1)$$

then I contains the substitutions leaving ξ_1 and ξ_2 fixed,

$$S^{-1}O^{-1}SO \equiv \Sigma^{-1}O^{-1}\Sigma O,$$

not all of which reduce to the identity. In the contrary case, Σ must, by § 147, have the form

$$\xi_i' = \omega\xi_i \qquad (i = 3, \ldots, m).$$

Hence I contains the product

$$S^{-1}C_1(\xi_1\xi_3)\,S(\xi_1\xi_3)C_1 : \begin{cases} \xi_1' = \omega\varkappa^{-1}\xi_1, & \xi_3' = \omega^{-1}\varkappa\xi_3, \\ \xi_2' = \xi_2, & \xi_i' = \xi_i \quad (i = 4, \ldots, m), \end{cases}$$

which does not reduce to the identity; for, if so, $\varkappa = \omega$ and S would, contrary to the hypothesis made in § 148, have the form

$$\xi_i' = \omega\xi_i \qquad (i = 1, 2, \ldots, m).$$

151. Theorem. *Except in the case $m = 3$, $p^s = 2$, the group I coincides with the group $\mathsf{H}_{m,p,s}$.*

The proofs of the theorems of §§ 149—150 hold for any value of $m \geqq 3$. Hence by a repeated application of these theorems, we finally reach in the group I a substitution $S \neq 1$ leaving $m - 2$ indices fixed and therefore of the form $O_{1,2}^{\varkappa,\sigma}$, we may assume. If it reduce to C_1C_2, when $p \neq 2$, its transformed by $O_{1,3}^{\alpha,\beta}$ gives the substitution

$$O_{1,3}^{\alpha^{p^s+1}-\beta^{p^s+1},-2\alpha\beta}C_1C_2,$$

so that I will contain an $O_{1,3}$ neither the identity nor C_1C_3. Indeed, by § 144, there exist solutions $\alpha \neq 0$, $\beta \neq 0$ in the $GF[p^{2s}]$, $p^s > 2$, of the equation $\alpha^{p^s+1} + \beta^{p^s+1} = 1$. Hence I contains a substitution $O_{1,2}^{\varkappa,\sigma}$ neither the identity nor C_1C_2. It follows then from § 144 that, for $p^s > 3$, I contains every substitution $O_{1,2}^{\alpha,\beta}$. Transforming by substitutions of the form $(\xi_i\xi_j)C_i$, we obtain in I every $O_{i,j}^{\alpha,\beta}$.

These substitutions suffice, except when $m \geqq 3$, $p^s = 2$, to generate the group $\mathsf{H}_{m,p,s}$. Indeed, by applying the formula

132) $$\qquad (O_{i,j}^{\alpha,\beta})^{-1}T_{i,\tau}O_{i,j}^{\alpha,\beta}T_{i,\tau}^{-1} \equiv O_{i,j}^{\alpha',\beta'}$$

where

$$\alpha' \equiv \alpha^{p^s+1} + \tau^{-1}\beta^{p^s+1}, \quad \beta' \equiv \alpha\beta(\tau^{-1}-1); \quad \tau^{p^s+1} = 1,$$

it follows from § 146 that every substitution of $G_{m,p,s}$ has the form h or $hT_{m,\varkappa}$ where h is generated from the $O_{i,j}^{\alpha,\beta}$ and has determinant unity. Hence the substitutions of $\mathsf{H}_{m,p,s}$ (of determinant unity) are of the form h.

THE HYPERORTHOGONAL AND RELATED LINEAR GROUPS. 143

For the case $p^s = 3$, we first prove that I contains the substitution $C_1 C_2$. We have shown that I contains an $O_{1,2}^{\alpha,\beta}$ not the identity and therefore $O_{1,2}^{\alpha',\beta'}$ given by 132). If $\beta \neq 0$, we can make $\alpha' = 0$; indeed, if α be not itself zero, we have in the $GF[3^2]$

$$\alpha^4 = \beta^4 = -1$$

and we need only take $\tau = -1$. But the square of $O_{1,2}^{0,\beta'}$ gives $C_1 C_2$ since $\beta'^4 = 1$ when $\alpha' = 0$. If, however, $\beta = 0$, then $\alpha \neq 1$. If $\alpha = -1$, we have at once $O_{1,2}^{\alpha,0} = C_1 C_2$. If $\alpha \neq \pm 1$, then the square of $O_{1,2}^{\alpha,0}$ gives $O_{1,2}^{\alpha^2,0} = C_1 C_2$.

Having $C_1 C_2$, I contains (as above) the substitution

$$O_{1,3}^{\lambda,\mu}, \quad \lambda \equiv \alpha^{p^s+1} - \beta^{p^s+1}, \quad \mu \equiv -2\alpha\beta \equiv \alpha\beta \pmod{3}.$$

Taking for α and β an arbitrary set of solutions of

$$\alpha^4 = -1, \quad \beta^4 = -1, \quad \text{whence } \alpha^4 + \beta^4 = 1,$$

we have $O_{1,3}^{0,\mu}$ where $\mu = \alpha\beta$ is an arbitrary solution of $\mu^4 = 1$. Hence I contains

$$O_{1,3}^{0,\mu} (O_{1,3}^{0,1})^{-1} = T_{1,\mu^3} T_{3,\mu}.$$

Transforming the latter by $O_{1,2}^{\alpha,\beta}$, we obtain by 132),

$$O_{1,2}^{\alpha',\beta'} T_{1,\mu^3} T_{3,\mu} \qquad \begin{pmatrix} \alpha' \equiv \alpha^4 + \mu\beta^4 \\ \beta' \equiv \alpha\beta(\mu - 1) \end{pmatrix}.$$

Hence I contains every such $O_{1,2}^{\alpha',\beta'}$. For $\alpha = 0$, $\beta^4 = 1$, we have $\alpha' = \mu$, $\beta' = 0$; for $\alpha^4 = -1$, $\beta^4 = -1$, we have $\alpha' = -1 - \mu$. We have therefore reached in the group I every $O_{1,2}^{\varkappa,\sigma}$ in which $\varkappa = \mu$, 0, $\pm 1 + \mu$, where μ is an arbitrary one of the four roots of $\mu^4 = 1$. Defining the $GF[3^2]$ by the irreducible quadratic congruence,

$$i^2 \equiv -1 \pmod{3},$$

we have $\varkappa = 0, \pm 1, \pm i, \pm 1 \pm i$. Hence $\varkappa$ takes every value in the $GF[3^2]$. We thus reach all 24 substitutions $O_{1,2}^{\varkappa,\sigma}$. It follows that I coincides with $H_{m,3,1}$.

For the case $p^s = 2$, we have in I a substitution $O_{1,2}^{\alpha,\beta} \neq 1$. By the result at the end of § 146, it must be one of the six substitutions

$$T_{1,\varrho} T_{2,\varrho^{-1}}, \quad (\xi_1 \xi_2) T_{1,\varrho} T_{2,\varrho^{-1}} \qquad (\varrho^3 = 1).$$

The transformed of the latter by $T_{1,\tau} T_{3,\tau^{-1}}$ gives

$$(\xi_1 \xi_2) T_{1,\varrho} T_{2,\varrho^{-1}} \cdot T_{1,\tau} T_{2,\tau^{-1}}.$$

Hence, in every case, I contains a substitution of the form

$$T_{1,\tau} T_{2,\tau^{-1}} \neq 1 \qquad (\tau^3 = 1)$$

144   CHAPTER V. THE HYPERORTHOGONAL etc.

Its reciprocal gives $T_{1,\tau}{-1}T_{2,\tau}$. If $m > 3$, I contains

$$W^{-1}T_{1,\tau}T_{4,\tau}{-1}W \equiv \begin{pmatrix} \tau & \tau+1 & \tau+1 \\ \tau+1 & \tau & \tau+1 \\ \tau+1 & \tau+1 & \tau \end{pmatrix} T_{4,\tau}{-1}$$

$$= T_{1,\tau}T_{2,\tau}{-1}T_{3,\tau}{-1}WT_{2,\tau}T_{3,\tau}T_{4,\tau}{-1},$$

where

$$W \equiv \begin{pmatrix} 1 & 1 & 1 \\ 1 & \tau & \tau^2 \\ 1 & \tau^2 & \tau \end{pmatrix} \qquad (\tau^2 = \tau + 1).$$

Hence I contains

$$T_{3,\tau}{-1}WT_{2,\tau} \equiv T_{3,\tau}{-1}T_{4,\tau}WT_{4,\tau}{-1}T_{2,\tau}$$

and therefore W. Hence I contains $W^2 = (\xi_2 \xi_3)$. Hence I coincides with $\mathsf{H}_{m,2,1}$ if $m > 3$.

152. Theorem. *The group $G_{m,p,s}$ is isomorphic with a subgroup of the linear group*[1]) *on $2m$ indices in the $GF[p^s]$ defined by a quadratic invariant*

$$\Psi \equiv \sum_{i=1}^{m} (x_i^2 + y_i^2 + \Theta x_i y_i).$$

Indeed, we may define the $GF[p^{2s}]$ by an equation of the form

$$I^2 - \Theta I + 1 = 0,$$

belonging to and irreducible in the $GF[p^s]$. Its roots I and $I^{p^s} \equiv I^{-1}$ belong to the $GF[p^{2s}]$. Set

$$\xi_i \equiv x_i + Iy_i, \quad \alpha_{ij} = a_{ij} + Ic_{ij} \qquad (i,j = 1, \ldots, m).$$

Then

$$\xi_i^{p^s} = x_i + I^{-1}y_i, \quad \xi_i^{p^s+1} = x_i^2 + y_i^2 + \Theta x_i y_i.$$

The invariant $\sum_{i=1}^{m} \xi_i^{p^s+1}$ becomes the quadratic form ψ. The general substitution of $G_{m,p,s}$,

$$S: \qquad \xi_i' = \sum_{j=1}^{m} \alpha_{ij}\xi_j \qquad (i = 1, \ldots, m)$$

takes the following form

$$\begin{cases} x_i' = \sum_{j=1}^{m}(a_{ij}x_j - c_{ij}y_j) \\ y_i' = \sum_{j=1}^{m}[c_{ij}x_j + (a_{ij} + \Theta c_{ij})y_j] \end{cases} \qquad (i = 1, \ldots, m).$$

---

1) Cf. Chapters VII and VIII. See also the note to § 139.

CHAPTER VI.

THE COMPOUNDS OF A LINEAR HOMOGENEOUS GROUP.[1]

153. It was shown in § 98 that the linear substitutions

$$A: \quad \xi'_i = \sum_{j=1}^{m} \alpha_{ij}\xi_j \quad (i=1,\ldots,m)$$

combine according to the law

$$A'' \equiv A'A: \quad \xi''_i = \sum_{j=1}^{m} \alpha''_{ij}\xi_j \quad (i=1,\ldots,m)$$

where

$$\alpha''_{ij} \equiv \sum_{k=1}^{m} \alpha_{ik}\alpha'_{kj} \quad (i,j=1,\ldots,m).$$

In Sylvester's umbral notation, the general q^{th} minor of the determinant $|\alpha_{ij}|$ is as follows:

$$\left|\begin{matrix} i_1 & i_2 \ldots i_q \\ j_1 & j_2 \ldots j_q \end{matrix}\right|_\alpha \equiv \begin{vmatrix} \alpha_{i_1 j_1} & \alpha_{i_1 j_2} & \ldots & \alpha_{i_1 j_q} \\ \alpha_{i_2 j_1} & \alpha_{i_2 j_2} & \ldots & \alpha_{i_2 j_q} \\ \cdot & \cdot & \cdot & \cdot \\ \alpha_{i_q j_1} & \alpha_{i_q j_2} & \ldots & \alpha_{i_q j_q} \end{vmatrix}.$$

The formula expressing the qth minors of $|\alpha''_{ij}|$ in terms of the qth minors of $|\alpha_{ij}|$ and of $|\alpha'_{ij}|$ is the following[2]:

$$133) \quad \left|\begin{matrix} i_1 \ldots i_q \\ j_1 \ldots j_q \end{matrix}\right|_{\alpha''} = \sum \left|\begin{matrix} i_1 \ldots i_q \\ l_1 \ldots l_q \end{matrix}\right|_\alpha \cdot \left|\begin{matrix} l_1 \ldots l_q \\ j_1 \ldots j_q \end{matrix}\right|_{\alpha'},$$

the summation extending over the $C_{m,q}$ combinations $l_1, l_2, \ldots, l_q$ of the m integers $1, 2, \ldots, m$ taken q at a time.

Consider the linear homogeneous substitutions on $C_{m,q}$ variables $Y_{l_1, l_2, \ldots, l_q}$

$$[\alpha]_q: \quad Y'_{i_1, i_2, \ldots, i_q} = \sum_{l_1, \ldots, l_q} \left|\begin{matrix} i_1 & i_2 \ldots i_q \\ l_1 & l_2 \ldots l_q \end{matrix}\right|_\alpha Y_{l_1, l_2, \ldots, l_q}$$

where the sets $(i_1, i_2, \ldots, i_q)$ and $(l_1, l_2, \ldots, l_q)$ take independently

---

[1] This chapter gives a new exposition of results published by the author in the following journals: *Bulletin of the Amer. Math. Soc.*, vol. 5 (1898), pp. 120—135; *Proceed. Lond. Math. Soc.*, vol. 30 (1898), pp. 70—98; *Transactions of the Amer. Math. Soc.*, vol. 1 (1900), pp. 91—96.

[2] Scott, Theory of determinants, p. 53.

146   CHAPTER VI.

the $C_{m,q}$ combinations q together of the integers $1, 2, \ldots, m$ and where we suppose

$$i_1 < i_2 < \ldots < i_q, \quad l_1 < l_2 < \ldots < l_q.$$

The determinant of the substitution $[\alpha]_q$ is called the qth compound of the determinant $\begin{vmatrix} 1 & 2 \ldots m \\ 1 & 2 \ldots m \end{vmatrix}_\alpha$ and equals[1]) the latter raised to the power $C_{m-1, q-1}$. In virtue of formula 133), we have the following formula of composition:

$$[\alpha'']_q = [\alpha']_q [\alpha]_q.$$

Hence if the substitutions $A \equiv (\alpha_{ij})$ form a group G_m, the substitutions $[\alpha]_q$ form a group $G_{m,q}$ called "the qth compound of the m-ary group G_m". We may therefore state the theorem:

Any linear homogeneous group is isomorphic with each of its compounds.

154. Theorem. — *The general linear homogeneous group $GLH(m, p^n)$ has $(d, 1)$ isomorphism with its qth compound, if d be the greatest common divisor of q and $p^n - 1$.*

We verify first that at least d substitutions of $G_m \equiv GLH(m, p^n)$ correspond to the identical substitution in its qth compound $G_{m,q}$. In fact, there exist in the $GF[p^n]$ exactly d marks δ for which $\delta^d = 1$ (§ 16). For every such mark δ, the substitution belonging to G_m,

$$(\alpha_{ij}) \equiv \begin{pmatrix} \delta & 0 \ldots 0 \\ 0 & \delta \ldots 0 \\ \cdot & \cdot \cdot \cdot \\ 0 & 0 \ldots \delta \end{pmatrix}$$

gives rise to the substitution $[\alpha]_q \equiv I$ in $G_{m,q}$.

To prove the inverse, consider the matrix J formed of certain coefficients of the substitution $[\alpha]_q$, in which $q < j \leq m$:

$$\begin{pmatrix} \begin{vmatrix} 1 & 2 \ldots q-1\,q \\ 1 & 2 \ldots q-1\,q \end{vmatrix} & -\begin{vmatrix} 1 & 2 \ldots q-1\,q \\ 1 & 2 \ldots q-1\,j \end{vmatrix} \cdots (-1)^q & \begin{vmatrix} 1 & 2 \ldots q-1\,q \\ 2 & 3 \ldots q \quad j \end{vmatrix} \\ -\begin{vmatrix} 1 & 2 \ldots q-1\,j \\ 1 & 2 \ldots q-1\,q \end{vmatrix} & \begin{vmatrix} 1 & 2 \ldots q-1\,j \\ 1 & 2 \ldots q-1\,j \end{vmatrix} \cdots (-1)^{q+1} & \begin{vmatrix} 1 & 2 \ldots q-1\,j \\ 2 & 3 \ldots q \quad j \end{vmatrix} \\ \cdot & \cdot \cdot \cdot & \cdot \\ (-1)^q\begin{vmatrix} 2 & 3 \ldots q \quad j \\ 1 & 2 \ldots q-1\,q \end{vmatrix} & (-1)^{q+1}\begin{vmatrix} 2 & 3 \ldots q \quad j \\ 1 & 2 \ldots q-1\,j \end{vmatrix} \cdots & \begin{vmatrix} 2 & 3 \ldots q \quad j \\ 2 & 3 \ldots q \quad j \end{vmatrix} \end{pmatrix}.$$

---

1) Muir, Theory of determinants, § 174.

Consider also the matrix A of determinant Δ,

$$A \equiv \begin{pmatrix} \alpha_{jj} & \alpha_{qj} & \ldots & \alpha_{1j} \\ \alpha_{jq} & \alpha_{qq} & \ldots & \alpha_{1q} \\ \vdots & & & \vdots \\ \alpha_{j1} & \alpha_{q1} & \ldots & \alpha_{11} \end{pmatrix}.$$

The composition of the matrices J and A gives the result

$$JA \equiv \begin{pmatrix} \Delta & 0 & \ldots & 0 \\ 0 & \Delta & \ldots & 0 \\ \vdots & & & \vdots \\ 0 & 0 & \ldots & \Delta \end{pmatrix}.$$

We seek those substitutions of G_m which correspond to the identity in $G_{m,q}$. Suppose, therefore, that $[\alpha]_q$ reduces to the identical substitution, so that the matrix J is the identity. In this case we have

$$\Delta^{q+1} = \Delta, \quad \alpha_{ik} = 0, \quad \alpha_{ii} = \Delta \quad (i, k = 1, 2 \ldots, q, j;\ i \neq k).$$

Taking in turn $j = q + 1, q + 2, \ldots, m$, we have the result

$$(\alpha_{ij}) \equiv \begin{pmatrix} \Delta & 0 & \ldots & 0 \\ 0 & \Delta & \ldots & 0 \\ \vdots & & & \vdots \\ 0 & 0 & \ldots & \Delta \end{pmatrix}.$$

Hence $\Delta \neq 0$ and therefore $\Delta^q = 1$.

155. Theorem. — *The special linear homogeneous group $SLH(m, p^n)$ has $(g, 1)$ isomorphism with its qth compound, if g denotes the greatest common divisor of m, q, $p^n - 1$.*

The proof is quite similar to that of the last section. The following m-ary substitution of determinant unity in the $GF[p^n]$,

$$\begin{pmatrix} \delta & 0 & \ldots & 0 \\ 0 & \delta & \ldots & 0 \\ \vdots & & & \vdots \\ 0 & 0 & \ldots & \delta \end{pmatrix}, \quad \delta^m = 1, \quad \delta^{p^n-1} = 1,$$

will give $[\alpha]_q = I$ only when $\delta^q = 1$. Hence must $\delta^g = 1$. The inverse is proven as above.

156. Theorem. — *The second compound of the general linear homogeneous group $GLH(m, p^n)$ leaves invariant the Pfaffian*

$$[1 \quad 2 \ldots m] \equiv \begin{vmatrix} Y_{12}, & Y_{13}, \ldots, & Y_{1m} \\ & Y_{23}, \ldots, & Y_{2m} \\ & & \vdots \\ & & Y_{m-1\,m} \end{vmatrix}.$$

The square of $[1\ 2\ \ldots m]$ is the skew-symmetric determinant

$$\Delta \equiv \begin{vmatrix} 0 & Y_{12} & Y_{13} & Y_{14} & \ldots & Y_{1m} \\ Y_{21} & 0 & Y_{23} & Y_{24} & \ldots & Y_{2m} \\ Y_{31} & Y_{32} & 0 & Y_{34} & \ldots & Y_{3m} \\ \ldots & \ldots & \ldots & \ldots & \ldots & \ldots \\ Y_{m1} & Y_{m2} & Y_{m3} & Y_{m4} & \ldots & 0 \end{vmatrix},$$

where $Y_{ij} \equiv -Y_{ji}$.

By § 100, $G \equiv GLH(m, p^n)$ is generated by the substitutions $B_{r,s,\lambda}$ and D_m. The corresponding substitutions of the second compound $G_{m,2}$ will therefore generate the latter group. To $B_{1,2,\lambda}$ and D_1 there correspond the respective substitutions of $G_{m,2}$:

$$\overline{B}_{1,2,\lambda}: \begin{cases} Y'_{12} = Y_{12}, & Y'_{1i} = Y_{1i} + \lambda Y_{2i} & (i = 3, \ldots, m) \\ Y'_{ij} = Y_{ij} & (i, j = 2, \ldots, m;\ i < j). \end{cases}$$

$$\overline{D}_1: \quad Y'_{1j} = D Y_{1j}\ (j = 2, \ldots, m), \quad Y'_{ij} = Y_{ij}\ (i, j = 2, \ldots, m).$$

But Δ is unaltered by an interchange of any two subscripts as 1 with 3; for, the resulting determinant may be derived from Δ by interchanging the first and third rows and the first and third columns. It therefore suffices to prove that Δ remains invariant, up to a multiplicative constant, upon applying the substitutions $\overline{B}_{1,2,\lambda}$ and $\overline{D}_1$. By inspection, $\overline{D}_1$ multiplies Δ by D^2. Also $\overline{B}_{1,2,\lambda}$ transforms Δ into the determinant

$$\begin{vmatrix} 0 & Y_{12} & Y_{13} + \lambda Y_{23} & Y_{14} + \lambda Y_{24} & \ldots & Y_{1m} + \lambda Y_{2m} \\ Y_{21} & 0 & Y_{23} & Y_{24} & \ldots & Y_{2m} \\ Y_{31} + \lambda Y_{32} & Y_{32} & 0 & Y_{34} & \ldots & Y_{3m} \\ \ldots & \ldots & \ldots & \ldots & \ldots & \ldots \\ Y_{m1} + \lambda Y_{m2} & Y_{m2} & Y_{m3} & Y_{m4} & \ldots & 0 \end{vmatrix}.$$

This reduces at once to Δ since $Y_{12} + Y_{21} = 0$.

157. Theorem. — *For m odd, the substitution $[\alpha]_2$ of the second compound gives rise to the substitution*

$$F'_i = \sum_{j=1}^{m} A_{ij} F_j \qquad (i = 1, \ldots, m)$$

upon the Pfaffians $F_j \equiv [1\ 2 \ldots j-1\ j+1 \ldots m]$, if A_{ij} denotes the minor[1]. *complementary to α_{ij} in the determinant $|\alpha_{ij}|$.*

---

1) Or the adjoint of α_{ij} without its prefixed sign.

THE COMPOUNDS OF A LINEAR HOMOGENEOUS GROUP. 149

Consider the Pfaffian F_j, j being a fixed integer $\lessgtr m$. By the last section, it is unchanged by the substitutions

$$\bar{B}_{r,s,\lambda} \quad (r, s = 1, 2, \ldots, j-1, j+1, \ldots, m).$$

Furthermore, $\bar{B}_{j,s,\lambda}$ alters no element of F_j and hence leaves F_j unchanged. Finally, we prove that $\bar{B}_{r,j,\lambda}$ replaces F_j by

$$F_j + (-1)^{j+r+1} \lambda F_r.$$

Indeed, $\bar{B}_{1,j,\lambda}$ replaces F_j by

$$\begin{vmatrix} Y_{12}+\lambda Y_{j2}, & Y_{13}+\lambda Y_{j3}, \ldots, & Y_{1j-1}+\lambda Y_{jj-1}, & Y_{1j+1}+\lambda Y_{jj+1}, \ldots, & Y_{1m}+\lambda Y_{jm} \\ & Y_{23} & , \ldots, & Y_{2j-1} & , & Y_{2j+1} & , \ldots, & Y_{2m} \\ & & & \cdots & & & & \\ & & & & Y_{j-1\,j+1} & , \ldots, & Y_{j-1\,m} \\ & & & & & \ldots, & Y_{j+1\,m} \\ & & & & & & Y_{m-1\,m} \end{vmatrix}$$

$$\equiv \begin{vmatrix} Y_{12}, & Y_{13}, \ldots, Y_{1j-1}, & Y_{1j+1}, \ldots, Y_{1m} \\ Y_{23}, \ldots, Y_{2j-1}, & Y_{2j+1}, \ldots, Y_{2m} \\ \cdots \\ & & Y_{m-1\,m} \end{vmatrix} + \lambda \begin{vmatrix} Y_{j2}, & Y_{j3}, \ldots, Y_{jm} \\ Y_{23}, \ldots, Y_{2m} \\ \cdots \\ & Y_{m-1\,m} \end{vmatrix}$$

$\equiv [1 \; 2 \; 3 \ldots j-1 \; j+1 \ldots m] + \lambda [j \; 2 \; 3 \ldots j-1 \; j+1 \ldots m]$
$\equiv F_j + (-1)^{j-2} \lambda F_1.$

Interchanging 1 with r, we see that $\bar{B}_{r,j,\lambda}$ replaces

by $\quad [r \; 2 \; 3 \ldots j-1 \; j+1 \ldots r-1 \; 1 \; r+1 \ldots m] \equiv -F_j$

$-F_j + (-1)^{j-2}\lambda[2 \; 3 \ldots r-1 \; 1 \; r+1 \ldots m] \equiv -F_j + (-1)^{j-2+r-2}\lambda F_r.$

Hence $B_{r,j,\lambda}$ induces upon the Pfaffians F_i the substitution

$$S_{rj}: \quad F_j' = F_j + (-1)^{j+r+1}\lambda F_r, \quad F_i' = F_i$$
$$(i = 1, \ldots, j-1, j+1, \ldots, m).$$

By inspection, $\bar{D}_j$ gives rise to the substitution

$$S_j: \quad F_j' = F_j, \quad F_i' = DF_i \quad (i = 1, \ldots, j-1, j+1, \ldots, m).$$

Our theorem is therefore true for the particular substitutions $B_{r,s,\lambda}$, D_1 which generate the group G_m.

To complete the proof of the theorem, we show that, if $S \equiv (\alpha_{ij})$ induces upon the F_i the substitution

150   CHAPTER VI.

$$\Sigma: \qquad F_i' = \sum_{j=1}^{m} A_{ij} F_j \qquad (i = 1, \ldots, m),$$

where A_{ij} is the minor of $|\alpha_{ij}|$ complementary to α_{ij}, the products $D_1 S$ and $B_{r,s,\lambda} S$ will induce upon the F_i the substitutions called for by the theorem. First, the product $D_1 S$ will induce upon the F_i the substitution

$$S_1 \Sigma: \qquad F_i' = A_{i1} F_1 + \sum_{j=2}^{m} D A_{ij} F_j \qquad (i = 1, \ldots, m).$$

The matrices of the two products $D_1 S$ and $S_1 \Sigma$ are respectively

$$\begin{pmatrix} D\alpha_{11} & \alpha_{12} & \ldots & \alpha_{1m} \\ D\alpha_{21} & \alpha_{22} & \ldots & \alpha_{2m} \\ \vdots & \vdots & & \vdots \\ D\alpha_{m1} & \alpha_{m2} & \ldots & \alpha_{mm} \end{pmatrix}, \quad \begin{pmatrix} A_{11} & DA_{12} & \ldots & DA_{1m} \\ A_{21} & DA_{22} & \ldots & DA_{2m} \\ \vdots & \vdots & & \vdots \\ A_{m1} & DA_{m2} & \ldots & DA_{mm} \end{pmatrix}.$$

Here the second matrix is derived from the first by the law expressed by our theorem.

Next, the product $B_{r,s,\lambda} S$ induces upon the F_i the substitution

$$S_{rs} \Sigma: \qquad F_i' = \sum_{k=1}^{m} A_{ik} F_k + (-1)^{s+r+1} \lambda A_{is} F_r \qquad (i = 1, \ldots, m).$$

The matrices of the two products $R_{r,s,\lambda} S$ and $S_{r,s} \Sigma$ are

$$\begin{pmatrix} \alpha_{11} & \alpha_{12} & \ldots & \alpha_{1s} + \lambda \alpha_{1r} & \ldots & \alpha_{1m} \\ \alpha_{21} & \alpha_{22} & \ldots & \alpha_{2s} + \lambda \alpha_{2r} & \ldots & \alpha_{2m} \\ \vdots & \vdots & & \vdots & & \vdots \\ \alpha_{m1} & \alpha_{m2} & \ldots & \alpha_{ms} + \lambda \alpha_{mr} & \ldots & \alpha_{mm} \end{pmatrix}, \quad \begin{pmatrix} A_{11} & A_{12} & \ldots & A_{1r} + (-1)^{s+r+1} \lambda A_{1s} & \ldots & A_{1m} \\ A_{21} & A_{22} & \ldots & A_{2r} + (-1)^{s+r+1} \lambda A_{2s} & \ldots & A_{2m} \\ \vdots & \vdots & & \vdots & & \vdots \\ A_{m1} & A_{m2} & \ldots & A_{mr} + (-1)^{s+r+1} \lambda A_{ms} & \ldots & A_{mm} \end{pmatrix}.$$

The second matrix is seen to be derived from the first according to the law expressed in the theorem.

Corollary. — The second compound of any linear homogeneous group G_m gives rise to a linear group on the m Pfaffians $F_1, \ldots, F_m$ which is identical with the $m-1^{st}$ compound of G_m.

158. We can establish in an analogous manner the theorem: *The linear substitution $[\alpha]_2$ of the second compound of any m-ary linear homogeneous group G_m, which corresponds to the substitution (α_{ij}) of G_m, effects upon the $C_{m,2}$ Pfaffians*

$$[i_1 i_2 \ldots i_{m-2}] \quad \begin{pmatrix} i_1, i_2, \ldots, i_{m-2} = 1, \ldots, m \\ i_1 < i_2 < \ldots < i_{m-2} \end{pmatrix}$$

a linear homogeneous substitution identical with the substitution $[\alpha]_{m-2}$ of the $(m-2)^{nd}$ compound of G_m.

The group induced by the second compound of G_m upon these Pfaffians is therefore the $(m-2)^{nd}$ compound of G_m.

159. Theorem. — *The qth and $m - q^{\text{th}}$ compounds of the special linear group $SLH(m, p^n)$ are holoedrically isomorphic.*

The theorem follows from § 155 since the greatest common divisor of m, q, $p^n - 1$ equals that of m, $m - q$, $p^n - 1$.

We proceed to set up the correspondence between the individual substitutions of the two groups. We may express the qth minors of the determinant $|A_{ij}|$, adjungate to $D \equiv |\alpha_{ij}|$, in terms of the $m - q^{\text{th}}$ minors of the latter determinant by the formula,

$$\begin{vmatrix} i_1 & i_2 \ldots i_q \\ j_1 & j_2 \ldots j_q \end{vmatrix}_A = D^{q-1} \begin{vmatrix} 1 & 2 \ldots i_1 - 1 & i_1 + 1 \ldots i_q - 1 & i_q + 1 \ldots m \\ 1 & 2 \ldots j_1 - 1 & j_1 + 1 \ldots j_q - 1 & j_q + 1 \ldots m \end{vmatrix}_\alpha.$$

Hence, if we write (for every $i_1 < i_2 < \cdots < i_q \leq m$)

$$Y_{1\ 2\ i_1 - 1\ i_1 + 1 \ldots i_q - 1\ i_q + 1 \ldots m} \equiv Z_{i_1\ i_2 \ldots i_q},$$

the general substitution $[\alpha]_{m-q}$ of the $m - q^{\text{th}}$ compound of the general m-ary linear group takes the form

$$Z'_{i_1\ i_2 \ldots i_q} = D^{1-q} \sum_{j_1 < j_2 \ldots < j_q}^{1 \ldots m} \begin{vmatrix} i_1 & i_2 \ldots i_q \\ j_1 & j_2 \ldots j_q \end{vmatrix}_A Z_{j_1\ j_2 \ldots j_q}.$$

If we take $D = 1$, this substitution belongs to the qth compound, being derived from the substitution (A_{ij}) of determinant

$$|A_{ij}| = |\alpha_{ij}|^{m-1} = D^{m-1} = 1.$$

Hence to $[\alpha]_{m-q}$, the $m - q^{\text{th}}$ compound of (α_{ij}) of determinant unity, corresponds $[A]_q$, the qth compound of (A_{ij}).

160. Theorem. — *The general Abelian group $GA(2m, p^n)$ is the largest $2m$-ary linear homogeneous group in the $GF[p^n]$ whose second compound has as a relative invariant the linear function of its $C_{m,2}$ variables Y_{ij},*

$$Z \equiv \sum_{l=1}^{m} Y_{2l-1\ 2l}.$$

It will be convenient to employ a notation for the general substitution S of $GA(2m, p^n)$ more compact than that of § 110, viz.,

$$S: \qquad \xi_i' = \sum_{j=1}^{2m} \alpha_{ij} \xi_j \qquad (i = 1, \ldots, 2m).$$

The Abelian conditions 76) then take the form (see § 112)

139) $\quad \sum_{l=1}^{m} \begin{vmatrix} \alpha_{2l-1\,j} & \alpha_{2l-1\,k} \\ \alpha_{2l\,j} & \alpha_{2l\,k} \end{vmatrix} = \mu \varepsilon_{jk} \equiv \begin{cases} \mu & \text{(if } k = j + 1 = \text{ even)} \\ 0 & \text{(unless } k = j + 1 = \text{ even)} \end{cases}$

$(j, k = 1, \ldots, 2m; j \leq k).$

These conditions may also be obtained by the method of § 129.

152 CHAPTER VI.

The corresponding substitution of the second compound is

$$[\alpha]_2: \quad Y'_{i_1 i_2} = \sum_{j_1 < j_2}^{j_1, j_2 = 1 \ldots 2m} \begin{vmatrix} \alpha_{i_1 j_1} & \alpha_{i_1 j_2} \\ \alpha_{i_2 j_1} & \alpha_{i_2 j_2} \end{vmatrix} Y_{j_1 j_2} \quad \left(\begin{matrix} i_1, i_2 = 1, \ldots, 2m \\ i_1 < i_2 \end{matrix} \right).$$

In virtue of (139), $[\alpha]_2$ transforms Z into

$$\sum_{j_1, j_2} \left\{ \sum_{l=1}^{m} \begin{vmatrix} \alpha_{2l-1 \, j_1} & \alpha_{2l-1 \, j_2} \\ \alpha_{2l \, j_1} & \alpha_{2l \, j_2} \end{vmatrix} \right\} Y_{j_1 j_2} = \mu \sum_{j_1, j_2} \varepsilon_{j_1 j_2} Y_{j_1 j_2} = \mu Z.$$

Inversely, if $[\alpha]_2$ transforms Z into μZ, the relations 139, follow.

161. Since the Abelian group $GA(2m, p^n)$ contains the substitution

$$T: \qquad \xi'_i = -\xi_i \qquad (i = 1, \ldots, 2m),$$

it is (by § 154) holoedrically or hemiedrically isomorphic with its second compound according as $p = 2$ or $p > 2$.

If S belong to the special Abelian group $SA(2m, p^n)$, so that $\mu = 1$, the corresponding substitution $[\alpha]_2$ of the second compound will leave Z absolutely invariant. Since S then has determinant unity (§ 114), $[\alpha]_2$ will leave absolutely invariant the Pfaffian $[1 \, 2 \ldots 2m]$ (§ 156). If in $SA(2m, p^n)$ we consider S and TS to be identical, we obtain the quotient-group $A(2m, p^n)$. The latter is therefore simply isomorphic with the second compound of $SA(2m, p^n)$. Applying § 119, we may state the theorem:

Except for $(2m, p^n) = (2, 2), (2, 3)$ and $(4, 2)$ the second compound of $SA(2m, p^n)$ is a simple group which leaves absolutely invariant the Pfaffian $[1 \, 2 \ldots 2m]$ and the linear function Z.

162. For $2m = 4$, $p > 2$, we introduce as new variables

$$Y \equiv \frac{1}{2}(Y_{12} - Y_{34}), \quad Z_1 \equiv \frac{1}{2}(Y_{12} + Y_{34}).$$

The general substitution $[\alpha]_2$ of the second compound of $SA(4, p^n)$ takes the form, in which the unaltered index Z_1 does not appear[1]),

	Y	Y_{13}	Y_{14}	Y_{23}	Y_{24}
$Y' =$	$2\begin{vmatrix}1\,2\\1\,2\end{vmatrix} - 1$	$\begin{vmatrix}1\,2\\1\,3\end{vmatrix}$	$\begin{vmatrix}1\,2\\1\,4\end{vmatrix}$	$\begin{vmatrix}1\,2\\2\,3\end{vmatrix}$	$\begin{vmatrix}1\,2\\2\,4\end{vmatrix}$
$Y'_{13} =$	$2\begin{vmatrix}1\,3\\1\,2\end{vmatrix}$	$\begin{vmatrix}1\,3\\1\,3\end{vmatrix}$	$\begin{vmatrix}1\,3\\1\,4\end{vmatrix}$	$\begin{vmatrix}1\,3\\2\,3\end{vmatrix}$	$\begin{vmatrix}1\,3\\2\,4\end{vmatrix}$
$Y'_{14} =$	$2\begin{vmatrix}1\,4\\1\,2\end{vmatrix}$	$\begin{vmatrix}1\,4\\1\,3\end{vmatrix}$	$\begin{vmatrix}1\,4\\1\,4\end{vmatrix}$	$\begin{vmatrix}1\,4\\2\,3\end{vmatrix}$	$\begin{vmatrix}1\,4\\2\,4\end{vmatrix}$
$Y'_{23} =$	$2\begin{vmatrix}2\,3\\1\,2\end{vmatrix}$	$\begin{vmatrix}2\,3\\1\,3\end{vmatrix}$	$\begin{vmatrix}2\,3\\1\,4\end{vmatrix}$	$\begin{vmatrix}2\,3\\2\,3\end{vmatrix}$	$\begin{vmatrix}2\,3\\2\,4\end{vmatrix}$
$Y'_{24} =$	$2\begin{vmatrix}2\,4\\1\,2\end{vmatrix}$	$\begin{vmatrix}2\,4\\1\,3\end{vmatrix}$	$\begin{vmatrix}2\,4\\1\,4\end{vmatrix}$	$\begin{vmatrix}2\,4\\2\,3\end{vmatrix}$	$\begin{vmatrix}2\,4\\2\,4\end{vmatrix}$

---

[1] In § 164 below, the second compound $[\alpha]_2$ of an arbitrary quaternary linear homogeneous substitution is written in matrix form.

THE COMPOUNDS OF A LINEAR HOMOGENEOUS GROUP. 153

For example, $[\alpha]_2$ replaces Y_{13} by the function

$$\begin{vmatrix}1&3\\1&2\end{vmatrix}(Y+Z_1) + \begin{vmatrix}1&3\\1&3\end{vmatrix}Y_{13} + \begin{vmatrix}1&3\\1&4\end{vmatrix}Y_{14} + \begin{vmatrix}1&3\\2&3\end{vmatrix}Y_{23} + \begin{vmatrix}1&3\\2&4\end{vmatrix}Y_{24} + \begin{vmatrix}1&3\\3&4\end{vmatrix}(Z_1-Y),$$

which becomes Y'_{13} of the table if we apply the Abelian relation

$$\begin{vmatrix}1&3\\1&2\end{vmatrix} + \begin{vmatrix}1&3\\3&4\end{vmatrix} \equiv \begin{vmatrix}\alpha_{11}&\alpha_{12}\\\alpha_{31}&\alpha_{32}\end{vmatrix} + \begin{vmatrix}\alpha_{13}&\alpha_{14}\\\alpha_{33}&\alpha_{34}\end{vmatrix} = 0.$$

Similarly, it replaces Y by the function

$$\frac{1}{2}\left\{\begin{vmatrix}1&2\\1&2\end{vmatrix} - \begin{vmatrix}3&4\\1&2\end{vmatrix}\right\}(Y+Z_1) + \frac{1}{2}\left\{\begin{vmatrix}1&2\\1&3\end{vmatrix} - \begin{vmatrix}3&4\\1&3\end{vmatrix}\right\}Y_{13} + \cdots$$

$$+ \frac{1}{2}\left\{\begin{vmatrix}1&2\\3&4\end{vmatrix} - \begin{vmatrix}3&4\\3&4\end{vmatrix}\right\}(Z_1-Y).$$

By means of the Abelian relations

$$\begin{vmatrix}3&4\\1&2\end{vmatrix} = 1 - \begin{vmatrix}1&2\\1&2\end{vmatrix}, \quad \begin{vmatrix}1&2\\3&4\end{vmatrix} = 1 - \begin{vmatrix}1&2\\1&2\end{vmatrix}, \quad \begin{vmatrix}3&4\\3&4\end{vmatrix} = \begin{vmatrix}1&2\\1&2\end{vmatrix}, \quad \begin{vmatrix}3&4\\1&3\end{vmatrix} = -\begin{vmatrix}1&2\\1&3\end{vmatrix}.$$

Hence Y is replaced by the function Y' given by the above table.

It is therefore a substitution on five indices leaving absolutely invariant the function

$$\Phi \equiv Z_1^2 - [1\ 2\ 3\ 4] \equiv Y^2 + Y_{13}Y_{24} - Y_{14}Y_{23}.$$

For $p > 2$, the simple group $A(4, p^n)$ is holoedrically isomorphic with a subgroup of the quinary linear group leaving the quadratic function Φ absolutely invariant.

This theorem and the results of § 163—165 find application in Chapters VII and VIII.

163. By § 155, the quaternary linear group of determinant unity $SLH(4, p^n) \equiv G'_4$ is holoedrically or hemiedrically isomorphic with its second compound $G'_{4,2}$ according as $p = 2$ or $p > 2$. By §§ 103—104, G'_4 has as maximal invariant subgroup the group generated by the substitution

$$M_\mu: \qquad \xi'_i = \mu \xi_i \qquad (i = 1, 2, 3, 4),$$

where μ is a primitive root of $\mu^d = 1$, d being the greatest common divisor of 4 and $p^n - 1$. The quotient-group $LF(4, p^n)$ is a simple group of order

$$\frac{1}{d}(p^{4n}-1)p^{3n}(p^{3n}-1)p^{2n}(p^{2n}-1)p^n.$$

To M_μ there corresponds in $G'_{4,2}$ the substitution which multiplies every index by μ^2 and therefore the identity if $p = 2$ or $p^n = 4l+3$; while, for $p^n = 4l+1$, it is the substitution T multiplying each of the six indices by -1. We may state the theorem:

154 CHAPTER VI.

For $p^n = 2^n$ or $p^n = 4l + 3$, $G'_{4,2}$ is a simple group holoedrically isomorphic with $LF(4, p^n)$. For $p^n = 4l + 1$, $G'_{4,2}$ has a maximal self-conjugate subgroup $\{I, T\}$ of order two, the quotient-group being holoedrically isomorphic with $LF(4, p^n)$. If $e = 1$ or 2 according as $p = 2$ or $p > 2$, the order of $G'_{4,2}$ is

$$\frac{1}{e}(p^{4n} - 1)p^{3n}(p^{3n} - 1)p^{2n}(p^{2n} - 1)p^n.$$

164. Theorem. — *The second compound $G_{4,2}$ of the general linear homogeneous group G_4 in the $GF[p^n]$ contains the substitution*

140) $Y'_{12} = \nu Y_{12},\ Y'_{13} = Y_{13},\ Y'_{14} = Y_{14},\ Y'_{23} = Y_{23},\ Y'_{24} = Y_{24},\ Y'_{34} = \nu^{-1}Y_{34}$

if, and only if, ν be a square in the field.

To the substitution (α_{ij}) of G_4 corresponds in $G_{4,2}$ the substitution $[\alpha]_2$:

	Y_{12}	Y_{13}	Y_{14}	Y_{23}	Y_{24}	Y_{34}
$Y'_{12} =$	$\begin{vmatrix}1\,2\\1\,2\end{vmatrix}$	$\begin{vmatrix}1\,2\\1\,3\end{vmatrix}$	$\begin{vmatrix}1\,2\\1\,4\end{vmatrix}$	$\begin{vmatrix}1\,2\\2\,3\end{vmatrix}$	$\begin{vmatrix}1\,2\\2\,4\end{vmatrix}$	$\begin{vmatrix}1\,2\\3\,4\end{vmatrix}$
$Y'_{13} =$	$\begin{vmatrix}1\,3\\1\,2\end{vmatrix}$	$\begin{vmatrix}1\,3\\1\,3\end{vmatrix}$	$\begin{vmatrix}1\,3\\1\,4\end{vmatrix}$	$\begin{vmatrix}1\,3\\2\,3\end{vmatrix}$	$\begin{vmatrix}1\,3\\2\,4\end{vmatrix}$	$\begin{vmatrix}1\,3\\3\,4\end{vmatrix}$
$Y'_{14} =$	$\begin{vmatrix}1\,4\\1\,2\end{vmatrix}$	$\begin{vmatrix}1\,4\\1\,3\end{vmatrix}$	$\begin{vmatrix}1\,4\\1\,4\end{vmatrix}$	$\begin{vmatrix}1\,4\\2\,3\end{vmatrix}$	$\begin{vmatrix}1\,4\\2\,4\end{vmatrix}$	$\begin{vmatrix}1\,4\\3\,4\end{vmatrix}$
$Y'_{23} =$	$\begin{vmatrix}2\,3\\1\,2\end{vmatrix}$	$\begin{vmatrix}2\,3\\1\,3\end{vmatrix}$	$\begin{vmatrix}2\,3\\1\,4\end{vmatrix}$	$\begin{vmatrix}2\,3\\2\,3\end{vmatrix}$	$\begin{vmatrix}2\,3\\2\,4\end{vmatrix}$	$\begin{vmatrix}2\,3\\3\,4\end{vmatrix}$
$Y'_{24} =$	$\begin{vmatrix}2\,4\\1\,2\end{vmatrix}$	$\begin{vmatrix}2\,4\\1\,3\end{vmatrix}$	$\begin{vmatrix}2\,4\\1\,4\end{vmatrix}$	$\begin{vmatrix}2\,4\\2\,3\end{vmatrix}$	$\begin{vmatrix}2\,4\\2\,4\end{vmatrix}$	$\begin{vmatrix}2\,4\\3\,4\end{vmatrix}$
$Y'_{34} =$	$\begin{vmatrix}3\,4\\1\,2\end{vmatrix}$	$\begin{vmatrix}3\,4\\1\,3\end{vmatrix}$	$\begin{vmatrix}3\,4\\1\,4\end{vmatrix}$	$\begin{vmatrix}3\,4\\2\,3\end{vmatrix}$	$\begin{vmatrix}3\,4\\2\,4\end{vmatrix}$	$\begin{vmatrix}3\,4\\3\,4\end{vmatrix}$

Consider the "partial substitution", possibly of determinant zero,

	Y_{23}	Y_{24}	Y_{34}
$Y'_{23} =$	$\begin{vmatrix}2\,3\\2\,3\end{vmatrix}$	$\begin{vmatrix}2\,3\\2\,4\end{vmatrix}$	$\begin{vmatrix}2\,3\\3\,4\end{vmatrix}$
141) $\quad Y'_{24} =$	$\begin{vmatrix}2\,4\\2\,3\end{vmatrix}$	$\begin{vmatrix}2\,4\\2\,4\end{vmatrix}$	$\begin{vmatrix}2\,4\\3\,4\end{vmatrix}$
$Y'_{34} =$	$\begin{vmatrix}3\,4\\2\,3\end{vmatrix}$	$\begin{vmatrix}3\,4\\2\,4\end{vmatrix}$	$\begin{vmatrix}3\,4\\3\,4\end{vmatrix}$

Its determinant is readily seen to equal

$$\begin{vmatrix} \alpha_{22} & \alpha_{23} & \alpha_{24} \\ \alpha_{32} & \alpha_{33} & \alpha_{34} \\ \alpha_{42} & \alpha_{43} & \alpha_{44} \end{vmatrix}^2.$$

THE COMPOUNDS OF A LINEAR HOMOGENEOUS GROUP. 155

If $[\alpha]_2$ be the particular substitution 140), the "partial substitution" 141) becomes
$$\begin{pmatrix} 1 \cdot 0 & 0 \\ 0 & 1 & 0 \\ 0 & 0 & \nu^{-1} \end{pmatrix}$$
of determinant ν^{-1}. Hence if 140) belong to $G_{4,2}$, ν must be a square in the field.

Inversely, if ν be a square, 140) is the second compound of the following substitution of determinant unity:
$$\begin{pmatrix} \nu^{1/2} & 0 & 0 & 0 \\ 0 & \nu^{1/2} & 0 & 0 \\ 0 & 0 & \nu^{-1/2} & 0 \\ 0 & 0 & 0 & \nu^{-1/2} \end{pmatrix}.$$

Note. — The second compound contains the substitution
$$Y'_{12}=\nu Y_{12}, \; Y'_{13}=\nu Y_{13}, \; Y'_{14}=Y_{14}, \; Y'_{23}=Y_{23}, \; Y'_{24}=\nu^{-1}Y_{24}, \; Y'_{34}=\nu^{-1}Y_{34}.$$
In fact, the latter is the second compound of the substitution
$$\begin{pmatrix} \nu & 0 & 0 & 0 \\ 0 & 1 & 0 & 0 \\ 0 & 0 & 1 & 0 \\ 0 & 0 & 0 & \nu^{-1} \end{pmatrix}.$$

165. Theorem. — *For $p=2$, every substitution of $G'_{4,2}$ satisfies the relation*

$$\begin{vmatrix} 1\,2 \\ 1\,2 \end{vmatrix}\begin{vmatrix} 3\,4 \\ 3\,4 \end{vmatrix} + \begin{vmatrix} 1\,2 \\ 1\,3 \end{vmatrix}\begin{vmatrix} 3\,4 \\ 2\,4 \end{vmatrix} + \begin{vmatrix} 1\,2 \\ 1\,4 \end{vmatrix}\begin{vmatrix} 3\,4 \\ 2\,3 \end{vmatrix} + \begin{vmatrix} 1\,3 \\ 1\,2 \end{vmatrix}\begin{vmatrix} 2\,4 \\ 3\,4 \end{vmatrix} + \begin{vmatrix} 1\,3 \\ 1\,3 \end{vmatrix}\begin{vmatrix} 2\,4 \\ 2\,4 \end{vmatrix}$$
$$+ \begin{vmatrix} 1\,3 \\ 1\,4 \end{vmatrix}\begin{vmatrix} 2\,4 \\ 2\,3 \end{vmatrix} + \begin{vmatrix} 1\,4 \\ 1\,2 \end{vmatrix}\begin{vmatrix} 2\,3 \\ 3\,4 \end{vmatrix} + \begin{vmatrix} 1\,4 \\ 1\,3 \end{vmatrix}\begin{vmatrix} 2\,3 \\ 2\,4 \end{vmatrix} + \begin{vmatrix} 1\,4 \\ 1\,4 \end{vmatrix}\begin{vmatrix} 2\,3 \\ 2\,3 \end{vmatrix} \equiv 1 \;(\mathrm{mod}\,2),$$

formed by multiplying each coefficient of the partial substitution 141) *by that coefficient of the matrix* $[\alpha]_2$ *which lies symmetrical to it. $G'_{4,2}$ does not contain the substitution* $M_1 \equiv (Y_{12}Y_{34})$.

The left member of our relation is seen to be the expansion of the expression
$$\begin{vmatrix} \alpha_{11} & \alpha_{12} & \alpha_{13} & \alpha_{14} \\ \alpha_{21} & \alpha_{22} & \alpha_{23} & \alpha_{24} \\ \alpha_{31} & \alpha_{32} & \alpha_{33} & \alpha_{34} \\ \alpha_{41} & \alpha_{42} & \alpha_{43} & \alpha_{44} \end{vmatrix} + 2\alpha_{44} \begin{vmatrix} \alpha_{11} & \alpha_{12} & \alpha_{13} \\ \alpha_{21} & \alpha_{22} & \alpha_{23} \\ \alpha_{31} & \alpha_{32} & \alpha_{33} \end{vmatrix}$$
and is therefore $\equiv 1 \;(\mathrm{mod}\,2)$, since $|\alpha_{ij}|=1$. The substitution M_1 does not satisfy the relation and so does not belong to the group $G'_{4,2}$.

CHAPTER VII.

LINEAR HOMOGENEOUS GROUP IN THE $GF[p^n]$, $p > 2$, DEFINED BY A QUADRATIC INVARIANT.[1]

166. Any quadratic form with coefficients in the $GF[p^n]$, $p > 2$,
$$f \equiv \alpha_{11}\xi_1^2 + 2\alpha_{12}\xi_1\xi_2 + \alpha_{22}\xi_2^2 + 2\alpha_{13}\xi_1\xi_3 + \cdots + \alpha_{mm}\xi_m^2$$
may, by using the notation $\alpha_{ji} \equiv \alpha_{ij}$, be written in the form
$$f \equiv \sum_{i,j}^{1,\ldots,m} \alpha_{ij}\xi_i\xi_j.$$
By the *determinant* (or discriminant) of f we mean
$$\Delta \equiv \begin{vmatrix} \alpha_{11} & \alpha_{12} & \cdots & \alpha_{1m} \\ \alpha_{21} & \alpha_{22} & \cdots & \alpha_{2m} \\ \cdot & \cdot & \cdot & \cdot \\ \alpha_{m1} & \alpha_{m2} & \cdots & \alpha_{mm} \end{vmatrix}.$$

167. Theorem. — *Upon applying to f a linear m-ary transformation of determinant D, the determinant Δ of f is multiplied by D^2.*

In view of § 100, it suffices to prove the theorem for the types of transformations considered in the cases 1^0 and 2^0 following.

1^0. Upon applying to f the transformation
$$\xi_1' = \xi_1 + \lambda\xi_2, \quad \xi_i' = \xi_i \qquad (i = 2, \ldots, m)$$
we obtain the function
$$f + 2\lambda \sum_{j=1}^m \alpha_{1j}\xi_2\xi_j + \lambda^2 \alpha_{11}\xi_2^2.$$
Its determinant is
$$\begin{vmatrix} \alpha_{11} & \alpha_{12}+\lambda\alpha_{11} & \alpha_{13} & \cdots & \alpha_{1m} \\ \alpha_{21}+\lambda\alpha_{11} & \alpha_{22}+\lambda\alpha_{21}+\lambda\alpha_{12}+\lambda^2\alpha_{11} & \alpha_{23}+\lambda\alpha_{13} & \cdots & \alpha_{2m}+\lambda\alpha_{1m} \\ \alpha_{31} & \alpha_{32}+\lambda\alpha_{31} & \alpha_{33} & \cdots & \alpha_{3m} \\ \cdot & \cdot & \cdot & & \cdot \\ \alpha_{m1} & \alpha_{m2}+\lambda\alpha_{m1} & \alpha_{m3} & \cdots & \alpha_{mm} \end{vmatrix}.$$
Multiply the first row by λ and subtract from the second row; afterwards multiply the first column by λ and subtract from the second. We obtain the original determinant $\Delta = |\alpha_{ij}|$.

---

[1] The results of this chapter were given by the author in the *American Journal of Mathematics*, vol. 21 (1899), pp. 193—256, and partially in earlier papers there cited. For the case $n = 1$, the order of the first orthogonal group was determined by Jordan, Traité, pp. 161—170.

2^0. Upon applying to f the transformation
$$\xi_1' = D\xi_1, \quad \xi_i' = \xi_i \qquad (i = 2, \ldots, m)$$
we obtain the function
$$D^2 \alpha_{11} \xi_1^2 + 2D \sum_{j=2}^{m} \alpha_{1j} \xi_1 \xi_j + \sum_{i,j}^{2,\ldots,m} \alpha_{ij} \xi_i \xi_j.$$
Its determinant is
$$\begin{vmatrix} D^2\alpha_{11} & D\alpha_{12} & \ldots & D\alpha_{1m} \\ D\alpha_{21} & \alpha_{22} & \ldots & \alpha_{2m} \\ \vdots & & & \\ D\alpha_{m1} & \alpha_{m2} & \ldots & \alpha_{mm} \end{vmatrix} \equiv D^2 |\alpha_{ij}|.$$

168. Theorem. — *A quadratic form f with coefficients in the $GF[p^n]$, $p > 2$, and of determinant $\Delta \neq 0$ can be reduced by a linear homogeneous substitution belonging to the field to the form*

142) $$\sum_{i=1}^{m} \alpha_i \xi_i^2 \qquad (each\ \alpha_i \neq 0).$$

Since $\Delta \neq 0$, the coefficients $\alpha_{11}, \alpha_{12}, \ldots, \alpha_{1m}$ are not all zero. If $\alpha_{11} = 0$, we may suppose that $\alpha_{12} \neq 0$, for example. Applying to f the substitution of determinant $-2\lambda \neq 0$,
$$\xi_1' = \lambda\xi_1 + \lambda\xi_2, \quad \xi_2' = \xi_1 - \xi_2, \quad \xi_i' = \xi_i \quad (i = 3, \ldots, m)$$
we obtain a form in which the coefficient of ξ_1^2 is $\alpha_{22} + 2\lambda\alpha_{12}$. Taking for λ any one of the $p^n - 2$ marks different from zero and from $-\alpha_{22}/2\alpha_{12}$, the coefficient of ξ_1^2 will be not zero. Whether α_{11} be zero or not, we thus obtain a form
$$f' \equiv \sum_{i,j}^{1,\ldots,m} \beta_{ij}\xi_i\xi_j \qquad (\beta_{11} \neq 0)$$
whose determinant Δ' is not zero by § 167.

Applying to f' the substitution
$$\xi_1' = \xi_1 - \frac{\beta_{12}}{\beta_{11}}\xi_2, \quad \xi_i' = \xi_i \qquad (i = 2, \ldots, m)$$
we obtain a form in which the coefficient of $\xi_1\xi_2$ is zero, while that ξ_1^2 remains $\beta_{11} \neq 0$. In a similar manner, we can make the coefficients of $\xi_1\xi_3, \ldots, \xi_1\xi_m$ all zero. In the resulting form
$$\beta_{11}\xi_1^2 + \sum_{i,j}^{2,\ldots,m} \gamma_{ij}\xi_i\xi_j,$$
the coefficients $\gamma_{22}, \gamma_{23}, \ldots, \gamma_{2m}$ are not all zero, since the determinant of the transformed form is not zero by § 167.

158   CHAPTER VII.

Proceeding with this form as we did with f, we reach a form
$$\beta_{11}\xi_1^2 + \delta_{11}\xi_2^2 + \sum_{i,j}^{3,\ldots,m}\varepsilon_{ij}\xi_i\xi_j$$
of determinant $\neq 0$. After $m-1$ such steps we reach the form 142).

169. Certain of the α_i in 142) are squares and the others are not-squares in the $GF[p^n]$. By applying a suitable substitution which interchanges the ξ_i, we may suppose that in the resulting form $\alpha_1, \ldots, \alpha_s$ are squares, say $a_1^2, \ldots, a_s^2$, while $\alpha_{s+1}, \ldots, \alpha_m$ are not-squares, say $\nu a_{s+1}^2, \ldots, \nu a_m^2$, ν being a particular not-square. Applying the substitution
$$\xi_i' = a_i^{-1}\xi_i \qquad (i = 1, \ldots, m)$$
our form is transformed into
$$f_s \equiv \sum_{i=1}^{s}\xi_i^2 + \nu\sum_{i=s+1}^{m}\xi_i^2.$$

Furthermore, we can transform f_s into f_{s+2} and *vice versa*. In fact, the substitution of determinant $\alpha^2 + \beta^2$
$$\xi_i' = \alpha\xi_i - \beta\xi_j, \quad \xi_j' = \beta\xi_i + \alpha\xi_j$$
transforms $\xi_i^2 + \xi_j^2$ into $(\alpha^2 + \beta^2)(\xi_i^2 + \xi_j^2)$. By § 64, α and β may be chosen in the $GF[p^n]$, $p > 2$, such that
$$\alpha^2 + \beta^2 = \nu.$$

We have therefore only two canonical forms, f_m and f_{m-1}. The latter form may be dropped if m be odd. Indeed, f_{m-1} can, for m odd, be transformed into
$$f_0 \equiv \nu(\xi_1^2 + \xi_2^2 + \cdots + \xi_m^2).$$
But the linear group leaving f_0 invariant leaves also
$$f_m \equiv \xi_1^2 + \xi_2^2 + \cdots + \xi_m^2$$
invariant. We may therefore state the theorem:

The group of all linear homogeneous m-ary substitutions in the $GF[p^n]$, $p > 2$, which leave invariant a quadratic form f belonging to the field and of determinant not zero, can be transformed by a linear homogeneous m-ary substitution belonging to the field into the group of all linear homogeneous m-ary substitutions in the $GF[p^n]$ which leave invariant
$$F_\mu \equiv \sum_{i=1}^{m-1}\xi_i^2 + \mu\xi_m^2,$$
where $\mu = 1$ for m odd, but $\mu = 1$ or a particular not-square ν for m even.

LINEAR GROUP WITH QUADRATIC INVARIANT. 159

170. The conditions that the substitution

$$S: \qquad \xi'_i = \sum_{j=1}^{m} \alpha_{ij} \xi_j \qquad (i = 1, \ldots, m)$$

shall leave F_μ invariant are the following:

143) $\quad \alpha_{1j}^2 + \alpha_{2j}^2 + \cdots + \alpha_{m-1\,j}^2 + \mu \alpha_{mj}^2 = \begin{cases} 1 & (j = 1, \ldots, m-1) \\ \mu & (j = m) \end{cases}$

144) $\quad \alpha_{1j}\alpha_{1k} + \cdots + \alpha_{m-1\,j}\alpha_{m-1\,k} + \mu \alpha_{mj}\alpha_{mk} = 0$
$\qquad\qquad\qquad\qquad\qquad\qquad (j, k = 1, \ldots, m;\ j \neq k).$

It follows readily that the inverse of S is

$$S^{-1}: \quad \begin{cases} \xi'_i = \sum_{j=1}^{m-1} \alpha_{ji} \xi_j + \mu \alpha_{mi} \xi_m & (i = 1, \ldots, m-1) \\ \xi'_m = \dfrac{1}{\mu} \sum_{j=1}^{m-1} \alpha_{jm} \xi_j + \alpha_{mm} \xi_m. \end{cases}$$

The determinant of S^{-1} is seen to equal the determinant D of S. Hence $D^2 = 1$, being the determinant of $S^{-1}S = I$.

Writing the relations 143) and 144) for the substitution S^{-1}, we obtain the following relations, which are evidently together equivalent to the set 143) and 144):

145) $\quad \alpha_{j1}^2 + \alpha_{j2}^2 + \cdots + \alpha_{j\,m-1}^2 + \dfrac{1}{\mu}\alpha_{jm}^2 = \begin{cases} 1 & (j = 1, \ldots, m-1) \\ 1/\mu & (j = m) \end{cases}$

146) $\quad \alpha_{j1}\alpha_{k1} + \cdots + \alpha_{j\,m-1}\alpha_{k\,m-1} + \dfrac{1}{\mu}\alpha_{jm}\alpha_{km} = 0$
$\qquad\qquad\qquad\qquad\qquad\qquad (j, k = 1, \ldots, m;\ j \neq k).$

171. The substitutions leaving F_μ invariant were proven to have determinant ± 1. Among them occur substitutions of determinant -1, as

$$C_i: \qquad \xi'_i = -\xi_i, \quad \xi'_j = \xi_j \qquad (j = 1, \ldots, m;\ j \neq i).$$

The group $\bar{O}_\mu(m, p^n)$ of all linear substitutions leaving F_μ invariant has therefore a subgroup of index two $O_\mu(m, p^n)$ composed of all linear m-ary substitutions in the $GF[p^n]$ of determinant unity which leave F_μ invariant. The latter substitutions will be called *orthogonal*.[1]) For $\mu = 1$, we have the *first orthogonal group* $O_1(m, p^n)$; for m even and $\mu = \nu$, we have the *second orthogonal group* $O_\nu(m, p^n)$.

---

1) This unusual restriction of the term orthogonal to substitutions of determinant $+1$ is done in the interest of the later terminology and notation. We will be concerned with such substitutions alone. If it became necessary to consider substitutions of determinant -1 which leave F_μ invariant, they might be designated *extended* (erweiterte) *orthogonal substitutions* and the group $\bar{O}_\mu(m, p^n)$ designated the extended orthogonal group.

160   CHAPTER VII.

172. Theorem. — *The order $\Omega_\mu(m, p^n)$ of $O_\mu(m, p^n)$ is, for m odd,*

$$(p^{n(m-1)} - 1)\, p^{n(m-2)} (p^{n(m-3)} - 1)\, p^{n(m-4)} \ldots (p^{2n} - 1)\, p^n,$$

and, for m even,

$$\left[p^{n(m-1)} \mp \varepsilon^{\frac{m}{2}} p^{n\left(\frac{m}{2}-1\right)} \right] (p^{n(m-2)} - 1)\, p^{n(m-3)} \ldots (p^{2n} - 1)\, p^n,$$

where the sign $\mp$ is $-$ or $+$ according as $\mu = 1$ or ν, and where $\varepsilon = \pm 1$ according to the form $4l \pm 1$ of p^n.

We notice first that the number of substitutions $S, S', \ldots$ of $O_\mu(m, p^n)$ which leave ξ_1 fixed is $\Omega_\mu(m-1, p^n)$. In fact, they have $\alpha_{11} = 1$, $\alpha_{12} = \alpha_{13} = \cdots = \alpha_{1m} = 0$, and therefore by 146) for $j = 1$,

$$\alpha_{k1} = 0 \qquad (k = 2, \ldots, m).$$

Hence they belong to the group $O_\mu(m-1, p^n)$ leaving invariant

$$\xi_2^2 + \xi_3^2 + \cdots + \xi_{m-1}^2 + \mu \xi_m^2.$$

Let T be a general substitution of $O_\mu(m, p^n)$ and let it replace ξ_1 by

$$\omega_1 = \sum_{k=1}^{m} \alpha_{1k} \xi_k,$$

where, by 145) for $j = 1$,

147) $\qquad \alpha_{11}^2 + \alpha_{12}^2 + \cdots + \alpha_{1\,m-1}^2 + \frac{1}{\mu} \alpha_{1m}^2 = 1.$

The $\Omega_\mu(m-1, p^n)$ substitutions $TS, TS', \ldots$ and no others of the group will replace ξ_1 by ω_1. If, therefore, $P_\mu(m, p^n)$ denote the number of distinct linear functions ω_1 by which the substitutions of $O_\mu(m, p^n)$ can replace ξ_1, we have for the order of the latter group,

$$\Omega_\mu(m, p^n) = P_\mu(m, p^n)\, \Omega_\mu(m-1, p^n).$$

This recursion formula gives

$$\Omega_\mu(m, p^n) = P_\mu(m, p^n)\, P_\mu(m-1, p^n) \ldots P_\mu(2, p^n),$$

since the identity is the only substitution of determinant unity on one index which leaves $\mu \xi_m^2$ invariant, so that $\Omega_\mu(1, p^n) = 1$.

It will be shown in §§ 174—180 that $P_\mu(k, p^n)$ equals the number of sets of solutions in the $GF[p^n]$ of the equation

$$\alpha_1^2 + \alpha_2^2 + \cdots + \alpha_{k-1}^2 + \frac{1}{\mu} \alpha_k^2 = 1,$$

and hence, by §§ 65—66, $P_\mu(k, p^n) =$

$$p^{n(k-1)} \mp \varepsilon^{\frac{k}{2}} p^{n\left(\frac{k}{2}-1\right)} \qquad (k \text{ even})$$

$$p^{n(k-1)} \pm \varepsilon^{\frac{k-1}{2}} p^{n(k-1)/2} \qquad (k \text{ odd})$$

the upper signs holding if $\mu = 1$, the lower signs if $\mu = \nu$, and ε denoting $+1$ or -1 according as -1 is a square or a not-square

LINEAR GROUP WITH QUADRATIC INVARIANT.   161

in the $GF[p^n]$. Whether the integer t be even or odd, we find that the product

$$P_\mu(2t+1, p^n) P_\mu(2t, p^n) = p^{nt}(p^{nt} \pm \varepsilon^t) \cdot p^{n(t-1)}(p^{nt} \mp \varepsilon^t)$$
$$\equiv (p^{2nt} - 1) p^{n(2t-1)}.$$

We derive at once the expression for $\Omega_\mu(m, p^n)$ given in the theorem.

173. Theorem. — *The orthogonal group $O_\mu(m, p^n)$ is generated by the substitutions*[1])

$$O_{i,j}^{\alpha,\beta}: \quad \begin{cases} \xi_i' = \alpha \xi_i + \beta \xi_j \\ \xi_j' = -\beta \xi_i + \alpha \xi_j \end{cases} \quad [\alpha^2 + \beta^2 = 1] \quad (i,j < m)$$

$$O_{i,m}^{\gamma,\delta}: \quad \begin{cases} \xi_i' = \gamma \xi_i + \delta \xi_m \\ \xi_m' = -\dfrac{\delta}{\mu} \xi_i + \gamma \xi_m \end{cases} \quad \left[\gamma^2 + \dfrac{1}{\mu}\delta^2 = 1\right] \quad (i < m)$$

with the following exceptions:

1°. *For $p^n = 5$, $m \geqq 3$, $\mu = 1$, we may take as the necessary additional generator*

$$R_{123}: \quad \begin{cases} \xi_1' = \xi_1 + \xi_2 + 2\xi_3, \\ \xi_2' = \xi_1 + 2\xi_2 + \xi_3, \\ \xi_3' = 2\xi_1 + \xi_2 + \xi_3. \end{cases} \quad R_{123}^2 = I.$$

2°. *For $p^n = 3$, $m \geqq 4$, $\mu = 1$, we may choose as the necessary additional generator*

$$W_{1234}: \quad \begin{cases} \xi_1' = \xi_1 - \xi_2 - \xi_3 - \xi_4, \\ \xi_2' = \xi_1 - \xi_2 + \xi_3 + \xi_4, \\ \xi_3' = \xi_1 + \xi_2 - \xi_3 + \xi_4, \\ \xi_4' = \xi_1 + \xi_2 + \xi_3 - \xi_4. \end{cases} \quad W_{1234}^3 = I.$$

3°. *For $p^n = 3$, $m \geqq 3$, $\mu = \nu \equiv -1$, we may take as the necessary additional generator*

$$V_{12m}: \quad \begin{cases} \xi_1' = \xi_1 - \xi_2 - \xi_m, \\ \xi_2' = \xi_1 - \xi_2 + \xi_m, \\ \xi_m' = -\xi_1 - \xi_2. \end{cases} \quad V_{12m}^3 = I.$$

For $m = 2$, the theorem is readily proven. If any orthogonal substitution replaces ξ_1 by $\gamma \xi_1 + \delta \xi_2$, then $S \equiv O_{1,2}^{\gamma,\delta} S_1$, where S_1 leaves ξ_1 fixed and is therefore the identity.

For $m = 3$, the theorem follows from §§ 174—179. For $m > 3$, it follows from § 180.

---

[1]) For simplicity we write only the indices altered by the substitution.

162   CHAPTER VII.

174. Theorem. — *If α_1, α_2, α_3 be any set of solutions in the $GF[p^n]$, $p > 2$, of the equation*
$$\alpha_1^2 + \alpha_2^2 + \frac{1}{\mu}\alpha_3^2 = 1,$$
there exists a substitution S, derived from the generators of § 173 which leave $\xi_1^2 + \xi_2^2 + \mu\xi_3^2$ invariant, such that S replaces ξ_1 by $\alpha_1\xi_1 + \alpha_2\xi_2 + \alpha_3\xi_3$.
The proposition follows if $1 - \alpha_1^2$ or $1 - \alpha_2^2$ be a square $\neq 0$ in the $GF[p^n]$. For example, if $1 - \alpha_2^2 = \tau^2$, then
$$\left(\frac{\alpha_1}{\tau}\right)^2 + \frac{1}{\mu}\left(\frac{\alpha_3}{\tau}\right)^2 = 1,$$
so that we may take
$$S \equiv O_{1,3}^{\frac{\alpha_1}{\tau},\frac{\alpha_3}{\tau}} O_{1,2}^{\tau,\alpha_2}.$$

The proposition will be true for α_1, α_2, α_3 if true for the quantities
where
$$\alpha_1' \equiv \alpha_1\varrho - \alpha_2\sigma, \quad \alpha_2' \equiv \alpha_1\sigma + \alpha_2\varrho, \quad \alpha_3' \equiv \alpha_3,$$
so that we have
$$\varrho^2 + \sigma^2 = 1,$$
148) $\quad\quad \alpha_1'^2 + \alpha_2'^2 + \frac{1}{\mu}\alpha_3'^2 = \alpha_1^2 + \alpha_2^2 + \frac{1}{\mu}\alpha_3^2 = 1.$

In fact, if the group contains a substitution S' which replaces ξ_1 by $\alpha_1'\xi_1 + \alpha_2'\xi_2 + \alpha_3'\xi_3$, it will contain the product $S \equiv O_{1,2}^{\varrho,-\sigma}S'$ which replaces ξ_1 by $\alpha_1\xi_1 + \alpha_2\xi_2 + \alpha_3\xi_3$.

175. Consider first the case in which -1 is a not-square in the $GF[p^n]$. By § 64, there are $p^n + 1$ sets of solutions ϱ, σ in the field of the equation $\varrho^2 + \sigma^2 = 1$. Not more than two of these sets of solutions give the same value to
$$\alpha_2' \equiv \alpha_1\sigma + \alpha_2\varrho.$$
Indeed, upon eliminating σ, we obtain a quadratic for ϱ. Hence α_2' takes at least $\frac{1}{2}(p^n + 1)$ distinct values. But, by § 67, there are exactly $\frac{1}{2}(p^n - 3)$ distinct marks $\eta \neq 0$ for which $\eta^2 - 1$ is a square[1]), so that $1 - \eta^2$ is a not-square. Hence there exist at least two values of α_2' for which $1 - \alpha_2'^2$ is a square or zero. If it be a square, our theorem follows from the previous section. There remains the case $\alpha_2'^2 = 1$, for which, by 148),
$$\alpha_1'^2 = -\frac{1}{\mu}\alpha_3'^2.$$
If $\mu = 1$, we have $\alpha_1' = \alpha_3' = 0$, since -1 is a not-square, and the required substitution is $S \equiv O_{1,2}^{0,\alpha'_2}$. If μ be a not-square, we may take $\mu = -1$, so that
$$\alpha_1' = \pm \alpha_3', \quad \alpha_2'^2 = 1.$$

---
1) Zero is not reckoned as a square.

LINEAR GROUP WITH QUADRATIC INVARIANT.   163

But the theorem is true for α_1', α_2', α_3' if true for the quantities

$$\alpha_1'' \equiv \alpha_1'\beta - \alpha_3'\gamma, \quad \alpha_2'' \equiv \alpha_2', \quad \alpha_3'' \equiv -\alpha_1'\gamma + \alpha_3'\beta,$$

where $\beta^2 - \gamma^2 = 1$. In fact, if S'' replaces ξ_1 by $\alpha_1''\xi_1 + \alpha_2''\xi_2 + \alpha_3''\xi_3$, then $O_{1,3}^{\beta,\gamma} S''$ will replace ξ_1 by $\alpha_1'\xi_1 + \alpha_2'\xi_2 + \alpha_3'\xi_3$. The $p^n - 1$ sets of solutions in the $GF[p^n]$ of the equation $\beta^2 - \gamma^2 = 1$ are given by

$$\beta = \frac{1}{2}(\tau + 1/\tau), \quad \mp \gamma = \frac{1}{2}(\tau - 1/\tau),$$

where τ runs through the series of marks $\neq 0$ of the field. Hence $\beta \mp \gamma$ may be given an arbitrary value $\tau \neq 0$ in the field. The theorem being evident if $\alpha_1' = 0$, we exclude this case. Then $\alpha_1'' \equiv \alpha_1' (\beta \mp \gamma)$ may be made to assume an arbitrary value except zero, and hence, if $p^n > 3$, a value for which $1 - \alpha_1''^2$ is a square in the field (§ 64). For $p^n = 3$, α_1', α_2', α_3' are each ± 1, so that we may evidently take

$$S = CV_{123}K,$$

where C and K are products formed from C_1, C_2, C_3. But, if C be the product of an odd number of the C_i, we note that

$$C_1 V_{123} = C_2 C_3 V_{123} C_1 C_2 C_3.$$

We may therefore assume that C and K are each products of an even number of the C_i and therefore derived from the given generators.

176. Suppose next that -1 is the square of a mark i of the $GF[p^n]$, while μ is a not-square. There exist $p^n + 1$ sets of solutions in the field of the equation

149) $$\beta^2 + \frac{1}{\mu}\gamma^2 = 1.$$

But the theorem is true for α_1, α_2, α_3 if true for

$$\alpha_1' \equiv \alpha_1, \quad \alpha_2' \equiv \beta\alpha_2 + \frac{\gamma}{\mu}\alpha_3, \quad \alpha_3' \equiv -\gamma\alpha_2 + \beta\alpha_3.$$

Indeed, if S' replaces ξ_1 by $\alpha_1'\xi_1 + \alpha_2'\xi_2 + \alpha_3'\xi_3$, then $O_{2,3}^{\beta,\gamma} S'$ will replace ξ_1 by $\alpha_1\xi_1 + \alpha_2\xi_2 + \alpha_3\xi_3$.

There are at least $\frac{1}{2}(p^n + 1)$ sets of solutions of 149) for which the values of α_2' are distinct; for, upon eliminating β, we obtain a quadratic for γ. But, by § 67, there exist only $\frac{1}{2}(p^n - 1)$ marks $i\xi$, and hence as many distinct marks ξ, for which

$$(i\xi)^2 + 1 \equiv 1 - \xi^2 = \text{not-square}.$$

Hence at least one set of solutions of 149) will make $1 - \alpha_2'^2$ a square or zero. If it be a square, the theorem follows from § 174. If it be zero, we have by 148),

$$\alpha_1'^2 = -\frac{1}{\mu}\alpha_3'^2.$$

Hence $\alpha_1' = \alpha_3' = 0$, $\alpha_2'^2 = 1$, so that we may take $S = O_{1,2}^{0,\,\alpha'_2}$.

177. For the case[1] -1 a square in the $GF[p^n]$ and $\mu = 1$, it follows from §§ 178—179 that $O_1(3, p^n)$ contains a subgroup of order at least $p^n(p^{2n}-1)$ generated by the substitutions $O_{i,j}^{\alpha,\beta}$, together with R_{123} if $p^n = 5$, all of determinant $+1$. But, by § 172, the order of $O_1(3, p^n)$ is $P_1(3, p^n) P_1(2, p^n)$. Here $P_1(2, p^n) = p^n - 1$, being the number of functions

$$\alpha_1 \xi_1 + \alpha_2 \xi_2 \qquad (\alpha_1^2 + \alpha_2^2 = 1)$$

by which the substitutions of $O_1(2, p^n)$ can replace ξ_1. Also

$$P_1(3, p^n) \overline{\leq} p^{2n} + p^n.$$

In fact, if a substitution of $O_1(3, p^n)$ replace ξ_1 by

then
$$\omega_1 \equiv \alpha_1 \xi_1 + \alpha_2 \xi_2 + \alpha_3 \xi_3,$$
150)
$$\alpha_1^2 + \alpha_2^2 + \alpha_3^2 = 1.$$

By § 66, this equation has $p^{2n} + p^n$ sets of solutions in the $GF[p^n]$, -1 being a square. The order of $O_1(3, p^n)$ is thus *at most* $(p^{2n} + p^n)(p^n - 1)$. From the two results it follows that this number equals the order of $O_1(3, p^n)$ and that for every set of solutions of 150) there exists a substitution of $O_1(3, p^n)$, derived from $O_{i,j}^{\alpha,\beta}$ and R, which replaces ξ_1 by ω_1.

178. Theorem. — *The first orthogonal group $O_1(3, p^n)$ contains a subgroup $O_1'(3, p^n)$ holoedrically isomorphic with the group $LF(2, p^n)$ of linear fractional substitutions of determinant unity.*

Let $-1 = i^2$, so that i belongs to the $GF[p^n]$ if and only if -1 be a square in that field. Introduce in place of ξ_1, ξ_2, ξ_3 the new indices

$$\eta_1 \equiv -i\xi_1, \quad \eta_2 \equiv \xi_2 - i\xi_3, \quad \eta_3 \equiv \xi_2 + i\xi_3,$$
so that
$$-\eta_1^2 + \eta_2 \eta_3 \equiv \xi_1^2 + \xi_2^2 + \xi_3^2.$$

---

[1] For a more direct treatment of this case, but one involving considerable calculation, see *Amer. Journal*, vol. 21, pp. 202—204, in which the proof of Jordan, Traité, pp. 164—166, for $n = 1$, is corrected and generalized.

The following substitution of determinant unity,

$$Y: \quad \begin{array}{c} \eta_1' = \\ \eta_2' = \\ \eta_3' = \end{array} \begin{array}{|ccc|} \hline \eta_1 \cdot & \eta_2 & \eta_3 \\ \alpha\delta + \beta\gamma & \alpha\gamma & \beta\delta \\ 2\alpha\beta & \alpha^2 & \beta^2 \\ 2\gamma\delta & \gamma^2 & \delta^2 \\ \hline \end{array} \quad (\alpha\delta - \beta\gamma) = 1$$

leaves $\eta_2\eta_3 - \eta_1^2$ absolutely invariant. Written in terms of the indices ξ_1, ξ_2, ξ_3, it takes the form

$$X \equiv \begin{pmatrix} \alpha\delta + \beta\gamma & i(\alpha\gamma + \beta\delta) & \alpha\gamma - \beta\delta \\ -i(\alpha\beta + \gamma\delta) & \frac{1}{2}(\alpha^2 + \beta^2 + \gamma^2 + \delta^2) & \frac{i}{2}(-\alpha^2 + \beta^2 - \gamma^2 + \delta^2) \\ \alpha\beta - \gamma\delta & \frac{i}{2}(\alpha^2 + \beta^2 - \gamma^2 - \delta^2) & \frac{1}{2}(\alpha^2 - \beta^2 - \gamma^2 + \delta^2) \end{pmatrix}.$$

It follows that X has determinant unity (§ 101) and leaves $\xi_1^2 + \xi_2^2 + \xi_3^2$ absolutely invariant. Giving to the substitution Y the notation

151) $$\begin{bmatrix} \alpha & \beta \\ \gamma & \delta \end{bmatrix} \equiv \begin{bmatrix} -\alpha & -\beta \\ -\gamma & -\delta \end{bmatrix} \quad (\alpha\delta - \beta\gamma = 1)$$

we readily verify the formula of composition

$$\begin{bmatrix} \alpha' & \beta' \\ \gamma' & \delta' \end{bmatrix} \begin{bmatrix} \alpha & \beta \\ \gamma & \delta \end{bmatrix} = \begin{bmatrix} \alpha\alpha' + \beta\gamma' & \alpha\beta' + \beta\delta' \\ \gamma\alpha' + \delta\gamma' & \gamma\beta' + \delta\delta' \end{bmatrix}.$$

The group of the substitutions X, being isomorphic with the group of the substitutions Y, is isomorphic with the group of the linear fractional substitutions 151). But Y and therefore X is the identity if and only if $\alpha = \delta = \pm 1$, $\beta = \gamma = 0$. Hence the isomorphism is holoedric.

If -1 be a square in the $GF[p^n]$, so that the coefficients of X belong to that field, the substitutions X form a group $O_1'(3, p^n)$, a subgroup of $O_1(3, p^n)$, which is holoedrically isomorphic with $LF(2, p^n)$.

If -1 be a not-square in the $GF[p^n]$, the coefficients of X will belong to the $GF[p^n]$ if we choose $\alpha, \beta, \gamma, \delta$ in the $GF[p^{2n}]$ such that α is conjugate (§ 73) with δ, β with $-\gamma$, with respect to the $GF[p^n]$. By § 144, the resulting substitutions 151) of determinant unity form a group holoedrically isomorphic with $LF(2, p^n)$. The corresponding substitutions X form a subgroup $O_1'(3, p^n)$ of $O_1(3, p^n)$.

In each case, the subgroup $O_1'(3, p^n)$ has the order $\frac{1}{2}p^n(p^{2n} - 1)$, since it is holoedrically isomorphic with $LF(2, p^n)$. We proceed to prove that this subgroup does not coincide with $O_1(3, p^n)$. In order that $O_{2,3}^{a,b}$ shall be of the form X, it is necessary and sufficient that

$$\beta = \gamma = 0, \quad \alpha\delta = 1, \quad \alpha^2 = a + bi, \quad \delta^2 = a - bi.$$

According to the definition of α, β, γ, δ in the above cases, the expressions
$$A \equiv \frac{1}{2}(\delta + \alpha), \quad B \equiv \frac{i}{2}(\delta - \alpha)$$
belong to the $GF[p^n]$. The above conditions then give
$$a = 2A^2 - 1, \quad b = 2AB, \quad A^2 + B^2 = 1,$$
so that $O_{2,3}^{a,b}$ must be the special substitution $Q_{2,3}^{A,B}$ defined in § 181. Any orthogonal substitution $O_{2,3}$ not of the form $Q_{2,3}$, and therefore not of the form X, will extend $O_1'(3, p^n)$ to a larger subgroup G of $O_1(3, p^n)$. The order of G is therefore *at least* $p^n(p^{2n} - 1)$. From the remarks at the end of § 177, it follows that G has exactly this order and hence coincides with $O_1(3, p^n)$.

179. We proceed to the proof that, if -1 be a square i^2 in the $GF[p^n]$, the group $O_1(3, p^n)$ is generated by the substitutions $O_{i,j}^{\alpha,\beta}$ together with R_{123} if $p^n = 5$. If $p^n > 5$, there exist (§ 64) marks β and τ in the $GF[p^n]$ such that
$$1 + \beta^2 = \tau^2 \qquad (\beta \neq 0, \ \tau \neq 0).$$
Then the product
$$O_{1,2}^{0,1} \, O_{2,3}^{\frac{1}{\tau}, \frac{\beta}{\tau}} \, O_{1,2}^{i\beta, -\tau} \, O_{2,3}^{\frac{1}{2}\left(\frac{1}{\tau}+\tau\right), \frac{i}{2}\left(\frac{1}{\tau}-\tau\right)}$$
$$\equiv \begin{pmatrix} 1 & i\beta & -\beta \\ -i\beta & 1+\frac{1}{2}\beta^2 & \frac{i}{2}\beta^2 \\ \beta & \frac{i}{2}\beta^2 & 1-\frac{1}{2}\beta^2 \end{pmatrix} \equiv \begin{bmatrix} 1 & \beta \\ 0 & 1 \end{bmatrix}.$$
But $\begin{bmatrix} \alpha & 0 \\ 0 & \alpha^{-1} \end{bmatrix}$, which is an $O_{2,3}$, transforms $\begin{bmatrix} 1 & \beta \\ 0 & 1 \end{bmatrix}$ into $\begin{bmatrix} 1 & \beta\alpha^2 \\ 0 & 1 \end{bmatrix}$. Furthermore,
$$\begin{bmatrix} 1 & \beta\alpha_1^2 \\ 0 & 1 \end{bmatrix} \begin{bmatrix} 1 & \beta\alpha_2^2 \\ 0 & 1 \end{bmatrix} = \begin{bmatrix} 1 & \beta(\alpha_1^2 + \alpha_2^2) \\ 0 & 1 \end{bmatrix}.$$
Since $\beta \neq 0$, we can (§ 64) find marks α_1 and α_2 in the field such that $\beta(\alpha_1^2 + \alpha_2^2) = \varkappa$, where $\varkappa$ is an arbitrary mark $\neq 0$. Also
$$\begin{bmatrix} 0 & 1 \\ -1 & 0 \end{bmatrix}^{-1} \begin{bmatrix} 1 & -\varkappa \\ 0 & 1 \end{bmatrix} \begin{bmatrix} 0 & 1 \\ -1 & 0 \end{bmatrix} = \begin{bmatrix} 1 & 0 \\ \varkappa & 1 \end{bmatrix}, \quad \begin{bmatrix} 0 & 1 \\ -1 & 0 \end{bmatrix} \equiv O_{1,3}^{-1,0}.$$
Hence, if $p^n > 5$, we have reached from the $O_{i,j}^{\alpha,\beta}$ the substitutions
$$\begin{bmatrix} 1 & \varkappa \\ 0 & 1 \end{bmatrix}, \quad \begin{bmatrix} 1 & 0 \\ \varkappa & 1 \end{bmatrix} \qquad (\varkappa \text{ arbitrary}).$$
By §§ 100 and 108, these substitutions generate the group $LF(2, p^n)$. Hence the $O_{i,j}^{\alpha,\beta}$ from which they were derived generate the isomorphic

group $O_1'(3, p^n)$. Then, by the last section, all the $O_{i,j}^{\alpha,\beta}$ generate $O_1(3, p^n)$. For $p^n = 5$, $i = 2$, we have

$$R_{123} \equiv \begin{pmatrix} 1 & 1 & 2 \\ 1 & 2 & 1 \\ 2 & 1 & 1 \end{pmatrix} = \begin{bmatrix} 2 & 1 \\ 0 & 3 \end{bmatrix} = \begin{bmatrix} 1 & 3 \\ 0 & 1 \end{bmatrix} \begin{bmatrix} 2 & 0 \\ 0 & 2^{-1} \end{bmatrix}.$$

Hence from R_{123} and $\begin{bmatrix} 2 & 0 \\ 0 & 2^{-1} \end{bmatrix} \equiv O_{2,3}^{-1,0}$ we reach $\begin{bmatrix} 1 & 3 \\ 0 & 1 \end{bmatrix}$. It follows as above that R_{123} and $O_{2,3}^{-1,0} \equiv C_2 C_3$ and $C_1 C_3$ generate $O_1'(3, 5)$. The latter is extended to $O_1(3, 5)$ by any $O_{i,j}^{0,\pm 1}$.

180. Theorem. — *If $\alpha_1, \alpha_2, \ldots, \alpha_m$ be any set of solutions in the $GF[p^n]$ of*

$$\alpha_1^2 + \alpha_2^2 + \cdots + \alpha_{m-1}^2 + \frac{1}{\mu}\alpha_m^2 = 1,$$

there exists a substitution S derived from the generators[1]) of § 173 which replaces ξ_1 by $\omega_1 \equiv \alpha_1 \xi_1 + \alpha_2 \xi_2 + \cdots + \alpha_m \xi_m$.

The proposition having been proven for $m = 2$ and $m = 3$, we will give a proof by induction from $m - 1$ to m, supposing $m > 3$.

Consider first the case in which every sum of three of the terms $\alpha_1^2, \alpha_2^2, \ldots, \alpha_{m-1}^2, \frac{1}{\mu}\alpha_m^2$ is zero. These terms must all be equal and therefore

$$m\alpha_1^2 = 1, \quad 3\alpha_1^2 = 0, \quad \mu = \text{square}.$$

Hence $p = 3$, while m is of the form $3k + 2$ or $3k + 1$.

If $m = 3k + 2$, we have $1 - \alpha_1^2 = \alpha_1^2 \neq 0$, so that the theorem is reduced by § 174 to the case of $m - 1$ indices.

If $m = 3k + 1$, we must have $\alpha_1^2 = 1$. But the product $O_{1,2}^{\alpha,\beta} S$ will replace ξ_1 by $\alpha_1' \xi_1 + \cdots + \alpha_m' \xi_m$, where

$$\alpha_1' \equiv \alpha\alpha_1 - \beta\alpha_2, \quad \alpha_2' \equiv \beta\alpha_1 + \alpha\alpha_2, \quad \alpha_j' \equiv \alpha_j \quad (j = 3, \ldots, m).$$

Of the $3^n \pm 1$ sets of values in the $GF[3^n]$ satisfying

$$\alpha^2 + \beta^2 = 1,$$

at most two give the same value to α_1' and hence at most four make $\alpha_1'^2 = 1$. Hence, if $n > 1$, we can avoid the case $\alpha_1^2 = 1$. For $p^n = 3$, we may take

---

1) For the case $p^n = 5$, $m \leqq 4$, $\mu = \text{not-square}$, it would appear that the generator R_{123} were necessary in addition to the $O_{i,j}^{\alpha,\beta}$. We can, however, express R_{123} in terms of the generators

$$O_{i,m}^{2,1}: \begin{cases} \xi_i' = 2\xi_i + \xi_m \\ \xi_m' = 3\xi_i + 2\xi_m \end{cases},$$

leaving invariant $\xi_1^2 + \xi_2^2 + \cdots + \xi_{m-1}^2 + 3\xi_m^2$. Indeed,

$$R_{123} = O_{1m} O_{2m} O_{3m}^{-1} O_{1m} O_{2m}^{-1} O_{3m}^{-1}.$$

168 CHAPTER VII.

$$S = CW_{1234}W_{1567} \ldots W_{13k-1\,3k\,3k+1}K,$$

where C and K denote products of an even number of the C_i [compare the end of § 175].

Suppose next that the above sums are not all zero, for example[1])

$$a_1^2 + a_2^2 + \frac{1}{\mu}a_m^2 \neq 0.$$

We have proven that, for every set of solutions of

152) $\qquad a^2 + \beta^2 + \dfrac{1}{\mu}\gamma^2 = 1,$

there exists a substitution Σ of the group,

$$\xi_1' = \alpha\xi_1 + \beta\xi_2 + \gamma\xi_m, \quad \xi_2' = \alpha'\xi_1 + \beta'\xi_2 + \gamma'\xi_m, \quad \xi_m' = \alpha''\xi_1 + \beta''\xi_2 + \gamma''\xi_m$$

which therefore satisfies the relation 152) and the following:

$$\overset{2}{\alpha'} + \overset{2}{\beta'} + \frac{1}{\mu}\overset{2}{\gamma'} = 1, \quad \alpha^2 + \overset{2}{\alpha'} + \mu\overset{2}{\alpha''} = 1, \quad \alpha\beta + \alpha'\beta' + \mu\alpha''\beta'' = 0, \text{ etc.}$$

If there be a substitution S' in our group which replaces ξ_1 by

$$\alpha_1'\xi_1 + \alpha_2'\xi_2 + \alpha_m'\xi_m + \sum_{j=3}^{m-1}\alpha_j\xi_j,$$

where

$$\alpha_1' = \alpha\alpha_1 + \beta\alpha_2 + \frac{\gamma}{\mu}\alpha_m,$$

$$\alpha_2' = \alpha'\alpha_1 + \beta'\alpha_2 + \frac{\gamma'}{\mu}\alpha_m,$$

$$\alpha_m' = \mu\alpha''\alpha_1 + \mu\beta''\alpha_2 + \gamma''\alpha_m,$$

then the group will contain $\Sigma S'$ which replaces ξ_1 by ω_1. The proposition is therefore true for the quantities α_j if true for α_1', α_2', α_m', α_3, α_4, ..., α_{m-1}. We may thus make our proof by induction from $m-1$ to m by showing that it is possible to choose α, β, γ among the sets of solutions of 152) in such a way that $\alpha_1' = 0$. We may suppose that $\alpha_1 \neq 0$, since otherwise the proposition is already proven.

If $\alpha_1^2 + \alpha_2^2 = 0$, then $\alpha_2 \neq 0$. From $\dfrac{1}{\mu}\alpha_m^2 = 1$, it follows that μ is a square, say $\mu = 1$. Then the values

$$\alpha = \frac{-\alpha_m}{2\alpha_1}, \quad \beta = \frac{-\alpha_m}{2\alpha_2}, \quad \gamma = 1$$

satisfy 152) and make $\alpha_1' = 0$.

If $\alpha_1^2 + \alpha_2^2 \neq 0$, the condition 152) combines with $\alpha_1' = 0$ to give a single condition for β and γ:

$$\left(\beta\alpha_2 + \frac{\gamma}{\mu}\alpha_m\right)^2 + \alpha_1^2\left(\beta^2 + \frac{1}{\mu}\gamma^2\right) = \alpha_1^2.$$

---

[1]) The treatment for a case like $\alpha_1^2 + \alpha_2^2 + \alpha_3^2 \neq 0$ is quite similar, taking $\mu = 1$.

Multiplying this by $\alpha_1^2 + \alpha_2^2$, it may be given the form

$$\left\{\beta(\alpha_1^2 + \alpha_2^2) + \frac{\alpha_2 \alpha_m}{\mu}\gamma\right\}^2 + \frac{\gamma^2 \alpha_1^2}{\mu}\left(\alpha_1^2 + \alpha_2^2 + \frac{\alpha_m^2}{\mu}\right) = \alpha_1^2(\alpha_1^2 + \alpha_2^2).$$

Since the coefficient of γ^2 is not zero, this equation has in the $GF[p^n]$ (by § 64) $p^n \pm 1$ sets of solutions for γ and

$$\left\{\beta(\alpha_1^2 + \alpha_2^2) + \frac{\alpha_2 \alpha_m}{\mu}\gamma\right\}$$

and hence as many sets of solutions β, γ.

The structure of the first and second orthogonal groups, §§ 181—198.

181. The group $O_\mu(m, p^n)$ contains the substitutions

$$O_{i,j}^{\varrho,\sigma}: \begin{cases} \xi_i' = \varrho\xi_i + \sigma\xi_j \\ \xi_j' = -\frac{\sigma}{\lambda}\xi_i + \varrho\xi_j \end{cases} \qquad \left(\varrho^2 + \frac{1}{\lambda}\sigma^2 = 1\right)$$

leaving $\xi_i^2 + \lambda\xi_j^2$ invariant, where $\lambda = 1$ if $i, j < m$, but $\lambda = \mu$ if $i < j = m$. For i and j fixed, while ϱ, σ take all possible values in the field such that $\varrho^2 + \frac{1}{\lambda}\sigma^2 = 1$, the substitutions $O_{i,j}^{\varrho,\sigma}$ form a subgroup denoted by $O_{i,j}$. Its substitutions are commutative since the following product is unaltered if we interchange ϱ with ϱ' and σ with σ':

$$O_{i,j}^{\varrho',\sigma'} O_{i,j}^{\varrho,\sigma}: \begin{cases} \xi_i' = \left(\varrho\varrho' - \frac{\sigma\sigma'}{\lambda}\right)\xi_i + (\varrho\sigma' + \varrho'\sigma)\xi_j \\ \xi_j' = -\left(\frac{\varrho\sigma' + \varrho'\sigma}{\lambda}\right)\xi_i + \left(\varrho\varrho' - \frac{\sigma\sigma'}{\lambda}\right)\xi_j. \end{cases}$$

By § 64, the order of O_{ij} is $p^n - \varepsilon_{ij}$, where $\varepsilon_{ij} = +1$ or -1 according as $-1/\lambda$ is a square or a not-square in the $GF[p^n]$.

The squares of the substitutions of $O_{i,j}$ form a commutative group $Q_{i,j}$, composed of the substitutions,

$$Q_{i,j}^{\varrho,\sigma} \equiv (O_{i,j}^{\varrho,\sigma})^2: \begin{cases} \xi_i' = (2\varrho^2 - 1)\xi_i + 2\varrho\sigma\xi_j, \\ \xi_j' = -\frac{2\varrho\sigma}{\lambda}\xi_i + (2\varrho^2 - 1)\xi_j. \end{cases}$$

The order of Q_{ij} is $\frac{1}{2}(p^n - \varepsilon_{ij})$. Indeed, the identity

$$Q_{i,j}^{\varrho,\sigma} \equiv Q_{i,j}^{\varrho',\sigma'}$$

holds if and only if $\varrho' = \pm \varrho, \sigma' = \pm \sigma$.

170 CHAPTER VII.

Let $O_{i,j}^{\varrho,\sigma}$ be a particular substitution which extends Q_{ij} to O_{ij}, the values ϱ, σ depending on λ. Consider the subgroup $O'_\mu(m, p^n)$ of $O_\mu(m, p^n)$ generated by the substitutions

$$Q_{i,j}^{\alpha,\beta}, \quad O_{i,j}^{\varrho,\sigma} \cdot O_{kl}^{\varrho,\sigma} \quad (i, j, k, l = 1, \ldots, m; \; i \neq j, k \neq l)$$

where α, β take all the values in the $GF[p^n]$ satisfying $\alpha^2 + \frac{1}{\lambda}\beta^2 = 1$, the generator R, W or V being added in the exceptional cases of § 173.

182. Theorem. — *The order of $O'_\mu(m, p^n)$ is at least half of the order of $O_\mu(m, p^n)$.*

By the theorem of § 173, every substitution of $O_\mu(m, p^n)$ has the form
$$S \equiv h_1 \, O_{i,j}^{\varrho,\sigma} \, h_2 \, O_{k,l}^{\varrho,\sigma} \, h_3 \ldots$$
where the h_i (and the h'_i, h'', h, ... below) are derived from the generators of $O'_\mu(m, p^n)$. For $m > 2$, $O_{1,2}^{\gamma,\delta}$ and $O_{1,2}^{\gamma,-\delta} \equiv O_{2,1}^{\gamma,\delta}$ are reciprocal. Hence
$$O_{i,j}^{\varrho,\sigma} = O_{i,j}^{\varrho,\sigma} \cdot O_{2,1}^{\varrho,\sigma} O_{1,2}^{\varrho,\sigma} = h_4 \, O_{1,2}^{\varrho,\sigma}.$$
Hence
$$S = h'_1 \, O_{1,2}^{\varrho,\sigma} h'_2 \, O_{1,2}^{\varrho,\sigma} h'_3 \ldots$$
Furthermore, $O_{1,2}^{\varrho,\sigma}$ is commutative with every $Q_{1,2}^{\alpha,\beta}$ and every $Q_{i,j}^{\alpha,\beta}$, i and $j > 2$. Since the square of $O_{1,k}^{\varrho,\sigma}$ is $Q_{1,k}^{\varrho,\sigma}$, whose reciprocal is $Q_{1,k}^{\varrho,-\sigma}$,
$$O_{1,2}^{\varrho,\sigma} Q_{1,k}^{\alpha,\beta} = O_{1,2}^{\varrho,\sigma} (O_{1,k}^{\varrho,\sigma})^2 Q_{1,k}^{\varrho,-\sigma} \cdot Q_{1,k}^{\alpha,\beta}$$
$$= O_{1,2}^{\varrho,\sigma} O_{1,k}^{\varrho,\sigma} \cdot Q_{1,k}^{\varrho,-\sigma} Q_{1,k}^{\alpha,\beta} \cdot O_{1,k}^{\varrho,\sigma} = h' \, O_{1,k}^{\varrho,\sigma} = h'' \, O_{1,2}^{\varrho,\sigma}.$$
Aside from the above exceptional cases, we may conclude that S is of the form h or else $h \cdot O_{1,2}^{\varrho,\sigma}$. We treat next the exceptional cases.

1^0. For $p^n = 5$, $m \geq 3$, $\mu = 1$, the additional generator is R_{123}, and the only $Q_{i,j}^{\alpha,\beta}$ are $Q_{i,j}^{0,\pm 1} \equiv C_i C_j$ and $Q_{i,j}^{\pm 1, 0} = I$. Since
$$O_{i,j}^{0,-1} \equiv T_{ij} C_i,$$
where $T_{ij} \equiv (\xi_i \xi_j)$, is not in Q_{ij}, it may be taken for $O_{i,j}^{\varrho,\sigma}$. To complete the proof that $S = h$ or $h O_{1,2}^{\varrho,\sigma}$, we note that
$$T_{12} C_1 \cdot R_{123} = C_1 C_3 R_{123} \cdot T_{23} C_2 \cdot T_{12} C_1 \cdot C_1 C_3 \cdot T_{12} C_1.$$

2^0. For $p^n = 3$, $m \geq 4$, $\mu = 1$, the additional generator is W_{1234}. The remarks of 1^0 apply here, if we replace the last formula by
$$T_{12} C_1 \cdot W_{1234} = C_1 C_2 W_{1234}^2 T_{12} C_1.$$

3^0. For $p^n = 3$, $m \geq 3$, $\mu = \nu = -1$, the additional generator is V_{12m} and the only $O_{i,m}^{\alpha,\beta}$ ($\alpha^2 - \beta^2 = 1$) are $C_i C_m$ and I, the only $Q_{i,m}^{\alpha,\beta}$ being I. We may take $O_{i,j}^{\varrho,\sigma} \equiv T_{ij} C_i$ ($i, j < m$) and $O_{i,m}^{\varrho,\sigma} \equiv C_i C_m$. To complete our proof, we use the formulae
$$T_{12} C_1 V_{12m} = C_1 C_2 V_{12m}^2 T_{12} C_1, \quad C_1 C_m V_{12m} = V_{12m}^2 T_{12} C_2.$$

LINEAR GROUP WITH QUADRATIC INVARIANT. 171

183. Theorem. — *The group $O_1'(3, p^n)$ just defined is identical with the subgroup of $O_1(3, p^n)$ of index two defined in § 178.*

It is only necessary to show that every $Q_{i,j}^{a,b}$ and every $O_{i,j}^{x,\sigma} O_{i,k}^{x,\sigma}$ are of the form X or, if we prefer, 151). We have

$$O_{1,2}^{x,\sigma} O_{2,3}^{x,\sigma} \equiv \begin{pmatrix} x & \sigma & 0 \\ -x\sigma & x^2 & \sigma \\ \sigma^2 & -x\sigma & x \end{pmatrix} \equiv \begin{bmatrix} \dfrac{x+1+i\sigma}{2} & -\dfrac{x-1+i\sigma}{2} \\ \dfrac{x-1-i\sigma}{2} & \dfrac{x+1-i\sigma}{2} \end{bmatrix},$$

$$T_{12} T_{13} \equiv \begin{bmatrix} \dfrac{1}{2}(1-i) & -\dfrac{1}{2}(1-i) \\ \dfrac{1}{2}(1+i) & \dfrac{1}{2}(1+i) \end{bmatrix}.$$

In particular, we reach $T_{12} T_{13}$ and $T_{12} T_{23}$. These suffice to transform $O_{1,2}^{x,\sigma} O_{2,3}^{x,\sigma}$ into $O_{2,3}^{x,\sigma} O_{3,1}^{x,\sigma}$ and $O_{3,1}^{x,\sigma} O_{1,2}^{x,\sigma}$. Transforming these products by $C_1 C_2$, $C_1 C_3$ and $C_2 C_3$, we obtain every $O_{i,j}^{x,\sigma} O_{k,l}^{x,\sigma}$, since C_i transforms $O_{i,j}^{x,\sigma}$ into $O_{j,i}^{x,\sigma}$. Transforming $Q_{2,3}^{a,b}$ (which is of the form X by § 178) by $T_{12} T_{13}$ and $T_{12} T_{23}$, we obtain every $Q_{i,j}^{a,b}$.

184. Theorem. — *The group $O_\nu'(3, p^n)$ is of index two under the second orthogonal group $O_\nu(3, p^n)$.*

Consider the substitutions, in which $\alpha^2 + \beta^2 = \nu$,

$$O: \begin{cases} \xi_1' = \alpha \xi_1 - \beta \xi_2 \\ \xi_2' = \beta \xi_1 + \alpha \xi_2 \end{cases}, \quad O^{-1}: \begin{cases} \xi_1' = \dfrac{\alpha}{\nu} \xi_1 + \dfrac{\beta}{\nu} \xi_2 \\ \xi_2' = -\dfrac{\beta}{\nu} \xi_1 + \dfrac{\alpha}{\nu} \xi_2 \end{cases}.$$

Since O transforms $\xi_1^2 + \xi_2^2 + \nu \xi_3^2$ into $\nu(\xi_1^2 + \xi_2^2 + \xi_3^2)$, it transforms the group $O_1(3, p^n)$ of the latter into the group $O_\nu(3, p^n)$ of the former. But O is commutative with $O_{1,2}^{0,\sigma}$. Hence if $O_{1,2}^{x,\tau}$ serves to extend the subgroup $O_1'(3, p^n)$ to $O_1(3, p^n)$, there exists a subgroup G of $O_\nu(3, p^n)$ which $O_{1,2}^{x,\tau}$ extends to the latter. We readily prove that G is identical with the $O_\nu'(3, p^n)$ defined in § 181. For example, O transform $O_{1,2}^{x,\tau} O_{1,2}^{0,\sigma} C_1 C_3$ into $O_{1,2}^{x,\tau} C_1 C_3$, where

$$\varrho \equiv (\alpha^2 - \beta^2)/\nu, \quad \sigma \equiv -2\alpha\beta/\nu, \quad \varrho^2 + \sigma^2 = 1.$$

Here $O_{1,2}^{\varrho,\sigma}$ is not a $Q_{1,2}$ since $(1+\varrho)/2 \equiv \alpha^2/\nu$ is a not-square. But $C_1 C_3$ is a $Q_{1,3}$ in $O_1(3, p^n)$, but not in $O_\nu(3, p^n)$. It follows that G contains the product $O_{1,2}^{x,\tau} C_1 C_3$, neither factor being a Q.

185. It will be shown in the following sections that $O_\mu'(m, p^n)$ is not identical with $O_\mu(m, p^n)$ in the cases $m = 4, 5, 6$ and therefore, by § 182, that its index under $O_\mu(m, p^n)$ is exactly two. By §§ 181—184, the same result is true for $m = 2$ and $m = 3$. For

172 CHAPTER VII.

various reasons it would seem that the same result holds true when $m > 6$, but no explicit investigation has yet been made. The developments in §§ 191—193 are made on the assumption that this index is 2. Moreover, if this conjecture prove false, very simple alterations would be necessary in the treatment.

186. We continue the investigations begun in § 163 on the senary group $G'_{4,2}$, whose substitutions leave absolutely invariant the Pfaffian [1 2 3 4], viz.,

$$F_4 \equiv Y_{12}Y_{34} - Y_{13}Y_{24} + Y_{14}Y_{23}.$$

Denote by G_6 the group of all substitutions of determinant unity in the $GF[p^n]$, $p > 2$, which leave F_4 absolutely invariant. We will prove that G_6 is holoedrically isomorphic with $O_\mu(6, p^n)$, where $\mu = 1$ or ν according as $p^n = 4l + 1$ or $4l + 3$. Hence G_6 has the order (§ 172)

153) $(p^{4n} - 1)p^{3n}(p^{3n} - 1)p^{2n}(p^{2n} - 1)p^n.$

It will therefore follow from the theorem of § 163 that $G_{4,2}$ is a subgroup of index two under the group G_6.

187. Let $p^n = 4l + 1$, so that -1 is the square of a mark i belonging to the $GF[p^n]$. We make the following transformation of indices:

154) $Y_{12} \equiv \xi_1 + i\xi_2, \quad -Y_{13} \equiv \xi_3 + i\xi_4, \quad Y_{14} \equiv \xi_5 + i\xi_6,$
   $Y_{34} \equiv \xi_1 - i\xi_2, \quad Y_{24} \equiv \xi_3 - i\xi_4, \quad Y_{23} \equiv \xi_5 - i\xi_6.$

Then F_4 takes the form

$$\Psi \equiv \sum_{i=1}^{6} \xi_i^2.$$

Hence G_6 is holoedrically isomorphic with $O_1(6, p^n)$. By § 164, the following substitution of G_6 (leaving four of the indices fixed):

$$Y'_{13} = \tau Y_{13}, \quad Y'_{24} = \tau^{-1} Y_{24}$$

belongs to the subgroup $G'_{4,2}$ if and only if τ be a square in the field. Expressed in the new indices, it has the form

155) $\xi'_3 = \dfrac{1}{2}(\tau + \tau^{-1})\xi_3 + \dfrac{i}{2}(\tau - \tau^{-1})\xi_4,$
   $\xi'_4 = -\dfrac{i}{2}(\tau - \tau^{-1})\xi_3 + \dfrac{1}{2}(\tau + \tau^{-1})\xi_4.$

For τ an arbitrary mark $\neq 0$ of the field, 155) may be written

156) $O_{3,4}^{\varrho,\sigma}$ $\varrho \equiv \dfrac{1}{2}(\tau + \tau^{-1}), \ \sigma \equiv \dfrac{i}{2}(\tau - \tau^{-1}), \ \varrho^2 + \sigma^2 = 1.$

For $\tau = t^2$, 155) may be expressed in the form

157) $\quad Q_{3,4}^{\gamma,\delta} \qquad \gamma \equiv \frac{1}{2}(t + t^{-1}), \quad \delta \equiv \frac{i}{2}(t - t^{-1}), \quad \gamma^2 + \delta^2 = 1.$

For τ a not-square, 155) is not of the form 157), since that would require $\gamma^2 = (\tau + 1)^2/4\tau$. It follows that to the subgroup $G'_{4,2}$ of G_6 of index two there corresponds a subgroup O of $O_1(6, p^n)$ of index two, where O is extended to $O_1(6, p^n)$ by any substitution $O_{3,4}^{\varrho,\sigma}$ not of the form $Q_{3,4}$. We proceed to prove that O is identical with the group $O'_1(6, p^n)$ defined in § 181. We first show that O contains all even substitutions on the six letters $\xi_1, \ldots, \xi_6$. Expressing the substitution

$(\xi_1 \xi_2 \xi_3): \quad \xi'_1 = \xi_2, \quad \xi'_2 = \xi_3, \quad \xi'_3 = \xi_1, \quad \xi'_i = \xi_i \qquad (i = 4, 5, 6)$

in terms of the indices Y_{ij}, it takes the form

	Y_{12}	Y_{13}	Y_{14}	Y_{23}	Y_{24}	Y_{34}
$Y'_{12} =$	$-i/2$	$-i/2$	0	0	$i/2$	$i/2$
$Y'_{13} =$	$-1/2$	$1/2$	0	0	$1/2$	$-1/2$
$Y'_{14} =$	0	0	1	0	0	0
$Y'_{23} =$	0	0	0	1	0	0
$Y'_{24} =$	$1/2$	$1/2$	0	0	$1/2$	$1/2$
$Y'_{34} =$	$-i/2$	$i/2$	0	0	$-i/2$	$i/2$

By inspection this substitution is the second compound of

$$\begin{pmatrix} \frac{1}{2}(1-i) & 0 & 0 & \frac{1}{2}(1-i) \\ 0 & \frac{1}{2}(1-i) & \frac{1}{2}(1-i) & 0 \\ 0 & -\frac{1}{2}(1+i) & \frac{1}{2}(1+i) & 0 \\ -\frac{1}{2}(1+i) & 0 & 0 & \frac{1}{2}(1+i) \end{pmatrix}$$

having determinant unity. Hence O contains the substitution $(\xi_1 \xi_2 \xi_3)$. In the transformation of indices 154), the pairs ξ_1 and ξ_2, ξ_3 and ξ_4, ξ_5 and ξ_6 enter symmetrically. Hence O contains the substitution $(\xi_i \xi_j \xi_k)$, two of the distinct integers i, j, k, each ≤ 6, being chosen from one of the above pairs. But the *linear* substitution denoted by $(\xi_1 \xi_2 \xi_3)$ transforms $(\xi_1 \xi_2 \xi_5)$ into $(\xi_1 \xi_5 \xi_3)$. Hence O contains every cyclic substitution $(\xi_r \xi_s \xi_t)$ on the six indices and therefore every even permutation of the six indices.[1]

---

1) Netto-Cole, The theory of substitutions, p. 35.

Having every $Q_{3,4}^{\gamma,\delta}$, O contains their transformed $Q_{i,j}^{\gamma,\delta}$ ($i \neq j$) by the even substitution $(\xi_3 \xi_i)(\xi_4 \xi_j)$. By § 164, Note, O contains the product $O_{1,2}^{\varrho,\sigma} O_{3,4}^{\varrho,\sigma}$ and therefore also every $O_{i,j}^{\varrho,\sigma} O_{k,l}^{\varrho,\sigma}$, where i, j, k, l are distinct. Hence O contains

$$O_{1,2}^{\varrho,\sigma} O_{3,4}^{\varrho,\sigma} \cdot O_{4,3}^{\varrho,\sigma} O_{1,5}^{\varrho,\sigma} \equiv O_{1,2}^{\varrho,\sigma} O_{1,5}^{\varrho,\sigma}$$

and therefore every $O_{i,j}^{\varrho,\sigma} O_{k,l}^{\varrho,\sigma}$ in which two of the subscripts are alike. For the case $p^n = 5$, $i = 2$, there is an additional generator, viz., R_{123}. Expressing R_{123} in the indices Y_{ij}, we obtain the substitution

$$\begin{pmatrix} 4 & 3 & 0 & 0 & 2 & 4 \\ 0 & 1 & 0 & 0 & 0 & 3 \\ 0 & 0 & 1 & 0 & 0 & 0 \\ 0 & 0 & 0 & 1 & 0 & 0 \\ 0 & 0 & 0 & 0 & 1 & 2 \\ 0 & 0 & 0 & 0 & 0 & 4 \end{pmatrix}.$$

By inspection, this is the second compound of the following substitution of determinant unity with coefficients mod 5:

$$\begin{pmatrix} 2 & 0 & 0 & 4 \\ 0 & 2 & 4 & 0 \\ 0 & 0 & 3 & 0 \\ 0 & 0 & 0 & 3 \end{pmatrix}.$$

The group O therefore contains all the generators of $O'_1(6, p^n)$. Since O is of index 2 under $O_1(6, p^n)$ and $O'_1(6, p^n)$ of index at most 2 under $O_1(6, p^n)$ (§ 182), it follows that $O \equiv O'_1(6, p^n)$. We have therefore, by § 163, the theorem:

For $p^n = 4l + 1$, the group $O'_1(6, p^n)$ has a maximal invariant subgroup $\{I, T \equiv C_1 C_2 \ldots C_6\}$ of order two, the quotient-group being holoedrically isomorphic with the simple group $LF(4, p^n)$. $O'_1(6, p^n)$ is of index two under the first orthogonal group $O_1(6, p^n)$ and is extended to it by any $O_{i,j}^{\varrho,\sigma}$ not a Q_{ij}.

188. Let $p^n = 4l + 3$, so that -1 is a not-square in the $GF[p^n]$. We make the following transformation of indices:

158) $\quad Y_{12} \equiv \xi_1 - \alpha \xi_2 + \beta \xi_4, \quad -Y_{13} \equiv \xi_3 - \beta \xi_2 - \alpha \xi_4, \quad Y_{14} \equiv \xi_5 - \xi_6,$
$\quad Y_{34} \equiv \xi_1 + \alpha \xi_2 - \beta \xi_4, \quad Y_{24} \equiv \xi_3 + \beta \xi_2 + \alpha \xi_4, \quad Y_{23} \equiv \xi_5 + \xi_6,$

where α and β is a suitable set of solutions in the field of

159) $\quad\quad\quad\quad \alpha^2 + \beta^2 = -1.$

LINEAR GROUP WITH QUADRATIC INVARIANT. 175

Under this transformation, F_4 takes the form
$$\Psi' \equiv \xi_1^2 + \xi_2^2 + \xi_3^2 + \xi_4^2 + \xi_5^2 - \xi_6^2.$$
Hence G_6 is holoedrically isomorphic with the second orthogonal group $O_{-1}(6, p^n)$. Reversing equations 158), we find

160) $\quad 2\xi_1 \equiv Y_{12} + Y_{34}, \; 2\xi_2 \equiv \alpha Y_{12} - \alpha Y_{34} - \beta Y_{13} - \beta Y_{24}, \; 2\xi_3 \equiv Y_{24} - Y_{13},$
$\quad\quad 2\xi_5 \equiv Y_{14} + Y_{23}, \; 2\xi_4 \equiv -\beta Y_{12} + \beta Y_{34} - \alpha Y_{13} - \alpha Y_{24}, \; 2\xi_6 \equiv Y_{23} - Y_{14}.$

The following substitution leaving F_4 invariant,
$$T_\tau: \quad\quad Y'_{14} = \tau Y_{14}, \quad Y'_{23} = \tau^{-1} Y_{23},$$
becomes, when expressed in the new indices 160),
$$\xi'_5 = \tfrac{1}{2}(\tau + \tau^{-1})\xi_5 - \tfrac{1}{2}(\tau - \tau^{-1})\xi_6,$$
$$\xi'_6 = -\tfrac{1}{2}(\tau - \tau^{-1})\xi_5 + \tfrac{1}{2}(\tau + \tau^{-1})\xi_6.$$

It is always an $O_{5,6}$, but is of the form $Q_{5,6}^{\gamma,\delta}$ if and only if τ be a square. It follows that $O_{-1}(6, p^n)$ contains a subgroup O' of index 2, which is the form taken by $G_{4,2}$ when expressed in terms of the ξ_i. The subgroup O' may be extended to $O_{-1}(6, p^n)$ by the substitution $C_5 C_6$, the new form of T_{-1}.

We proceed to prove that O' is identical with the subgroup $O'_{-1}(6, p^n)$ defined in § 181. Expressing the orthogonal substitution $O_{2,4}^{\varrho,\sigma}$ in the indices Y_{ij}, we obtain the substitution, denoted for the moment $\overline{O}_{2,4}^{\varrho,\sigma}$:

	Y_{12}	Y_{13}	Y_{14}	Y_{23}	Y_{24}	Y_{34}
$Y'_{12} =$	$\tfrac{1}{2}(1+\varrho)$	$-\tfrac{1}{2}\sigma$	0	0	$-\tfrac{1}{2}\sigma$	$\tfrac{1}{2}(1-\varrho)$
$Y'_{13} =$	$\tfrac{1}{2}\sigma$	$\tfrac{1}{2}(1+\varrho)$	0	0	$-\tfrac{1}{2}(1-\varrho)$	$-\tfrac{1}{2}\sigma$
$Y'_{14} =$	0	0	1	0	0	0
$Y'_{23} =$	0	0	0	1	0	0
$Y'_{24} =$	$\tfrac{1}{2}\sigma$	$-\tfrac{1}{2}(1-\varrho)$	0	0	$\tfrac{1}{2}(1+\varrho)$	$-\tfrac{1}{2}\sigma$
$Y'_{34} =$	$\tfrac{1}{2}(1-\varrho)$	$\tfrac{1}{2}\sigma$	0	0	$\tfrac{1}{2}\sigma$	$\tfrac{1}{2}(1+\varrho)$

For $\varrho \equiv 2\gamma^2 - 1$, $\sigma \equiv 2\gamma\delta$, whence $O_{2,4}^{\varrho,\sigma} \equiv Q_{2,4}^{\gamma,\delta}$, we see that $\overline{O}_{2,4}^{\varrho,\sigma}$ is the second compound of the substitution of determinant $(\gamma^2 + \delta^2)^2 = 1$

$$\begin{Bmatrix} \gamma & 0 & 0 & \delta \\ 0 & \gamma & -\delta & 0 \\ 0 & \delta & \gamma & 0 \\ -\delta & 0 & 0 & \gamma \end{Bmatrix}.$$

176   CHAPTER VII.

As shown above, T_{-1} corresponds to $C_5 C_6$. The product $\bar{O}_{2,4}^{\varrho,\sigma} T_{-1}$ is seen to be the second compound of

$$\begin{pmatrix} x & 0 & 0 & -\frac{\sigma}{2x} \\ 0 & -x & -\frac{\sigma}{2x} & 0 \\ 0 & \frac{\sigma}{2x} & -x & 0 \\ \frac{\sigma}{2x} & 0 & 0 & x \end{pmatrix} \quad \left[x^2 \equiv -\frac{1}{2}(1+\varrho)\right].$$

But x belongs to the $GF[p^n]$, in which -1 is a not-square, if and only if $\frac{1}{2}(1+\varrho)$ is a not-square, which occurs if and only if $O_{2,4}^{\varrho,\sigma}$ is not of the form $Q_{2,4}^{\varrho,\sigma}$. Hence O' contains $O_{2,4}^{\varrho,\sigma} C_5 C_6$ if $O_{2,4}^{\varrho,\sigma}$ is not a $Q_{2,4}$, but not in the contrary case. As shown above, O' contains every $Q_{2,4}^{\gamma,\delta}$. To prove that O' contains all the generators of $O'_{-1}(6, p^n)$, it evidently remains only to prove that O' contains all even substitutions on $\xi_1, \xi_2, \xi_3, \xi_4, \xi_5$, and, if $p^n = 3$, also $V_{1,2,6}$.

Expressing the linear substitution $(\xi_1 \xi_2 \xi_3)$ in the indices Y_{ij}, we get

$$\begin{pmatrix} \frac{1}{2}(\alpha-\beta^2) & \frac{1}{2}(\alpha-\beta-\alpha\beta) & 0 & 0 & -\frac{1}{2}(\alpha+\beta+\alpha\beta) & -\frac{1}{2}(\alpha-\beta^2) \\ -\frac{1}{2}(1+\alpha\beta) & -\frac{1}{2}(\beta+\alpha^2) & 0 & 0 & \frac{1}{2}(\beta-\alpha^2) & -\frac{1}{2}(1-\alpha\beta) \\ 0 & 0 & 1 & 0 & 0 & 0 \\ 0 & 0 & 0 & 1 & 0 & 0 \\ \frac{1}{2}(1-\alpha\beta) & -\frac{1}{2}(\beta+\alpha^2) & 0 & 0 & \frac{1}{2}(\beta-\alpha^2) & \frac{1}{2}(1+\alpha\beta) \\ \frac{1}{2}(\alpha+\beta^2) & -\frac{1}{2}(\alpha+\beta-\alpha\beta) & 0 & 0 & \frac{1}{2}(\alpha-\beta+\alpha\beta) & -\frac{1}{2}(\alpha+\beta^2) \end{pmatrix}$$

This substitution is the second compound of

161) $\quad \begin{pmatrix} x & 0 & 0 & y \\ 0 & z & w & 0 \\ 0 & -W & Z & 0 \\ -Y & 0 & 0 & X \end{pmatrix}$

having determinant unity, where

$$x^2 = \frac{\alpha-\beta-\alpha\beta}{2}, \quad z = \frac{\alpha-\beta^2}{2x}, \quad w = \frac{\alpha-\beta-\alpha\beta}{2x}, \quad y = \frac{x(\alpha-\beta^2)}{\alpha-\beta-\alpha\beta},$$

$$X = \frac{x(1+\alpha\beta)}{\alpha-\beta-\alpha\beta}, \quad Z = \frac{-\beta-\alpha^2}{2x}, \quad W = \frac{1+\alpha\beta}{2x}, \quad Y = \frac{-x(\beta+\alpha^2)}{\alpha-\beta-\alpha\beta}.$$

In order that 161) shall belong to the $GF[p^n]$, it is necessary and sufficient that x be a mark $\neq 0$ of the field. We proceed to prove that, for every set of solutions in the field of $\alpha^2 + \beta^2 = -1$, the expression

$$s \equiv \tfrac{1}{2}(\alpha - \beta - \alpha\beta)$$

is a square in the field or else zero.[1]) Eliminating β between the two equations, we find
$$1 + \alpha^2 + \left(\frac{\alpha - 2s}{1+\alpha}\right)^2 = 0,$$
or
$$(1 + \alpha + \alpha^2)^2 - 4\alpha s + 4s^2 = 0$$
or
$$(1 + \alpha + \alpha^2 + 2s)^2 = 4s(1+\alpha)^2.$$

Hence will s be a square.[2]) Solving, we find
$$x \equiv s^{\tfrac{1}{2}} = \tfrac{1}{2}(1 + \alpha - \beta).$$

The linear substitution $(\xi_2\xi_4\xi_5)$ expressed in the indices Y_{ij} takes the following form, say V:

$$\begin{pmatrix}
\tfrac{1}{2}(1+\alpha\beta) & \tfrac{1}{2}\alpha^2 & \tfrac{1}{2}\beta & \tfrac{1}{2}\beta & \tfrac{1}{2}\alpha^2 & \tfrac{1}{2}(1-\alpha\beta) \\
-\tfrac{1}{2}\beta^2 & \tfrac{1}{2}(1-\alpha\beta) & \tfrac{1}{2}\alpha & \tfrac{1}{2}\alpha & -\tfrac{1}{2}(1+\alpha\beta) & \tfrac{1}{2}\beta^2 \\
\tfrac{1}{2}\alpha & -\tfrac{1}{2}\beta & \tfrac{1}{2} & -\tfrac{1}{2} & -\tfrac{1}{2}\beta & -\tfrac{1}{2}\alpha \\
\tfrac{1}{2}\alpha & -\tfrac{1}{2}\beta & -\tfrac{1}{2} & \tfrac{1}{2} & -\tfrac{1}{2}\beta & -\tfrac{1}{2}\alpha \\
-\tfrac{1}{2}\beta^2 & -\tfrac{1}{2}(1+\alpha\beta) & \tfrac{1}{2}\alpha & \tfrac{1}{2}\alpha & \tfrac{1}{2}(1-\alpha\beta) & \tfrac{1}{2}\beta^2 \\
\tfrac{1}{2}(1-\alpha\beta) & -\tfrac{1}{2}\alpha^2 & -\tfrac{1}{2}\beta & -\tfrac{1}{2}\beta & -\tfrac{1}{2}\alpha^2 & \tfrac{1}{2}(1+\alpha\beta)
\end{pmatrix}.$$

The product VE will be simpler than V, if we take as E:
$Y'_{12}=Y_{12}-\beta Y_{23}$, $Y'_{13}=Y_{13}-\alpha Y_{23}$, $Y'_{14}=Y_{14}-Y_{23}-\beta Y_{12}+\beta Y_{34}-\alpha Y_{13}-\alpha Y_{24}$
$Y'_{34}=Y_{34}+\beta Y_{23}$, $Y'_{24}=Y_{24}-\alpha Y_{23}$, $Y'_{23}=Y_{23}$,
which is recognized as the second compound of
$$E' = \begin{pmatrix} 1 & -\alpha & \beta & 0 \\ 0 & 1 & 0 & 0 \\ 0 & 0 & 1 & 0 \\ 0 & -\beta & -\alpha & 1 \end{pmatrix}.$$

---

1) The case $s = 0$ requires $1 + \alpha + \alpha^2 = 0$, and may thus be avoided.
2) For $p^n = 3$, 7 or 11, there exist solutions of $\alpha^2 + \beta^2 = -1$ for which is an *arbitrary* square in the field. Is this always true?

DICKSON, Linear Groups.

The product VE has indeed the simple form $U \equiv$

$$\begin{pmatrix} \frac{1}{2} & -\frac{1}{2} & \beta & 0 & -\frac{1}{2} & \frac{1}{2} \\ \frac{1}{2} & \frac{1}{2} & \alpha & 0 & -\frac{1}{2} & -\frac{1}{2} \\ 0 & 0 & 2 & 0 & 0 & 0 \\ \alpha/2 & -\beta/2 & -\frac{1}{2} & 1/2 & -\beta/2 & -\alpha/2 \\ 1/2 & -1/2 & \alpha & 0 & 1/2 & -1/2 \\ 1/2 & 1/2 & -\beta & 0 & 1/2 & 1/2 \end{pmatrix}$$

which is seen by inspection to be the second compound of

$$U' \equiv \begin{pmatrix} 1 & 0 & 0 & 1 \\ \frac{1}{2}(\alpha-\beta) & 1/2 & -1/2 & \frac{1}{2}(\alpha+\beta) \\ -\frac{1}{2}(\alpha+\beta) & 1/2 & 1/2 & \frac{1}{2}(\alpha-\beta) \\ -1 & 0 & 0 & 1 \end{pmatrix}.$$

Hence $V \equiv UE^{-1}$ is the second compound of $V' \equiv U'E'^{-1}$:[1])

$$\begin{pmatrix} \frac{1}{2} & \frac{1}{2}(\alpha-\beta) & -\frac{1}{2}(\alpha+\beta) & \frac{1}{2} \\ \frac{1}{2}(\alpha-\beta) & \frac{1}{2} & -\frac{1}{2} & \frac{1}{2}(\alpha+\beta) \\ -\frac{1}{2}(\alpha+\beta) & \frac{1}{2} & \frac{1}{2} & \frac{1}{2}(\alpha-\beta) \\ -\frac{1}{2} & \frac{1}{2}(\alpha+\beta) & \frac{1}{2}(\alpha-\beta) & \frac{1}{2} \end{pmatrix}.$$

Having the linear substitutions $(\xi_1 \xi_2 \xi_3)$ and $(\xi_2 \xi_4 \xi_5)$, O' contains every even substitution on $\xi_1, \ldots, \xi_5$. It will suffice to prove this for *literal* substitutions (123), etc. Transforming (245) by (123) and by $(123)^2$, we reach (345) and (145). We then get

$$(124) = (154)(245), \quad (314) = (132)^{-1}(124)(132),$$
$$(12)(34) = (124)(134), \quad (12)(45) = (12)(34) \cdot (354).$$

But (123), (12)(34) and (12)(45) generate the alternating group on five letters (Cf. §§ 265—266).

For $p^n = 3$, $O'_{-1}(6, p^n)$ requires an additional generator V_{126}. Expressing the latter in terms of the indices Y_{ij} defined by 158), where we may now take $\alpha = \beta = +1$, it becomes $\overline{V}$:

---

1) The reciprocal of E' is given by changing the signs of α and β.

	Y_{12}	Y_{13}	Y_{14}	Y_{23}	Y_{24}	Y_{34}
$Y'_{12} =$	1	1	1	-1	1	-1
$Y'_{13} =$	1	-1	1	-1	1	0
$Y'_{14} =$	1	1	-1	-1	1	0
$Y'_{23} =$	-1	-1	-1	-1	-1	0
$Y'_{24} =$	1	1	1	-1	-1	0
$Y'_{34} =$	-1	0	0	0	0	0

having determinant unity. $\overline{V}$ is seen to be the second compound of

$$\begin{Bmatrix} -1 & 0 & -1 & 1 \\ 0 & -1 & -1 & -1 \\ -1 & -1 & 0 & 0 \\ 1 & -1 & 0 & 0 \end{Bmatrix}$$

of determinant $+1$. Hence O' contains V_{126} when $p^n = 3$.

Since O' contains the group $O'_{-1}(6, p^n)$ but is of index 2 under $O_{-1}(6, p^n)$, it follows from § 182 that $O' \equiv O'_{-1}(6, p^n)$. Applying § 163, we have the theorem:

For $p^n = 4l + 3$, the group $O'_{-1}(6, p^n)$ is holoedrically isomorphic with the simple group $LF(4, p^n)$ and is of index two under the second orthogonal group $O_{-1}(6, p^n)$, being extended to it by $C_5 C_6$.

189. Theorem. — *The subgroup $O'_1(5, p^n)$ is of index two under $O_1(5, p^n)$ and is holoedrically isomorphic with the simple Abelian group $A(4, p^n)$.*

By § 161, $A(4, p^n)$, $p > 2$ is holoedrically isomorphic with the second compound $A_{4,2}$ of the quaternary special Abelian group. $A_{4,2}$ leaves absolutely invariant the Pfaffian [1234] and $Y_{12} + Y_{34}$. By the introduction of the new indices (§ 162)

$$Y \equiv \tfrac{1}{2}(Y_{12} - Y_{34}), \quad Z_1 \equiv \tfrac{1}{2}(Y_{12} + Y_{34}); \quad Y_{12} \equiv Z_1 + Y, \quad Y_{34} \equiv Z_1 - Y,$$

$A_{4,2}$ takes a form not involving Z_1 and so becomes a quinary group Q leaving absolutely invariant the quadratic function

$$\Phi \equiv Y^2 + Y_{13} Y_{24} - Y_{14} Y_{23}.$$

The group G of all quinary linear substitutions of determinant unity which leave Φ absolutely invariant will be proven holoedrically isomorphic with $O_1(5, p^n)$ and therefore (§ 172) of order

$$(p^{4n} - 1) p^{3n} (p^{2n} - 1) p^n.$$

180   CHAPTER VII.

Since Q is holoedrically isomorphic with $A(4, p^n)$, its order is, by § 115, half of that of $O_1(5, p^n)$. To complete the proof of the theorem, we then show that Q is holoedrically isomorphic with a subgroup of $O_1(5, p^n)$ containing all the generators of the group $O_1'(5, p^n)$ defined in § 181.

Let first -1 be the square of a mark i of the $GF[p^n]$. Set

162) $\quad Y \equiv i\xi_2, \quad -Y_{13} \equiv \xi_3 + i\xi_4, \quad Y_{14} \equiv \xi_5 + i\xi_6,$
$\quad\quad\quad Y_{24} \equiv \xi_3 - i\xi_4, \quad Y_{23} \equiv \xi_5 - i\xi_6,$
whence
$$-\Phi \equiv \xi_2^2 + \xi_3^2 + \xi_4^2 + \xi_5^2 + \xi_6^2.$$

Hence G is holoedrically isomorphic with $O_1(5, p^n)$. Proceeding[1]) as in § 187, we find that Q' (Q expressed in the indices ξ_i) contains the substitution $O_{3,4}^{\varrho,\sigma}$ if and only if it be a $Q_{3,4}$, also that Q' contains $(\xi_3 \xi_5 \xi_6)$. The latter with $(\xi_2 \xi_3 \xi_4)$ will generate all even substitutions on $\xi_2, \ldots, \xi_6$ by the preceding section. But $(\xi_2 \xi_3 \xi_4)$ expressed in the indices Y_{ij} is

	Y_{12}	Y_{13}	Y_{14}	Y_{23}	Y_{24}	Y_{34}
$Y_{12}' =$	$1/2$	$-i/2$	0	0	$i/2$	$1/2$
$Y_{13}' =$	$-1/2$	$-i/2$	0	0	$-i/2$	$1/2$
$Y_{14}' =$	0	0	1	0	0	0
$Y_{23}' =$	0	0	0	1	0	0
$Y_{24}' =$	$-1/2$	$i/2$	0	0	$i/2$	$1/2$
$Y_{34}' =$	$1/2$	$i/2$	0	0	$-i/2$	$1/2$

This is seen to be the second compound of the following special Abelian substitution:

$$\begin{pmatrix} \frac{1}{2}(1-i) & 0 & 0 & -\frac{1}{2}(1+i) \\ 0 & \frac{1}{2}(1+i) & \frac{1}{2}(1-i) & 0 \\ 0 & -\frac{1}{2}(1+i) & \frac{1}{2}(1-i) & 0 \\ \frac{1}{2}(1-i) & 0 & 0 & \frac{1}{2}(1+i) \end{pmatrix}.$$

It follows that Q' contains every $Q_{i,j}^{\gamma,\delta}$ ($i, j = 2, \ldots 6$; $i \neq j$). Also Q' contains $O_{3,4}^{\varrho,\sigma} O_{5,6}^{\varrho,\sigma}$ and hence every $O_{i,j}^{\varrho,\sigma} O_{k,l}^{\varrho,\sigma}$. For $p^n = 5$, we take $i = 2$. Expressing the additional generator R_{345} in the indices Y_{ij}, we reach the substitution (mod 5)

---

[1]) Comparing the transformations of indices 154) and 162), we note that they are identical as far as ξ_3, ξ_4, ξ_5 and ξ_6 are concerned.

$$\begin{pmatrix} 1 & 0 & 0 & 0 & 0 & 0 \\ 0 & -1 & 3 & 3 & 1 & 0 \\ 0 & 0 & 1 & 0 & 2 & 0 \\ 0 & 0 & 0 & 1 & 2 & 0 \\ 0 & 0 & 0 & 0 & -1 & 0 \\ 0 & 0 & 0 & 0 & 0 & 1 \end{pmatrix}$$

which is the second compound of the special Abelian substitution

$$\begin{pmatrix} 3 & 1 & 0 & 0 \\ 0 & 2 & 0 & 0 \\ 0 & 0 & 3 & 1 \\ 0 & 0 & 0 & 2 \end{pmatrix}.$$

Hence Q' coincides with $O_1'(5, p^n)$.

Consider next the case -1 a not-square in the $GF[p^n]$. Set

163) $\quad Y \equiv \xi_5, \quad Y_{13} \equiv \xi_1 - \alpha\xi_2 + \beta\xi_4, \quad -Y_{14} \equiv \xi_3 - \beta\xi_2 - \alpha\xi_4,$
$\quad\quad\quad Y_{24} \equiv \xi_1 + \alpha\xi_2 - \beta\xi_4, \quad Y_{23} \equiv \xi_3 + \beta\xi_2 + \alpha\xi_4,$

where $\alpha^2 + \beta^2 = -1$. Then Φ becomes $\xi_1^2 + \xi_2^2 + \xi_3^2 + \xi_4^2 + \xi_5^2$. Hence G_6 is holoedrically isomorphic with $O_1(5, p^n)$. Reversing equations 163), we get

164) $\quad 2\xi_1 = Y_{13} + Y_{24}, \quad 2\xi_2 = \alpha Y_{13} - \alpha Y_{24} - \beta Y_{14} - \beta Y_{23}, \quad \xi_5 = Y,$
$\quad\quad\quad 2\xi_3 = Y_{23} - Y_{14}, \quad 2\xi_4 = -\beta Y_{13} + \beta Y_{24} - \alpha Y_{14} - \alpha Y_{23}.$

As in § 188, we find that Q_1 (Q expressed in the indices ξ_i) contains every $Q_{2,4}^{\gamma,\delta}$ and the linear substitution $(\xi_1 \xi_2 \xi_3)$ and consequently also $Q_{1,4}^{\gamma,\delta}$, the transformed of the former by the latter.

Expressing in the indices ξ_i the following substitution of G_6,

$K: \quad\quad Y_{13}' = -Y_{13}, \quad Y_{24}' = -Y_{24},$

we get $C_1 C_4 O_{2,4}^{\alpha^2-\beta^2, 2\alpha\beta}$. This $O_{2,4}$ is not a $Q_{2,4}^{x,y}$ since $2x^2 - 1 = \alpha^2 - \beta^2$ requires $x^2 = -\beta^2$. But $C_1 C_4 = Q_{1,4}^{0,1}$ belongs to Q_1. Since K does not belong to Q (§ 164), it follows that $O_{2,4}^{\varrho,\sigma}$

$$\varrho \equiv \alpha^2 - \beta^2, \quad \sigma \equiv 2\alpha\beta,$$

extends Q_1 to $O_1(5, p^n)$. If $\overline{O}_{1,3}^{\varrho,\sigma}$ denotes $O_{1,3}^{\varrho,\sigma}$ when expressed in the indices Y_{ij}, we find that the product $K\overline{O}_{1,3}^{\varrho,\sigma}$ has the form

182 CHAPTER VII.

$$\begin{array}{r|cccc} & Y_{13} & Y_{14} & Y_{23} & Y_{24} \\ \hline Y'_{13} = & \beta^2 & -\alpha\beta & \alpha\beta & -\alpha^2 \\ Y'_{14} = & -\alpha\beta & -\beta^2 & -\alpha^2 & -\alpha\beta \\ Y'_{23} = & \alpha\beta & -\alpha^2 & -\beta^2 & \alpha\beta \\ Y'_{24} = & -\alpha^2 & -\alpha\beta & \alpha\beta & \beta^2 \end{array}$$

and is the second compound of the special Abelian substitution

$$\begin{pmatrix} \beta & \alpha & 0 & 0 \\ \alpha & -\beta & 0 & 0 \\ 0 & 0 & \beta & -\alpha \\ 0 & 0 & -\alpha & -\beta \end{pmatrix}.$$

Hence Q_1 contains $O_{2,4}^{0,\sigma} O_{1,3}^{0,\sigma}$. We next show that Q_1 contains the linear substitution $(\xi_2 \xi_4 \xi_5)$, so that with $(\xi_1 \xi_2 \xi_3)$ Q_1 will contain all even substitutions on $\xi_1, \ldots, \xi_5$. Expressing $(\xi_2 \xi_4 \xi_5)$ in the indices Y_{ij}, we get

$$\begin{pmatrix} \frac{1}{2} & \frac{\alpha}{2} & \frac{-\beta}{2} & \frac{-\beta}{2} & \frac{-\alpha}{2} & \frac{1}{2} \\ \frac{\beta}{2} & \frac{1+\alpha\beta}{2} & \frac{\alpha^2}{2} & \frac{\alpha^2}{2} & \frac{1-\alpha\beta}{2} & \frac{-\beta}{2} \\ \frac{\alpha}{2} & \frac{-\beta^2}{2} & \frac{1-\alpha\beta}{2} & \frac{-1-\alpha\beta}{2} & \frac{\beta^2}{2} & \frac{-\alpha}{2} \\ \frac{\alpha}{2} & \frac{-\beta^2}{2} & \frac{-1-\alpha\beta}{2} & \frac{1-\alpha\beta}{2} & \frac{\beta^2}{2} & \frac{-\alpha}{2} \\ \frac{-\beta}{2} & \frac{1-\alpha\beta}{2} & \frac{-\alpha^2}{2} & \frac{-\alpha^2}{2} & \frac{1+\alpha\beta}{2} & \frac{\beta}{2} \\ \frac{1}{2} & \frac{-\alpha}{2} & \frac{\beta}{2} & \frac{\beta}{2} & \frac{\alpha}{2} & \frac{1}{2} \end{pmatrix}$$

which is the second compound of the special Abelian substitution

$$\begin{pmatrix} \frac{1}{2} & \frac{-1}{2} & \frac{-1}{2}(\alpha-\beta) & \frac{1}{2}(\alpha+\beta) \\ \frac{1}{2} & \frac{1}{2} & \frac{1}{2}(\alpha+\beta) & \frac{1}{2}(\alpha-\beta) \\ \frac{-1}{2}(\alpha-\beta) & \frac{1}{2}(\alpha+\beta) & \frac{1}{2} & \frac{-1}{2} \\ \frac{1}{2}(\alpha+\beta) & \frac{1}{2}(\alpha-\beta) & \frac{1}{2} & \frac{1}{2} \end{pmatrix}.$$

For $p^n = 3$, $O'_1(5, p^n)$ requires an additional generator W_{1234}. For $\alpha = \beta = 1$, the following substitution

$$Y'_{13} = Y_{13} + Y_{23}, \quad Y'_{14} = Y_{14} + Y_{24}$$

when expressed in the indices $\xi_1, \ldots, \xi_5$ becomes (mod 3)

$$\begin{pmatrix} 1 & 2 & 2 & 2 & 0 \\ 1 & 1 & 2 & 1 & 0 \\ 1 & 1 & 1 & 2 & 0 \\ 1 & 2 & 1 & 1 & 0 \\ 0 & 0 & 0 & 0 & 1 \end{pmatrix} \equiv W_{1234}(\xi_2 \xi_3 \xi_4).$$

In every case it follows that Q_1 coincides with $O'_1(5, p^n)$.

190. Theorem.[1] — If $p^n = 4l + 1$, $O'_\nu(6, p^n)$ is holoedrically isomorphic with the simple group $HA(4, p^{2n})$. If $p^n = 4l + 3$, $O'_1(6, p^n)$ has the maximal invariant subgroup $\{I, C_1 C_2 C_3 C_4 C_5 C_6\}$ of order 2, the quotient-group being holoedrically isomorphic with $HA(4, p^{2n})$. In either case, $O'_\mu(6, p^n)$ is of index 2 under $O_\mu(6, p^n)$ and is extended to the latter by any substitution O_{ij} not a Q_{ij}.

Consider the group H' of quaternary hyperabelian substitutions in the $GF[p^{2n}]$ of determinant unity. It has the order

$$h' \equiv (p^{4n} - 1) p^{3n} (p^{3n} + 1) p^{2n} (p^{2n} - 1) p^n.$$

The special Abelian group $SA(4, p^n)$ is a subgroup of H'. Denote their second compound groups by $A_{4,2}$ and $H_{4,2}$ respectively. By § 161, $A_{4,2}$ leaves absolutely invariant the functions

$$F_4 \equiv Y_{12} Y_{34} - Y_{13} Y_{24} + Y_{14} Y_{23}, \quad Z \equiv Y_{12} + Y_{34}.$$

For an arbitrary mark $\omega \neq 0$ in the $GF[p^{2n}]$, the substitution

$$\Omega \equiv \begin{pmatrix} \omega & 0 & 0 & 0 \\ 0 & \omega^{-p^n} & 0 & 0 \\ 0 & 0 & \omega^{-1} & 0 \\ 0 & 0 & 0 & \omega^{p^n} \end{pmatrix}$$

is hyperabelian and of determinant unity. Its second compound is

$$\Omega': \begin{cases} Y'_{12} = \omega^{-p^n+1} Y_{12}, & Y'_{13} = Y_{13}, & Y'_{14} = \omega^{p^n+1} Y_{14}, \\ Y'_{34} = \omega^{p^n-1} Y_{34}, & Y'_{24} = Y_{24}, & Y'_{23} = \omega^{-p^n-1} Y_{23}. \end{cases}$$

Taking $p > 2$, we introduce in place of Y_{12}, Y_{34} the new indices

165) $\qquad \xi_1 \equiv \frac{1}{2}(Y_{12} - Y_{34}), \quad \xi_6 \equiv \frac{-1}{2J}(Y_{12} + Y_{34}),$

where J is a mark of the $GF[p^{2n}]$ satisfying the equation

166) $\qquad\qquad\qquad J^{p^n-1} = -1.$

---

[1] *Bulletin Amer. Math. Soc.*, May, 1900; *Transactions*, July, 1900.

Reversing relations 165), we find

167) $\quad Y_{12} \equiv \xi_1 - J\xi_6, \ldots Y_{34} \equiv -\xi_1 - J\xi_6.$

Written in the new indices, the substitution Ω' becomes

$$\Omega'': \begin{cases} \xi_1' = \Theta\xi_1 + J\varrho\xi_6, & Y_{13}' = Y_{13}, & Y_{14}' = \omega^{p^n+1}Y_{14}, \\ \xi_6' = \dfrac{1}{J}\varrho\xi_1 + \Theta\xi_6, & Y_{24}' = Y_{24}, & Y_{23}' = \omega^{-p^n-1}Y_{23}, \end{cases}$$

where

$$\Theta \equiv \frac{1}{2}(\omega^{p^n-1} + \omega^{-p^n+1}), \quad \varrho \equiv \frac{1}{2}(\omega^{p^n-1} - \omega^{-p^n+1}).$$

The coefficients Θ, $J\varrho$, ϱ/J belong to the $GF[p^n]$ since

$$\Theta^{p^n} \equiv \Theta, \quad J^{p^n} = -J, \quad \varrho^{p^n} = -\varrho.$$

Hence Ω'' belongs to the $GF[p^n]$ and has determinant unity.

If, in § 162, we set $Y \equiv \xi_1$, $Z_1 \equiv -J\xi_6$, we obtain the present transformation of indices 165). Hence, if we express any substitution $[\alpha]_2$ of $A_{4,2}$ in terms of the new indices, we obtain a substitution, not involving ξ_6, the matrix of whose coefficients is given in § 162. Hence $A_{4,2}$ is transformed into a group A'' of substitutions belonging to the $GF[p^n]$ which do not involve ξ_6 and which leave absolutely invariant

$$\xi_1^2 + Y_{13}Y_{24} - Y_{14}Y_{23}.$$

In order that A'' shall contain the substitution

$$A: \quad Y_{14}' = \omega^{p^n+1}Y_{14}, \quad Y_{23}' = \omega^{-p^n-1}Y_{23},$$

it is necessary and sufficient (by § 164) that ω^{p^n+1} be a square in the $GF[p^n]$ and hence that ω be a square in the $GF[p^{2n}]$,

$$\omega^{(p^n+1)(p^n-1)/2} = +1.$$

Hence the group G'', given by the extension of A'' by Ω'', will contain

$$K \equiv \Omega''A^{-1}: \quad \xi_1' = \Theta\xi_1 + J\varrho\xi_6, \quad \xi_6' = \frac{1}{J}\varrho\xi_1 + \Theta\xi_6,$$

if and only if ω be a square in the $GF[p^{2n}]$. Now K leaves $-\xi_1^2 + J^2\xi_6^2$ invariant and is therefore an O_{16}. We proceed to prove that, *if ω be a square in the $GF[p^{2n}]$, every K is a $Q_{1,6}^{\alpha,\beta}$ and every $Q_{1,6}^{\alpha,\beta}$ is a K, where α, β belong to the $GF[p^n]$*. Let, in fact,

168) $\quad \alpha \equiv \dfrac{1}{2}(\omega^{(p^n-1)/2} + \omega^{(-p^n+1)/2}), \quad \beta \equiv \dfrac{1}{2}J(\omega^{(p^n-1)/2} - \omega^{(-p^n+1)/2}).$

Since $\omega^{(p^{2n}-1)/2} = 1$, we see that α and β belong to the $GF[p^n]$. Also

169) $\quad\quad\quad\quad \alpha^2 - \beta^2/J^2 = 1$

170) $\quad 2\alpha\beta = \dfrac{1}{2}J(\omega^{p^n-1} - \omega^{-p^n+1}), \quad 2\alpha^2 - 1 = \dfrac{1}{2}(\omega^{p^n-1} + \omega^{-p^n+1}).$

Hence K has the form $Q_{1,6}^{\alpha,\beta}$, where α, β are defined by 168).

LINEAR GROUP WITH QUADRATIC INVARIANT. 185

Given, inversely, a $Q_{1,6}^{\alpha,\beta}$, where α, β are marks of the $GF[p^n]$ satisfying 169), we can determine a square ω in the $GF[p^{2n}]$ which satisfies 170). In fact, 170) may be written in the solved form

$$\omega^{p^n-1} = 2\alpha^2 - 1 + 2\alpha\beta/J = \left(\alpha + \frac{\beta}{J}\right)^2, \quad \omega^{-p^n+1} = \left(\alpha - \frac{\beta}{J}\right)^2,$$

of which the second follows from the first in virtue of 169). That the first can be satisfied by a square ω in the $GF[p^{2n}]$ follows from the relation

$$\left(\alpha + \frac{\beta}{J}\right)^{p^n+1} = \left(\alpha + \frac{\beta}{J^{p^n}}\right)\left(\alpha + \frac{\beta}{J}\right) = \alpha^2 - \frac{\beta^2}{J^2} = 1.$$

For ω a not-square in the $GF[p^{2n}]$, Ω'' is the product of an O_{16}, not a Q_{16} in the $GF[p^n]$, by the substitution A, neither factor belonging separately to G''.

Under the transformation of indices 165), F_4 becomes

$$\Psi \equiv -\xi_1^2 + J^2\xi_6^2 - Y_{13}Y_{24} + Y_{14}Y_{23},$$

where, by 166), J^2 belongs to the $GF[p^n]$ but is a not-square in it. We introduce in place of the Y_{ij} new indices such that

$$Y_{13}Y_{24} - Y_{14}Y_{23} \equiv \xi_2^2 + \xi_3^2 + \xi_4^2 + \xi_5^2.$$

Then Ψ becomes $J^2\xi_6^2 - \sum_{i=1}^{5}\xi_i^2$. Therefore, by § 189, A'' will be transformed into $O_1'(5, p^n)$.

For -1 the square of a mark i in the $GF[p^n]$, we may take

$$Y_{13} \equiv \xi_2 + i\xi_3, \quad Y_{24} \equiv \xi_2 - i\xi_3, \quad -Y_{14} \equiv \xi_4 + i\xi_5, \quad Y_{23} \equiv \xi_4 - i\xi_5.$$

As in § 187, A becomes an $O_{4,5}$, which is a $Q_{4,5}$ if and only if ω be a square in the $GF[p^{2n}]$. Hence G'' is isomorphic with a subgroup of $O_\nu(6, p^n)$. The subgroup contains every $Q_{1,6}$ and every $O_{16}O_{45}$, neither factor a Q, but does not contain the separate factors.

For -1 a not-square, so that $p^n = 4l + 3$, we may take ω so that

$$\omega^{p^n+1} = -1, \quad \omega^{(p^{2n}-1)/2} = -1.$$

Then A multiplies Y_{14} and Y_{23} by -1. The required transformation of indices, transforming G'' into a subgroup of $O_1(6, p^n)$, is the following:

$$Y_{13} \equiv \xi_2 - \alpha\xi_4 + \beta\xi_5, \quad -Y_{14} \equiv \xi_3 - \beta\xi_4 - \alpha\xi_5,$$
$$Y_{24} \equiv \xi_2 + \alpha\xi_4 - \beta\xi_5, \quad Y_{23} \equiv \xi_3 + \beta\xi_4 + \alpha\xi_5. \quad (\alpha^2 + \beta^2 = -1)$$

As in § 189, A becomes in the new indices $C_3C_4 O_{4,5}^{\alpha^2-\beta^2, 2\alpha\beta}$, the last factor being not of the form $Q_{4,5}$, while $C_3C_4 \equiv Q_{3,4}^{0,1}$ belongs to $O_1'(5, p^n)$. Hence G'' is isomorphic with a subgroup of $O_1(6, p^n)$. The subgroup contains every Q_{16} and every $O_{16}O_{45}$, neither factor being a Q, but does not contain the factors separately.

186   CHAPTER VII.

It follows that G'' is holoedrically isomorphic with $O'_\nu(6, p^n)$ or $O'_1(6, p^n)$ according as $p^n = 4l+1$ or $4l+3$. But, for $p > 2$, the order of the second compound $H_{4,2}$ of H' is $\frac{1}{2}h'$ and therefore equals that of $O'_\mu(6, p^n)$. Hence G'', $H_{4,2}$ and $O'_\mu(6, p^n)$ are holoedrically isomorphic.

By § 132, we pass from H' to the quotient-group $HA(4, p^{2n})$ by making the substitutions $T_\varkappa$ (106) correspond to the identity. The corresponding substitutions of $H_{4,2}$ are the identity I if $p^n = 4l+1$, but are I and the substitution T changing the signs of the six indices if $p^n = 4l+3$. Hence $O'_\nu(6, p^n)$ is holoedrically isomorphic with $HA(4, p^{2n})$ if $p^n = 4l+1$; while, for $p^n = 4l+3$, $O'_1(6, p^n)$ has the maximal invariant subgroup $\{I, C_1C_2C_3C_4C_5C_6\}$ of order 2, the quotient-group being isomorphic with $HA(4, p^{2n})$.

191. We proceed to determine the structure of the orthogonal subgroups $O'_\mu(m, p^n)$, $m \geqq 7$. Every m-ary linear homogeneous substitution is commutative with

$$C \equiv C_1C_2 \ldots C_m: \qquad \xi'_i = -\xi_i \qquad (i = 1, \ldots, m).$$

C belongs to the group $O'_\mu(m, p^n)$ only when m is even and $\mu = 1$ (see § 185). Suppose that $O'_\mu(m, p^n)$ has a self-conjugate subgroup G containing a substitution S neither the identity I nor C:

$$S: \qquad \xi'_i = \sum_{j=1}^{m} \alpha_{ij}\xi_j \qquad (i = 1, \ldots, m).$$

Suppose first that S reduces to the form

171) $\qquad \xi'_i = \alpha_{ii}\xi_i \qquad (i = 1, \ldots, m)$

where $\alpha_{ii}^2 = 1$. Then S is merely a product of an even number of the C_i, in which certain ones as C_k are lacking since $S \neq C$. If $\mu = \nu$ and therefore m even by hypothesis, we may suppose that both C_m and C_k ($k < m$) are lacking, since C_iC_m does not belong to $O'_\nu(m, p^n)$. But if $S \equiv C_iC_jC_rC_s\ldots$, its transformed by $T_{ij}T_{ik}$ (always in the main group) gives $S' \equiv C_kC_jC_rC_s\ldots$, so that G contains the product

$$S'S^{-1} \equiv C_kC_i.$$

From it we obtain in G the substitution C_1C_2 and are thus led to the case treated in § 193.

Suppose, on the contrary, that S is not of the form 171). We may assume that $\alpha_{12}, \alpha_{13}, \ldots, \alpha_{1m}$ are not all zero. In fact, either S or its reciprocal will have at least one $\alpha_{ij} \neq 0$ in which $i < j$. Transforming the one or the other by $T_{1j}T_{1i}$, if $j < m$, we obtain a substitution in G which replaces ξ_1 by

$$\alpha_{ii}\xi_1 + \alpha_{ij}\xi_i + \cdots.$$

If $j = m$, we transform S by $T_{1k}T_{1i}$ (k not 1, i or m) and obtain a substitution in G which replaces ξ_1 by

$$\alpha_{ii}\xi_1 + \cdots + \alpha_{im}\xi_m.$$

From the resulting substitution S in which $\alpha_{12}, \alpha_{13}, \ldots, \alpha_{1m}$ are not all zero, we derive a substitution S_1 belonging to G and having $\alpha_{11}^2 + \alpha_{12}^2 \neq 1$. We get S_1 immediately if $\alpha_{11}^2 + \alpha_{1j}^2 \neq 1$ for $j = 2, 3, \ldots$, or $m-1$. In the contrary case, we have

172) $\qquad \alpha_{11}^2 + \alpha_{12}^2 = 1, \quad \alpha_{12}^2 = \alpha_{13}^2 = \cdots = \alpha_{1m-1}^2.$

If $\alpha_{12} = 0$, then $\alpha_{11}^2 = 1$ and therefore $\alpha_{1m} = 0$ by 147), contrary to the assumption that $\alpha_{12}, \alpha_{13}, \ldots, \alpha_{1m}$ are not all zero. Hence $\alpha_{12} \neq 0$. Transforming S by a suitable product of the C_i, we can take

$$\alpha_{12} = \alpha_{13} = \cdots = \alpha_{1m-1} \neq 0.$$

Transforming[1]) the resulting substitution by $O_{2,3}^{\alpha,\beta}$, we obtain a substitution which replaces ξ_1 by

$$\alpha_{11}\xi_1 + (\alpha\alpha_{12} + \beta\alpha_{13})\xi_2 + (-\beta\alpha_{12} + \alpha\alpha_{13})\xi_3 + \alpha_{14}\xi_4 + \cdots + \alpha_{1m}\xi_m.$$

If $p^n > 5$, we can determine α and β in the $GF[p^n]$ such that

$$\alpha^2 + \beta^2 = 1, \quad \alpha_{11}^2 + (\alpha\alpha_{12} + \beta\alpha_{13})^2 \neq 1.$$

Indeed, since $\alpha_{12} = \alpha_{13} \neq 0$, and $\alpha_{11}^2 + \alpha_{12}^2 = 1$, the second condition becomes $2\alpha\beta \neq 0$. But, of the $p^n - \varepsilon$ sets of solutions in the $GF[p^n]$ of the first condition, where $\varepsilon = \pm 1$ according as -1 is a square or a not-square in the field, only four sets of solutions have either α or β equal zero. Hence, if $p^n > 5$, there exist other solutions.

For $p^n = 3$, we transform S, in which

$$\alpha_{11} = 0, \quad \alpha_{12} = \alpha_{13} = \alpha_{14} = \alpha_{15} = \pm 1,$$

by W_{3245}^2 and obtain a substitution in G which replaces ξ_1 by

$$\pm \xi_3 + \alpha_{16}\xi_6 + \cdots + \alpha_{1m}\xi_m,$$

for which therefore $\alpha_{11}^2 + \alpha_{12}^2 = 0$.

For $p^n = 5$, S has $\alpha_{11} = 0$, $\alpha_{12}^2 = \alpha_{13}^2 = \alpha_{14}^2 = 1$ in virtue of 172). Transforming S by a product of the C_i, we may take

$$-\alpha_{12} = \alpha_{13} = \alpha_{14} = 1.$$

The resulting substitution is transformed by R_{234} into a substitution of G which replaces ξ_1 by $2\xi_2 + 2\xi_3 + \alpha_{15}\xi_5 + \cdots$, for which

$$\alpha_{11}^2 + \alpha_{12}^2 = 4.$$

---

[1] If the transformer does not belong to $O'_\mu(m, p^n)$, we afterwards transform by $O_{4,5}^{\alpha,\beta}$. Since the product $O_{2,3}^{\alpha,\beta} O_{4,5}^{\alpha,\beta}$ belongs to the main group, the transformed substitution will belong to G. A like remark is to be understood throughout this section.

Taking the reciprocal of the substitution of G which has
$$\alpha_{11}^2 + \alpha_{12}^2 \neq 1,$$
we obtain in G a substitution S in which $\alpha_{11}^2 + \alpha_{21}^2 \neq 1$. Then G contains the product
$$S_1 \equiv S^{-1} C_1 C_2 S \cdot C_1 C_2 \equiv S_\alpha C_1 C_2,$$
where $S_\alpha \equiv S^{-1} C_1 C_2 S$ is of period two and has the form
$$\xi_i' = \xi_i - 2\alpha_{i1}\left(\sum_{j=1}^{m-1} \alpha_{j1}\xi_j + \mu \alpha_{m1}\xi_m\right) - 2\alpha_{i2}\left(\sum_{j=1}^{m-1} \alpha_{j2}\xi_j + \mu \alpha_{m2}\xi_m\right)$$
$$(i = 1, 2, \ldots, m).$$

S_1 is not the identity since S would then be commutative with $C_1 C_2$ and would therefore break up into the product of
$$\xi_1' = \alpha_{11}\xi_1 + \alpha_{12}\xi_2, \quad \xi_2' = \alpha_{21}\xi_1 + \alpha_{22}\xi_2 \quad (\alpha_{11}^2 + \alpha_{21}^2 = 1)$$
by a substitution on $\xi_3, \ldots, \xi_m$.

We readily obtain the transformed $S_{\alpha'}$ of S_α by an orthogonal substitution O, in which $i, j < k$:
$$\xi_i' = \alpha \xi_i + \beta \xi_j + \gamma \xi_k, \quad \xi_j' = \alpha' \xi_i + \beta' \xi_j + \gamma' \xi_k, \quad \xi_k' = \alpha'' \xi_i + \beta'' \xi_j + \gamma'' \xi_k,$$
where by 145)
173) $\qquad \alpha^2 + \beta^2 + \dfrac{1}{\lambda}\gamma^2 = 1 \quad (\lambda = 1 \text{ if } k < m; \ \lambda = \mu \text{ if } k = m).$

We have $S_{\alpha'} \equiv (SO)^{-1} C_1 C_2 (SO)$. But $S' \equiv SO$ has the coefficients
$$\begin{cases} \alpha_{il}' \equiv \alpha \alpha_{il} + \beta \alpha_{jl} + \gamma \alpha_{kl} \\ \alpha_{jl}' \equiv \alpha' \alpha_{il} + \beta' \alpha_{jl} + \gamma' \alpha_{kl} \qquad (l = 1, 2, \ldots, m) \\ \alpha_{kl}' \equiv \alpha'' \alpha_{il} + \beta'' \alpha_{jl} + \gamma'' \alpha_{kl} \end{cases}$$
$$\alpha_{sl}' \equiv \alpha_{sl} \qquad (s = 1, \ldots, m; \ s \neq i, j, k).$$

If $\alpha_{i1}^2 + \alpha_{j1}^2 + \lambda \alpha_{k1}^2 \neq 0$, we can find solutions in the $GF[p^n]$ of 173), which make $\alpha_{i1}' = 0$. We suppose $\alpha_{i1} \neq 0$, the transformation of S_α being unnecessary if α_{i1} be already zero. Eliminating α from 173) and
174) $\qquad \alpha \alpha_{i1} + \beta \alpha_{j1} + \gamma \alpha_{k1} = 0,$
we find the single condition on β and γ,
175) $\qquad \beta^2(\alpha_{i1}^2 + \alpha_{j1}^2) + 2\beta\gamma \alpha_{j1}\alpha_{k1} + \gamma^2\left(\alpha_{k1}^2 + \dfrac{1}{\lambda}\alpha_{i1}^2\right) = \alpha_{i1}^2.$

If $\alpha_{i1}^2 + \alpha_{j1}^2 = 0$, so that $\alpha_{j1} \neq 0$ and $\alpha_{k1} \neq 0$, this equation determines β, when γ is assigned any value $\neq 0$ in the field. Then 174) determines α in the field. But, if $\alpha_{i1}^2 + \alpha_{j1}^2$ be $\neq 0$, we multiply it into 175), which then takes the form
$$\{\beta(\alpha_{i1}^2 + \alpha_{j1}^2) + \gamma \alpha_{j1}\alpha_{k1}\}^2 + \dfrac{1}{\lambda}\alpha_{i1}^2(\alpha_{i1}^2 + \alpha_{j1}^2 + \lambda \alpha_{k1}^2)\gamma^2 = \alpha_{i1}^2(\alpha_{i1}^2 + \alpha_{j1}^2).$$

LINEAR GROUP WITH QUADRATIC INVARIANT.

The coefficient of γ^2 being not zero, this equation has solutions for
$$\{\beta(\alpha_{i1}^2 + \alpha_{j1}^2) + \gamma \alpha_{j1}\alpha_{k1}\}, \quad \gamma$$
in the field and hence solutions β, γ.

Transforming S_α by the orthogonal substitution
$$\xi'_i = \alpha \xi_i + \beta \xi_j + \gamma \xi_l + \delta \xi_k, \quad \xi'_j = \alpha' \xi_i + \beta' \xi_j + \gamma' \xi_l + \delta' \xi_k, \ldots$$

176) $\qquad \alpha^2 + \beta^2 + \gamma^2 + \dfrac{1}{\lambda}\delta^2 = 1,$

we obtain as above a substitution $S_{\bar{\alpha}}$ in which
$$\bar{\alpha}_{i1} \equiv \alpha \alpha_{i1} + \beta \alpha_{j1} + \gamma \alpha_{l1} + \delta \alpha_{k1}.$$

We proceed to show that solutions of 176) exist in the $GF[p^n]$ which make $\bar{\alpha}_{i1} = 0$. We may suppose that $\alpha_{i1}, \alpha_{j1}, \alpha_{l1}$ are not zero, since otherwise the result follows by inspection. If either of the sums
$$\alpha_{i1}^2 + \alpha_{j1}^2 + \alpha_{l1}^2, \quad \alpha_{i1}^2 + \alpha_{j1}^2 + \lambda \alpha_{k1}^2$$
be not zero, the problem is solved as above. If both sums be zero, then
$$\alpha_{l1}^2 = \lambda \alpha_{k1}^2, \quad \lambda = \text{square} = 1, \quad \alpha_{i1}^2 + \alpha_{j1}^2 + \alpha_{k1}^2 = 0.$$

Then the following set of solutions of 176) will make $\bar{\alpha}_{i1}$ zero:
$$\alpha = -\alpha_{j1}/\alpha_{l1}, \quad \beta = \alpha_{i1}/\alpha_{l1}, \quad \gamma = -\alpha_{k1}/\alpha_{l1}, \quad \delta = 1.$$

192. Transforming $S_1 \equiv S_\alpha C_1 C_2$ by $O_{534m}, O_{634m}, \ldots, O_{m-134m}$ in succession, we obtain in G a substitution S' in which $\alpha'_{51}, \alpha'_{61}, \ldots, \alpha'_{m-11}$, are all zero. Then by 143),

Also
$$\alpha'^2_{11} + \alpha'^2_{21} + \alpha'^2_{31} + \alpha'^2_{41} + \mu \alpha'^2_{m1} = 1.$$
$$\alpha'^2_{11} + \alpha'^2_{21} \equiv \alpha^2_{11} + \alpha^2_{21} \not\equiv 1.$$

Hence $\alpha'^2_{31} + \alpha'^2_{41} + \mu \alpha'^2_{m1} \not\equiv 0$, so that we can transform S' by a suitable O_{34m} into $S'' \equiv S_{\alpha''} C_1 C_2$ in which
$$\alpha''_{41} = \alpha''_{51} = \alpha''_{61} = \cdots = \alpha''_{m-11} = 0.$$

Transforming S'' by O_{j456} ($j = 7, 8, \ldots, m-1$) in succession, we can obtain a substitution $S_2 \equiv S_\beta C_1 C_2$ which leaves $\xi_7, \xi_8, \ldots, \xi_{m-1}$ fixed and has $\beta_{41} = \beta_{51} = \beta_{61} = 0$. If $\beta_{42}, \beta_{52}, \beta_{62}$ are all not zero, we transform S_2 by O_{456} and obtain a substitution S'_2 in which we can make $\beta'_{62} = 0$ except in the case[1])
$$\beta_{42}^2 + \beta_{52}^2 + \beta_{62}^2 = 0, \quad \beta_{31} \not\equiv 0, \quad \beta_{m1} \not\equiv 0.$$

---

[1]) If $\beta_{31} = 0$ or $\beta_{m1} = 0$, we transform S_2 by O_{6345} or O_{6m45} and make $\beta'_{62} = 0$.

CHAPTER VII.

In the latter case, we transform S_2 by $O_{3,m,4,5,6}^{a,b,c,d,e}$ and require that
$$\beta'_{62} \equiv a\beta_{32} + b\beta_{m2} + c\beta_{42} + d\beta_{52} + e\beta_{62} = 0,$$
$$\beta'_{61} \equiv a\beta_{31} + b\beta_{m1} = 0.$$
Taking
$$a = -b\beta_{m1}/\beta_{31}, \quad c = -d\beta_{52}/\beta_{42},$$
the second condition becomes an identity and the first takes the form
$$e\beta_{62} = b\beta, \quad \beta \equiv \frac{\beta_{m1}}{\beta_{31}}\beta_{32} - \beta_{m2}.$$
The further condition $a^2 + \frac{1}{\mu}b^2 + c^2 + d^2 + e^2 = 1$ then becomes

177) $\qquad d^2\left(1 + \frac{\beta_{52}^2}{\beta_{42}^2}\right) + b^2\left(\frac{1}{\mu} + \frac{\beta_{m1}^2}{\beta_{31}^2} + \frac{\beta^2}{\beta_{62}^2}\right) = 1.$

Since $\beta_{11}^2 + \beta_{21}^2 \neq 1$, $\beta_{11}^2 + \beta_{21}^2 + \beta_{31}^2 + \mu\beta_{m1}^2 = 1$, it follows that

178) $\qquad\qquad\qquad \frac{1}{\mu} + \frac{\beta_{m1}^2}{\beta_{31}^2} \neq 0.$

The coefficient of b^2 is zero for at most two values of β_{62}. In view of 178), these two values of β_{62} can be avoided, if $p^n > 3$, by an earlier transformation of S_2 by O_{456}, an operation not affecting the previous argument. Also the coefficient of d^2 is not zero. Hence 177) has solutions d, b in the field. The conditions $\beta'_{62} = \beta'_{61} = 0$ can thus be satisfied.

For $p^n = 3$, 178) requires $\mu = 1$. The coefficient of b^2 in 177) is then zero only when $\beta \neq 0$. If $\beta \neq 0$, we can determine a and b (each $\neq 0$) such that
$$a\beta_{32} + b\beta_{m2} = 1, \quad a\beta_{31} + b\beta_{m1} = 0.$$
Since $a^2 = b^2 = 1$, $\mu = 1$, the remaining conditions become
$$c\beta_{42} + d\beta_{52} + e\beta_{62} = -1, \quad c^2 + d^2 + e^2 = -1.$$
These are satisfied modulo 3 by taking $c = \beta_{42}$, $d = \beta_{52}$, $e = 0$.

193. We have thus reached in G a substitution Σ which leaves fixed $\xi_6, \xi_7, \ldots, \xi_{m-1}$ and which is not the identity. If
$$\Sigma \equiv C_1 C_2 C_3 C_4 C_5 C_m,$$
we obtain from it the substitution $C_1 C_2$ as at the beginning of § 191. From the known structure of the subgroup $O'_\mu(6, p^n)$, it follows that G contains all the substitutions of this subgroup. Transforming these by suitable even substitutions on the ξ_i, we obtain all the generators of $O'_\mu(m, p^n)$, with which G therefore coincides.

194. In stating our results concerning the structure of the orthogonal groups on $m \neq 4$ indices, we introduce permanent notations for the simple groups reached. For the first orthogonal group the case $m=4$ is shown in §§ 195—196 to be quite exceptional. We denote by $FO(m, p^n)$ the first orthogonal subgroup $O'_1(m, p^n)$, when m is odd, and the quotient-group of $O'_1(m, p^n)$ by its maximal self-conjugate subgroup $\{I, C\}$, when m is even and > 4. By § 72, $FO(m, p^n)$ has the order

$$FO[m, p^n] \equiv \frac{1}{2}(p^{n(m-1)}-1)p^{n(m-2)}(p^{n(m-3)}-1)p^{n(m-4)}\ldots(p^{2n}-1)p^n$$

for m odd; while, for m even, $m > 4$,

$$FO[m, p^n] \equiv \frac{1}{4}\left[p^{n(m-1)} - \varepsilon^{\frac{m}{2}} p^{n\left(\frac{m}{2}-1\right)}\right](p^{n(m-2)}-1)p^{n(m-3)}\ldots(p^{2n}-1)p^n.$$

The second orthogonal group on an even number $m > 4$ of indices has a simple subgroup[1] $SO(m, p^n)$, previously denoted by $O'_\nu(m, p^n)$ of order

$$SO[m, p^n] \equiv \frac{1}{2}\left[p^{n(m-1)} + \varepsilon^{\frac{m}{2}} p^{n\left(\frac{m}{2}-1\right)}\right](p^{n(m-2)}-1)p^{n(m-3)}\ldots(p^{2n}-1)p^n.$$

It will be shown in §§ 197—198 that this result holds true for $m=4$[2]. In both places, ε equals ± 1 according to the form $4l \pm 1$ of p^n.

Theorem. — *The first orthogonal group $O_1(m, p^n)$ has for m even and > 4 the factors of composition 2, $FO[m, p^n]$, 2 and for m odd the factors of composition 2, $FO[m, p^n]$, the case $m=3$, $p^n=3$ being exceptional. The second orthogonal group $O_\nu(m, p^n)$ on an even number $m > 2$ of indices has the factors of composition 2, $SO[m, p^n]$. The orthogonal groups on 2 indices are commutative groups.*

195. In virtue of the identity

$$\xi_1^2 + \xi_2^2 + \cdots + \xi_s^2 - \xi_{s+1}^2 - \xi_{s+2}^2 - \cdots - \xi_{2s}^2 \equiv \sum_{i=1}^{s}(\xi_i - \xi_{s+i})(\xi_i + \xi_{s+i}),$$

it follows from § 169 that the group[3] L_{s, p^n} of $2s$-ary linear homogeneous substitutions of determinant unity in the $GF[p^n]$, $p > 2$, which leave $\sum_{i=1}^{s} X_i Y_i$ invariant is holoedrically isomorphic with $O_1(2s, p^n)$ if -1 be a square in the $GF[p^n]$, $p > 2$, or if -1 be a not-square while s is even, but is isomorphic with $O_\nu(2s, p^n)$ if -1 be a not-square while s is odd. In particular, L_{2, p^n} is, for $p > 2$, holoedrically isomorphic with $O_1(4, p^n)$. In determining the structure of L_{2, p^n} we do not exclude the case $p = 2$.

---

[1] In view of the not-square factor in its invariant, it first appeared in the literature with the notation $NS(m, p^n)$.

[2] This result is readily verified for the case $p^n = 3$ not treated in §§ 197—198.

[3] The structure of this group was first determined by the author without making use of its isomorphism with orthogonal groups, *Proc. Lond. Math. Soc.*, vol. 30, pp. 70—98.

192   CHAPTER VII.

196. Theorem. — *The factors of composition of L_{2,p^n} are*

(if $p > 2$) $2, \; \frac{1}{2}(p^{2n}-1)p^n, \; \frac{1}{2}(p^{2n}-1)p^n, \; 2,$

(if $p = 2$) $2, \; (2^{2n}-1)2^n, \; (2^{2n}-1)2^n,$

except when $p^n = 2$ or 3, when the composite numbers 6 and 12 respectively are to be replaced by their prime factors.

To determine the quaternary substitutions leaving $\xi_1 \eta_1 + \xi_2 \eta_2$ absolutely invariant, consider the two pairs of equations[1]),

179) $\xi_1 + \varkappa \xi_2 = 0, \quad \eta_2 - \varkappa \eta_1 = 0,$

180) $\xi_1 + \varkappa \eta_2 = 0, \quad \xi_2 - \varkappa \eta_1 = 0.$

The most general quaternary linear homogeneous substitution, leaving invariant the pair of equations 179), for every value of $\varkappa$ in the field, is readily seen to be

181) $\begin{cases} \xi_1' = \alpha \xi_1 + \gamma \eta_2, & \xi_2' = -\gamma \eta_1 + \alpha \xi_2, \\ \eta_1' = \delta \eta_1 - \beta \xi_2, & \eta_2' = \beta \xi_1 + \delta \eta_2, \end{cases}$

having the determinant $(\alpha \delta - \beta \gamma)^2$. For it we have

$\xi_1' + \varkappa \xi_2' = \alpha(\xi_1 + \varkappa \xi_2) + \gamma(\eta_2 - \varkappa \eta_1),$
$\eta_2' - \varkappa \eta_1' = \beta(\xi_1 + \varkappa \xi_2) + \delta(\eta_2 - \varkappa \eta_1).$

The group of the substitutions 181) is therefore simply isomorphic with the binary group on the variables $\xi_1 + \varkappa \xi_2$ and $\eta_2 - \varkappa \eta_1$. Since the transposition $(\xi_2 \eta_2)$ transforms the pair of equations 179) into the pair 180), we obtain the most general linear homogeneous substitution, leaving invariant the pair of equations 180), for every $\varkappa$, if we transform the set of substitutions 181) by $(\xi_2 \eta_2)$, giving the set

182) $\begin{cases} \xi_1' = A\xi_1 + C\xi_2, & \xi_2' = B\xi_1 + D\xi_2, \\ \eta_1' = D\eta_1 - B\eta_2, & \eta_2' = -C\eta_1 + A\eta_2. \end{cases}$

The product of any substitution 181) by any substitution 182) gives

183)

	ξ_1	η_1	ξ_2	η_2
$\xi_1' =$	αA	$-\gamma C$	αC	γA
$\eta_1' =$	$-\beta B$	δD	$-\beta D$	$-\delta B$
$\xi_2' =$	αB	$-\gamma D$	αD	γB
$\eta_2' =$	βA	$-\delta C$	βC	δA

---

[1]) They give the two sets of generators on the ruled surface $\xi_1 \eta_1 + \xi_2 \eta_2 = 0$.

The same result holds if the substitutions be compounded in reverse order, so that the substitutions are commutative. Further, the only substitutions belonging to both of the sets 181) and 182) are seen to be

184) $\quad \xi_1' = \alpha \xi_1, \quad \eta_1' = \alpha \eta_1, \quad \xi_2' = \alpha \xi_2, \quad \eta_2' = \alpha \eta_2.$

The substitution 181) leaves $\xi_1 \eta_1 + \xi_2 \eta_2$ absolutely invariant if and only if $\alpha \delta - \beta \gamma = 1$. Hence there are $(p^{2n} - 1) p^n$ such substitutions. It follows that there are

$$\{(p^{2n} - 1) p^n\}^2, \qquad \text{(if } p = 2\text{)}$$
$$\frac{1}{2} \{(p^{2n} - 1) p^n\}^2 \qquad \text{(if } p > 2\text{)}$$

distinct substitutions 183) for which

185) $\qquad \alpha \delta - \beta \gamma = 1, \quad AD - BC = 1.$

The substitution $T_{2,\varkappa}$, defined in § 114, will be of the form 183) only if

$$\alpha A = \delta D = 1, \quad \alpha D = \varkappa, \quad \delta A = \varkappa^{-1}, \quad \beta = \gamma = B = C = 0.$$

Therefore $A = \alpha^{-1}, \; D = \varkappa \alpha^{-1}, \; \delta = \varkappa^{-1} \alpha$, so that

$$\alpha \delta - \beta \gamma = \varkappa^{-1} \alpha^2 \quad AD - BC = \varkappa \alpha^{-2}.$$

It will thus satisfy the relations 185) only when $\varkappa$ is a square in the $GF[p^n]$. Hence there are at least $\{(p^{2n} - 1) p^n\}^2$ substitutions 183) which satisfy the single relation

186) $\qquad (\alpha \delta - \beta \gamma)(AD - BC) = 1.$

For $p > 2$, L_{2, p^n} is holoedrically isomorphic with $O_1(4, p^n)$ and therefore, by § 172, has the order $(p^{3n} - p^n)(p^{2n} - 1) p^n$. Hence L_{2, p^n} is composed of the substitutions 183) alone. Those of these substitutions which satisfy 185) form a subgroup L_{2, p^n}' of index two. It is extended to the main group L_{2, p^n} by a substitution $T_{2,\varkappa}$.

For $p = 2$, the substitutions 183) which satisfy 185) form a subgroup $L_{2, 2^n}'$ of index two under $L_{2, 2^n}$. In fact, by § 204, the order of $L_{2, 2^n}$ is $2(2^{2n} - 1)^2 2^{2n}$, which is double the order of $L_{2, 2^n}'$. The transposition $(\xi_1 \eta_1)$ serves to extend $L_{2, 2^n}'$ to $L_{2, 2^n}$; for, if 183) reduces to the form $(\xi_1 \eta_1)$, then $\alpha A = \alpha C = \alpha B = 0$, $\alpha D = 1$, whence $A = C = B = 0$.

For either $p > 2$ or $p = 2$, the group L_{2, p^n}' of the substitutions 183) satisfying 185) has an invariant subgroup formed of the substitutions 181) which satisfy the relation $\alpha \delta - \beta \gamma = 1$. The quotient-group is holoedrically isomorphic with the simple group $LF(2, p^n)$. Indeed, it is clearly the quotient-group of the group of substitutions 182) satisfying $AD - BC = 1$ by the group of the substitutions 184), $\alpha^2 = 1$, common to the two sets 181) and 182) under the conditions 185).

CHAPTER VII.

197. Theorem. — *For $p^n > 3$, the second orthogonal group $O_\nu(4, p^n)$ is holoedrically isomorphic with the group E_{4,p^n} of quaternary linear substitutions in the $GF[p^n]$ of determinant unity which leave absolutely invariant the function*

$$f \equiv \xi_1 \eta_1 + \xi_2 \eta_2 + \xi_1^2 + \lambda^2 \eta_1^2,$$

in which $q \equiv \xi_1 \eta_1 + \xi_1^2 + \lambda^2 \eta_1^2$ *is irreducible in the field.*

For $p^n = 3$, the theorem necessarily fails, since q then becomes $(\xi_1 - \eta_1)^2$. For $p^n > 3$, there exists a quaternary substitution in the $GF[p^n]$ which transforms the invariant of the orthogonal group,

$$\Phi \equiv \xi_1^2 + \xi_2^2 + \xi_3^2 + \nu \xi_4^2 \qquad (\nu = \text{not-square})$$

into the function $f_1 \equiv \xi_1 \eta_1 + \xi_2 \eta_2 + \lambda \xi_1^2 + \lambda \eta_1^2$. But, for any p^n, f_1 is transformed into $\lambda^{-1} f$ by the substitution $\xi_1' = \lambda^{-1} \xi_1$, $\xi_2' = \lambda^{-1} \xi_2$.

If -1 be a not-square in the $GF[p^n]$, we may take $\nu = -1$. Then the substitution of determinant $\alpha\beta$,

$$\xi_1 = \alpha(\xi_1 - \eta_1), \quad \xi_2 = \beta(\xi_1 + \eta_1), \quad \xi_3 = \frac{1}{2}(\eta_2 + \xi_2), \quad \xi_4 = \frac{1}{2}(\eta_2 - \xi_2)$$

converts Φ into the function

$$(2\beta^2 - 2\alpha^2)\xi_1 \eta_1 + \xi_2 \eta_2 + (\alpha^2 + \beta^2)(\xi_1^2 + \eta_1^2).$$

Of the $p^n - 1$ sets of solutions in the $GF[p^n]$, $p > 2$, of $2\beta^2 - 2\alpha^2 = 1$, two sets make $\alpha\beta = 0$.[1] Hence there are $p^n - 3$ substitutions which reduce Φ to f_1. The irreducibility of q follows from that of $\xi_1^2 + \xi_2^2$.

If $-1 = I^2$, where I belongs to the $GF[p^n]$, the substitution

$$\xi_1 = \frac{1}{2}(\xi_2 + \eta_2), \quad \xi_2 = \frac{I}{2}(-\xi_2 + \eta_2), \quad \xi_3 = \alpha(\xi_1 - \eta_1), \quad \xi_4 = \beta(\xi_1 + \eta_1)$$

of determinant $I\alpha\beta$ transforms Φ into the function

$$\xi_2 \eta_2 + (2\nu\beta^2 - 2\alpha^2)\xi_1 \eta_1 + (\alpha^2 + \nu\beta^2)(\xi_1^2 + \eta_1^2).$$

Of the $p^n + 1$ sets of solutions in the $GF[p^n]$, $p > 2$, of $2\nu\beta^2 - 2\alpha^2 = 1$, two sets make $\alpha\beta = 0$. Hence there are $p^n - 1$ substitutions which transform Φ into f_1. The irreducibility of q now follows from that of $\xi_3^2 + \nu \xi_4^2$.

198. Theorem. — *Whether $p = 2$ or $p > 2$, the group E_{4,p^n} contains a subgroup E'_{4,p^n} of index two which is holoedrically isomorphic with $LF(2, p^{2n})$. According as $p = 2$ or $p > 2$, E'_{4,p^n} is extended to E_{4,p^n} by $(\xi_1 \eta_1)$ or $T_{2,\nu}$.*

---

[1] According as 2 is a square or a not-square, the solutions are given by $\alpha = 0$ or $\beta = 0$ respectively.

Let σ be a root of the equation

$$\sigma^2 + \sigma + \lambda^2 = 0,$$

which is irreducible in the $GF[p^n]$ in virtue of the irreducibility of q. The second root is therefore $\sigma^{p^n} \equiv \bar{\sigma}$, so that $\sigma\bar{\sigma} = \lambda^2$. The substitution

Z: $\qquad X = \xi_1 - \sigma\eta_1, \quad Y = \xi_1 - \bar{\sigma}\eta_1$

transforms the function $F \equiv XY + \xi_2\eta_2$ into f. Let $\alpha, \beta, \gamma, \delta$ be any set of marks in the $GF[p^{2n}]$ subject to the condition $\alpha\delta - \beta\gamma = 1$. Then F is absolutely invariant under the substitution [see 181)]

U: $X' = \alpha X + \gamma\eta_2, \quad Y' = \delta Y - \beta\xi_2, \quad \xi_2' = -\gamma Y + \alpha\xi_2, \quad \eta_2' = \beta X + \delta\eta_2.$

If we regard[1]) $\xi_1, \eta_1, \xi_2, \eta_2$ to be arbitrary marks of the $GF[p^n]$, Y will be conjugate to X with respect to the $GF[p^n]$, while F will be absolutely invariant under the following substitution conjugate to U [see 182)]

$\bar{U}$: $X' = \bar{\delta}X - \bar{\beta}\xi_2, \quad Y' = \bar{\alpha}Y + \bar{\gamma}\eta_2, \quad \xi_2' = -\bar{\gamma}X + \bar{\alpha}\xi_2, \quad \eta_2' = \bar{\beta}Y + \bar{\delta}\eta_2.$

If therefore the product $U\bar{U}$ be expressed in terms of the indices $\xi_1, \eta_1, \xi_2, \eta_2$, the resulting substitution W will leave f absolutely invariant and have its coefficients in the $GF[p^n]$. To give the explicit form of W, let U and $\bar{U}$ become U_1 and $\bar{U}_1$ when written in the indices ξ_i, η_i. Since the reciprocal of Z is

Z^{-1}: $\quad (\bar{\sigma} - \sigma)\xi_1 = \bar{\sigma}X - \sigma Y, \quad (\bar{\sigma} - \sigma)\eta_1 = X - Y,$

we find for U_1 the substitution

	ξ_1	η_1	ξ_2	η_2
$(\bar{\sigma} - \sigma)\xi_1' =$	$\sigma\alpha - \sigma\delta$	$\lambda^2\delta - \lambda^2\alpha$	$\sigma\beta$	$\bar{\sigma}\gamma$
$(\bar{\sigma} - \sigma)\eta_1' =$	$\alpha - \delta$	$\bar{\sigma}\delta - \sigma\alpha$	β	γ
$\xi_2' =$	$-\gamma$	$\bar{\sigma}\gamma$	α	0
$\eta_2' =$	β	$-\sigma\beta$	0	δ

The coefficients of $\bar{U}_1$ are conjugate to the corresponding coefficients of U_1. The product $W \equiv U_1\bar{U}_1$ is readily found to be the substitution

---

[1]) This interpretation is not a necessary one in view of the later explicit calculations.

196   CHAPTER VII.

	ξ_1	η_1	ξ_2	η_2
$\xi_1' =$	S	R	$\varkappa(\sigma\beta\overline{\alpha} - \overline{\sigma}\overline{\beta}\alpha)$	$\varkappa(\overline{\sigma}\gamma\overline{\delta} - \sigma\overline{\gamma}\delta)$
$\eta_1' =$	P	Q	$\varkappa(\beta\overline{\alpha} - \overline{\beta}\alpha)$	$\varkappa(\gamma\overline{\delta} - \delta\overline{\gamma})$
$\xi_2' =$	$-\overline{\alpha}\gamma - \alpha\overline{\gamma}$	$\sigma\alpha\overline{\gamma} + \overline{\sigma}\overline{\alpha}\gamma$	$\alpha\overline{\alpha}$	$-\gamma\overline{\gamma}$
$\eta_2' =$	$\beta\overline{\delta} + \overline{\beta}\delta$	$-\sigma\beta\overline{\delta} - \overline{\sigma}\overline{\beta}\delta$	$-\beta\overline{\beta}$	$\delta\overline{\delta}$

where $\varkappa^{-1} \equiv \overline{\sigma} - \sigma$ and

$$S \equiv \varkappa(\overline{\sigma}\alpha\overline{\delta} - \sigma\delta\overline{\alpha} + \overline{\sigma}\gamma\overline{\beta} - \sigma\beta\overline{\gamma}), \quad R \equiv \varkappa(\lambda^2\delta\overline{\alpha} - \lambda^2\alpha\overline{\delta} + \sigma^2\beta\overline{\gamma} - \overline{\sigma}^2\overline{\beta}\gamma),$$
$$Q \equiv \varkappa(\overline{\sigma}\alpha\overline{\delta} - \sigma\alpha\overline{\delta} - \overline{\sigma}\gamma\overline{\beta} + \sigma\beta\overline{\gamma}), \quad P \equiv \varkappa(\alpha\overline{\delta} - \delta\overline{\alpha} + \overline{\beta}\gamma - \beta\overline{\gamma}).$$

Since every coefficient of W equals its own conjugate with respect to the $GF[p^n]$, W belongs to that field.

As in § 196, the substitutions U form a group $\{U\}$ holoedrically isomorphic with the group of binary linear substitutions of determinant unity in the $GF[p^{2n}]$. The substitutions W form an isomorphic group $\{W\}$ leaving f absolutely invariant and therefore a subgroup of E_{4,p^n}. Indeed, if we take $U \sim W$ and $U' \sim W'$, then to UU' will correspond

$$U_1 U_1' \cdot \overline{U_1 U_1'} \equiv U_1 U_1' \overline{U_1} \overline{U_1'} = U_1 \overline{U_1} \cdot U_1' \overline{U_1'} \equiv WW',$$

since the set of substitutions U is commutative with the set $\overline{U}$ by § 196. Moreover, an identity $U\overline{U} = U'\overline{U}'$ or $U'^{-1}U = \overline{U}'\overline{U}^{-1}$ requires $U' = U$ or CU, where C merely changes the signs of the four indices. In fact, the groups $\{U\}$ and $\{\overline{U}\}$ have in common only the identity and C. Hence CU is the only substitution in addition to U which corresponds to the product $W \equiv U_1 \overline{U}_1 \equiv CU_1 \cdot C\overline{U}_1$. It follows that the quotient-group of $\{U\}$ by $\{I, C\}$ is holoedrically isomorphic both with the simple linear fractional group $LF(2, p^{2n})$ and with the group $\{W\}$. In particular, the order of $\{W\}$ is $\frac{1}{2}(p^{4n} - 1)p^{2n}$ or $(2^{4n} - 1) 2^{2n}$ according as $p > 2$ or $p = 2$. For $p > 2$, $p^n > 3$, E_{4,p^n} has the order $(p^{3n} + p^n)(p^{2n} - 1)p^n$, being holoedrically isomorphic with $O_\nu(4, p^n)$, whose order is given in § 172. For $p = 2$, $E_{4,2^n}$ is holoedrically isomorphic with the group leaving $\xi_1\eta_1 + \xi_2\eta_2 + \lambda\xi_1^2 + \lambda\eta_1^2$ absolutely invariant, whose order is shown in § 204 to be $2(2^{4n} - 1) 2^{2n}$. Hence $\{W\}$ is of index 2 under E_{4,p^n}.

According as $p > 2$ or $p = 2$, $\{W\}$ is extended to E_{4,p^n} by $T_{2,\varkappa}$ or $(\xi_1\eta_1)$, where $\varkappa$ is any not-square in the $GF[p^n]$. It is only necessary to show that these substitutions are not of the form W. If $(\xi_1\eta_1)$ were of the form W, then $\gamma\overline{\gamma} = \beta\overline{\beta} = 0$, $S = Q = 0$. Hence $\beta = \gamma = 0$, $\overline{\sigma}\alpha\overline{\delta} = \sigma\delta\overline{\alpha}$, $\overline{\sigma}\overline{\alpha}\delta = \sigma\alpha\overline{\delta}$. Hence would $\sigma\delta\overline{\alpha}$ and $\sigma\alpha\overline{\delta}$

CHAPTER VIII. LINEAR HOMOGENEOUS GROUP IN THE $GF[2^n]$ etc. 197

and consequently also their product belong to the $GF[p^n]$. But $\sigma^2 \alpha \bar{\alpha} \delta \bar{\delta}$ belongs to that field only when α or δ vanishes, so that $\alpha \delta - \beta \gamma = 0$.

If W reduce to the form $T_{2,\varkappa}$, then $\alpha \bar{\alpha} = \varkappa$, $\delta \bar{\delta} = \varkappa^{-1}$, $\beta = \gamma = 0$, $S = Q = 1$, $R = P = 0$. By the latter, $\bar{\alpha} \delta = \alpha \bar{\delta}$. Then $S = I$ gives $\delta \bar{\alpha} = 1$. But $\alpha \delta = \alpha \delta - \beta \gamma = 1$. Hence $\alpha = \bar{\alpha}$, so that $\varkappa = \alpha^2$, α belonging to the $GF[p^n]$.

CHAPTER VIII.

LINEAR HOMOGENEOUS GROUP IN THE $GF[2^n]$ DEFINED BY A QUADRATIC INVARIANT.

199. Theorem. — *If a quadratic form with coefficients in the $GF[2^n]$*

$$f \equiv \sum_{i \leq j}^{i,j=1,\ldots,m} \alpha_{ij} \xi_i \xi_j$$

can not be expressed in the field as a quadratic form in fewer than m linear homogeneous functions of $\xi_1, \ldots, \xi_m$, it can be reduced by a linear homogeneous substitution belonging to the field to one of the canonical forms

$$F \equiv \xi_1 \xi_2 + \xi_3 \xi_4 + \cdots + \xi_{m-2} \xi_{m-1} + \xi_m^2 \qquad (m \text{ odd})$$
$$F_\lambda \equiv \xi_1 \xi_2 + \xi_3 \xi_4 + \cdots + \xi_{m-3} \xi_{m-2} + \xi_{m-1} \xi_m + \lambda \xi_{m-1}^2 + \lambda \xi_m^2 \quad (m \text{ even})$$

where λ is zero or is a particular one of the values λ' for which

$$Q \equiv \xi_{m-1} \xi_m + \lambda' \xi_{m-1}^2 + \lambda' \xi_m^2$$

is irreducible in the $GF[2^n]$.

We first prove that, if $m \geqq 3$, f can be transformed into a quadratic form having $\alpha_{11} = 0$. If every α_{ij} $(i,j = 1, \ldots, m;\ i < j)$ were zero, f would reduce modulo 2 to the form

$$(\sqrt{\alpha_{11}} \xi_1 + \sqrt{\alpha_{22}} \xi_2 + \cdots + \sqrt{\alpha_{mm}} \xi_m)^2.$$

This being contrary to our hypothesis, we may assume that $\alpha_{23} \neq 0$, for example. We may also suppose that $\alpha_{22} \neq 0$, since otherwise the transformed of f by $(\xi_1 \xi_2)$ would have $\alpha_{11} = 0$. The terms of f which involve ξ_2 may be written thus,

$$\alpha_{22} \xi_2^2 + \xi_2 (\alpha_{12} \xi_1 + \alpha_{23} \xi_3 + \alpha_{24} \xi_4 + \cdots + \alpha_{2m} \xi_m).$$

198   CHAPTER VIII.

Hence the inverse of the following substitution,
$$\xi_3' = \alpha_{12}\xi_1 + \alpha_{23}\xi_3 + \alpha_{24}\xi_4 + \cdots + \alpha_{2m}\xi_m,$$
$$\xi_i' = \xi_i \qquad (i = 1, \ldots, m;\ i \neq 3),$$
will transform f into
$$\alpha_{22}\xi_2^2 + \xi_2\xi_3 + \Sigma \beta_{ij}\xi_i\xi_j,$$
summed for $i, j = 1, 3, 4, \ldots, m;\ i \leqq j$. Applying the substitution
$$\xi_2' = \xi_2 + \lambda\xi_1, \quad \xi_i' = \xi_i, \qquad (i = 1, 3, 4, \ldots, m)$$
we obtain as the new coefficient of ξ_1^2 the function $\alpha_{22}\lambda^2 + \beta_{11}$, which may be made to vanish by determining λ.

We may therefore suppose that $\alpha_{11} = 0$ in our original function f. Since the α_{1j} are not all zero, we may assume that $\alpha_{12} \neq 0$. Applying to f the inverse of the substitution
$$\xi_2' = \alpha_{12}\xi_2 + \alpha_{13}\xi_3 + \cdots + \alpha_{1m}\xi_m, \quad \xi_i' = \xi_i \quad (i = 1, 3, 4, \ldots, m)$$
we obtain the function
$$\xi_1\xi_2 + \sum_{i \leqq j}^{i,j=2,\ldots,m} \gamma_{ij}\xi_i\xi_j.$$

Replacing $\xi_1 + \gamma_{22}\xi_2 + \gamma_{23}\xi_3 + \cdots + \gamma_{2m}\xi_m$ by ξ_1, we get
$$f' \equiv \xi_1\xi_2 + \sum_{i \leqq j}^{i,j=3,\ldots,m} \gamma_{ij}\xi_i\xi_j.$$

Similarly, if $m \geqq 5$, we can transform f' into
$$\xi_1\xi_2 + \xi_3\xi_4 + \sum_{i \leqq j}^{i,j=5,\ldots,m} \varepsilon_{ij}\xi_i\xi_j.$$

If m be odd, we reach ultimately the form
$$\xi_1\xi_2 + \xi_3\xi_4 + \cdots + \xi_{m-2}\xi_{m-1} + \varkappa\xi_m^2.$$

Applying to it the substitution which replaces ξ_m by $\varkappa^{-1/2}\xi_m$, we obtain F.

If m be even, we reach ultimately the form
$$\Theta \equiv \xi_1\xi_2 + \xi_3\xi_4 + \cdots + \xi_{m-3}\xi_{m-2} + \alpha\xi_{m-1}^2 + \beta\xi_{m-1}\xi_m + \gamma\xi_m^2.$$

If $\alpha\xi_{m-1}^2 + \beta\xi_{m-1}\xi_m + \gamma\xi_m^2$ be reducible in the $GF[2^n]$, i. e., be the product of two linear homogeneous functions of ξ_{m-1} and ξ_m, an evident substitution will reduce Θ to F_0. In the contrary case, α, β, γ are certainly distinct from zero, so that the substitution

$\xi'_{m-1} = \alpha^{-1/2}\xi_{m-1}$, $\xi'_m = \alpha^{1/2}\beta^{-1}\xi_m$, $\xi'_i = \xi_i$ $(i=1,\ldots,m-2)$

will belong to the $GF[2^n]$. It transforms Θ into

187) $\quad \xi_1\xi_2 + \xi_3\xi_4 + \cdots + \xi_{m-3}\xi_{m-2} + \xi^2_{m-1} + \xi_{m-1}\xi_m + \delta\xi^2_m$

δ being such a mark that the equation

188) $\quad\quad\quad\quad\quad\quad \xi^2 + \xi + \delta = 0$

is irreducible in the $GF[2^n]$. It follows from 188) that

$$\xi^{2^n} = \xi + \delta + \delta^2 + \delta^4 + \cdots + \delta^{2^{n-1}}.$$

Hence 188) has a root ξ in the $GF[2^n]$ if and only if

$$\delta + \delta^2 + \cdots + \delta^{2^{n-1}} = 0.$$

The left member being its own square in the $GF[2^n]$ and hence either 0 or 1, it follows that 188) is irreducible in that field if and only if

189) $\quad\quad\quad\quad \delta + \delta^2 + \delta^4 + \cdots + \delta^{2^{n-1}} = 1.$

Applying to the quadratic form 187) the transformation

$\xi'_{m-1} = \xi_{m-1} + \lambda\xi_m$, $\xi'_i = \xi_i$, $(i=1,\ldots,m;\ i \neq m-1)$

the constant δ is replaced by

$$\delta' \equiv \delta + \lambda + \lambda^2,$$

which is therefore a root of 189). Giving to λ all possible values in the $GF[2^n]$, we obtain the 2^{n-1} roots of 189). Indeed, if in the $GF[2^n]$,

$$\delta + \lambda + \lambda^2 = \delta + \lambda_1 + \lambda_1^2,$$

we must have $\lambda_1 = \lambda$ or $\lambda + 1$. Hence all irreducible quadratic forms in two variables of the $GF[2^n]$ can be transformed linearly into each other. Applying, finally, the transformation

$\xi'_{m-1} = \delta^{\frac{1}{4}}\xi_{m-1}$, $\xi'_m = \delta^{-\frac{1}{4}}\xi'_m$, $\xi'_i = \xi_i$ $(i=1,\ldots,m-2)$

187) becomes $F_{\delta^{\frac{1}{2}}}$.

200. Changing the notation used in exhibiting F, the canonical quadratic form for an odd number $2m+1$ of indices may be written

$$\Psi \equiv \xi_0^2 + \xi_1\eta_1 + \xi_2\eta_2 + \cdots + \xi_m\eta_m.$$

The conditions upon the coefficients of the substitution S:

200 CHAPTER VIII.

$$\begin{cases} \xi'_i = \varkappa_i \xi_0 + \sum_{j=1}^{m} (\alpha_{ij}\xi_j + \gamma_{ij}\eta_j) & (i = 0, 1, \ldots, m) \\ \eta'_i = \sigma_i \xi_0 + \sum_{j=1}^{m} (\beta_{ij}\xi_j + \delta_{ij}\eta_j) & (i = 1, \ldots, m) \end{cases}$$

in order that it leave Ψ absolutely invariant are seen to be the special Abelian relations[1]) 76) for $\mu = 1$ together with the following:

190) $\quad \sum_{i=1}^{m}(\varkappa_i\beta_{ik} + \sigma_i\alpha_{ik}) = 0, \quad \sum_{i=1}^{m}(\varkappa_i\delta_{ik} + \sigma_i\gamma_{ik}) = 0, \quad (k=1,2,\ldots,m)$

191) $\quad \alpha_{0j}^2 = \sum_{i=1}^{m} \alpha_{ij}\beta_{ij}, \quad \gamma_{0j}^2 = \sum_{i=1}^{m}\gamma_{ij}\delta_{ij}, \quad \varkappa_0^2 + \sum_{i=1}^{m}\varkappa_i\sigma_i = 1.$

It follows from § 114 that every set of solutions $\alpha_{ij}, \beta_{ij}, \gamma_{ij}, \delta_{ij}$ in the $GF[2^n]$ of the relations $76)_{\mu=1}$ leads to a special Abelian substitution

$$\Sigma: \quad \xi'_i = \sum_{j=1}^{m}(\alpha_{ij}\xi_j + \gamma_{ij}\eta_j), \quad \eta'_i = \sum_{j=1}^{m}(\beta_{ij}\xi_j + \delta_{ij}\eta_j) \quad (i=1,\ldots,m)$$

whose determinant Δ is unity.[2])

The determinant of the coefficients of the $2m$ quantities $\varkappa_i, \sigma_i$ in the $2m$ equations 190) is seen to equal Δ. Hence, since $\Delta \neq 0$,

$$\varkappa_i = \sigma_i = 0 \qquad (i = 1, \ldots, m).$$

It follows that S takes the form

$$S': \begin{cases} \xi'_i = \sum_{j=1}^{m}(\alpha_{ij}\xi_j + \gamma_{ij}\eta_j), \quad \eta'_i = \sum_{j=1}^{m}(\beta_{ij}\xi_j + \delta_{ij}\eta_j) \quad i=1,\ldots,m \\ \xi'_0 = \xi_0 + \sum_{j=1}^{m}\left\{\left(\sum_{i=1}^{m}\alpha_{ij}\beta_{ij}\right)^{1/2}\xi_j + \left(\sum_{i=1}^{m}\gamma_{ij}\delta_{ij}\right)^{1/2}\eta_j\right\}, \end{cases}$$

the coefficients of S' being subject to the Abelian conditions 76) only. The group of the substitutions S is therefore holoedrically isomorphic with the special Abelian group $SA(2m, 2^n)$ of the substitutions Σ. The structure of the latter group is given in § 117.

201. Changing the notation employed in exhibiting the function F_λ, the canonical quadratic form for $2m$ indices may be written

$$f_\lambda \equiv \sum_{i=1}^{m} \xi_i \eta_i + \lambda \xi_m^2 + \lambda \eta_m^2.$$

---

[1]) Since $p = 2$, we have $-1 = +1$ in the field.
[2]) For a direct proof that $\Delta \neq 0$, see *American Journal*, vol. 21, p. 244.

We study the group G_λ of $2m$-ary linear substitutions in the $GF[2^n]$,

192) $\quad S: \quad \xi'_i = \sum_{j=1}^{m}(\alpha_{ij}\xi_j + \gamma_{ij}\eta_j), \quad \eta'_i = \sum_{j=1}^{m}(\beta_{ij}\xi_j + \delta_{ij}\eta_j)$
$\hspace{8cm} (i = 1, \ldots, m)$

which leave f_λ absolutely invariant. The conditions upon the coefficients of S are the Abelian relations[1]) 76), for $\mu = 1$, together with

193) $\quad \begin{cases} \sum_{i=1}^{m} \alpha_{ij}\beta_{ij} + \lambda\alpha_{mj}^2 + \lambda\beta_{mj}^2 = \begin{matrix} 0 \\ \lambda \end{matrix} & \begin{matrix}(j < m) \\ (j = m),\end{matrix} \\ \sum_{i=1}^{m} \gamma_{ij}\delta_{ij} + \lambda\gamma_{mj}^2 + \lambda\delta_{mj}^2 = \begin{matrix} 0 \\ \lambda \end{matrix} & \begin{matrix}(j < m) \\ (j = m).\end{matrix} \end{cases}$

Since S must be an Abelian substitution in the $GF[2^n]$, its reciprocal is obtained by replacing α_{ij}, β_{ij}, γ_{ij}, δ_{ij} by respectively δ_{ji}, β_{ji}, γ_{ji}, α_{ji}. Writing for S^{-1} the conditions 76) and 193), we obtain the equivalent set of conditions 78), for $\mu = 1$, and

194) $\quad \begin{cases} \sum_{i=1}^{m} \delta_{ji}\beta_{ji} + \lambda\delta_{jm}^2 + \lambda\beta_{jm}^2 = \begin{matrix} 0 \\ \lambda \end{matrix} & \begin{matrix}(j < m) \\ (j = m),\end{matrix} \\ \sum_{i=1}^{m} \gamma_{ji}\alpha_{ji} + \lambda\gamma_{jm}^2 + \lambda\alpha_{jm}^2 = \begin{matrix} 0 \\ \lambda \end{matrix} & \begin{matrix}(j < m) \\ (j = m).\end{matrix} \end{cases}$

Among the simplest substitutions leaving f_λ invariant occur
$M_i \equiv (\xi_i \eta_i) \quad (i = 1, \ldots, m); \quad N_{i,j,\varkappa}, \quad R_{i,j,\varkappa}, \quad Q_{i,j,\varkappa}, \quad T_{i,\varkappa}, \quad P_{ij}$
$\hspace{7cm}(i, j < m \text{ if } \lambda = \lambda')$

$L: \quad \xi'_m = \eta_m, \quad \eta'_m = \xi_m + \lambda^{-1}\eta_m \hspace{2cm} (\text{if } \lambda = \lambda');$
$N_{m,j,\varkappa}: \quad \xi'_m = \xi_m + \varkappa\eta_j, \quad \xi'_j = \xi_j + \lambda\varkappa^2\eta_j + \varkappa\eta_m$
$R_{m,j,\varkappa}: \quad \eta'_m = \eta_m + \varkappa\xi_j, \quad \eta'_j = \eta_j + \lambda\varkappa^2\xi_j + \varkappa\xi_m$
$Q_{m,j,\varkappa}: \quad \xi'_m = \xi_m + \varkappa\xi_j, \quad \eta'_j = \eta_j + \lambda\varkappa^2\xi_j + \varkappa\eta_m$
$Q_{j,m,\varkappa}: \quad \eta'_m = \eta_m + \varkappa\eta_j, \quad \xi'_j = \xi_j + \lambda\varkappa^2\eta_j + \varkappa\xi_m$

which reduce, when $\lambda = 0$, to the $N_{m,j,\varkappa}$, $R_{m,j,\varkappa}$, etc., defined in § 114.

According as $\lambda = 0$ or $\lambda = \lambda'$, G_λ is called *the first or the second hypoabelian group*[2]). The name arises from the fact that G_λ is a *subgroup* of the special Abelian group $SA(2m, 2^n)$.

---

[1]) This also follows from the fact that the invariance of f_λ implies that of its polar. Hence, if $p = 2$, G_λ leaves invariant $\sum_{i=1}^{m}(\xi_{i1}\eta_{i2} + \xi_{i2}\eta_{i1})$, where ξ_{i1}, η_{i1} and ξ_{i2}, η_{i2} are sets of cogredient variables.

[2]) For the case $n = 1$, these groups were studied at length by Jordan, Traité des substitutions, pp. 195—213 and p. 440. For general n, they were set up and investigated by the author in the papers, *Quarterly Journal*, 1898, pp. 1—16; *Bulletin of the Amer. Math. Soc.*, 1898, pp. 495—510; *Proceed. Lond. Math. Soc.*, vol. 30, pp. 70—98; *American Journal*, 1899, pp. 222—243.

CHAPTER VIII.

202. Theorem. — *If $m > 1$, G_λ may be generated by the substitutions*[1])

195) $\qquad M_i, \quad N_{i,j,\varkappa} \quad (i,j = 1, \ldots, m; \; \varkappa \text{ arbitrary in the field}).$

We note that M_i transforms $N_{j,i,\varkappa}$ into $Q_{j,i,\varkappa}$ and $Q_{i,j,\varkappa}$ into $R_{i,j,\varkappa}$. Further, for $i,j < m$ if $\lambda = \lambda'$, we have

$$P_{ij} \equiv Q_{j,i,1}^{-1} Q_{i,j,1} Q_{j,i,1},$$
$$T_{i,\mu} T_{j,\mu} \equiv M_i M_j P_{ij} R_{i,j,\mu^{-1}} N_{i,j,\mu} R_{i,j,\mu^{-1}}.$$
$$T_{i,\mu^2} \equiv T_{i,\mu} T_{j,\mu} \cdot M_j^{-1} T_{i,\mu} T_{j,\mu} M_j.$$

But every mark of the $GF[2^n]$ may be expressed as a square μ^2. Except in the case $m = 2$, $\lambda = \lambda'$, we thus reach every $T_{i,\varkappa}$. In the latter case, we derive every $T_{1,\varkappa}$ from the formula

196) $\qquad N_{m,1,\varkappa} Q_{m,1,\varkappa^{-1}\lambda^{-1}} N_{m,1,\varkappa} = L M_1 M_m T_{1,\lambda\varkappa^2}.$

Taking first $\varkappa = \lambda^{-1/2}$, we find that L may be derived from the substitutions 195). Applying 196) again, we reach every $T_{1,\lambda\varkappa^2}$.

To prove that every substitution S satisfying the relations $78)_{\mu=1}$ and 194) can be derived from the substitutions 195), we first set up a substitution T derived from them which, like S, replaces ξ_1 by

$$f_1 \equiv \sum_{j=1}^{m} (\alpha_{1j}\xi_j + \gamma_{1j}\eta_j)$$

where by 194),

197) $\qquad \sum_{j=1}^{m} \alpha_{1j}\gamma_{1j} + \lambda\alpha_{1m}^2 + \lambda\gamma_{1m}^2 = 0.$

a) If $\alpha_{11} \neq 0$, we may take as T the product

$$T_{1,\alpha_{11}} Q_{1,2,\alpha_{12}} N_{2,1,\gamma_{12}} \ldots Q_{1,m,\alpha_{1m}} N_{m,1,\gamma_{1m}},$$

since it replaces ξ_1 by

$$\alpha_{11}\xi_1 + \alpha_{11}^{-1}(\alpha_{12}\gamma_{12} + \cdots + \alpha_{1m}\gamma_{1m} + \lambda\alpha_{1m}^2 + \lambda\gamma_{1m}^2)\eta_1 + \sum_{j=2}^{m}(\alpha_{1j}\xi_j + \gamma_{1j}\eta_j) \equiv f_1.$$

b) If $\alpha_{11} = 0$, $\gamma_{11} \neq 0$, we may take for T the product

$$T_{1,\gamma_{11}^{-1}} Q_{2,1,\gamma_{12}} R_{2,1,\alpha_{12}} \ldots Q_{m,1,\gamma_{1m}} R_{m,1,\alpha_{1m}} \cdot M_1 M_m,$$

which replaces ξ_1 by

$$\gamma_{11}\eta_1 + \gamma_{11}^{-1}(\alpha_{12}\gamma_{12} + \cdots + \alpha_{1m}\gamma_{1m} + \lambda\alpha_{1m}^2 + \lambda\gamma_{1m}^2)\xi_1 + \sum_{j=2}^{m}(\alpha_{1j}\xi_j + \gamma_{1j}\eta_j) \equiv f_1.$$

---

[1]) The structure of G_λ being evident from § 203 if $m = 1$, we exclude this case henceforth.

c) If $\alpha_{1j} = \gamma_{1j} = 0$ $(j = 1, \ldots, k-1)$, but α_{1k} and γ_{1k} not both zero, we may, for $k < m$, proceed as in case a) or b) and obtain a substitution T' which replaces ξ_k by f_1 and is derived from the substitutions 195). We then take $T = T' P_{1k}$.

d) If $\alpha_{1j} = \gamma_{1j} = 0$ $(j = 1, \ldots, m-1)$, the proof given in c) applies if $\lambda = 0$, since then P_{1m} is generated by the substitutions 195) of G_λ. For $\lambda = \lambda'$, this case cannot exist, since the equation

$$\alpha_{1m}\gamma_{1m} + \lambda'\alpha_{1m}^2 + \lambda'\gamma_{1m}^2 = 0$$

requires $\alpha_{1m} = \gamma_{1m} = 0$ on account of the irreducibility of Q. Then would $f_1 \equiv 0$.

It follows that $S = TS_1$, where S_1 leaves ξ_1 fixed but is a substitution belonging to G_λ. Let S_1 replace η_1 by

$$f' \equiv \sum_{j=1}^{m} (\beta_{1j}\xi_j + \delta_{1j}\eta_j),$$

where, by 78), $\mu = 1$, and 194),

198) $\qquad \delta_{11} = 1, \sum_{j=1}^{m} \beta_{1j}\delta_{1j} + \lambda\beta_{1m}^2 + \lambda\delta_{1m}^2 = 0.$

The product

$$S' \equiv R_{2,1,\beta_{12}} Q_{2,1,\delta_{12}} \ldots R_{m,1,\beta_{1m}} Q_{m,1,\delta_{1m}}$$

replaces ξ_1 by ξ_1 and η_1 by

$$(\beta_{12}\delta_{12} + \cdots + \beta_{1m}\delta_{1m} + \lambda\beta_{1m}^2 + \lambda\delta_{1m}^2)\xi_1 + \eta_1 + \sum_{j=2}^{m}(\beta_{1j}\xi_j + \delta_{1j}\eta_j),$$

which equals f' since the coefficient of ξ_1 equals β_{11} by 198).

We may therefore set $S_1 = S'S_2$, where S_2 is a substitution of G_λ which leaves ξ_1 and η_1 fixed. Then by 78),

$$\alpha_{i1} = \beta_{i1} = \gamma_{i1} = \delta_{i1} = 0 \qquad (i = 2, \ldots, m).$$

The relations holding between $\alpha_{ij}, \beta_{ij}, \gamma_{ij}, \delta_{ij}$ $(i, j = 2, \ldots, m)$ are seen to be the relations 78) and 194) when $m-1$ is written for m. Proceeding with S_2 as we did with S, etc., we find ultimately that $S = T'\Sigma$, where T' is derived from the substitutions 195), while Σ is a substitution of G_λ which affects only ξ_m and η_m,

$\Sigma: \qquad \xi_m' = \alpha\xi_m + \gamma\eta_m, \quad \eta_m' = \beta\xi_m + \delta\eta_m.$

The conditions 78), 193) and 194) become, for $m = 1$,

199) $\quad \alpha\delta + \beta\gamma = 1, \quad \alpha\beta + \lambda\alpha^2 + \lambda\beta^2 = \lambda, \quad \gamma\delta + \lambda\gamma^2 + \lambda\delta^2 = \lambda,$
200) $\qquad\qquad\qquad \delta\beta + \lambda\delta^2 + \lambda\beta^2 = \lambda, \quad \gamma\alpha + \lambda\gamma^2 + \lambda\alpha^2 = \lambda.$

Combining 199) with 200), we may replace 200) by

201) $\qquad\qquad \beta(\alpha + \delta) = \gamma(\alpha + \delta) = \lambda(\alpha + \delta)^2.$

Suppose first that $\alpha + \delta \neq 0$. By 201), Σ becomes

202) $\quad O_m^{\alpha,\delta} : \begin{cases} \xi_m' = \alpha \xi_m + \lambda(\alpha + \delta)\eta_m \\ \eta_m' = \lambda(\alpha + \delta)\xi_m + \delta \eta_m \end{cases} \quad (\alpha\delta + \lambda^2\alpha^2 + \lambda^2\delta^2 = 1).$

Suppose next that $\beta + \gamma \neq 0$. Applying the above procedure to
$$\Sigma_1 \equiv \Sigma M_m: \quad \xi_m' = \beta \xi_m + \delta \eta_m, \quad \eta_m' = \alpha \xi_m + \gamma \eta_m,$$
it follows that $\Sigma_1 = O_m^{\beta,\gamma}$. Hence $\Sigma = O_m^{\beta,\gamma} M_m$.

Suppose finally that $\alpha + \delta = \beta + \gamma = 0$. Conditions 199) and 201) become
$$\alpha^2 + \beta^2 = 1, \quad \alpha\beta = 0,$$
so that $\Sigma = I$ or M_m.

In every case, $\Sigma = O_m^{\alpha,\delta}$ or $\Sigma = O_m^{\beta,\gamma} M_m$. If $\lambda = 0$, $O_m^{\alpha,\delta} \equiv T_{m,\alpha}$ and the theorem is proven. If $\lambda = \lambda'$, λ' being suitably chosen, we prove in the next section that every $O_m^{\alpha,\delta}$ is a power of $L \equiv O_m^{0,\lambda-1}$ and may therefore be derived from the substitutions 195).

203. Let ϱ be a primitive root of $\varrho^{2^n+1} = 1$. It will satisfy an equation belonging to and irreducible in the $GF[2^n]$,
$$\varrho^2 + \Theta\varrho + 1 = 0.$$
If we set $\Theta = \lambda^{-1}$, $\varrho = \xi_m/\eta_m$, we find that $\lambda \xi_m^2 + \lambda \eta_m^2 + \xi_m \eta_m$ is irreducible in and belongs to the $GF[2^n]$. Changing the variable from ϱ to $\sigma \equiv \lambda\varrho$, we obtain for σ the irreducible equation

203) $\quad\quad\quad\quad\quad \sigma^2 + \sigma + \lambda^2 = 0.$

Since the roots of 203) are σ and σ^{2^n}, we have $\sigma + \sigma^{2^n} = 1$.

We make the transformation of indices:

204) $\quad \xi_m \equiv \lambda^{3/2}\sigma^{-1} Y_{12} + \sigma\lambda^{-1/2} Y_{34}, \quad \eta_m \equiv \lambda^{1/2}(Y_{12} + Y_{34}).$

Solving, we find, for $p = 2$,

205) $\quad Y_{12} \equiv \lambda^{1/2}\xi_m + \sigma\lambda^{-1/2}\eta_m, \quad Y_{34} \equiv \lambda^{1/2}\xi_m + \lambda^{3/2}\sigma^{-1}\eta_m.$

Then
$$\xi_m \eta_m + \lambda \xi_m^2 + \lambda \eta_m^2 \equiv Y_{12} Y_{34}.$$

The substitution 202) takes the form

206) $\quad Y_{12}' = \tau Y_{12}, \quad Y_{34}' = \tau^{-1} Y_{34},$

where
$$\tau \equiv \alpha + (\alpha + \delta)\sigma, \quad \tau^{-1} \equiv \delta + (\alpha + \delta)\sigma,$$
$$\tau\tau^{-1} \equiv \alpha\delta + \sigma(\alpha+\delta)^2 + \sigma^2(\alpha+\delta)^2 \equiv \alpha\delta + \lambda^2(\alpha^2+\delta^2) = 1.$$

We have $\tau^{2^n+1} = 1$ since (mod 2),
$$\tau^{2^n} = \alpha + (\alpha+\delta)\sigma^{2^n} = \alpha + (\alpha+\delta)(\sigma+1) = \delta + (\alpha+\delta)\sigma = \tau^{-1}.$$

In particular, $L \equiv O_m^{0,\lambda-1}$ takes the form

207) $\quad\quad\quad\quad Y_{12}' = \varrho Y_{12}, \quad Y_{34}' = \varrho^{-1} Y_{34} \quad\quad\quad (\varrho \equiv \sigma\lambda^{-1}).$

LINEAR HOMOGENEOUS GROUP IN THE $GF[2^n]$ etc.   205

The substitutions 206) are evidently powers of 207), ϱ being a primitive root of $x^{2^n+1} = 1$. Hence the substitutions 202) are powers of L.

Inversely, every substitution 206) for which $\tau^{2^n+1} = 1$ may be transformed by 205) into a substitution 202) of the $GF[2^n]$. In fact,
$$\alpha + \delta = \tau + \tau^{-1}, \quad \alpha = \tau + (\tau + \tau^{-1})\sigma, \quad \delta = \tau^{-1} + (\tau + \tau^{-1})\sigma,$$
so that $\alpha + \delta$ belongs to the $GF[2^n]$ and likewise α since
$$\alpha^{2^n} = \tau^{2^n} + (\tau + \tau^{-1})\sigma^{2^n} = \tau^{-1} + (\tau + \tau^{-1})(\sigma + 1) = \alpha.$$

The number of substitutions 206) is $2^n + 1$. The number of substitutions 202) is therefore $2^n + 1$. Furthermore $M_m \equiv (\xi_m \eta_m)$ takes the form
$$Y'_{12} = \varrho Y_{34}, \quad Y'_{34} = \varrho^{-1} Y_{12} \qquad (\varrho \equiv \sigma \lambda^{-1}).$$

We have therefore a new proof of the results at the end of § 202.

It is worth while to verify independently that the number of substitutions 202) is $2^n \pm 1$ according as $\lambda = \lambda'$ or $\lambda = 0$. We have only to determine the number of sets of solutions in the $GF[2^n]$ of
208) $$\alpha\delta + \lambda^2 \alpha^2 + \lambda^2 \delta^2 = 1.$$
The result for the case $\lambda = 0$ being evident, we suppose that $\lambda = \lambda'$. The left member of 208) vanishes only when $\alpha = \delta = 0$; for, otherwise,
$$(1 + \omega + \lambda^2 \omega^{-1})^2 = 0, \quad \omega^2 \equiv \lambda^2 \alpha/\delta$$
would be reducible in the field, contrary to the irreducibility of 203). Hence each of the $2^{2n} - 1$ sets of marks α_1, δ_1, not both zero, in the $GF[2^n]$ will make
$$\alpha_1 \delta_1 + \lambda^2 \alpha_1^2 + \lambda^2 \delta_1^2 = \varkappa^2 \neq 0.$$
Then will $\alpha_1/\varkappa$, $\delta_1/\varkappa$ be a set of solutions of 208), and inversely every set of solutions of 208) may be so obtained. Hence, if $\lambda = \lambda'$, the number of distinct sets of solutions is $(2^{2n} - 1)/(2^n - 1)$.

204. We can now readily determine the order $\Omega_{m,n}^{(\lambda)}$ of G_λ. The number of distinct linear functions f_1 by which the substitutions of G_λ can replace ξ_1 is $P_{m,n}^{(\lambda)} - 1$, if $P_{m,n}^{(\lambda)}$ denotes the number of sets of solutions in the $GF[2^n]$ of 197). For $m > 1$, the pair of equations
$$\alpha_{11}\gamma_{11} = \tau, \quad \sum_{j=2}^{m} \alpha_{1j}\gamma_{1j} + \lambda \alpha_{1m}^2 + \lambda \gamma_{1m}^2 = \tau$$
has $(2^{n+1} - 1) P_{m-1,n}^{(\lambda)}$ sets of solutions when $\tau = 0$ and has
$$(2^n - 1)(2^{n(2m-2)} - P_{m-1,n}^{(\lambda)})$$
sets of solutions when τ runs through the series of marks $\neq 0$ of the $GF[2^n]$. We have therefore the recursion formula $(m > 1)$
$$P_{m,n}^{(\lambda)} = 2^n P_{m-1,n}^{(\lambda)} + (2^n - 1) 2^{n(2m-2)}.$$

According as $\lambda = 0$ or $\lambda = \lambda'$, the number of sets of solutions of
$$\alpha_{11}\gamma_{11} + \lambda a_{11}^2 + \lambda \gamma_{11}^2 = 0$$
is $P_{1,n}^{(0)} \equiv 2^{n+1} - 1$ or $P_{1,n}^{(\lambda')} \equiv 1$. We find by simple induction,
$$P_{s,n}^{(0)} - 1 = (2^{ns} - 1)(2^{n(s-1)} + 1), \quad P_{s,n}^{(\lambda')} - 1 = (2^{ns} + 1)(2^{n(s-1)} - 1).$$

The number of distinct linear functions f' is $2^{n(2m-2)}$. In fact, 198) determines β_{11} in terms of β_{1j}, δ_{1j} ($j = 2, \ldots, m$), so that the latter may be chosen arbitrarily in the $GF[2^n]$.

It follows therefore, from § 202, that
$$\Omega_{m,n}^{(\lambda)} = (P_{m,n}^{(\lambda)} - 1) 2^{2n(m-1)} \Omega_{m-1,n}^{(\lambda)} \qquad (m > 1).$$

By § 203, we have the initial values
$$\Omega_{1,n}^{(0)} = 2(2^n - 1), \quad \Omega_{1,n}^{(\lambda')} = 2(2^n + 1).$$

We now readily obtain the formulae
$$\Omega_{m,n}^{(0)} = 2(2^{nm} - 1)(2^{2n(m-1)} - 1)2^{2n(m-1)}(2^{2n(m-2)} - 1)2^{2n(m-2)} \ldots (2^{2n} - 1)2^{2n},$$
$$\Omega_{m,n}^{(\lambda')} = 2(2^{nm} + 1)(2^{2n(m-1)} - 1)2^{2n(m-1)}(2^{2n(m-2)} - 1)2^{2n(m-2)} \ldots (2^{2n} - 1)2^{2n}.$$

205. Theorem. — *Those substitutions of G_λ which satisfy the further relation*
$$209) \quad I(\alpha, \beta, \gamma, \delta) \equiv \sum_{i,j}^{1,\ldots,m} \alpha_{ij}\delta_{ij} + \lambda^2(\alpha_{mm}^2 + \beta_{mm}^2 + \gamma_{mm}^2 + \delta_{mm}^2) = m$$

form a subgroup of index 2 which any M_i extends to G_λ. If $m > 2$, this subgroup is identical with the group generated as follows:
$$J_\lambda \equiv \{M_i M_j,\ N_{i,j,\varkappa}\} \quad (i, j = 1, \ldots, m;\ \varkappa\ \textit{arbitrary in field}).$$

If $m = 2$, it is identical with the group
$$J_\lambda \equiv \{M_1 M_2,\ N_{2,1,1},\ T_{1,\varkappa},\ Q_{2,1,1}\}.$$

We first prove that every substitution of J_λ satisfies 209). To do this, it suffices to show that, if Σ be any substitution of G_λ which satisfies 209), the products $M_i M_j \Sigma$, $N_{i,j,\varkappa} \Sigma$ will also satisfy 209), the case $m = 2$, being treated later. Let Σ have the form 192).

a) If the product $M_j \Sigma$ be expressed in the form
$$210) \quad \xi_i' = \sum_{j=1}^m (\alpha_{ij}'\xi_j + \gamma_{ij}'\eta_j), \quad \eta_i' = \sum_{j=1}^m (\beta_{ij}'\xi_j + \delta_{ij}'\eta_j) \quad (i = 1, \ldots, m)$$
we have
$$\alpha_{ij}' = \gamma_{ij}, \quad \gamma_{ij}' = \alpha_{ij}, \quad \beta_{ij}' = \delta_{ij}, \quad \delta_{ij}' = \beta_{ij} \quad (i = 1, \ldots, m)$$
$$\alpha_{ik}' = \alpha_{ik}, \quad \beta_{ik}' = \beta_{ik}, \quad \gamma_{ik}' = \gamma_{ik}, \quad \delta_{ik}' = \delta_{ik} \quad \binom{i = 1, \ldots, m}{k = 1, \ldots, m;\ k \neq j}.$$

Hence

$$I(\alpha',\beta',\gamma',\delta') \equiv \sum_{k \neq j}^{i,k=1,\ldots,m} \alpha_{ik}\delta_{ik} + \sum_{i=1}^{m} \gamma_{ij}\beta_{ij} + \lambda^2(\alpha_{mm}^2 + \beta_{mm}^2 + \gamma_{mm}^2 + \delta_{mm}^2)$$

$$= I(\alpha,\beta,\gamma,\delta) + \sum_{i=1}^{m}(\gamma_{ij}\beta_{ij} - \alpha_{ij}\delta_{ij}) = m+1,$$

upon applying 209) and 76). Hence $M_j\Sigma$ does not satisfy 209), while $M_iM_j\Sigma$ does.

b) If the product $N_{m,j,\varkappa}\Sigma$ be expressed in the form 210), we have

$$\alpha'_{rs} = \alpha_{rs}, \qquad \beta'_{rs} = \beta_{rs} \qquad (r,s=1,\ldots,m)$$
$$\gamma'_{rs} = \gamma_{rs}, \qquad \delta'_{rs} = \delta_{rs} \qquad (r,s=1,\ldots,m; \; s \neq m, j)$$
$$\gamma'_{rm} = \gamma_{rm} + \varkappa\alpha_{rj}, \; \gamma'_{rj} = \gamma_{rj} + \varkappa\alpha_{rm} + \lambda\varkappa^2\alpha_{rj} \qquad (r=1,\ldots,m)$$
$$\delta'_{rm} = \delta_{rm} + \varkappa\beta_{rj}, \; \delta'_{rj} = \delta_{rj} + \varkappa\beta_{rm} + \lambda\varkappa^2\beta_{rj} \qquad (r=1,\ldots,m).$$

Hence $I(\alpha',\beta',\gamma',\delta')$ equals

$$\sum_{s \neq m,j}^{r,s=1,\ldots,m} \alpha_{rs}\delta_{rs} + \sum_{r=1}^{m}\alpha_{rm}(\delta_{rm}+\varkappa\beta_{rj}) + \sum_{r=1}^{m}\alpha_{rj}(\delta_{rj}+\varkappa\beta_{rm}+\lambda\varkappa^2\beta_{rj})$$
$$+ \lambda^2\{\alpha_{mm}^2 + \beta_{mm}^2 + (\gamma_{mm}+\varkappa\alpha_{mj})^2 + (\delta_{mm}+\varkappa\beta_{mj})^2\}$$
$$= I(\alpha,\beta,\gamma,\delta) + \varkappa\sum_{r=1}^{m}(\alpha_{rm}\beta_{rj}+\alpha_{rj}\beta_{rm}) + \lambda\varkappa^2\left(\sum_{r=1}^{m}\alpha_{rj}\beta_{rj}+\lambda\alpha_{mj}^2+\lambda\beta_{mj}^2\right).$$

But the last two sums are zero by 76) and 193).

c) An analogous proof holds for $N_{i,j,\varkappa}$ $(i,j<m)$, the above terms involving $\lambda\varkappa^2$ not being present.

d) Since the substitutions $Q_{i,j,\varkappa}$, $R_{i,j,\varkappa}$ $(i,j=1,\ldots,m)$ and P_{ij}, $T_{i,\varkappa}$ $(i,j<m$, if $\lambda = \lambda')$ may be expressed as a product of the $N_{i,j,\varkappa}$ and an even number of the M_i (§ 202), the products $T_{i,\varkappa}\Sigma$, $Q_{i,j,\varkappa}\Sigma$, etc., will satisfy 209) if Σ does.

Inversely, every substitution S of G_λ which satisfies 209) *belongs to J_λ.* In fact, by the proof given in § 202, S is of one of the two forms K, KM_m, where K is derived from[1] M_iM_j, $Q_{i,j,\varkappa}$, $N_{i,j,\varkappa}$, $R_{i,j,\varkappa}$ $(i,j=1,\ldots,m)$; P_{ij}, $T_{i,\varkappa}$ $(i,j<m$ if $\lambda=\lambda')$. Since S shall satisfy 209), it is not of the form KM_m. It remains to show that these substitutions M_iM_j, $Q_{i,j,\varkappa}$, ..., $T_{i,\varkappa}$ belong to J_λ.

For $m>2$, J_λ contains $Q_{i,j,\varkappa}$, the transformed of $N_{i,j,\varkappa}$ by M_jM_k $(k \neq i,j)$; also $R_{i,j,\varkappa}$ and $Q_{j,i,\varkappa}$, the transformed of $N_{i,j,\varkappa}$ and $Q_{i,j,\varkappa}$ respectively by M_iM_j. Applying the formulae at the beginning of § 202, we reach P_{ij} and $T_{i,\mu}T_{j,\mu}$ $(i,j<m$, if $\lambda=\lambda')$. Then J_λ contains $T_{i,\mu}T_{j,\mu^{-1}}$, the transformed of the latter by M_mM_j. The product of the two gives T_{i,μ^2}.

---

[1] By 196), L and therefore every $O_m^{\alpha,\delta}$ is derived from M_1M_2, $Q_{2,1,\varkappa}$ and $N_{2,1,\varkappa}$.

208 CHAPTER VIII.

For $m=2$, J_λ contains $M_1 M_2$, $T_{1,\varkappa}$, $N_{2,1,\varkappa}$, $R_{2,1,\varkappa}$, $Q_{2,1,\varkappa}$, $Q_{1,2,\varkappa}$.
If $\lambda = 0$, J_λ contains $P_{12} \equiv Q_{2,1,1}^{-1} Q_{1,2,1} Q_{2,1,1}$.

The fact that $M_1 M_2$ and $N_{1,2,\varkappa}$ do not generate J_0, for $m=2$, follows readily from § 196. Since $M_1 M_2$ transforms $N_{1,2,\varkappa}$ into $R_{1,2,\varkappa}$, every substitution derived from the two former may be given the form V or $VM_1 M_2$, where V is derived from $N_{1,2,\varkappa}$ and $R_{1,2,\varkappa}$. The latter two are of the form 181). Hence the group of the substitutions V is a subgroup of the group of the substitutions 181) having the order
$$v \equiv (2^{2n}-1) 2^n.$$
Hence $M_1 M_2$ and the $N_{1,2,\varkappa}$ generate a group whose order is at most $2v$. But $2v < (2^{2n}-1)^2 2^{2n}$, the order of J_0 for $m=2$.

It follows similarly from §§ 197—198 that $M_1 M_2$ and $N_{2,1,\varkappa}$ do not generate $J_{\lambda'}$ for $m=2$. This result may be shown directly for the case $n=1$, when $J_{\lambda'}$ has the order 60 (§ 204). In fact, setting $M \equiv M_1 M_2$, $N \equiv N_{2,1,1}$, $R \equiv R_{2,1,1} = M^{-1} NM$, the group generated by M and N contains only ten distinct substitutions:

$$I, M, N, R, NM, RM, RN, NR, NRM, RNM.$$

For $m=2$, the structure of J_0 was determined in § 196 and that of $J_{\lambda'}$ in §§ 197—198.

206. Theorem. — *The senary first hypoabelian group J_0 in the $GF[2^n]$ is a simple group holoedrically isomorphic with $LF(4, 2^n)$.*

We obtained in § 163 a senary group $G'_{4,2}$, leaving absolutely invariant
$$Y_{12} Y_{34} - Y_{13} Y_{24} + Y_{14} Y_{23},$$
which is holoedrically isomorphic with the simple group $LF(4, 2^n)$. To identify $G'_{4,2}$ with J_0 ($m=3$), we set
$$Y_{12} = \xi_1, \quad Y_{13} = \xi_2, \quad Y_{14} = \xi_3, \quad Y_{23} = \eta_3, \quad Y_{24} = \eta_2, \quad Y_{34} = \eta_1.$$
The general substitution $[\alpha]_2$ of $G'_{4,2}$, given in § 164, may be written

	ξ_1	ξ_2	ξ_3	η_3	η_2	η_1
$\xi'_1 =$	α_{11}	α_{12}	α_{13}	γ_{13}	γ_{12}	γ_{11}
$\xi'_2 =$	α_{21}	α_{22}	α_{23}	γ_{23}	γ_{22}	γ_{21}
$\xi'_3 =$	α_{31}	α_{32}	α_{33}	γ_{33}	γ_{32}	γ_{31}
$\eta'_3 =$	β_{31}	β_{32}	β_{33}	δ_{33}	δ_{32}	δ_{31}
$\eta'_2 =$	β_{21}	β_{22}	β_{23}	δ_{23}	δ_{22}	δ_{21}
$\eta'_1 =$	β_{11}	β_{12}	β_{13}	δ_{13}	δ_{12}	δ_{11}

In this form the notation agrees with that employed in § 201 for the substitutions of J_0. In view of § 165, the above general substitution of $G'_{4,2}$ must satisfy the relation (mod 2)
$$\alpha_{11}\delta_{11}+\alpha_{12}\delta_{12}+\alpha_{13}\delta_{13}+\alpha_{21}\delta_{21}+\alpha_{22}\delta_{22}+\alpha_{23}\delta_{23}+\alpha_{31}\delta_{31}+\alpha_{32}\delta_{32}+\alpha_{33}\delta_{33} \equiv 1.$$
But this is relation 209) for $\lambda = 0$, $m = 3$, which defines the subgroup J_0 of the first hypoabelian group. Hence $G'_{4,2} \equiv J_0$.

207. Theorem. — *The senary second hyperabelian group $J_{\lambda'}$ in the $GF[2^n]$ is a simple group, holoedrically isomorphic with $HA(4, 2^{2n})$.*

We begin as in § 190, but make the following transformation of indices, including the transformation 204) for $m = 3$:

$$\xi_1 = Y_{14}, \quad \xi_2 = Y_{13}, \quad \eta_1 = Y_{23}, \quad \eta_2 = Y_{24},$$
$$\xi_3 \equiv \lambda^{3/2}\sigma^{-1}Y_{12} + \sigma\lambda^{-1/2}Y_{34}, \quad \eta_3 = \lambda^{1/2}(Y_{12} + Y_{34}).$$

The invariant of the second compound group is transformed thus:

$$Y_{12}Y_{34} - Y_{13}Y_{24} + Y_{14}Y_{23} = \xi_1\eta_1 + \xi_2\eta_2 + \xi_3\eta_3 + \lambda\xi_3^2 + \lambda\eta_3^2.$$

If we take
$$\tau \equiv \omega^{-2^n+1}, \quad \tau^{2^n+1} \equiv \omega^{-2^{2n}+1} = 1,$$

the substitution 206) becomes in the new indices a substitution 202) with coefficients in the $GF[2^n]$. In particular, if ω be a suitable primitive root of the $GF[2^{2n}]$, τ will be the primitive root ϱ of $x^{2^n+1} = 1$. We thus reach, by 207), the substitution L.

We next express in the new indices the general substitution $[\alpha]_2$, given in § 164, of the second compound $A_{4,2}$ of the group of quaternary Abelian substitutions of determinant unity in the $GF[2^n]$. For example, it will replace $\xi_2 \equiv Y_{13}$ by

$$\begin{vmatrix}1&3\\1&2\end{vmatrix}(\lambda^{1/2}\xi_3 + \sigma\lambda^{-1/2}\eta_3) + \begin{vmatrix}1&3\\1&3\end{vmatrix}\xi_2 + \begin{vmatrix}1&3\\1&4\end{vmatrix}\xi_1 + \begin{vmatrix}1&3\\2&3\end{vmatrix}\eta_1 + \begin{vmatrix}1&3\\2&4\end{vmatrix}\eta_2$$
$$+ \begin{vmatrix}1&3\\3&4\end{vmatrix}(\lambda^{1/2}\xi_3 + \lambda^{3/2}\sigma^{-1}\eta_3).$$

Here the coefficient of ξ_3 is $\equiv 0$ (mod 2) and that of η_3 is $\lambda^{-1/2}\begin{vmatrix}1&3\\1&2\end{vmatrix}$, since

$$\begin{vmatrix}1&3\\1&2\end{vmatrix} + \begin{vmatrix}1&3\\3&4\end{vmatrix} \equiv \begin{vmatrix}\alpha_{11}&\alpha_{12}\\\alpha_{31}&\alpha_{32}\end{vmatrix} + \begin{vmatrix}\alpha_{13}&\alpha_{14}\\\alpha_{33}&\alpha_{34}\end{vmatrix} = 0$$

by one of the Abelian conditions, while $\lambda^2\sigma^{-1} + \sigma + 1 \equiv 0$ by 203). Proceeding in this manner, we find that $[\alpha]_2$ takes the form

	ξ_1	ξ_2	ξ_3	η_3	η_2	η_1
$\xi_1' =$	$\begin{vmatrix}1&4\\1&4\end{vmatrix}$	$\begin{vmatrix}1&4\\1&3\end{vmatrix}$	0	$\lambda^{-\frac{1}{2}}\begin{vmatrix}1&4\\1&2\end{vmatrix}$	$\begin{vmatrix}1&4\\2&4\end{vmatrix}$	$\begin{vmatrix}1&4\\2&3\end{vmatrix}$
$\xi_2' =$	$\begin{vmatrix}1&3\\1&4\end{vmatrix}$	$\begin{vmatrix}1&3\\1&3\end{vmatrix}$	0	$\lambda^{-\frac{1}{2}}\begin{vmatrix}1&3\\1&2\end{vmatrix}$	$\begin{vmatrix}1&3\\2&4\end{vmatrix}$	$\begin{vmatrix}1&3\\2&3\end{vmatrix}$
211) $\xi_3' =$	$\lambda^{-\frac{1}{2}}\begin{vmatrix}1&2\\1&4\end{vmatrix}$	$\lambda^{-\frac{1}{2}}\begin{vmatrix}1&2\\1&3\end{vmatrix}$	1	$\lambda^{-1}\begin{vmatrix}1&2\\1&2\end{vmatrix} + \lambda^{-1}$	$\lambda^{-\frac{1}{2}}\begin{vmatrix}1&2\\2&4\end{vmatrix}$	$\lambda^{-\frac{1}{2}}\begin{vmatrix}1&2\\2&3\end{vmatrix}$
$\eta_3' =$	0	0	0	1	0	0
$\eta_2' =$	$\begin{vmatrix}2&4\\1&4\end{vmatrix}$	$\begin{vmatrix}2&4\\1&3\end{vmatrix}$	0	$\lambda^{-\frac{1}{2}}\begin{vmatrix}2&4\\1&2\end{vmatrix}$	$\begin{vmatrix}2&4\\2&4\end{vmatrix}$	$\begin{vmatrix}2&4\\2&3\end{vmatrix}$
$\eta_1' =$	$\begin{vmatrix}2&3\\1&4\end{vmatrix}$	$\begin{vmatrix}2&3\\1&3\end{vmatrix}$	0	$\lambda^{-\frac{1}{2}}\begin{vmatrix}2&3\\1&2\end{vmatrix}$	$\begin{vmatrix}2&3\\2&4\end{vmatrix}$	$\begin{vmatrix}2&3\\2&3\end{vmatrix}$

210   CHAPTER VIII.

To prove that this substitution satisfies relation 209) for $m = 3$, consider it to be expressed in the notation used for the general substitution $[\alpha]_2$ of § 206. The condition 209) built for the substitution 211) therefore becomes

$$\begin{vmatrix} 1\,4 \\ 1\,4 \end{vmatrix}\begin{vmatrix} 2\,3 \\ 2\,3 \end{vmatrix} + \begin{vmatrix} 1\,4 \\ 1\,3 \end{vmatrix}\begin{vmatrix} 2\,3 \\ 2\,4 \end{vmatrix} + \begin{vmatrix} 1\,3 \\ 1\,4 \end{vmatrix}\begin{vmatrix} 2\,4 \\ 2\,3 \end{vmatrix} + \begin{vmatrix} 1\,3 \\ 1\,3 \end{vmatrix}\begin{vmatrix} 2\,4 \\ 2\,4 \end{vmatrix} + \begin{vmatrix} 1\,2 \\ 1\,2 \end{vmatrix}^2 \equiv 3 \pmod{2}.$$

The left member may be written (mod 2):

$$\alpha_{11}\begin{vmatrix} 0 & \alpha_{23} & \alpha_{24} \\ \alpha_{32} & \alpha_{33} & \alpha_{34} \\ \alpha_{42} & \alpha_{43} & \alpha_{44} \end{vmatrix} + \alpha_{31}\begin{vmatrix} 0 & \alpha_{13} & \alpha_{14} \\ \alpha_{22} & \alpha_{23} & \alpha_{24} \\ \alpha_{42} & \alpha_{43} & \alpha_{44} \end{vmatrix} + \alpha_{41}\begin{vmatrix} 0 & \alpha_{13} & \alpha_{14} \\ \alpha_{22} & \alpha_{23} & \alpha_{24} \\ \alpha_{32} & \alpha_{33} & \alpha_{34} \end{vmatrix} + \alpha_{11}^2\alpha_{22}^2 + \alpha_{12}^2\alpha_{21}^2.$$

Upon expanding according to the elements of the first column the determinant on the left of the following identity

$$\begin{vmatrix} \alpha_{11} & \alpha_{12} & \alpha_{13} & \alpha_{14} \\ \alpha_{21} & \alpha_{22} & \alpha_{23} & \alpha_{24} \\ \alpha_{31} & \alpha_{32} & \alpha_{33} & \alpha_{34} \\ \alpha_{41} & \alpha_{42} & \alpha_{43} & \alpha_{44} \end{vmatrix} \equiv 1,$$

we obtain the first three terms in the above expression together with

$$\alpha_{21}\begin{vmatrix} \alpha_{12} & \alpha_{13} & \alpha_{14} \\ \alpha_{32} & \alpha_{33} & \alpha_{34} \\ \alpha_{42} & \alpha_{43} & \alpha_{44} \end{vmatrix} + \alpha_{11}\alpha_{22}\begin{vmatrix} \alpha_{33} & \alpha_{34} \\ \alpha_{43} & \alpha_{44} \end{vmatrix} + \alpha_{31}\alpha_{12}\begin{vmatrix} \alpha_{23} & \alpha_{24} \\ \alpha_{43} & \alpha_{44} \end{vmatrix} + \alpha_{41}\alpha_{12}\begin{vmatrix} \alpha_{23} & \alpha_{24} \\ \alpha_{33} & \alpha_{34} \end{vmatrix}.$$

It remains to show that the sum of these terms together with $\alpha_{11}^2\alpha_{22}^2 + \alpha_{12}^2\alpha_{21}^2$ is zero. Upon applying the Abelian relations (mod 2),

$$\begin{vmatrix} \alpha_{33} & \alpha_{34} \\ \alpha_{43} & \alpha_{44} \end{vmatrix} = \begin{vmatrix} \alpha_{11} & \alpha_{12} \\ \alpha_{21} & \alpha_{22} \end{vmatrix}, \quad \begin{vmatrix} \alpha_{32} & \alpha_{34} \\ \alpha_{42} & \alpha_{44} \end{vmatrix} = \begin{vmatrix} \alpha_{12} & \alpha_{14} \\ \alpha_{22} & \alpha_{24} \end{vmatrix}, \quad \begin{vmatrix} \alpha_{32} & \alpha_{33} \\ \alpha_{42} & \alpha_{43} \end{vmatrix} = \begin{vmatrix} \alpha_{12} & \alpha_{13} \\ \alpha_{22} & \alpha_{23} \end{vmatrix},$$

$$\begin{vmatrix} \alpha_{23} & \alpha_{24} \\ \alpha_{43} & \alpha_{44} \end{vmatrix} = \begin{vmatrix} \alpha_{21} & \alpha_{22} \\ \alpha_{41} & \alpha_{42} \end{vmatrix}, \quad \begin{vmatrix} \alpha_{23} & \alpha_{24} \\ \alpha_{33} & \alpha_{34} \end{vmatrix} = \begin{vmatrix} \alpha_{21} & \alpha_{22} \\ \alpha_{31} & \alpha_{32} \end{vmatrix}, \quad \begin{vmatrix} \alpha_{13} & \alpha_{14} \\ \alpha_{23} & \alpha_{24} \end{vmatrix} = \begin{vmatrix} \alpha_{31} & \alpha_{32} \\ \alpha_{41} & \alpha_{42} \end{vmatrix},$$

the sum is seen to be congruent to zero (mod 2). The substitutions 211) therefore belong to $J_{\lambda'}$. Their number equals the order

$$(2^{4n} - 1)\,2^{3n}(2^{2n} - 1)\,2^n$$

of the quaternary Abelian group $SA(4, 2^n)$ (§ 115), which was shown above to be holoedrically isomorphic with the group of the substitutions 211) leaving η_3 fixed. We prove in the next section that this number equals the total number of substitutions belonging to $J_{\lambda'}$ ($m = 3$) and leaving η_3 fixed. It follows that the substitutions 211) include the following substitutions of $J_{\lambda'}$ not altering η_m:

$$M_1 M_2, \quad N_{1,2,\varkappa}, \quad N_{m,1,\varkappa}, \quad N_{m,2,\varkappa}, \quad Q_{m,1,\varkappa}.$$

LINEAR HOMOGENEOUS GROUP IN THE $GF[2^n]$ etc. 211

These substitutions must therefore belong to the group C, the second compound of $HA(4, 2^{2n})$ when expressed in the indices ξ_i, η_i. Also C contains L and therefore also $M_1 M_m$ by formula 196). Hence C contains all the generators of $J_{\lambda'}$ ($m = 3$). But the order of C, being equal to that of $HA(4, 2^{2n})$, is

$$(2^{4n} - 1) 2^{3n} (2^{3n} + 1) 2^{2n} (2^{2n} - 1) 2^n,$$

which equals the order of $J_{\lambda'}$ ($m = 3$). Hence $J_{\lambda'} \equiv C$.

208. Theorem. — *If $m = 3$, the number of substitutions of $J_{\lambda'}$ which leave ξ_m fixed is*

$$(2^{2n} + 1) 2^{2n} (2^{2n} - 1)^2 2^{2n}.$$

If a substitution S of $J_{\lambda'}$ does not alter ξ_m and replaces η_m by

$$f_m \equiv \sum_{j=1}^{m} (\beta_{mj} \xi_j + \delta_{mj} \eta_j)$$

we must have, in virtue of the relations 78) and 194),

212) $\qquad \delta_{mm} = 1, \sum_{j=1}^{m} \beta_{mj} \delta_{mj} + \lambda \beta_{mm}^2 = 0.$

We proceed to prove, inversely, that if β_{mj}, δ_{mj} be any set of solutions in the $GF[2^n]$ of 212) there exists a substitution Σ in' $J_{\lambda'}$ which leaves ξ_m fixed and replaces η_m by f_m.

If $\beta_{mj} = \delta_{mj} = 0$ ($j = 1, \ldots, m-1$), then $\beta_{mm} = 0$ or λ^{-1}. Hence we may take as Σ the identity or $M_2 M_m L$ respectively.

In the contrary case, let $\beta_{m2} \neq 0$, for example. Then $J_{\lambda'}$ contains a first hypoabelian substitution T leaving ξ_m and η_m fixed and replacing η_2 by

$$\beta_{m1} \xi_1 + \delta_{m1} \eta_1 + \beta_{m2} \xi_2 + \delta \eta_2, \quad \delta \equiv \delta_{m2} + \frac{1}{\beta_{m2}} (\beta_{mm} + \lambda \beta_{mm}^2),$$

since $\beta_{m1} \delta_{m1} + \beta_{m2} \delta = 0$ in virtue of 212). Then we may take

$$\Sigma = Q_{2, m, \varkappa} T Q_{2, m, 1} \qquad \varkappa \equiv \beta_{mm} / \beta_{m2}.$$

For $m = 3$, the number of sets of solutions in the $GF[2^n]$ of 212):

$$\beta_{m1} \delta_{m1} + \beta_{m2} \delta_{m2} + \beta_{mm} + \lambda \beta_{mm}^2 = 0$$

is $(2^{2n} + 1) 2^{2n}$. Indeed, there are 2^{n-1} distinct values in the $GF[2^n]$ of

$$\tau \equiv \beta_{mm} + \lambda \beta_{mm}^2.$$

By § 204, $\beta_{m1} \delta_{m1} + \beta_{m2} \delta_{m2} = \tau$ has $2^{3n} + 2^{2n} - 2^n$ sets of solutions if $\tau = 0$; while, if τ have any one of the $2^{n-1} - 1$ possible values $\neq 0$, it has

212   CHAPTER VIII.

$$\frac{2^{4n}-(2^{3n}+2^{2n}-2^n)}{2^n-1} \equiv 2^n(2^{2n}-1)$$

sets of solutions, and therefore in all

$$2^{3n}+2^{2n}-2^n+(2^{n-1}-1)2^n(2^{2n}-1) \equiv 2^{4n-1}+2^{2n-1}$$

sets of solutions. But each value of τ furnishes two values of β_{mm}.

209. Theorem. — *The hypoabelian groups J_λ on $2m > 6$ indices are simple.*

Let K be a self-conjugate subgroup of J_λ containing a substitution

$$S: \quad \xi_i' = \sum_{j=1}^m (\alpha_{ij}\xi_j + \gamma_{ij}\eta_j), \quad \eta_i' = \sum_{j=1}^m (\beta_{ij}\xi_j + \delta_{ij}\eta_j) \quad (i=1,\ldots,m)$$

not the identity I. We first prove that K *contains a substitution $\neq I$ which multiplies ξ_1 by a constant.* Let S replace ξ_1 by

$$f_1 \equiv \sum_{j=1}^m (\alpha_{1j}\xi_j + \gamma_{1j}\eta_j),$$

where by 194),

213) $$\sum_{j=1}^m \alpha_{1j}\gamma_{1j} + \lambda\alpha_{1m}^2 + \lambda\gamma_{1m}^2 = 0.$$

If $f_1 \neq \alpha_{11}\xi_1$, we have one of the following three sub-cases.

a) $\gamma_{11} \neq 0$. Then J_λ contains the product

$$T \equiv T_{1,\gamma_{11}^{-1}} R_{2,1,\alpha_{12}} Q_{2,1,\gamma_{12}} \ldots R_{m,1,\alpha_{1m}} Q_{m,1,\gamma_{1m}}$$

which replaces ξ_1 by $\gamma_{11}^{-1}\xi_1$ and η_1 by the function

$$\gamma_{11}^{-1}(\alpha_{12}\gamma_{12} + \cdots + \alpha_{1m}\gamma_{1m} + \lambda\alpha_{1m}^2 + \lambda\gamma_{1m}^2)\xi_1 + \gamma_{11}\eta_1 + \sum_{j=2}^m (\alpha_{1j}\xi_j + \gamma_{1j}\eta_j).$$

This equals f_1, since the coefficient of ξ_1 is congruent to α_{11} modulo 2, in virtue of 213). Hence K contains $S_1 \equiv T^{-1}ST$, which replaces ξ_1 by $\gamma_{11}^{-1}\eta_1$.

If J_λ contains a substitution T_1 which leaves ξ_1 and η_1 fixed and is not commutative with S_1, K will contain the product

$$S_1^{-1} \cdot T_1^{-1} S_1 T_1 \neq I$$

which leaves ξ_1 fixed. Suppose on the contrary that S_1 is commutative with every substitution of J_λ which leaves ξ_1 and η_1 fixed. Among the latter are $R_{2,3,\varkappa}$ and $Q_{3,2,\varkappa}$. If we equate the two expressions by which $S_1 R_{2,3,\varkappa}$ and $R_{2,3,\varkappa} S_1$ replace η_3, we find

$$\eta_3' + \varkappa \xi_2' = \eta_3' + (\ )\xi_2 + (\ )\xi_3.$$

Similarly, if S_1 be commutative with $Q_{3,2,\varkappa}$, we have

$$\xi_3' + \varkappa \xi_2' = \xi_3' + (\ )\xi_2 + (\ )\eta_3.$$

Hence $\xi_2' = (\ )\xi_2$. Transforming S_1 by P_{12} we obtain a substitution $\neq I$ which multiplies ξ_1 by a constant and belongs to K.

b) Let $\gamma_{11} = 0$, $\alpha_{12} = \alpha_{13} = \cdots = \alpha_{1m-1} = 0$ and, if $\lambda = 0$, also $\alpha_{1m} = 0$. If $\lambda = \lambda'$, we must have $\alpha_{1m} = \gamma_{1m} = 0$, since 213) reduces to

$$\alpha_{1m}\gamma_{1m} + \lambda'\alpha_{1m}^2 + \lambda'\gamma_{1m}^2 = 0,$$

whereas Q (§ 199) is irreducible in the field. Since $f_1 \neq \alpha_{11}\xi_1$, we cannot have $\gamma_{12} = \gamma_{13} = \cdots = \gamma_{1m-1}$ together with $\gamma_{1m} = 0$, if $\lambda = 0$. Transforming S by a suitable P_{2j} ($j < m$, if $\lambda = \lambda'$), we reach a substitution S' having $\gamma_{12} \neq 0$ and belonging to K. Transforming S' by $M_2 M_3$, we reach a substitution of K in which $\gamma_{11} = 0$, $\alpha_{12} \neq 0$ [case c)].

c) Let $\gamma_{11} = 0$, $\alpha_{12}, \ldots, \alpha_{1m-1}$, α_{1m} be not all zero if $\lambda = 0$; let $\gamma_{11} = 0$, $\alpha_{12}, \ldots, \alpha_{1m-1}$ be not all zero if $\lambda = \lambda'$. Transforming S by a suitable P_{2j}, we reach a substitution S' of K having $\alpha_{12} \neq 0$. Then J_λ contains

$$T \equiv T_{2,\alpha_{12}} Q_{2,1,\alpha_{11}} \quad Q_{2,3,\alpha_{13}} N_{2,3,\gamma_{13}} \ldots Q_{2,m,\alpha_{1m}} N_{2,m,\gamma_{1m}}$$

which does not alter ξ_1 but replaces ξ_2 by

$$\alpha_{11}\xi_1 + \alpha_{12}\xi_2 + \alpha_{12}^{-1}(\alpha_{13}\gamma_{13} + \cdots + \alpha_{1m}\gamma_{1m} + \lambda\alpha_{1m}^2 + \lambda\gamma_{1m}^2)\eta_2$$
$$+ \sum_{j=3}^{m}(\alpha_{1j}\xi_j + \gamma_{1j}\eta_j).$$

Since $\gamma_{11} = 0$, this reduces to f_1 in virtue of 213). Hence K contains S_1, the transformed of S' by T, which replaces ξ_1 by ξ_2.

If S_1 be commutative with both $R_{3,1,\varkappa}$ and $R_{3,2,\varkappa}$, it merely multiplies ξ_3 by a constant, so that its transform by P_{13} gives the required substitution. In fact, $S_1 R_{3,j,\varkappa}$ and $R_{3,j,\varkappa} S_1$ replace η_j by respectively

$$\eta_j' + \lambda\xi_3', \quad \eta_j' + (\ )\xi_3 + (\ )\xi_j.$$

In the contrary case, K contains the two products

$$S_1^{-1} \cdot R_{3,j,\varkappa}^{-1} S_1 R_{3,j,\varkappa} \qquad (j=1,2)$$

which leave ξ_1 fixed and do not both reduce to the identity.

Next, K contains a substitution $\neq I$ leaving ξ_1 and η_1 fixed. We have previously reached in K a substitution $S \neq I$ which replaces ξ_1 by $\alpha\xi_1$. Let it replace η_1 by $\sum_{j=1}^{m}(\beta_{1j}\xi_j + \delta_{1j}\eta_j)$. By an Abelian relation 78), $\delta_{11} = \alpha^{-1}$. By 194), we have

$$214) \qquad \sum_{j=1}^{m}\beta_{1j}\delta_{1j} + \lambda\beta_{1m}^2 + \lambda\delta_{1m}^2 = 0.$$

214   CHAPTER VIII.

a) Let $\beta_{11} = 0$, $\beta_{1j} = \delta_{1j} = 0$ $(j = 2, \ldots, m-1)$, and, if $\lambda = 0$, also $\beta_{1m} = \delta_{1m} = 0$. If $\lambda = \lambda'$, then must $\beta_{1m} = \delta_{1m} = 0$ by 214). Evidently $S = T_{1,\alpha} S_1$, where S_1 leaves ξ_1 and η_1 fixed. By the Abelian relations 78), S_1 involves only the indices ξ_i, η_i $(i = 2, \ldots, m)$. If S_1 be not commutative with every substitution Σ_1 of J_λ which does not involve ξ_1, η_1, then K will contain a product

$$S^{-1} \Sigma_1^{-1} S \Sigma_1 \equiv S_1^{-1} \Sigma_1^{-1} S_1 \Sigma_1 \neq I,$$

which leaves ξ_1 and η_1 fixed. In the contrary case, S_1 is commutative with $R_{2,3,\varkappa}$ and $Q_{3,2,\varkappa}$, so that, as shown above, S_1 will replace ξ_2 by $\varrho \xi_2$. Since S_1 is to be commutative with $M_2 M_3$ also, it will replace η_2 by $\varrho \eta_2$. Hence, by an Abelian relation, $\varrho^2 = 1$; whence $\varrho = 1$. Transforming S by P_{12}, we obtain a substitution $\neq I$ which leaves ξ_1 and η_1 fixed and belongs to K.

b) Let $\beta_{11} = 0$, β_{1j}, δ_{1j} $(j = 2, \ldots, m)$ be not all zero if $\lambda = 0$, but let $\beta_{11} = 0$, β_{1j}, δ_{1j} $(j = 2, \ldots, m-1)$ be not all zero if $\lambda = \lambda'$. Then by § 202, J_λ contains a substitution T, affecting only

which replaces ξ_2 by ξ_i, η_i $(i = 2, \ldots, m)$,

$$\sum_{j=2}^m (\beta_{1j} \xi_j + \delta_{1j} \eta_j), \quad \sum_{j=2}^m \beta_{1j} \delta_{1j} + \lambda \beta_{1m}^2 + \lambda \delta_{1m}^2 = 0.$$

Hence K contains S_1, the transformed of S by T. S_1 replaces ξ_1 by $\alpha \xi_1$ and η_1 by $\alpha^{-1} \eta_1 + \xi_2$.

If J_λ contains a substitution V, leaving ξ_1, η_1, ξ_2 fixed, which is not commutative with S_1, K will contain

$$S_1^{-1} \cdot V^{-1} S_1 V \neq I,$$

which leaves ξ_1 and η_1 fixed.

In the contrary case, S_1 will be commutative with $R_{2,3,\lambda}$ and $R_{m,3,\lambda}$ and $M_3 M_m$. Equating the two functions by which $S_1 R_{2,3,\varkappa}$ and $R_{2,3,\varkappa} S_1$ replace η_2, we find $\xi_3' = (\ )\xi_3 + (\ )\xi_2$. Equating the two functions by which $S_1 R_{m,3,\varkappa}$ and $R_{m,3,\varkappa} S_1$ replace η_m, we find that $\xi_3' = (\ )\xi_3 + (\ )\xi_m$. Hence $\xi_3' = \varrho \xi_3$. Since S_1 is to be commutative with $M_3 M_m$, $\eta_3' = \varrho \eta_3$. Then $\varrho = 1$. Transforming S_1 by P_{13}, we have a substitution $\neq I$ in K which leaves ξ_1 and η_1 fixed.

c) Let $\beta_{11} \neq 0$. We can determine a substitution S' of K of form similar to that of S but having also $\delta_{12} \neq 0$. In fact, if $\lambda = 0$, the products $\beta_{1j} \delta_{1j}$ $(j = 2, \ldots, m)$ are not all zero by 214). Transforming by a suitable P_{2i}, we have $\beta_{12} \delta_{12} \neq 0$. If $\lambda = \lambda'$, the same result follows unless $\beta_{1j} = \delta_{1j} = 0$ $(j = 2, \ldots, m-1)$, in which case either $\beta_{1m} \neq 0$ or $\delta_{1m} \neq 0$ by 214). In the latter case, we can take

LINEAR HOMOGENEOUS GROUP IN THE $GF[2^n]$ etc.   215

$\delta_{1m} \neq 0$, transforming by $M_2 M_m$ if necessary. Transforming the resulting substitution of the form

$$\xi_1' = \alpha \xi_1, \quad \eta_1' = \beta_{11} \xi_1 + \alpha^{-1} \eta_1 + \beta_{1m} \xi_m + \delta_{1m} \eta_m, \ldots \quad (\beta_{11} \neq 0, \delta_{1m} \neq 0)$$

by the substitution $R_{m, 2, \varkappa}$, we obtain a similar substitution having in η_1' the additional term $\varkappa \delta_{1m} \xi_2$.

Recurring to S', in which $\delta_{12} \neq 0$, we transform it by T_2, δ_{12}^{-1} and obtain a substitution S_1 of K having the form

$$\xi_1' = \alpha \xi_1, \quad \eta_1' = \beta_{11} \xi_1 + \alpha^{-1} \eta_1 + \beta_{12} \xi_2 + \eta_2 + \sum_{j=3}^{m} (\beta_{1j} \xi_j + \delta_{1j} \eta_j), \ldots$$

Consider the following product, leaving ξ_1, η_1, ξ_2 fixed,

$$W \equiv Q_{3, 2, \delta_{13}} R_{3, 2, \beta_{13}} \ldots Q_{m, 2, \delta_{1m}} R_{m, 2, \beta_{1m}}.$$

It replaces η_2 by the function

$$\eta_2 + (\beta_{13} \delta_{13} + \cdots + \beta_{1m} \delta_{1m} + \lambda \beta_{1m}^2 + \lambda \delta_{1m}^2) \xi_2 + \sum_{j=3}^{m} (\beta_{1j} \xi_j + \delta_{1j} \eta_j),$$

in which the coefficient of ξ_2 equals $\beta_{11} \alpha^{-1} + \beta_{12}$ by 214), since $\delta_{12} = 1$ and $\delta_{11} = \alpha^{-1}$. Hence W transforms S_1 into the substitution S_2:

$$\xi_1' = \alpha \xi_1, \quad \eta_1' = \beta_{11} \xi_1 + \alpha^{-1} \eta_1 + \eta_2 + \beta_{11} \alpha^{-1} \xi_2, \ldots$$

Let $\mu \equiv \beta_{11} \alpha^{-1} \neq 0$. If among the substitutions $Q_{3, 2, \mu} N_{2, 3, 1}$, $T_{2, \mu} M_2 M_3$, etc., of J_λ, leaving ξ_1, η_1 and $\mu \xi_2 + \eta_2$ invariant, there exists one, say V, which is not commutative with S_2, then K contains

$$S_2^{-1} \cdot V^{-1} S_2 V \neq I,$$

which leaves ξ_1 and η_1 fixed. In the contrary case, we find, on equating the functions by which $S_2 Q_{3, 2, \mu} N_{2, 3, 1}$ and $Q_{3, 2, \mu} N_{2, 3, 1} S_2$ replace ξ_2, that

$$\eta_3' = (\alpha_{22} + \mu \alpha_{23} + \mu \gamma_{22}) \eta_3 + \alpha_{23} \eta_2 + \mu \alpha_{23} \xi_2.$$

By one of the relations 194), we find $\alpha_{23} = 0$. Then, if S_2 be also commutative with $T_{2, \mu} M_2 M_3$, we must have $\xi_3' = \xi_3$, $\eta_3' = \eta_3$.

In proving that K contains a substitution $S \neq I$ which leaves ξ_1 and η_1 fixed, we assumed the existence of the indices

$$\xi_i, \quad \eta_i \qquad (i = 1, 2, 3, m)$$

only. But, by the relations 78) and 194) S is a hypoabelian substitution on the indices ξ_i, η_i ($i = 2, \ldots, m$). Hence, if $m > 4$, a repetition of the previous argument shows that K contains a substitution $\neq I$ involving only the indices ξ_i, η_i ($i = 3, \ldots, m$). After $m - 3$ such steps, we reach in K a substitution $\neq I$ and affecting only six indices ξ_i, η_i ($i = m - 2, m - 1, m$). In view of the simplicity of the senary hypoabelian groups, K will contain all the substitutions of J_λ will affect only the last six indices, and, in particular, $M_i M_j$, $N_{i,j,\varkappa}$ $(i, j = m-2, m-1, m)$. Transforming the latter by suitable substitutions P_{rs} $(r, s < m,$ if $\lambda = \lambda')$, we reach all the generators of J_λ. Hence $K \equiv J_\lambda$, so that J_λ is a simple group.

In view of the importance of the subgroups J_0 and $J_{\lambda'}$ of the first and second hypoabelian groups respectively, they will be designated by the more explicit notation $FH(2m, 2^n)$ and $SH(2m, 2^n)$. They are both simple when $m \geqq 3$. The second is simple and the first is composite for $m = 2$ (§§ 196—198).

210. MISCELLANEOUS EXERCISES UPON CHAPTERS I—VIII.

1. Every m-ary linear homogeneous substitution in the $GF[2]$ leaves invariant the function $s_1 + s_2 + \cdots + s_m$, where s_r denotes the sum of the products of the m indices taken r at a time.

2. An m-ary linear homogeneous substitution in the $GF[p^n]$ of determinant D multiplies by D the function of the indices

$$\Psi \equiv \begin{vmatrix} \xi_1 & \xi_2 & \cdots & \xi_m \\ \xi_1^{p^n} & \xi_2^{p^n} & \cdots & \xi_m^{p^n} \\ \cdot & \cdot & \cdot & \cdot \\ \xi_1^{p^{n(m-1)}} & \xi_2^{p^{n(m-1)}} & \cdots & \xi_m^{p^{n(m-1)}} \end{vmatrix}.$$

Hence Ψ is a relative invariant of the group $GLH(m, p^n)$.

3. The structure of the m-ary linear homogeneous group in the $GF[2^n]$ which leaves $\xi_1^2 + \xi_2^2 + \cdots + \xi_m^2$ absolutely invariant may be derived from that of the special linear group $SLH(m-1, 2^n)$.
   [Take as new indices $X \equiv \xi_1 + \xi_2 + \cdots + \xi_m$ and $\xi_2, \xi_3, \ldots, \xi_m$].

4. Those substitutions of the hyperorthogonal group $G_{m,2,n}$ (§ 143) whose coefficients all belong to the $GF[2^n]$ form a group G, a subgroup of the group of Ex. 3. Prove that G is generated by the binary substitutions
$$\xi_i' = \alpha \xi_i + (\alpha+1)\xi_j, \quad \xi_j' = (\alpha+1)\xi_i + \alpha \xi_j,$$
and that G is a solvable group of order $2^{nm(m-1)/2}$.

5. Consider the group C of $2m$-ary substitutions in the $GF[p^n]$, $p > 2$,

$$S: \quad \xi_i' = \sum_{j=1}^{m}(\alpha_{ij}\xi_j + \gamma_{ij}\eta_j), \quad \eta_i' = \sum_{j=1}^{m}(\beta_{ij}\xi_j + \delta_{ij}\eta_j) \quad (i = 1, \ldots, m)$$

common to the special Abelian and orthogonal groups. Being Abelian its reciprocal is obtained by replacing α_{ij}, γ_{ij}, β_{ij}, δ_{ij} by δ_{ji}, $-\gamma_{ji}$, $-\beta_{ji}$, α_{ji} respectively. Being orthogonal, its reciprocal is obtained by replacing the former by α_{ji}, β_{ji}, γ_{ji}, δ_{ji}. Hence must

c) $\qquad \alpha_{ji} = \delta_{ji}, \quad \beta_{ji} = -\gamma_{ji} \qquad (i, j = 1, \ldots, m).$

The conditions that an arbitrary substitution S, for which c) hold, shall be orthogonal are the same as the conditions that it shall be a special Abelian substitution.

6. The most general $2m$-ary substitution commutative with the special Abelian substitution $M \equiv M_1 M_2 \ldots M_m$ has the form

$$S: \quad \xi_i' = \sum_{j=1}^{m}(\alpha_{ij}\xi_j+\gamma_{ij}\eta_j), \quad \eta_i'=\sum_{j=1}^{m}(-\gamma_{ij}\xi_j+\alpha_{ij}\eta_j) \quad (i=1,\ldots,m).$$

The group in the $GF[p^n]$, $p>2$, commutative with M is identical with C of Ex. 5.

7. Setting $-1 = I^2$, $X_i \equiv \xi_i + I\eta_i$, $A_{ij} \equiv \alpha_{ij} - I\gamma_{ij}$, S of Ex. 6 becomes

$$\Sigma: \quad X_i' = \sum_{j=1}^{m} A_{ij} X_j \quad (i=1,\ldots,m).$$

If -1 be a not-square in the $GF[p^n]$, we may pass, inversely, from an arbitrary substitution Σ in the $GF[p^{2n}]$ to a substitution S in the $GF[p^n]$ by equating the coefficients of I^0 and I. Σ leaves invariant the function

$$\sum_{i=1}^{m} X_i^{p^n+1} = \sum_{i=1}^{m}(\xi_i+I\eta_i)(\xi_i-I\eta_i) = \sum_{i=1}^{m}(\xi_i^2+\eta_i^2).$$

Hence, if p^n be of the form $4l+1$, the group C is simply isomorphic with the hyperorthogonal group $G_{m,p,n}$. If -1 be a square in the $GF[p^n]$, we introduce the further indices $Y_i \equiv \xi_i - I\eta_i$, $B_{ij} \equiv \alpha_{ij} + I\gamma_{ij}$, when S becomes

$$\Sigma_1: \quad X_i' = \sum_{j=1}^{m} A_{ij} X_j, \quad Y_i' = \sum_{j=1}^{m} B_{ij} Y_j \quad (i=1,\ldots,m)$$

leaving invariant $\sum_{i=1}^{m} X_i Y_i$. Inversely, from every substitution Σ_1 we derive a substitution of the form S. The group of "dualistic" substitutions is simply isomorphic with $GLH(m, p^n)$, since the B_{ij} are determined in terms of the A's.

8. The simple group $A(4, p^n)$, $p > 2$, of order

$$\frac{1}{2}(p^{4n}-1)(p^{2n}-1)p^{4n}$$

contains just two sets of conjugate substitutions of period 2. The one set contains $\frac{1}{2}(p^{2n}+1)p^{2n}$ substitutions conjugate with $T_{1,-1}$. Those of the other set are conjugate with $M_1 M_2$ and are in number $\frac{1}{2}p^{3n}(p^{2n}+1)(p^n \mp 1)$ according to the form $4l \mp 1$ of p^n.

9. The group of all quaternary linear homogeneous substitutions in the $GF[p^n]$ which leave absolutely invariant the functions $\xi_1\eta_1 + \xi_2\eta_2$ and $\xi_1 + \eta_1$ has a subgroup of index 4 holoedrically isomorphic with $LF(2, p^n)$.

10. The squares of the substitutions of the first orthogonal group $O_1(m, p^n)$ generate the subgroup $O'_1(m, p^n)$ of § 181.

11. To the subgroup E'_{4,p^n} of E_{4,p^n} corresponds, for $p > 2$, the subgroup $O'_\nu(4, p^n)$ of $O_\nu(4, p^n)$ defined in § 181.

12. In order that $\lambda_1 \xi_1^{p^s+1} + \lambda_2 \xi_2^{p^s+1}$ shall be capable of transformation into $\mu(\xi_1^{p^s+1} + \xi_2^{p^s+1})$ by a binary linear substitution with coefficients in the $GF[p^{2s}]$, it is necessary and sufficient that the ratio λ_1/λ_2 shall belong to the $GF[p^s]$.

CHAPTER IX.

LINEAR GROUPS WITH CERTAIN INVARIANTS OF DEGREE $q > 2$.

211. Consider the group G_3 of substitutions in an arbitrary field

$$S \begin{cases} x'_i = \sum_{j=1}^{r}(L_{ij}x_j + M_{ij}y_j + N_{ij}z_j) \\ y'_i = \sum_{j=1}^{r}(\lambda_{ij}x_j + \mu_{ij}y_j + \nu_{ij}z_j) \qquad (i=1,\ldots,r) \\ z'_i = \sum_{j=1}^{r}(l_{ij}x_j + m_{ij}y_j + n_{ij}z_j) \end{cases}$$

which leave absolutely invariant the function of degree $q = 3$

$$\Phi_3 \equiv \sum_{i=1}^{r} x_i y_i z_i.$$

It will be convenient to employ a symbol, analogous to a determinant,

$$\begin{Bmatrix} A & B & C \\ \alpha & \beta & \gamma \\ a & b & c \end{Bmatrix} \equiv A\beta c + A\gamma b + B\alpha c + B\gamma a + C\alpha b + C\beta a.$$

The conditions that S shall leave Φ_3 absolutely invariant are then

215) $\quad \sum_{i=1}^{r} L_{ij}\lambda_{ij}l_{ij} = 0, \quad \sum_{i=1}^{r} M_{ij}\mu_{ij}m_{ij} = 0, \quad \sum_{i=1}^{r} N_{ij}\nu_{ij}n_{ij} = 0,$

LINEAR GROUPS WITH CERTAIN INVARIANTS OF DEGREE $q > 2$. 219

216) $\quad \sum_{i=1}^{r}(L_{ij}\lambda_{ij}l_{ik}+L_{ij}l_{ij}\lambda_{ik}+\lambda_{ij}l_{ij}L_{ik}) = 0,$

217) $\quad \sum_{i=1}^{r}(L_{ij}\lambda_{ij}m_{ik}+L_{ij}l_{ij}\mu_{ik}+\lambda_{ij}l_{ij}M_{ik}) = 0,$

218) $\quad \sum_{i=1}^{r}(L_{ij}\lambda_{ij}n_{ik}+L_{ij}l_{ij}\nu_{ik}+\lambda_{ij}l_{ij}N_{ik}) = 0,$

219) $\quad \sum_{i=1}^{r}\begin{Bmatrix}L_{ij}L_{ik}L_{it}\\ \lambda_{ij}\ \lambda_{ik}\ \lambda_{it}\\ l_{ij}\ l_{ik}\ l_{it}\end{Bmatrix}=0,\ \sum_{i=1}^{r}\begin{Bmatrix}L_{ij}L_{ik}M_{it}\\ \lambda_{ij}\ \lambda_{ik}\ \mu_{it}\\ l_{ij}\ l_{ik}\ m_{it}\end{Bmatrix}=0,\ \sum_{i=1}^{r}\begin{Bmatrix}L_{ij}L_{ik}N_{it}\\ \lambda_{ij}\ \lambda_{ik}\ \nu_{it}\\ l_{ij}\ l_{ik}\ n_{it}\end{Bmatrix}=0,$

220) $\quad \sum_{i=1}^{r}\begin{Bmatrix}L_{ij}M_{ik}N_{it}\\ \lambda_{ij}\ \mu_{ik}\ \nu_{it}\\ l_{ij}\ m_{ik}\ n_{it}\end{Bmatrix}=\begin{matrix}1\\ 0\end{matrix}\qquad \begin{matrix}\text{(if } j=k=t)\\ \text{(unless } j=k=t),\end{matrix}$

where, throughout, $i, j, k = 1, \ldots, r$, while $k \neq j$ in 216) and 219), and $t \neq j, k$ in the first of the relations 219); together with relations derived from 216), 217), 218) and 219) upon interchanging L, λ, l with M, μ, m or with N, ν, n. But relation 216) must also hold for $k = j$, being then derived from the first one of set 215) upon multiplying the latter by 3. Similarly 219) must hold for $k = j$, being then derived from 216), 217), 218). Lastly, the first of relations 219) must hold for $t = k = j$, being then derived from the first of the set 215) upon multiplying by 6. Hence the above conditions must hold for $i, j, k = 1, \ldots, r$ independently.

Let j be any fixed integer $\leq r$ and consider the $3r$ equations 216), 217), 218) for $k = 1, \ldots, r$. Taking as unknowns the $3r$ products

221) $\qquad\qquad L_{ij}\lambda_{ij},\quad L_{ij}l_{ij},\quad \lambda_{ij}l_{ij} \qquad (i = 1, \ldots, r),$

the determinant of their coefficients is seen to equal the determinant of S and is therefore not zero by hypothesis. Hence the products 221) are all zero. From the analogous conditions,

222) $\qquad\qquad M_{ij}\mu_{ij} = M_{ij}m_{ij} = \mu_{ij}m_{ij} = 0 \qquad (i, j = 1, \ldots, r),$
223) $\qquad\qquad N_{ij}\nu_{ij} = N_{ij}n_{ij} = \nu_{ij}n_{ij} = 0 \qquad (i, j = 1, \ldots, r).$

Expanding the symbols in 219) according to the last columns and applying a similar reasoning to the resulting equations, we find

224) $\quad L_{ij}\lambda_{ik} + L_{ik}\lambda_{ij} = L_{ij}l_{ik} + L_{ik}l_{ij} = \lambda_{ij}l_{ik} + \lambda_{ik}l_{ij} = 0.$

We obtain similar identities 225) and 226) between the M, μ, m and the N, ν, n. From 220) for $j \neq k$ and the following of type 219),

220 CHAPTER IX. LINEAR GROUPS WITH CERTAIN INVARIANTS etc.

$$\sum_{i=1}^{r}\begin{Bmatrix} L_{ij}M_{ik}M_{it}\\ \lambda_{ij}\ \mu_{ik}\ \mu_{it}\\ l_{ij}\ m_{ik}\ m_{it}\end{Bmatrix}=0,\quad \sum_{i=1}^{r}\begin{Bmatrix} L_{ij}M_{ik}L_{it}\\ \lambda_{ij}\ \mu_{ik}\ \lambda_{it}\\ l_{ij}\ m_{ik}\ l_{it}\end{Bmatrix}=0,$$

each set holding for $t=1,\ldots,r$, we derive as above

227) $\quad L_{ij}\mu_{ik}+M_{ik}\lambda_{ij}=L_{ij}m_{ik}+M_{ik}l_{ij}=\lambda_{ij}m_{ik}+\mu_{ik}l_{ij}=0.$

By a similar process, we get, for $k\neq j$,

228) $\quad L_{ij}v_{ik}+N_{ik}\lambda_{ij}=L_{ij}n_{ik}+N_{ik}l_{ij}=\lambda_{ij}n_{ik}+v_{ik}l_{ij}=0,$

229) $\quad N_{ij}\mu_{ik}+M_{ik}v_{ij}=N_{ij}m_{ik}+M_{ik}n_{ij}=v_{ij}m_{ik}+\mu_{ik}n_{ij}=0.$

212. Theorem. — *The group G_3 is generated by the substitutions*

230) $\quad (x_i y_i),\quad (x_i z_i),\quad P_{ij}\equiv(x_i x_j)(y_i y_j)(z_i z_j).$

together with the substitutions of the type

231) $\quad x_i'=L_i x_i,\quad y_i'=\mu_i y_i,\quad z_i'=n_i z_i,\quad L_i\mu_i n_i=1,\quad (i=1,\ldots,r).$

Let S denote any given substitution of G_3. We can determine a suitable product Σ of the substitutions 230) such that $\Sigma S\equiv S_1$ will have the coefficient $L_{11}\neq 0$. Then by 221), 224), 227), 228), we find

$\lambda_{11}=l_{11}=0,\ \lambda_{1k}=l_{1k}=0,\ \mu_{1k}=m_{1k}=0,\ v_{1k}=n_{1k}=0\ (k=2,\ldots,r).$

Hence S_1 replaces y_1 and z_1 by the respective functions

$$\mu_{11}y_1+v_{11}z_1,\quad m_{11}y_1+n_{11}z_1.$$

The product $\Sigma_1 S_1\equiv S_2$, where Σ_1 is the identity if $\mu_{11}\neq 0$ but $\Sigma_1\equiv(y_1 z_1)$ if $\mu_{11}=0$, will be of the form S_1 with the new coefficient $\mu_{11}\neq 0$. Then by 222), 225), 227) and 229), we find

$$M_{11}=m_{11}=0,\quad M_{1k}=0,\quad L_{1j}=0,\quad N_{1j}=0\quad (j,k=2,\ldots,r).$$

Hence must $n_{11}\neq 0$ and therefore $N_{11}=v_{11}=0$ by 223). Hence S_2 replaces x_1, y_1, z_1 by $L_{11}x_1, \mu_{11}y_1, n_{11}z_1$ respectively. Also

$$S=\Sigma^{-1}\Sigma_1^{-1}S_2.$$

Since the determinant of S_2 is not zero, the coefficients L_{2j}, $M_{2j}, N_{2j}\ (j=2,\ldots,r)$ are not all zero. We may therefore determine a suitable product Σ' of the substitutions 230), in which $i,j>1$, such that $\Sigma'S_2\equiv S_3$ will have $L_{22}\neq 0$. Proceeding as above, we find that $S=\Sigma''S_4$, where Σ'' is derived from the substitutions 230), while S_4 merely multiplies $x_1, y_1, z_1, x_2, y_2, z_2$ by constants. After r such steps, we reach a substitution of the form 231).

Corollary. — Any substitution leaving Φ_3 invariant may be expressed as a product AB, where A is of the form 231) and B is derived from the substitutions 230).

213. The preceding methods may be employed[1]) to investigate the group G_q of linear substitutions S on rq indices with coefficients in an arbitrary field which leave absolutely invariant the function

$$\Phi_q \equiv \sum_{i=1}^{r} \xi_{i1}\xi_{i2} \ldots \xi_{iq}.$$

For $q > 2$, it is seen that $S = AB$, where A merely multiplies each index ξ_{ij} by a constant, while B is a permutation on the indices ξ_{ij} having the imprimitive systems[2])

232) $\xi_{11}, \xi_{12}, \ldots, \xi_{1q}; \quad \xi_{21}, \xi_{22}, \ldots, \xi_{2q}; \ldots; \quad \xi_{r1}, \xi_{r2}, \ldots, \xi_{rq}.$

The substitutions A form a commutative group which is transformed into itself by every substitution B and is therefore self-conjugate under G_q. The quotient-group is the group of the substitutions B. The latter has a self-conjugate subgroup R formed by the direct product of r symmetric groups, the general one being on the q letters $\xi_{i1}, \xi_{i2}, \ldots, \xi_{iq}$; the quotient-group $\{L\}/R$ is a symmetric group on r letters, viz., the r sets 232). The structure of the group G_q, $q > 2$, is therefore completely determined. The result is essentially different from that for the case $q = 2$ (see § 195).

CHAPTER X.

CANONICAL FORM AND CLASSIFICATION OF LINEAR SUBSTITUTIONS.

Canonical form of linear homogeneous substitutions[3]), §§ 214—216.

214. Consider a substitution with coefficients in the $GF[p^n]$,

$$S: \qquad \xi'_i = \sum_{j=1}^{m} \alpha_{ij}\xi_j \qquad (i = 1, \ldots, m).$$

In order that S shall multiply by a constant K the linear function

$$\eta \equiv \sum_{i=1}^{m} \lambda_i \xi_i,$$

we must have

$$\sum_{i,j}^{1,\ldots,m} \lambda_i \alpha_{ij}\xi_j \equiv K \sum_{j=1}^{m} \lambda_j \xi_j$$

or

$$\sum_{i=1}^{m} \lambda_i \alpha_{ij} = K\lambda_j \qquad (j = 1, \ldots, m).$$

---

1) *Proceed. Lond. Math. Soc.*, vol. 30, pp. 200—208. On pp. 203—204 the numerical factors are incorrect; C should equal $t_1!\,t_2!\ldots t_\tau!$ The proof however is valid.

2) B replaces the indices of any set $\xi_{i1}, \xi_{i2}, \ldots, \xi_{iq}$ by indices all in one set.

3) For $n = 1$, the results are due to Jordan, Traité, pp. 114—126. The simple proof by induction of the fundamental theorem is due to the author, *American Journal*, vol. 22, pp. 121—137.

222   CHAPTER X.

Hence K must be a root of the *characteristic equation*

$$\Delta(K) \equiv \begin{vmatrix} \alpha_{11}-K & \alpha_{12} & \ldots & \alpha_{1m} \\ \alpha_{21} & \alpha_{22}-K & \ldots & \alpha_{2m} \\ \vdots & \vdots & & \vdots \\ \alpha_{m1} & \alpha_{m2} & \ldots & \alpha_{mm}-K \end{vmatrix} = 0.$$

Corresponding to each root K, we may determine at least one set of solutions λ_i of the above linear equations and hence one invariant function η.

If $\Delta(K) = 0$ has m distinct roots $K_1, K_2, \ldots, K_m$ (not necessarily in the initial $GF[p^n]$), we reach m linear functions $\eta_1, \eta_2, \ldots, \eta_m$, which S multiplies by) $K_1, K_2, \ldots, K_m$ respectively. These functions are linearly independent with respect to the variables ξ_i. For, if constants μ_i exist such that

$$\mu_1\eta_1 + \mu_2\eta_2 + \cdots + \mu_m\eta_m \equiv 0,$$

we have on applying the substitutions $S, S^2, \ldots, S^{m-1}$ the further identities

$$K_1\mu_1\eta_1 + K_2\mu_2\eta_2 + \cdots + K_m\mu_m\eta_m \equiv 0,$$
$$K_1^2\mu_1\eta_1 + K_2^2\mu_2\eta_2 + \cdots + K_m^2\mu_m\eta_m \equiv 0,$$
$$\cdots \cdots \cdots \cdots \cdots \cdots \cdots \cdots \cdots \cdots$$
$$K_1^{m-1}\mu_1\eta_1 + K_2^{m-1}\mu_2\eta_2 + \cdots + K_m^{m-1}\mu_m\eta_m \equiv 0.$$

But the determinant

$$\begin{vmatrix} 1 & 1 & \ldots & 1 \\ K_1 & K_2 & \ldots & K_m \\ \vdots & \vdots & & \vdots \\ K_1^{m-1} & K_2^{m-1} & \ldots & K_m^{m-1} \end{vmatrix} \overset{i,j=1\ldots m}{=} \prod_{i<j} (K_i - K_j) \neq 0.$$

Hence
$$\mu_1 = \mu_2 = \cdots = \mu_m = 0.$$

Introducing the linear functions η_i as new indices in place of the ξ_i, the substitution S takes the canonical form

$$S': \qquad \eta_i' = K_i\eta_i \qquad (i=1,\ldots,m).$$

If we take in place of η_1 a suitable multiple of η_1, we may suppose the reduction of S to S' to be accomplished by a transformation of indices of determinant unity.

Suppose, however, that the roots of $\Delta(K) = 0$ are not all distinct. Let

CANONICAL FORM AND CLASSIFICATION OF LINEAR SUBSTITUTIONS. 223

$$\Delta(K) \equiv [F_k(K)]^\alpha [F_l(K)]^\beta \ldots \quad (k\alpha + l\beta + \cdots \equiv m)$$

where $F_k(K)$, $F_l(K)$, ... are the distinct factors of $\Delta(K)$ which belong to and are irreducible in the $GF[p^n]$. Designate the roots of $F_k(K) = 0$, and of $F_l(L) = 0$, etc., by the notations

$$K_0, \quad K_1 \equiv K_0^{p^n}, \quad K_2 \equiv K_0^{p^{2n}}, \ldots, K_{k-1} \equiv K_0^{p^{n(k-1)}};$$
$$L_0, \quad L_1 \equiv L_0^{p^n}, \quad L_2 \equiv L_0^{p^{2n}}, \ldots, L_{l-1} \equiv L_0^{p^{n(l-1)}}; \ldots$$

Theorem. — *By a suitable transformation of indices, S can be reduced to a canonical form of the following type:*

$$\eta'_{i\,1} = K_i \eta_{i\,1} \quad , \quad \eta'_{ij} = K_i(\eta_{ij} + \eta_{i\,j-1}) \quad (j = 2, \ldots, a_1)$$
$$\eta'_{i\,a_1+1} = K_i \eta_{i\,a_1+1} \quad , \quad \eta'_{i\,a_1+j} = K_i(\eta_{i\,a_1+j} + \eta_{i\,a_1+j-1}) \quad (j = 2, \ldots, a_2)$$
$$\eta'_{i\,a_2+a_1+1} = K_i \eta_{i\,a_2+a_1+1}, \quad \eta'_{i\,a_2+a_1+j} = K_i(\eta_{i\,a_2+a_1+j} + \eta_{i\,a_2+a_1+j-1}) \quad (j = 2, \ldots, a_3)$$
$$\cdots$$
$$(i = 0, 1, \ldots, k-1)$$

$$\xi'_{i\,1} = L_i \xi_{i\,1} \quad , \quad \xi'_{ij} = L_i(\xi_{ij} + \xi_{i\,j-1}) \quad (j = 2, \ldots, b_1)$$
$$\xi'_{i\,b_1+1} = L_i \xi_{i\,b_1+1} \quad , \quad \xi'_{i\,b_1+j} = L_i(\xi_{i\,b_1+j} + \xi_{i\,b_1+j-1}) \quad (j = 2, \ldots, b_2)$$
$$\cdots$$
$$(i = 0, 1, \ldots, l-1)$$
$$\cdots$$

where $a_1 + a_2 + a_3 + \cdots = \alpha$, $b_1 + b_2 + \cdots = \beta$, ...; and where the indices have the properties:

1) *The indices η_{0s} ($s = 1, \ldots, \alpha$) are linear homogeneous functions of the initial indices ξ_i having as coefficients polynomials in K_0 with coefficients in the $GF[p^n]$;*

2) *The indices η_{is} are conjugate to the η_{0s}, being obtained by replacing K_0 by K_i in the coefficients of η_{0s};*

3) *The indices ζ_{0s} ($s = 1, \ldots, \beta$) are linear homogeneous functions of the indices ξ_i whose coefficients are polynomials in L_0 with coefficients in the $GF[p^n]$;*

4) *The indices ξ_{is} are obtained from the ξ_{0s} by replacing L_0 by L_i; etc.*

5) *The $k\alpha$ indices η_{is} ($i = 0, 1, \ldots, k-1$; $s = 1, \ldots, \alpha$) may be replaced by $k\alpha$ linear homogeneous functions y_{is} of the initial indices ξ_i with coefficients in the $GF[p^n]$, such that S replaces each y_{is} by a linear homogeneous function of the y_{is} with coefficients in the $GF[p^n]$;*

6) *The $l\beta$ indices ζ_{is} may be replaced by an equal number of linear homogeneous functions z_{is} of the ξ_i with coefficients in the $GF[p^n]$, such that S replaces each by a linear homogeneous function of the z_{is} with coefficients in the field; etc.*

224   CHAPTER X.

For the case $\alpha = \beta = \cdots = 1$, we obtained above the canonical form
$$\eta'_{i1} = K_i \eta_{i1} \qquad (i = 0, 1, \ldots, k-1)$$
$$\zeta'_{i1} = L_i \zeta_{i1} \qquad (i = 0, 1, \ldots, l-1)$$
where $\eta_{01} = f(\xi_1, \ldots, \xi_m; K_0)$ and $\eta_{i1} = f(\xi_1, \ldots, \xi_m; K_i)$, and, similarly, ζ_{i1} are conjugate with ζ_{01}. The new indices therefore have the properties 1)—4).

We will prove the general theorem by induction, supposing it true for every substitution belonging to the $GF[p^n]$ whose characteristic determinant has no irreducible factors other than $F_k(K)$, $F_l(K), \ldots$, and has these to a degree at most $\alpha - 1, \beta, \ldots$ respectively. We will prove that the theorem is true for any substitution S for which these factors occur to the degree $\alpha, \beta, \ldots$ respectively, where $\alpha > 1$.

Corresponding to the distinct roots $K_0, K_1, \ldots, K_{k-1}$ of $F_k(K) = 0$, we obtain as above a set of linearly independent conjugate functions $\lambda_0, \lambda_1, \ldots, \lambda_{k-1}$ which S multiplies by $K_0, K_0, \ldots, K_{k-1}$ respectively. We may introduce these in place of an equal number of the original indices, e. g., $\xi_{m-k+1}, \ldots, \xi_m$. The substitution S then takes the form

$$S' \begin{cases} \lambda'_i = K_i \lambda_i & (i = 0, 1, \ldots, k-1) \\ \xi'_i = \sum_{j=1}^{m} \beta_{ij} \xi_j + \sum_{j=0}^{k-1} \gamma_{ij} \lambda_j & (i = 1, 2, \ldots, m-k). \end{cases}$$

The coefficients β_{ij} belong to the $GF[p^n]$. Indeed, we may set
$$\lambda_i \equiv X_0 + K_i X_1 + K_i^2 X_2 + \cdots + K_i^{k-1} X_{k-1},$$
$$(i = 0, 1, \ldots, k-1)$$
where the X_i are linear functions of the ξ_i with coefficients in the $GF[p^n]$. Since the λ_i are linearly independent, the X_i must be linearly independent functions of the ξ_i. Since
$$|K_i^j| \equiv \prod_{r<s}^{0,\ldots,k-1}(K_r - K_s) \neq 0,$$
the X_j can be expressed as linear functions of the λ_i. Taking the X_j as new indices in place of the λ_i, S' takes the form S'', a substitution on the indices X_j and ξ_i with coefficients in the $GF[p^n]$. But S'' replaces ξ_i by
$$\sum_{j=1}^{m-k} \beta_{ij} \xi_j + \sum_{j=0}^{k-1} \delta_{ij} X_j$$
for $i = 1, \ldots, m-k$. Since these functions belong to the field for arbitrary ξ_j and X_j, the coefficients β_{ij}, δ_{ij} must belong to the field.

CANONICAL FORM AND CLASSIFICATION OF LINEAR SUBSTITUTIONS. 225

Since the determinant of a linear substitution is not altered by a linear transformation of indices (§ 101), the determinant of S' equals the determinant of S:

$$K_0 K_1 \ldots K_{k-1} \cdot |\beta_{ij}| = D.$$

We may, therefore, consider the following substitution in the $GF[p^n]$:

$$S_1: \quad \xi_i' = \sum_{j=1}^{m-k} \beta_{ij} \xi_j \qquad (i = 1, \ldots, m-k)$$

of determinant $\neq 0$. Also, the characteristic determinant $\Delta(K)$ of S equals that of the transformed substitution S', viz.:

$$\Delta(K) = \prod_{i=0}^{k-1} (K - K_i) \begin{vmatrix} \beta_{11} - K & \beta_{12} & \cdots \\ \beta_{21} & \beta_{22} - K & \cdots \\ \cdots & \cdots & \cdots \end{vmatrix}.$$

Hence, the characteristic determinant of S_1 is

$$\frac{\Delta(K)}{F_k(K)} \equiv [F_k(K)^{a-1}[F_l(K)]^\beta \ldots$$

Hence, by hypothesis, S_1 can be reduced to a canonical form of the above type. Applying the same transformation of indices to S', it takes the form $\bar{S}$:

$$\lambda_i' = K_i \lambda_i \qquad (i = 0, \ldots, k-1)$$

$$\eta_{01}' = K_0 \eta_{01} + \sum_{i=0}^{k-1} \alpha_{1i} \lambda_i, \quad \eta_{0j}' = K_0(\eta_{0j} + \eta_{0\,j-1}) + \sum_{i=0}^{k-1} \alpha_{ji} \lambda_i \quad (j = 2, \ldots, a_1)$$

$$\eta_{0\,a_1+1}' = K_0 \eta_{0\,a_1+1} + \Sigma \beta_{1i} \lambda_i, \quad \eta_{0\,a_1+j}' = K_0(\eta_{0\,a_1+j} + \eta_{0\,a_1+j-1}) + \Sigma \beta_{ji} \lambda_i \quad (j = 2, \ldots, a_2)$$

$$\zeta_{01}' = L_0 \zeta_{01} + \Sigma \alpha_{1i}' \lambda_i, \quad \zeta_{0j}' = L_0(\zeta_{0j} + \zeta_{0\,j-1}) + \Sigma \alpha_{ji}' \lambda_i \quad (j = 2, \ldots, b_1)$$

the expression for η_{is}' being derived from that for η_{0s}' by replacing K_0 by K_i; the expression for ζ_{is}' from ζ_{0s}' upon replacing L_0 by L_i, etc.

To simplify the form of $\bar{S}$, introduce as new indices

$$Y_{0s} \equiv \eta_{0s} + \sum_{i=0}^{k-1} A_{si} \lambda_i \qquad (s = 1, \ldots, \alpha),$$

$$Z_{0s} \equiv \zeta_{0s} + \sum_{i=0}^{k-1} B_{si} \lambda_i \qquad (s = 1, \ldots, \beta), \ldots$$

and their conjugate functions $Y_{is}, Z_{is}, \ldots$ Then $\bar{S}$ replaces Y_{01}. Y_{02}, Y_{03} by

226   CHAPTER X.

$$K_0 Y_{01} + \alpha_{10}\lambda_0 + \sum_{i=1}^{k-1}[\alpha_{1i} + (K_i - K_0)A_{1i}]\lambda_i,$$

$$K_0(Y_{02} + Y_{01}) + \sum_{i=0}^{k-1}[\alpha_{2i} - K_0 A_{1i} + (K_i - K_0)A_{2i}]\lambda_i,$$

$$K_0(Y_{03} + Y_{02}) + \sum_{i=0}^{k-1}[\alpha_{3i} - K_0 A_{2i} + (K_i - K_0)A_{3i}]\lambda_i,$$

respectively. By choice of the A_{ji}, we can make the terms in brackets all zero; those of the first sum by choice of $A_{11}, \ldots, A_{1\,k-1}$, those of the second by choice of $A_{10}, A_{21}, \ldots, A_{2\,k-1}$, those of the third by choice of $A_{20}, A_{31}, \ldots, A_{3\,k-1}$. A like result holds for the remaining Y_{0s} $(s = 1, \ldots, \alpha)$.

$\overline{S}$ replaces $Z_{01}, Z_{02}, \ldots$ by respectively

$$L_0 Z_{01} + \sum_{i=0}^{k-1}[\alpha'_{1i} + (K_i - L_0)B_{1i}]\lambda_i,$$

$$L_0(Z_{02} + Z_{01}) + \sum_{i=0}^{k-1}[\alpha'_{2i} - L_0 B_{1i} + (K_i - L_0)B_{2i}]\lambda_i, \ldots$$

Since $K_i - L_0 \neq 0$, the coefficients of λ_i may be made to vanish by choice of the B_{1i}. Hence, $\overline{S}$ takes the form S_2:

$$\lambda'_i = K_i \lambda_i,$$

$Y'_{i1} = K_i Y_{i1} + \varphi(K_i) \cdot \lambda_i, \quad Y'_{ij} = K_i(Y_{ij} + Y_{i\,j-1}) \quad (j = 2, \ldots, a_1)$

$Y'_{i\,a_1+1} = K_i Y_{i\,a_1+1} + \Psi(K_i) \cdot \lambda_i, \quad Y'_{i\,a_1+j} = K_i(Y_{i\,a_1+j} + Y_{i\,a_1+j-1})$

$$(j = 2, \ldots, a_2)$$

. . . .

$$(i = 0, 1, \ldots, k-1)$$

$Z'_{i1} = L_i Z_{i1}, \quad\quad Z'_{ij} = L_i(Z_{ij} + Z_{i\,j-1}) \quad (j = 2, \ldots, b_1)$

$Z'_{i\,b_1+1} = L_i Z_{i\,b_1+1}, \quad Z'_{i\,b_1+j} = L_i(Z_{i\,b_1+j} + Z_{i\,b_1+j-1})$

$$(j = 2, \ldots, b_2)$$

. . . .

$$(i = 0, 1, \ldots, l-1)$$

. . . .

If the constants $\varphi(K_i), \psi(K_i), \chi(K_i), \ldots$ are all zero, no further reduction is necessary. If any two are not zero, as φ and ψ, suppose for definiteness that $a_1 \leqq a_2$, and introduce in place of $Y_{i1}, \ldots, Y_{i a_1}$ the new indices

$$\overline{Y}_{ij} \equiv Y_{ij} - \frac{\varphi}{\psi} Y_{i\,a_1+j}. \quad\quad (j = 1, \ldots, a_1).$$

CANONICAL FORM AND CLASSIFICATION OF LINEAR SUBSTITUTIONS. 227

The substitution S_2 replaces $\overline{Y}_{i1}$, $\overline{Y}_{ij}$ $(j = 2, \ldots, a_1)$ by respectively
$$K_i \overline{Y}_{i1}, \quad K_i(\overline{Y}_{ij} + \overline{Y}_{ij-1}) \quad (j = 2, \ldots, a_1).$$
Hence, the introduction of the $\overline{Y}_{ij}$ has the effect of setting $\varphi = 0$ in S_2. Proceeding similarly, we can suppose that $\varphi, \psi, \chi, \ldots$ are all zero but one, say $\psi \neq 0$. In the latter case, we set
$$\psi(K_i) \cdot \lambda_i \equiv K_i y_{i\,a_1}$$
and find for S_2 the canonical form

$$\begin{aligned}
Y'_{i1} &= K_i Y_{i1}, & Y'_{ij} &= K_i(Y_{ij} + Y_{ij-1}), \quad (j = 2, \ldots, a_1)\\
y'_{i\,a_1} &= K_i y_{i\,a_1}, & Y'_{i\,a_1+1} &= K_i(Y_{i\,a_1+1} + y_{i\,a_1}),\\
& & Y'_{i\,a_1+j} &= K_i(Y_{i\,a_1+j} + Y_{i\,a_1+j-1}),\\
& & & \quad (j = 2, \ldots, a_2)\\
Y'_{i\,a_1+a_2+1} &= K_i Y_{i\,a_1+a_2+1}, & Y'_{i\,a_1+a_2+j} &= K_i(Y_{i\,a_1+a_2+j} + Y_{i\,a_1+a_2+j-1}),\\
& & & \quad (j = 2, \ldots, a_3)\\
& \ldots \ldots \ldots \ldots \ldots \ldots \ldots \ldots \ldots \ldots \ldots\\
Z'_{i1} &= L_i Z_{i1}, & Z'_{ij} &= L_i(Z_{ij} + Z_{ij-1}), \quad (j = 2, \ldots, b_1)\\
& \ldots \ldots \ldots \ldots \ldots \ldots \ldots \ldots \ldots \ldots \ldots
\end{aligned}$$

In every case we reach a canonical form of the type given in the theorem, for which the indices Y_{is} have the properties 1) and 2). But the indices Z_{ij} are linear functions of the ξ_i with coefficients which certainly involve L_i and apparently[1]) also K_i. If the K_i be involved, we proceed as follows. From the canonical form actually reached, $S = YS_1$, where Y is the partial substitution on the indices Y_{ij}, not altering the indices Z_{ij}, etc., while S_1 does not involve the indices Y_{ij}, but affects the Z_{ij}, etc. Setting
$$Y_{is} \equiv y_s + y'_s K_i + y''_s K_i^2 + y_s^{(k-1)} K_i^{k-1},$$
$$(s = 1, \ldots, \alpha;\ i = 0, \ldots, k-1)$$
where the y's are linear functions of the ξ_i with coefficients in the $GF[p^n]$, we can evidently introduce the y's as new indices in place of the Y_{is}, so that Y takes the form of a substitution belonging to the $GF[p^n]$ and affecting only $k\alpha$ indices. Likewise, by introducing in place of the Z_{ij}, etc., an equal number of linear functions z_{ij}, etc., belonging to the $GF[p^n]$, it is possible to give to S_1 the form of a substitution in the field and affecting only $m - k\alpha$ indices. Its characteristic determinant is $[F_i(K)]^\beta \ldots$ Hence, by the hypothesis made for the induction, S_1 can be reduced by a linear transformation T to a canonical form

---

1) By the considerations in the text, we may dispense with the difficult proof, analogous to that of Jordan, Traité, pp. 121—122, that the Z_{ij} do not involve K_i, but the single imaginary L_i.

$$\xi'_{i1} = L_i \xi_{i1}, \quad \xi'_{ij} = L_i(\xi_{i1} + \xi_{ij-1}), \qquad (j = 2, \ldots, b_1)$$

where the ξ_{ij} are linear functions of the ξ_i with coefficients involving the imaginary L_i only. As the transformation $\mathcal{T}$ does not alter the indices which Y affects, we obtain the desired canonical form.

215. Consider as an example the substitution in the $GF[p^n]$, p^n of the form $4l - 1$,

$$S: \quad \xi'_1 = -2\xi_2 - \xi_4, \quad \xi'_2 = \xi_1, \quad \xi'_3 = \xi_2, \quad \xi'_4 = \xi_3,$$

having the characteristic determinant

$$\Delta(K) \equiv (K^2 + 1)^2,$$

where $K^2 + 1$ is irreducible in the field. A root of $i^2 = -1$ belongs to the $GF[p^{2n}]$ but not to the $GF[p^n]$. The functions which S multiplies by i and $-i$ are readily found to be respectively

$$\lambda_1 \equiv -ix_1 + x_2 - ix_3 + x_4, \quad \lambda_2 \equiv ix_1 + x_2 + ix_3 + x_4.$$

Introducing λ_1, λ_2 in place of the indices x_2, x_3, S takes the form

$$x'_1 = x_4 - \lambda_1 - \lambda_2, \quad x'_4 = -x_1 + i/2\,\lambda_1 - i/2\,\lambda_2, \quad \lambda'_1 = i\lambda_1, \quad \lambda'_2 = -i\lambda_2.$$

The partial substitution of determinant unity,

$$x'_1 = x_4, \quad x'_4 = -x_1$$

multiplies $y_1 \equiv x_1 - ix_4$ by i and multiplies $y_2 \equiv x_1 + ix_4$ by $-i$. Introducing y_1 and y_2 as new indices in place of x_1 and x_4, S takes the form

$$y'_1 = iy_1 - \frac{1}{2}\lambda_1 - \frac{3}{2}\lambda_2, \quad y'_2 = -iy_2 - \frac{1}{2}\lambda_2 - \frac{3}{2}\lambda_1,$$
$$\lambda'_1 = i\lambda_1, \quad \lambda'_2 = -i\lambda_2.$$

Introducing as new indices,

$$\bar{\lambda}_1 \equiv i/2\,\lambda_1, \quad \bar{\lambda}_2 \equiv -i/2\,\lambda_2, \quad \bar{y}_1 \equiv y_1 + \frac{3i}{4}\lambda_2, \quad \bar{y}_2 \equiv y_2 - \frac{3i}{4}\lambda_1,$$

S takes the canonical form

$$\bar{\lambda}'_1 = i\bar{\lambda}_1, \quad \bar{y}'_1 = i(\bar{y}_1 + \bar{\lambda}_1), \quad \bar{\lambda}'_2 = -i\bar{\lambda}_2, \quad \bar{y}'_2 = -i(\bar{y}_2 + \bar{\lambda}_2),$$

where $\bar{\lambda}_1$ and $\bar{\lambda}_2$ are conjugate linear functions of $\xi_1, \xi_2, \xi_3, \xi_4$, and likewise for $\bar{y}_1, \bar{y}_2$.

216. Theorem. — *Two linear homogeneous substitutions S and T in the $GF[p^n]$ on the indices $\xi_1, \xi_2, \ldots, \xi_m$ have the same canonical form C if, and only if, T is the transformed of S by a linear homogeneous substitution W in the $GF[p^n]$ on the same indices.*

If $T = W^{-1}SW$, then S can be reduced to $\mathcal{T}$ by the introduction of new indices defined by the transformation W and therefore S and T have the same canonical form.

Suppose, inversely, that two substitutions S and T in the $GF[p^n]$ on the indices ξ_i can be reduced to the same canonical form by the respective transformations S' and T'. Let T' denote the transformation from the indices $\xi_1, \ldots, \xi_m$ to the indices $\eta_{is}, \zeta_{is}, \ldots$, where

$$\eta_{is} \equiv Y_s + Y_s' K_i + Y_s'' K_i^2 + \cdots + Y_s^{(k-1)} K_i^{k-1}$$
$$(s = 1, \ldots, \alpha;\ i = 0, 1, \ldots, k-1)$$
$$\zeta_{is} \equiv Z_s + Z_s' L_i + Z_s'' L_i^2 + \cdots + Z_s^{(l-1)} L_i^{l-1}$$
$$(s = 1, \ldots, \beta;\ i = 0, 1, \ldots, l-1),$$

$Y_s, Y_s', \ldots, Z_s, Z_s', \ldots$ being linearly independent linear functions of the ξ_i with coefficients in the $GF[p^n]$. Denote by τ the transformation of indices from $\eta_{is}, \zeta_{is}, \ldots$ to $Y_s, Y_s', \ldots, Z_s, \ldots$ By hypothesis, T' transforms T into the canonical form C. Let τ transform C into C_τ. Then $T'\tau$ is a substitution in the $GF[p^n]$ which transforms T into C_τ, likewise in the $GF[p^n]$. Similarly, let S' denote the transformation from the indices $\xi_1, \ldots, \xi_m$ to the indices $\overline{\eta}_{is}, \overline{\zeta}_{is}, \ldots$, where

$$\overline{\eta}_{is} \equiv \overline{Y}_s + \overline{Y}_s' K_i + \cdots,\quad \overline{\zeta}_{is} \equiv \overline{Z}_s + \overline{Z}_s' L_i + \cdots,\quad \ldots$$

Denote by σ the transformation of indices from $\overline{\eta}_{is}, \overline{\zeta}_{is}, \ldots$ to $\overline{Y}_s, \ldots, \overline{Z}_s, \ldots$ By hypothesis, S' transforms S into the canonical form $\overline{C}$, which in the same substitution on the indices $\overline{\eta}_{is}, \overline{\zeta}_{is}, \ldots$ that C is on the indices $\eta_{is}, \zeta_{is}, \ldots$ Let σ transform $\overline{C}$ into $\overline{C}_\sigma$. Then, if R be the substitution in the $GF[p^n]$ which transforms $Y_s, \ldots, Z_s, \ldots$ into $\overline{Y}_s, \ldots, \overline{Z}_s, \ldots$ respectively, then

$$\overline{C}_\sigma = R^{-1} C_\tau R.$$

It follows that the product $T'\tau R(S'\sigma)^{-1}$ is a substitution on the indices ξ_i with coefficients in the $GF[p^n]$ which transforms T into S.

§§ 217—220.
Substitutions commutative with a given linear substitution [1]).

217. Let the given linear homogeneous substitution S on m indices ξ_i with coefficients in the $GF[p^n]$ be brought to its canonical form S_1. For definiteness, suppose there are three sets of new indices,

$$\eta_{ij}\ (i=0,1,\ldots,k-1;\ j=1,\ldots,\alpha);\quad \zeta_{ij}\ (i=0,\ldots,l-1;\ j=1,\ldots,\beta);$$
$$\psi_{ij}\ (i=0,\ldots,q-1;\ j=1,\ldots,\gamma);$$

where

$$\alpha \equiv a_1 + a_2 + \cdots + a_{r+1},\quad \beta \equiv b_1 + b_2 + \cdots + b_{s+1},\quad \gamma \equiv c_1 + c_2 + \cdots + c_{t+1}.$$

---
1) *Amer. Journ.*, vol. 22, pp. 121—137; *Proceed. Lond. Math. Soc.*, vol. 32, pp. 165—170.

230 CHAPTER X.

In order to express more compactly the canonical form S_1, we let a, b, c denote an arbitrary one of the respective sets of integers

a) $1, a_1+1, a_1+a_2+1, \ldots, a_1+a_2+\cdots+a_r+1;$
b) $1, b_1+1, b_1+b_2+1, \ldots, b_1+b_2+\cdots+b_s+1;$
c) $1, c_1+1, c_1+c_2+1, \ldots, c_1+c_2+\cdots+c_t+1.$

Also let A denote any integer $\leqq \alpha$ not an a, B any integer $\leqq \beta$ not a b, C any integer $\leqq \gamma$ not a c. The canonical form S_1 may now be written as follows:

$$\eta'_{ia} = K_i \eta_{ia}, \quad \eta'_{iA} = K_i \eta_{iA} + K_i \eta_{iA-1} \quad (i=0,1,\ldots,k-1)$$
$$\zeta'_{ib} = L_i \zeta_{ib}, \quad \zeta'_{iB} = L_i \zeta_{iB} + L_i \zeta_{iB-1} \quad (i=0,1,\ldots,l-1)$$
$$\psi'_{ic} = Q_i \psi_{ic}, \quad \psi'_{iC} = Q_i \psi_{cC} + Q_i \psi_{iC-1} \quad (i=0,1,\ldots,q-1).$$

An arbitrary linear homogeneous substitution T_1 on these indices replaces η_{ij} by a linear function

233) $\qquad \Sigma D^{ij}_{\varkappa u} \eta_{\varkappa u} + \Sigma E^{ij}_{\lambda v} \zeta_{\lambda v} + \Sigma F^{ij}_{\mu w} \psi_{\mu w},$

where (as henceforth) the summation indices have the series of values

$\varkappa = 0,1,\ldots,k-1; \quad \lambda = 0,1,\ldots,l-1; \quad \mu = 0,1,\ldots,q-1;$
$u = 1,\ldots,\alpha; \qquad v = 1,\ldots,\beta; \qquad w = 1,\ldots,\gamma.$

In order that T_1 be commutative with S_1 it is necessary that 233) involve only the indices η_{iu} $(u=1,\ldots,\alpha)$. Equating the functions by which $T_1 S_1$ and $S_1 T_1$ replace η_{ia}, we get

$$K_i(\Sigma D^{ia}_{\varkappa u} \eta_{\varkappa u} + \Sigma E^{ia}_{\lambda v} \zeta_{\lambda v} + \Sigma F^{ia}_{\mu w} \psi_{\mu w})$$
$$\equiv \Sigma D^{ia}_{\varkappa u} K_\varkappa \eta_{\varkappa u} + \Sigma E^{ia}_{\lambda v} L_\lambda \zeta_{\lambda v} + \Sigma F^{ia}_{\mu w} Q_\mu \psi_{\mu w}$$
$$+ \sum_{\varkappa, A} D^{ia}_{\varkappa A} K_\varkappa \eta_{\varkappa A-1} + \sum_{\lambda, B} E^{ia}_{\lambda B} L_\lambda \zeta_{\lambda B-1} + \sum_{\mu, C} F^{ia}_{\mu C} Q_\mu \psi_{\mu C-1}.$$

Equating the coefficients of the η's and ζ's in this identity, we get

$$K_i D^{ia}_{\varkappa u} = K_\varkappa D^{ia}_{\varkappa u} \qquad (u \neq A-1)$$
$$K_i D^{ia}_{\varkappa A-1} = K_\varkappa D^{ia}_{\varkappa A-1} + K_\varkappa D^{ia}_{\varkappa A}$$
$$K_i E^{ia}_{\lambda v} = L_\lambda E^{ia}_{\lambda v} \qquad (v \neq B-1)$$
$$K_i E^{ia}_{\lambda B-1} = L_\lambda E^{ia}_{\lambda B-1} + L_\lambda E^{ia}_{\lambda B}.$$

Since $K_i \neq L_\lambda$, the third equation gives $E^{ia}_{\lambda b-1} = 0$, where b is any integer > 1 of the set b). If $b-1$ is a B, the fourth equation gives $E^{ia}_{\lambda b-2} = 0$. In the contrary case, $b-2 \neq B-1$, and the third equation gives $E^{ia}_{\lambda b-2} = 0$. Similarly, according as $b-2$ is or is not a B, the fourth or third equation gives $E^{ia}_{\lambda b-3} = 0$. Proceeding in this manner, we find that every $E^{ia}_{\lambda v} = 0$ $(\lambda = 0,\ldots,l-1;\ v=1,\ldots,\beta)$.

By a similar argument, the first and second equations give

$$D^{ia}_{\varkappa u} = 0 \quad (\varkappa \neq i,\ u = 1,\ldots,\alpha), \quad D^{ia}_{iA} = 0.$$

Equating the coefficients of the ψ's in the above identity, we find analogously that every $F^{ia}_{\mu w} = 0$. Hence T_1 replaces η_{ia} by

$$\sum_{a'} D^{ia}_{ia'} \eta_{ia'} \quad (a' = 1,\ a_1 + 1,\ a_1 + a_2 + 1, \ldots).$$

Consider any a such that $a + 1$ is an A and equate the functions by which $T_1 S_1$ and $S_1 T_1$ replace $\eta_{i\,a+1}$. Among the relations occur

$$\begin{cases} K_i D^{i\,a+1}_{\varkappa u} = K_\varkappa D^{i\,a+1}_{\varkappa u} & (\varkappa \neq i,\ u \neq A - 1) \\ K_i D^{i\,a+1}_{\varkappa A - 1} = K_\varkappa D^{i\,a+1}_{\varkappa A - 1} + K_\varkappa D^{i\,a+1}_{\varkappa A} & (\varkappa \neq i) \end{cases}$$

$$\begin{cases} K_i E^{i\,a+1}_{\lambda v} = L_\lambda E^{i\,a+1}_{\lambda v} & (v \neq B - 1) \\ K_i E^{i\,a+1}_{\lambda B - 1} = L_\lambda E^{i\,a+1}_{\lambda B - 1} + L_\lambda E^{i\,a+1}_{\lambda B} \end{cases}$$

$$\begin{cases} K_i F^{i\,a+1}_{\mu w} = Q_\mu F^{i\,a+1}_{\mu w} & (w \neq C - 1) \\ K_i F^{i\,a+1}_{\mu C - 1} = Q_\mu F^{i\,a+1}_{\mu C - 1} + Q_\mu F^{i\,a+1}_{\mu C}. \end{cases}$$

From these three pairs of equations we find (as above) respectively

$$D^{i\,a+1}_{\varkappa u} = 0 \ (\varkappa \neq i), \quad E^{i\,a+1}_{\lambda v} = 0, \quad F^{i\,a+1}_{\mu w} = 0.$$

Hence T_1 replaces $\eta_{i\,a+1}$ by a function of the η_{iu} only.

Considering any a such that $a + 1$ and $a + 2$ are of the set A, we find by the same method that T_1 replaces $\eta_{i\,a+2}$ by a function of the η_{iu} only. We readily verify that, if T_1 replaces $\eta_{i\,a+d}$ by a function of the η_{iu} only, the same will hold for $\eta_{i\,a+d+1}$. Since the series $a,\ a+1,\ a+2,\ a+3, \ldots$ yields every integer, we have proven that T_1 must replace each η_{ij} by a function of the η_{iu} only, if T_1 shall be commutative with S_1.

Similarly, T_1 must replace each ζ_{ij} by a function of the ζ_{iv} only and each ψ_{ij} by a function of the ψ_{iw} only.

When we return from the indices η_{ij}, ζ_{ij}, ψ_{ij} to the initial indices $\xi_1, \ldots, \xi_m$, S_1 becomes, by hypothesis, a substitution S having its coefficients in the $GF[p^n]$. Under what conditions will T, T_1 in the indices ξ_i, have its coefficients in the $GF[p^n]$? We have shown that T_1 replaces η_{ij} by a function of the form $\sum_{u=1}^{\alpha} D^{ij}_{iu} \eta_{iu}$. Recurring to the properties 1) and 2), § 214, of the indices η_{ij}, we must have as the D^{ij}_{iu} certain polynomials in the quantity K_i with coefficients in the $GF[p^n]$, such that

$$D^{ij}_{iu} \equiv D^{0j}_{0u}(K_i) = [D^{0j}_{0u}(K_0)]^{p^{ni}} \equiv (D^{0j}_{0u})^{p^{ni}}.$$

232 CHAPTER X.

Similar remarks hold for the indices ζ_{ij} and ψ_{ij}. We may now state our results in the following form:

Theorem. — *To determine the most general linear homogeneous substitution T on m indices with coefficients in the $GF[p^n]$ which shall be commutative with a particular one S, we apply the transformation of indices which reduces S to its canonical form S_1 and T to some form T_1. Then S_1 may be expressed as a product*

$$S_1 \equiv \eta_0 \eta_1 \ldots \eta_{k-1} \zeta_0 \zeta_1 \ldots \zeta_{l-1} \psi_0 \psi_1 \ldots \psi_{q-1}$$

where each substitution η_i, ζ_i, ψ_i is defined thus:

$\eta_i:$ $\eta'_{ia} = K_i \eta_{ia},$ $\eta'_{iA} = K_i \eta_{iA} + K_i \eta_{iA-1}$ (for every a, A)
$\zeta_i:$ $\zeta'_{ib} = L_i \zeta_{ib},$ $\zeta'_{iB} = L_i \zeta_{iB} + L_i \zeta_{iB-1}$ (for every b, B)
$\psi_i:$ $\psi'_{ic} = Q_i \psi_{ic},$ $\psi'_{iC} = Q_i \psi_{iC} + Q_i \psi_{iC-1}$ (for every c, C).

The most general T_1 must be expressible as a product

$$T_1 \equiv \mathsf{H}_0 \mathsf{H}_1 \ldots \mathsf{H}_{k-1} \mathsf{Z}_0 \mathsf{Z}_1 \ldots \mathsf{Z}_{l-1} \Psi_0 \Psi_1 \ldots \Psi_{q-1},$$

where the individual substitutions have the forms:

$\mathsf{H}_i:$ $\eta'_{ij} = \sum_{u=1}^{\alpha} \delta_{ju}^{p^{ni}} \eta_{iu}$ $(j = 1, \ldots, \alpha)$

$\mathsf{Z}_i:$ $\zeta'_{ij} = \sum_{v=1}^{\beta} \varrho_{jv}^{p^{ni}} \zeta_{iv}$ $(j = 1, \ldots, \beta)$

$\Psi_i:$ $\psi'_{ij} = \sum_{w=1}^{\gamma} \sigma_{jw}^{p^{ni}} \psi_{iw}$ $(j = 1, \ldots, \gamma)$,

the coefficients δ_{ju}, ϱ_{jv}, σ_{jw} being polynomials in K_0, L_0, Q_0, respectively, with coefficients in the $GF[p^n]$. Furthermore, H_0 must be commutative with η_0, Z_0 with ζ_0, Ψ_0 with ψ_0.

Inversely, if these conditions on H_i, Z_i, Ψ_i be satisfied, then the substitution T corresponding to the product T_1 will be commutative with S and will have its coefficients in the $GF[p^n]$.

218. In order that the substitutions H_0 and η_0 be commutative, it is necessary and sufficient that, for every a, A and A',

234) $\delta_{aA} = 0,$ $\delta_{A-1\,a-1} = 0$ $(a > 1),$ $\delta_{A-1\,a} = 0,$ $\delta_{A-1\,A'-1} = \delta_{AA'}.$

Indeed, $\eta_0 \mathsf{H}_0$ and $\mathsf{H}_0 \eta_0$ replace η_{0a} by the same function only if every $\delta_{aA} = 0$. In order that they shall replace η_{0A} by the same function, we must have

$$\sum_{A'} \delta_{AA'} \eta_{0\,A'-1} = \sum_{u=1}^{\alpha} \delta_{A-1\,u} \eta_{0u}.$$

If u is not of the form $A'-1$, it must be of the form $a-1$ or else α.

CANONICAL FORM AND CLASSIFICATION OF LINEAR SUBSTITUTIONS. 233

To take an example, let $r = 2$ and $a_1 = 3$, $a_2 = 3$, $a_3 = 2$. Then
$\delta_{1A} = \delta_{4A} = \delta_{7A} = 0$ $(A = 2, 3, 5, 6, 8)$; $\delta_{2u} = \delta_{5u} = 0$ $(u = 3, 6, 8)$
$\delta_{AA'} = \delta_{A-1\,A'-1}$ $(A, A' = 2, 3, 5, 6, 8)$.
Setting $\overset{.}{\eta}_{0u} \equiv \eta_u$, we find that H_0 has the following form [1]):

	η_1	η_2	η_3	η_4	η_5	η_6	η_7	η_8
$\eta_1' =$	δ_{11}			δ_{14}				
$\eta_2' =$	δ_{21}	δ_{11}		δ_{24}	δ_{14}		δ_{27}	
$\eta_3' =$	δ_{31}	δ_{21}	δ_{11}	δ_{34}	δ_{24}	δ_{14}	δ_{37}	δ_{27}
$\eta_4' =$	δ_{41}			δ_{44}				
$\eta_5' =$	δ_{51}	δ_{41}		δ_{54}	δ_{44}		δ_{57}	
$\eta_6' =$	δ_{61}	δ_{51}	δ_{41}	δ_{64}	δ_{54}	δ_{44}	δ_{67}	δ_{57}
$\eta_7' =$	δ_{71}			δ_{74}			δ_{77}	
$\eta_8' =$	δ_{81}	δ_{71}		δ_{84}	δ_{74}		δ_{87}	δ_{77}

Its determinant is readily seen to equal
$$\delta_{77}^2 \begin{vmatrix} \delta_{11} & \delta_{14} \\ \delta_{41} & \delta_{44} \end{vmatrix}^3.$$

In the general case, H_0 is seen to take the form:

	$\eta_{0\,1}$	$\eta_{0\,2}$	$\eta_{0\,3}$	$\ldots \eta_{0\,a_1}$	$\eta_{0\,a_1+1}$	$\eta_{0\,a_1+2}$	$\ldots \eta_{0\,a_1+a_2}$
$\eta'_{0\,1} =$	δ_{11}	0	0	$\ldots 0$	$\delta_{1\,a_1+1}$	0	$\ldots 0$
$\eta'_{0\,2} =$	δ_{21}	δ_{11}	0	$\ldots 0$	$\delta_{2\,a_1+1}$	$\delta_{1\,a_1+1}$	$\ldots 0$
$\eta'_{0\,3} =$	δ_{31}	δ_{21}	δ_{11}	$\ldots 0$	$\delta_{3\,a_1+1}$	$\delta_{2\,a_1+1}$	$\ldots 0$
$\eta'_{0\,a_1} =$	$\delta_{a_1\,1}$	$\delta_{a_1-1\,1}$	$\delta_{a_1-2\,1}$	$\ldots \delta_{11}$	$\delta_{a_1\,a_1+1}$	$\delta_{a_1-1\,a_1+1}$	$\ldots \delta''$
$\eta'_{0\,a_1+1} =$	$\delta_{a_1+1\,1}$	0	0	$\ldots 0$	$\delta_{a_1+1\,a_1+1}$	0	$\ldots 0$
$\eta'_{0\,a_1+2} =$	$\delta_{a_1+2\,1}$	$\delta_{a_1+1\,1}$	0	$\ldots 0$	$\delta_{a_1+2\,a_1+1}$	$\delta_{a_1+1\,a_1+1}$	$\ldots 0$
$\eta'_{0\,a_1+a_2} =$	$\delta_{a_1+a_2\,1}$	$\delta_{a_1+a_2-1\,1}$	$\delta_{a_1+a_2-2\,1}$	$\ldots \delta'$	$\delta_{a_1+a_2\,a_1+1}$	$\delta_{a_1+a_2-1\,a_1+1}$	$\ldots \delta_{a_1+1\,a_1+1}$

If $a_1 = a_2$, $\delta' = \delta_{a_1+1\,1}$ and $\delta'' = \delta_{1\,a_1+1}$. If $a_1 > a_2$, we have
$$\delta' \equiv \delta_{a_1+a_2\,a_1} = \delta_{a_1+1\,a_1-a_2+1} = 0, \quad \delta'' \equiv \delta_{a_1\,a_1+a_2} = \delta_{a_1-a_2+1\,a_1+1},$$
and $\delta_{1\,a_1+1} = \delta_{2\,a_1+1} = \cdots = \delta_{a_1-a_2\,a_1+1} = 0$. Finally, if $a_1 < a_2$, we have
$$\delta'' = 0, \quad \delta' = \delta_{a_2+1\,1}, \quad \delta_{a_1+1\,1} = \cdots = \delta_{a_2\,1} = 0.$$

---

[1]) δ_{17}, δ_{47}, δ_{83}, δ_{86} are zero, being equal to δ_{28}, δ_{58}, δ_{72}, δ_{75} respectively.

The matrix of the coefficients of H_0 is made up of $(r+1)^2$ rectangles, of which the general one R_{ij} is of height a_i and of base a_j. Let t be the smaller of the integers i, j or their common value if $i = j$. Then R_{ij} includes at its left or bottom a square array S_t of coefficients a_t to a side. The coefficients in its diagonal are all equal; likewise those in any parallel to the diagonal. All the coefficients in R_{ij} which lie above or to the right of the diagonal of the square S_t are zeros.

219. The results of § 218 will be applied only in such simple cases that the determinant D of H_0 can be simplified by inspection. It will therefore be sufficient to state without proof[1]) the simplest expression which can be given to D. Our notations may be fixed so that $a_1 \geqq a_2 \geqq a_3 \geqq \cdots \geqq a_{r+1}$. Let

$$a_1 = a_2 = \cdots = a_{\lambda_1} \equiv A_1, \quad a_{\lambda_1+1} = \cdots = a_{\lambda_1+\lambda_2} \equiv A_2, \ldots,$$
$$a_{\lambda_1+\lambda_2+\cdots+\lambda_{\tau-1}+1} = \cdots = a_{\lambda_1+\cdots+\lambda_\tau} \equiv A_\tau,$$

where
$$\lambda_1 + \lambda_2 + \cdots + \lambda_\tau \equiv r+1.$$

The determinant D equals $D_{\lambda_1}^{A_1} D_{\lambda_2}^{A_2} \ldots D_{\lambda_\tau}^{A_\tau}$, where, if $(i,j) \equiv \delta_{ij}$,

$$D_{\lambda_1} \equiv \begin{vmatrix} (1,1) & (1, A_1+1) & (1, 2A_1+1) & \ldots & (1, \lambda_1 A_1 - A_1 + 1) \\ (A_1+1, 1) & (A_1+1, A_1+1) & (A_1+1, 2A_1+1) & \ldots & (A_1+1, \lambda_1 A_1 - A_1 + 1) \\ \cdot & \cdot & \cdot & & \cdot \\ (\lambda_1 A_1 - A_1 + 1, 1) & (\lambda_1 A_1 - A_1 + 1, 1) & \ldots & (\lambda_1 A_1 - A_1 + 1, \lambda_1 A_1 - A_1 + 1) \end{vmatrix}$$

$$D_{\lambda_2} \equiv \begin{vmatrix} (\lambda_1 A_1+1, \lambda_1 A_1 + 1) & (\lambda_1 A_1+1, \lambda_1 A_1 + A_2 + 1) & \ldots & (\lambda_1 A_1+1, \lambda_1 A_1 + \lambda_2 A_2 - A_2 + 1) \\ \cdot & \cdot & & \cdot \\ (\lambda_1 A_1+\lambda_2 A_2 - A_2 + 1, \lambda_1 A_1 + 1) & & \ldots & (\lambda_1 A_1+\lambda_2 A_2 - A_2+1, \lambda_1 A_1+\lambda_2 A_2 - A_2+1) \end{vmatrix}$$

$$D_{\lambda_3} \equiv \begin{vmatrix} (\lambda_1 A_1+\lambda_2 A_2+1, \lambda_1 A_1+\lambda_2 A_2+1) & \ldots & (\lambda_1 A_1+\lambda_2 A_2+1, \lambda_1 A_1+\lambda_2 A_2+\lambda_3 A_3-A_3+1) \\ \cdot & & \cdot \\ (\lambda_1 A_1+\lambda_2 A_2+\lambda_3 A_3-A_3+1, \lambda_1 A_1+\lambda_2 A_2+1) & \ldots & (\lambda_1 A_1+\lambda_2 A_2+\lambda_3 A_3-A_3+1, \lambda_1 A_1+\lambda_2 A_2+\lambda_3 A_3-A_3+1) \end{vmatrix}$$

Since the coefficients δ_{ij} are functions of K_0, a root of an equation of degree k belonging to and irreducible in the $GF[p^n]$, the number of sets of values for the λ_σ^2 coefficients entering D_{λ_σ} for which this determinant is not zero is (§ 99)

$$\Omega(\lambda_\sigma, p^{nk}) \equiv (p^{nk\lambda_\sigma} - 1)(p^{nk\lambda_\sigma} - p^{nk}) \ldots (p^{nk\lambda_\sigma} - p^{nk\lambda_\sigma - nk}).$$

Excluding the coefficients of H_0 which are always zero, there remains the following number of distinct coefficients δ_{ij}:

---

1) A method of proof is given by the author in the *American Journal*, vol. 22, pp. 133—134.

CANONICAL FORM AND CLASSIFICATION OF LINEAR SUBSTITUTIONS. 235

$$\omega \equiv (a_1 + a_2 + a_3 + \cdots + a_{r+1}) + (2a_2 + a_3 + \cdots + a_{r+1})$$
$$+ (3a_3 + a_4 + \cdots + a_{r+1}) + \cdots + (\overline{r+1}\, a_{r+1})$$

the qth parenthesis giving the number of such δ_{ij} in the qth row of rectangles. On account of the equalities among the a's, we find

$$\omega \equiv A_1 \lambda_1^2 + A_2 \lambda_2 (\lambda_2 + 2\lambda_1) + A_3 \lambda_3 (\lambda_3 + 2\lambda_1 + 2\lambda_2) + \cdots$$
$$+ A_\tau \lambda_\tau (\lambda_\tau + 2\lambda_1 + \cdots + 2\lambda_{\tau-1}).$$

Excluding also the $\lambda_1^2 + \lambda_2^2 + \cdots + \lambda_\tau^2$ coefficients in the determinants D_{λ_σ}, there remains the following number of wholly arbitrary δ_{ij}:

$$\Omega \equiv \sum_{\sigma=1}^{\tau} \lambda_\sigma^2 (A_\sigma - 1) + 2 A_2 \lambda_2 \lambda_1 + 2 A_3 \lambda_3 (\lambda_2 + \lambda_1) + \cdots$$
$$+ 2 A_\tau \lambda_\tau (\lambda_{\tau-1} + \cdots + \lambda_2 + \lambda_1).$$

Each one of these Ω coefficients may take p^{nk} values. The total number of substitutions H_0 is therefore

$$f(a_1, \ldots, a_{r+1}, k, p^n) \equiv \Omega(\lambda_1, p^{nk})\, \Omega(\lambda_2, p^{nk}) \ldots \Omega(\lambda_\tau, p^{nk}) \cdot p^{nk\Omega}.$$

The total number of m-ary linear homogeneous substitutions T in the $GF[p^n]$ commutative with a particular one S, whose canonical form is expressed in the notations of § 217, is given by the product[1])

$$f(a_1, \ldots, a_{r+1}, k, p^n) \cdot f(b_1, \ldots, b_{s+1}, l, p^n) \cdot f(c_1, \ldots, c_{t+1}, q, p^n) \cdots$$

Recurring to the above example, $a_1 = 3$, $a_2 = 3$, $a_3 = 2$, we have

$$f(a_1, a_2, a_3, k, p^n) \equiv (p^{2nk} - 1)(p^{2nk} - p^{nk}) \cdot (p^{nk} - 1) \cdot p^{17nk},$$

as is directly evident from the form of H_0 and its determinant.

220. As an important example, suppose that S has the canonical form

$$\eta_i' = K_i \eta_i \qquad (i = 0, 1, \ldots, k-1)$$
$$\zeta_i' = L_i \zeta_i \qquad (i = 0, 1, \ldots, l-1)$$
$$\cdots \cdots \cdots \cdots \cdots \cdots \cdots \cdots \cdots \cdots$$
$$\psi_i' = Q_i \psi_i \qquad (i = 0, 1, \ldots, q-1).$$

The most general substitution T_1 commutative with S replaces η_0, $\zeta_0, \ldots, \psi_0$ by $\varkappa(K_0)\eta_0$, $\lambda(L_0)\zeta_0, \ldots, \varrho(Q_0)\psi_0$ respectively, in which the coefficients of the functions $\varkappa, \lambda, \ldots, \varrho$ belong to the $GF[p^n]$. If $K, L, \ldots, Q$ be primitive roots of the Galois fields of orders p^{nk}, $p^{nl}, \ldots, p^{nq}$ respectively, we may set

$$\varkappa(K_0) \equiv K^d, \quad \lambda(L_0) \equiv L^e, \ldots, \varrho(Q_0) \equiv Q^f.$$

---

1) This result is in accord with that of Jordan, who treats the case $n = 1$. His method of proof is merely illustrated by the consideration of a particular example, Traité, pp. 128—136. Moreover, it does not give the explicit form of the commutative substitutions.

If, upon returning to the initial indices ξ_i upon which S is a substitution with coefficients in the $GF[p^n]$, T_1 shall become a substitution with coefficients in that field, T_1 must have the form

$$\eta'_i = K^{d p^{ni}} \eta_i \qquad (i = 0, 1, \ldots, k-1)$$
$$\xi'_i = L^{e p^{ni}} \xi_i \qquad (i = 0, 1, \ldots, l-1)$$
$$\cdots\cdots\cdots\cdots\cdots\cdots\cdots\cdots\cdots\cdots$$
$$\psi'_i = Q^{f p^{ni}} \psi_i \qquad (i = 0, 1, \ldots, q-1).$$

Distribution of the substitutions of the general linear homogeneous group into complete sets of conjugate substitutions, §§ 221—223.

221. The substitutions of the group $G_m \equiv GLH(m, p^n)$ are to be classified into complete sets of conjugate substitutions and the number of substitutions in each set determined. Although a complete solution of this problem is furnished by the preceding general theorems, their generality and complexity make it desirable to consider in detail the special cases $m = 3$ and $m = 4$.

The classification employed is based upon the canonical forms of the substitutions of G_m. These in turn depend upon the characteristic determinants of the substitutions (α_{ij}), viz.,

$$\Delta(\lambda) \equiv \begin{vmatrix} \alpha_{11}-\lambda & \alpha_{12} & \ldots & \alpha_{1m} \\ \alpha_{21} & \alpha_{22}-\lambda & \ldots & \alpha_{2m} \\ \cdots & \cdots & & \cdots \\ \alpha_{m1} & \alpha_{m2} & \ldots & \alpha_{mm}-\lambda \end{vmatrix}$$
$$\equiv (-1)^m \{\lambda^m - \alpha_1 \lambda^{m-1} - \alpha_2 \lambda^{m-2} - \cdots - \alpha_{m-1}\lambda - \alpha_m\}.$$

Furthermore, G_m contains a substitution in whose characteristic determinant the coefficients $\alpha_1, \alpha_2, \ldots, \alpha_m$ are any preassigned marks of the $GF[p^n]$ such that $\alpha_m \neq 0$. The required substitution is

$$(\alpha_{ij}) \equiv \begin{pmatrix} \alpha_1 & \alpha_2 & \alpha_3 & \ldots & \alpha_{m-1} & \alpha_m \\ 1 & 0 & 0 & \ldots & 0 & 0 \\ 0 & 1 & 0 & \ldots & 0 & 0 \\ 0 & 0 & 1 & \ldots & 0 & 0 \\ \cdots & \cdots & \cdots & & \cdots & \cdots \\ 0 & 0 & 0 & \ldots & 1 & 0 \end{pmatrix}.$$

222. Consider first the group G_3 of order

$$N \equiv (p^{3n} - 1)(p^{3n} - p^n)(p^{3n} - p^{2n}).$$

By §§ 214—215, every linear homogeneous substitution in the $GF[p^n]$ on $m = 3$ indices can be reduced by a linear ternary transformation

CANONICAL FORM AND CLASSIFICATION OF LINEAR SUBSTITUTIONS. 237

(not necessarily in the $GF[p^n]$) to one of the following five types of canonical forms:

A: $\quad x' = \lambda x, \quad y' = \lambda^{p^n} y, \quad z' = \lambda^{p^{2n}} z$

B: $\quad x' = \mu x, \quad y' = \mu^{p^n} y, \quad z' = \alpha z$

C: $\quad x' = \alpha x, \quad y' = \beta y, \quad z' = \gamma z$

D: $\quad x' = \alpha x, \quad y' = \beta y, \quad z' = \beta(z+y)$

E: $\quad x' = \alpha x, \quad y' = \alpha(y+x), \quad z' = \alpha(z+y)$,

where λ satisfies a cubic equation and μ a quadratic equation each belonging to and irreducible in the $GF[p^n]$, while α, β, γ denote marks $\neq 0$ of the $GF[p^n]$.

Upon replacing λ by λ^{p^n} or by $\lambda^{p^{2n}}$, we obtain from A a substitution conjugate with A. Any other replacement of λ leads to a substitution not conjugate with A (§ 102, Corollary), since its characteristic determinant differs from that of A. Hence the type A includes $\frac{1}{3}(p^{3n} - p^n)$ distinct sets of conjugate substitutions, those in different sets being not conjugate under G_3.

Let S be a substitution of G_3 having the canonical form A, where λ is a definite mark of the $GF[p^{3n}]$ not in the $GF[p^n]$. If a substitution T of G_3 be commutative with S and if we apply to T the same transformation of indices which reduces S to the form A, then (§ 220) T will take the form

$$x' = \sigma^r x, \quad y' = \sigma^{rp^n} y, \quad z' = \sigma^{rp^{2n}} z,$$

where σ is a primitive root of the $GF[p^{3n}]$ and r is some positive integer $\leq p^{3n} - 1$. Hence S is commutative with exactly $p^{3n} - 1$ substitutions of G_3, so that S is one of $N \div (p^{3n} - 1)$ conjugate substitutions within G_3. The total number of substitutions of G_3 reducible to the canonical forms A is therefore

a) $\qquad \frac{1}{3}(p^{3n} - p^n)(p^{3n} - p^n)(p^{3n} - p^{2n}).$

Type B includes $\frac{1}{2}(p^{2n} - p^n)(p^n - 1)$ distinct sets of conjugate substitutions. In fact, the replacement of μ by μ^{p^n} leads to a substitution conjugate with B, while any other replacement of μ or any change in α leads to a substitution not conjugate with B. A substitution of G_3 commutative with a particular substitution reducible to a type B has the canonical form

238   CHAPTER X.

$$x' = \varrho^r x, \quad y' = \varrho^{rp^n} y, \quad z' = \delta z,$$

where ϱ is a primitive root of the $GF[p^{2n}]$ and δ belongs to the $GF[p^n]$, r being an integer $\lessgtr p^{2n} - 1$. The number of such substitutions is $(p^{2n}-1)(p^n-1)$. Hence the total number of substitutions of G_3 reducible to the canonical forms B is

b) $\qquad \frac{1}{2}(p^{2n} - p^n)(p^n - 1)(p^{3n} - 1)p^{3n}.$

Type C includes $p^n - 1$ canonical forms with $\alpha = \beta = \gamma$; $(p^n - 1)(p^n - 2)$ canonical forms with $\alpha = \beta \neq \gamma$; a like number with $\alpha = \gamma \neq \beta$; a like number with $\beta = \gamma \neq \alpha$; and $(p^n - 1)(p^n - 2)(p^n - 3)$ with α, β, γ all distinct. By a suitable transformation of indices the multipliers α, β, γ in C are permuted in an arbitrary manner. We have therefore the following numbers of distinct sets of conjugate canonical substitutions C:

$p^n - 1$ of type C_1 with $\alpha = \beta = \gamma$;

$(p^n - 1)(p^n - 2)$ of type C_2 with only two equal multipliers, say $\alpha = \beta \neq \gamma$;

$\frac{1}{6}(p^n - 1)(p^n - 2)(p^n - 3)$ of type C_3 with all three multipliers distinct.

The most general substitution of G_3 commutative with C_3 is

$$x' = ax, \quad y' = by, \quad z' = cz \qquad (a, b, c \text{ in the } GF[p^n]).$$

Hence C_3 is one of $N \div (p^n - 1)^3$ conjugate substitutions within G_3. The most general substitution of G_3 commutative with C_2 is

$$x' = ax + by, \quad y' = cx + dy, \quad z' = ez.$$

Hence C_2 is one of $N \div (p^{2n} - 1)(p^{2n} - p^n)(p^n - 1)$ conjugate substitutions. Finally, C_1 is commutative with every substitution of G_3 and thus is conjugate only with itself. The total number of substitutions of G_3 reducible to the canonical forms C is thus

c) $\qquad \begin{aligned} &(p^n - 1) + (p^{3n} - 1)(p^n - 2)p^{2n} \\ &\quad + \frac{1}{6}(p^n - 2)(p^n - 3)(p^{3n} - 1)(p^n + 1)p^{3n}.\end{aligned}$

Of the substitutions of type D, there are $p^n - 1$ with $\alpha = \beta$ and $(p^n - 1)(p^n - 2)$ with $\alpha \neq \beta$, no two being conjugate under G_3. A

CANONICAL FORM AND CLASSIFICATION OF LINEAR SUBSTITUTIONS. 239

substitution D with $\alpha = \beta$ is commutative only with the $p^{3n}(p^n-1)^2$ substitutions of G_3

$$x' = dy + ex, \quad y' = ay, \quad z' = by + az + cx \quad (a, b, c, d, e \text{ in the } GF[p^n]).$$

A substitution D with $\alpha \neq \beta$ is commutative only with the $p^n(p^n-1)^2$ substitutions of G_3

$$x' = ex, \quad y' = ay, \quad z' = by + az.$$

The total number of substitutions of G_3 reducible to the types D is thus

d) $\quad (p^n-1)(p^{3n}-1)(p^n+1) + (p^n-1)(p^n-2)(p^{3n}-1)(p^n+1)p^{2n}.$

No two of the p^n-1 substitutions of type E are conjugate under G_3. Each is commutative only with the $p^{2n}(p^n-1)$ substitutions of G_3

$$x' = ax, \quad y' = bx + ay, \quad z' = cx + by + az.$$

The number of substitutions reducible to the canonical forms E is

e) $\quad (p^n-1)(p^{3n}-1)(p^{2n}-1)p^n.$

A check on the above enumeration of the substitutions of G_3 consists is verifying that the sum of the numbers a), b), c), d), e) equals the order N of G_3.

223. Consider next the group[1]) G_4 of order

$$N \equiv (p^{4n}-1)(p^{4n}-p^n)(p^{4n}-p^{2n})(p^{4n}-p^{3n}).$$

By § 221, G_4 contains a substitution in whose characteristic determinant $\Delta(\lambda) \equiv \lambda^4 - \alpha_1\lambda^3 - \alpha_2\lambda^2 - \alpha_3\lambda - \alpha_4$ the coefficients $\alpha_1, \alpha_2, \alpha_3, \alpha_4$ are arbitrary marks of the $GF[p^n]$, $\alpha_4 \neq 0$. According to the possible factorizations of $\Delta(\lambda)$ in the $GF[p^n]$, we distinguish the cases: I) irreducible; II) linear factor and irreducible cubic; III) two distinct irreducible quadratic factors; IV) equal irreducible quadratic factors; V) irreducible quadratic and two distinct linear factors; VI) irreducible quadratic and two equal linear factors; VII)—XI) four linear factors, according to the number of equal factors. Denote by λ_t, μ_t marks of the $GF[p^{nt}]$ not in the $GF[p^{n\tau}]$, $\tau < t$. For simplicity, the subscript unity is omitted from the marks $\alpha, \beta, \gamma, \delta$ of the $GF[p^n]$. The types of canonical forms of the substitutions of G_4 may be exhibited in the following complete list:

---

[1]) Cf. T. M. Putnam, *Amer. Journ. Math.*, vol. XXIII, pp. 41—48. For the author's treatment of the case $n=3$, *ibid*, pp. 37—40.

CHAPTER X.

Type	Canonical substitutions[1]				Number M of distinct canonical forms
I	$\lambda_4 x$	$\lambda_4^{p^n} y$	$\lambda_4^{p^{2n}} z$	$\lambda_4^{p^{3n}} w$	$\frac{1}{4}(p^{4n} - p^{2n})$
II	$\lambda_3 x$	$\lambda_3^{p^n} y$	$\lambda_3^{p^{2n}} z$	$\lambda_1 w$	$\frac{1}{3}(p^{3n} - p^n)(p^n - 1)$
III	$\lambda_2 x$	$\lambda_2^{p^n} y$	$\mu_2 z$	$\mu_2^{p^n} w$	$\frac{1}{8}(p^{2n} - p^n)(p^{2n} - p^n - 2)$
IV_1	$\lambda_2 x$	$\lambda_2(y+x)$	$\lambda_2^{p^n} z$	$\lambda_2^{p^n}(w+z)$	$\frac{1}{2}(p^{2n} - p^n)$
IV_2	$\lambda_2 x$	$\lambda_2 y$	$\lambda_2^{p^n} z$	$\lambda_2^{p^n} w$	$\frac{1}{2}(p^{2n} - p^n)$
V	$\lambda_1 x$	$\mu_1 y$	$\lambda_2 z$	$\lambda_2^{p^n} w$	$\frac{1}{4}(p^{2n} - p^n)(p^n - 1)(p^n - 2)$
VI_1	$\lambda_1 x$	$\lambda_1(y+x)$	$\lambda_2 z$	$\lambda_2^{p^n} w$	$\frac{1}{2}(p^{2n} - p^n)(p^n - 1)$
VI_2	$\lambda_1 x$	$\lambda_1 y$	$\lambda_2 z$	$\lambda_2^{p^n} w$	$\frac{1}{2}(p^{2n} - p^n)(p^n - 1)$
VII	αx	βy	γz	δw	$\frac{1}{24}(p^n - 1)(p^n - 2)(p^n - 3)(p^n - 4)$
VIII_1	αx	βy	γz	$\gamma(w+z)$	$\frac{1}{2}(p^n - 1)(p^n - 2)(p^n - 3)$
VIII_2	αx	βy	γz	γw	$\frac{1}{2}(p^n - 1)(p^n - 2)(p^n - 3)$
IX_1	αx	βy	$\beta(z+y)$	$\beta(w+z)$	$(p^n - 1)(p^n - 2)$
IX_2	αx	βy	$\beta(z+y)$	βw	$(p^n - 1)(p^n - 2)$
IX_3	αx	βy	βz	βw	$(p^n - 1)(p^n - 2)$
X_1	αx	$\alpha(y+x)$	$\alpha(z+y)$	$\alpha(w+z)$	$p^n - 1$
X_2	αx	$\alpha(y+x)$	$\alpha(z+y)$	αw	$p^n - 1$
X_3	αx	$\alpha(y+x)$	αz	$\alpha(w+z)$	$p^n - 1$
X_4	αx	$\alpha(y+x)$	αz	αw	$p^n - 1$
X_5	αx	αy	αz	αw	$p^n - 1$
XI_1	αx	$\alpha(y+x)$	γz	$\gamma(w+z)$	$\frac{1}{2}(p^n - 1)(p^n - 2)$
XI_2	αx	$\alpha(y+x)$	γz	γw	$(p^n - 1)(p^n - 2)$
XI_3	αx	αy	γz	γw	$\frac{1}{2}(p^n - 1)(p^n - 2)$

---

[1] The notation αx, βy, γz, $\gamma(w+z)$, for example, is used for the substitution
$$x' = \alpha x, \quad y' = \beta y, \quad z' = \gamma z, \quad w' = \gamma(w+z).$$

CANONICAL FORM AND CLASSIFICATION OF LINEAR SUBSTITUTIONS. 241

Table giving the form and number C of the substitutions of the group G_4 commutative with the various types of canonical forms:

I	λx	$\lambda p^n y$	$\lambda p^{2n} z$	$\lambda p^{3n} w$	$p^{4n}-1$
II	μx	$\mu p^n y$	$\mu p^{2n} z$	aw	$(p^{3n}-1)(p^n-1)$
III	ϱx	$\varrho p^n y$	σz	$\sigma p^n w$	$(p^{2n}-1)^2$
IV$_1$	ϱx	$\sigma x + \varrho y$	$\varrho p^n z$	$\sigma p^n z + \varrho p^n w$	$(p^{2n}-1)p^{2n}$
IV$_2$	$\sigma x + \varrho y$	$\varkappa x + \tau y$	$\sigma p^n z + \varrho p^n w$	$\varkappa p^n z + \tau p^n w$	$(p^{4n}-1)(p^{4n}-p^{2n})$
V	ax	by	ϱz	$\varrho p^n w$	$(p^{2n}-1)(p^n-1)^2$
VI$_1$	ax	$bx + ay$	ϱz	$\varrho p^n w$	$(p^{2n}-1)(p^{2n}-p^n)$
VI$_2$	$ax + by$	$cx + dy$	ϱz	$\varrho p^n w$	$(p^{2n}-1)^2(p^{2n}-p^n)$
VII	ax	by	cz	dw	$(p^n-1)^4$
VIII$_1$	ax	by	cz	$dz + cw$	$(p^n-1)^3 p^n$
VIII$_2$	ax	by	$cz + dw$	$ez + fw$	$(p^{2n}-p^n)(p^{2n}-1)(p^n-1)^2$
IX$_1$	ax	by	$cy + bz$	$dy + cz + bw$	$(p^n-1)^2 p^{2n}$
IX$_2$	ax	by	$cy + bz + ew$	$fy + dw$	$(p^n-1)^3 p^{3n}$
IX$_3$	ax	$by + cz + dw$	$ey + fz + gw$	$hy + iz + jw$	$(p^{3n}-1)(p^{2n}-1)(p^n-1)^2 p^{3n}$
X$_1$	ax	$bx + ay$	$cx + by + az$	$dx + cy + bz + aw$	$(p^n-1) p^{3n}$
X$_2$	ax	$bx + ay$	$cx + by + az + ew$	$fx + dw$	$(p^n-1)^2 p^{4n}$
X$_3$	$ax + ez$	$bx + ay + fz + ew$	$gx + cz$	$hx + gy + dz + cw$	$(p^{2n}-1)(p^{2n}-p^n) p^{4n}$
X$_4$	ax	$bx + ay + fz + ew$	$gx + cz + kw$	$hx + dz + lw$	$(p^{2n}-1)(p^{2n}-p^n)(p^n-1) p^{5n}$
X$_5$			arbitrary		N
XI$_1$	ax	$bx + ay$	cz	$dz + cw$	$(p^n-1)^2 p^{2n}$
XI$_2$	ax	$bx + ay$	$cz + dw$	$ez + fw$	$(p^{2n}-1)(p^{2n}-p^n)(p^n-1) p^n$
XI$_3$	$ax + by$	$gx + hy$	$cz + dw$	$ez + fw$	$(p^{2n}-1)^2(p^{2n}-p^n)^2$

Here λ belongs to the $GF[p^{4n}]$, μ to the $GF[p^{3n}]$, ϱ, σ, $\varkappa$, τ to the $GF[p^{2n}]$, and $a, b, c, \ldots, j$ belong to the $GF[p^n]$. If M denote the number of distinct canonical forms in a general type, and C the number of substitutions of G_4 commutative with each, the number of substitutions of G_4 reducible to that type is MN/C. The sum of these numbers is found to equal N, the total number of the substitutions of G_4.

CHAPTER XI.

OPERATORS AND CYCLIC SUBGROUPS OF THE SIMPLE GROUP $LF(3, p^n)$.[1]

224. By § 108 the group $G \equiv LF(3, p^n)$ of all substitutions of determinant 1,

$$S:\quad x' = \frac{\alpha_{11}x + \alpha_{12}y + \alpha_{13}}{\alpha_{31}x + \alpha_{32}y + \alpha_{33}},\quad y' = \frac{\alpha_{21}x + \alpha_{22}y + \alpha_{23}}{\alpha_{31}x + \alpha_{32}y + \alpha_{33}},\quad |\alpha_{ij}| = 1,$$

in which the coefficients α_{ij} belong to the $GF[p^n]$, is a simple group of order

$$N \equiv \frac{1}{d}(p^{3n} - 1)(p^{2n} - 1)p^{3n},$$

where d is the greatest common divisor of 3 and $p^n - 1$, so that

$$d = 1,\text{ if } p^n = 3^n \text{ or } 3l - 1;\quad d = 3,\text{ if } p^n = 3l + 1.$$

The equation $\tau^3 = 1$ has in the $GF[p^n]$ a single root $\theta = 1$, if $d = 1$; but has three roots θ, θ^2, $\theta^3 \equiv 1$, if $d = 3$. Hence, if $d = 1$, there is a single homogeneous substitution of determinant unity

$$\Sigma:\qquad \xi_i' = \alpha_{i1}\xi_1 + \alpha_{i2}\xi_2 + \alpha_{i3}\xi_3 \qquad (i = 1, 2, 3)$$

which, when taken fractionally, leads to the non-homogeneous substitution S. If $d = 3$, let Θ denote the homogeneous substitution of determinant unity which multiplies each index by θ. Then there are exactly the three homogeneous substitutions of determinant unity, Σ, $\Theta\Sigma \equiv \Sigma\Theta$, $\Theta^2\Sigma \equiv \Sigma\Theta^2$:

$$\Theta^r\Sigma:\qquad \xi_i' = \theta^r(\alpha_{i1}\xi_1 + \alpha_{i2}\xi_2 + \alpha_{i3}\xi_3) \qquad (i = 1, 2, 3),$$

which, when taken fractionally, lead to the non-homogeneous substitution S. Combining the two cases, we may employ the group of ternary linear homogeneous substitutions of determinant unity in place of the group G provided we consider to be identical the d substitutions Σ, $\Theta\Sigma$ and $\Theta^2\Sigma$. Under this convention concerning the homogeneous substitutions, we employ henceforth the homogeneous notation for the substitutions of the group G.

225. Any substitution of G can be reduced by a linear ternary transformation of indices (not necessarily in the $GF[p^n]$ and not necessarily of determinant unity) to one of the canonical forms A, B, C, D, E of § 222. In the present case, the determinants of $A, \ldots, E$ must be unity.

---

[1] For $n = 1$, Burnside, *Proceed. Lond. Math. Soc.*, vol. 26, pp. 58—106; for general n, Dickson, *Amer. Journ.*, vol. 22, pp. 231—252, where certain errors in Burnside's paper are pointed out.

If two substitutions S and T of the group G have the same canonical form, there exists (§ 216) a ternary homogeneous substitution W belonging to the $GF[p^n]$ such that $T = W^{-1}SW$. It remains to consider whether or not there exists a ternary homogeneous substitution W_1 belonging to the $GF[p^n]$ *and having determinant unity* such that W_1 transforms S into T. If the canonical form be A, B, C or D, such a W_1 will be shown to exist; while for the canonical form E such a W_1 does not always exist.

It is first shown that any one of the types A, B, C, D can be transformed into itself by a substitution V of determinant equal to an arbitrary mark $\neq 0$ of the $GF[p^n]$ and obeying the same laws in regard to the conjugacy of its indices as does the canonical form in question. For type A we may take as V the substitution

$$x' = \sigma^r x, \quad y' = \sigma^{rp^n} y, \quad z' = \sigma^{rp^{2n}} z,$$

where σ is a primitive root of the $GF[p^{3n}]$ so that $\tau \equiv \sigma^{1+p^n+p^{2n}}$ is a primitive root of the $GF[p^n]$. The determinant of V is thus τ^r, which by suitable choice of r may be made equal to an arbitrary mark $\neq 0$ of the $GF[p^n]$. For types B and C we may take V to be

$$x' = x, \quad y' = y, \quad z' = \tau^r z.$$

For type D we may take as V the substitution

$$x' = \tau^r x, \quad y' = y, \quad z' = z.$$

Let W have the determinant w and choose V so that its determinant is w^{-1}. We may take as the required substitution W_1 the product $V_1 W$, where V_1 is the form taken by V when expressed in the initial indices. In fact V_1 and W have their coefficients in the $GF[p^n]$, while the product $V_1 W$ transforms S into T and has the determinant $w^{-1} \cdot w = 1$. Hence, if two substitutions of G have the same canonical form A, B, C, or D, they are conjugate within the group G.

For type E there arise two cases. If $d = 1$, so that 3 is prime to $p^n - 1$, every mark of the $GF[p^n]$ is a cube (§ 63, Corollary). Hence an integer r may be determined so that τ^{3r} shall be an arbitrary mark $\neq 0$ in the field. Hence the above argument holds if we choose as V the substitution

$$x' = \tau^r x, \quad y' = \tau^r y, \quad z' = \tau^r z.$$

For $d = 3$, only $\frac{1}{3}(p^n - 1)$ of the marks $\neq 0$ of the $GF[p^n]$ are cubes. Their products by β and β^2 will be not-cubes, if β be any particular not-cube. We can therefore determine V', of determinant a cube, such that T is the transformed of S by the substitution $V_1'W \equiv W'$ belonging to the $GF[p^n]$ and having as determinant one of the three marks $1, \beta, \beta^2$. Consider the three substitutions of G

E_r: $\quad x' = x, \quad y' = y + \beta^r x, \quad z' = z + y \qquad (r = 0, 1, 2).$

The following substitution of determinant β:

R: $\quad x' = \beta x, \quad y' = y, \quad z' = z$

transforms E_1 into E_0 and E_2 into E_1. If E has determinant unity, it is identical with E_0 in the group G. It follows from the proof above that any substitution T of G, which can be transformed into E_0 by a linear substitution W belonging to the $GF[p^n]$, can be transformed into E_0 by a similar substitution W' of determinant β^t ($t = 0, 1$ or 2). Also R^{-t} transforms E_0 into E_t. Hence T is transformed into E_t by the product $W'R^{-t}$ which belongs to the $GF[p^n]$ and has determinant unity. Hence every substitution of G of canonical form E is conjugate within G to one of the types E_0, E_1, E_2.

We next prove that no two of the types E_0, E_1, E_2 are conjugate within G, i. e., by means of a substitution of determinant unity. The most general ternary homogeneous substitution which transforms E_0 into E_1 is seen to be

$$x' = \beta^{-1} cx, \quad y' = cy + bx, \quad z' = cz + by + ax,$$

of determinant $\beta^{-1} c^3$, which can not be made unity. Transforming the latter by R^{-1}, we obtain the most general substitution which transforms E_1 into E_2, viz.,

$$x' = \beta^{-1} cx, \quad y' = cy + \beta bx, \quad z' = cz + by + \beta ax,$$

of determinant $\beta^{-1} c^3 \neq 1$. Finally, by § 102, E_0 can not be transformed into ΘE_1, nor E_1 into ΘE_2, by a linear substitution. The results now proven may be stated in the explicit form:

Every substitution of G can be reduced by a ternary linear homogeneous transformation to one of the canonical forms

A: $\quad x' = \lambda x, \quad y' = \lambda^{p^n} y, \quad z' = \lambda^{p^{2n}} z \qquad \left(\lambda^{p^{2n} + p^n + 1} = 1\right)$

B: $\quad x' = \mu x, \quad y' = \mu^{p^n} y, \quad z' = \mu^{-p^n - 1} z$

C: $\quad x' = \alpha x, \quad y' = \beta y, \quad z' = \gamma z \qquad (\alpha \beta \gamma = 1)$

D: $\quad x' = \alpha^{-2} x, y' = \alpha y, \quad z' = \alpha(z + y)$

E_0: $\quad x' = x, \quad y' = y + x, \quad z' = z + y$

E_1: $\quad x' = x, \quad y' = y + \beta x, \; z' = z + y$ (β not-cube in $GF[p^n]$)

E_2: $\quad x' = x, \quad y' = y + \beta^2 x, \; z' = z + y,$

in which λ satisfies a cubic and μ a quadratic equation each belonging to and irreducible in the $GF[p^n]$, while α, β, γ belong to the $GF[p^n]$.

OPERATORS AND CYCLIC SUBGROUPS etc. 245

Of the substitutions of G reducible to the forms A and B, those and only those are conjugate within G which are reducible to the same form A or to the same form B. Every other substitution of G is conjugate within G to one of the types C, D, E_0, E_1, E_2 and no two of the latter types are conjugate within G.

226. Type A. The substitution of determinant unity
$$x' = \alpha_1 x + \alpha_2 y + z, \quad y' = x, \quad z' = y$$
has the characteristic determinant
$$\Delta(\lambda) \equiv -\lambda^3 + \alpha_1 \lambda^2 + \alpha_2 \lambda + 1.$$
Hence α_1 and α_2 may be chosen in the $GF[p^n]$ so that a root λ of $\Delta(\lambda) = 0$ is a primitive root of the equation

235) $\qquad \lambda^{p^{2n}+p^n+1} = 1.$

The order of the corresponding substitution A is the least integer m for which
$$\lambda^m = \lambda^{m p^n} = \lambda^{m p^{2n}},$$
i. e., for which $m(p^n-1)$ is a multiple of $p^{2n}+p^n+1$. But the greatest common divisor of p^n-1 and $p^{2n}+p^n+1$ is also that of p^n-1 and 3 and therefore equals d. The order m is consequently $\frac{1}{d}(p^{2n}+p^n+1)$.

Moreover, the roots of any irreducible cubic of the form $\Delta(\lambda) = 0$ may be written λ^s, λ^{sp^n}, $\lambda^{sp^{2n}}$, so that the corresponding substitution is the sth power of the substitution just considered. Hence the orders of all substitutions having irreducible characteristic determinants are factors of $\frac{1}{d}(p^{2n}+p^n+1)$.

Consider a substitution S of G of canonical form A for which λ is a primitive root of equation 235). By § 220, the only substitutions of G which are commutative with S have, simultaneously with the canonical form A of S, the canonical form
$$x' = \sigma^r x, \quad y' = \sigma^{r p^n} y, \quad z' = \sigma^{r p^{2n}} z \quad \left(\sigma^{r(1+p^n+p^{2n})} = 1\right)$$
where σ is a primitive root of the $GF[p^{3n}]$. Hence $r(1+p^n+p^{2n})$ must be divisible by $p^{3n}-1$ and therefore r divisible by p^n-1. Setting $r = \varrho(p^n-1)$,
$$\sigma^r = (\sigma^{p^n-1})^\varrho = \lambda^{t\varrho},$$
since σ^{p^n-1} is a primitive root of 235) and hence equal to some power t of λ. The only substitutions of G which are commutative with S are therefore the powers of S. It follows that S is one of a set of

246   CHAPTER XI.

$$s \equiv \frac{N}{1/d\,(p^{2n}+p^n+1)}$$

distinct conjugate substitutions, N being the order of G.

The only distinct powers of S which have the same characteristic determinant as S are evidently S, S^{p^n} and $S^{p^{2n}}$. To each set of three substitutions such as S^r, S^{rp^n}, $S^{rp^{2n}}$ contained in the cyclic group generated by S and all belonging to the same characteristic determinant, there corresponds a set of s distinct conjugate substitutions. Hence there exist in G

$$\frac{1}{3}\left[\frac{1}{d}(p^{2n}+p^n+1)-1\right]$$

such sets of s conjugate substitutions. It follows that G contains in all

236) $$\frac{dN}{3(p^{2n}+p^n+1)}\left[\frac{1}{d}(p^{2n}+p^n+1)-1\right]$$

substitutions not the identity whose orders are factors of

$$\frac{1}{d}(p^{2n}+p^n+1).$$

Hence G contains $\frac{dN}{3(p^{2n}+p^n+1)}$ *distinct conjugate cyclic subgroups of order*
$$\frac{1}{d}(p^{2n}+p^n+1).$$

227. Type B. Since G contains substitutions in whose characteristic determinant $-\lambda^3+\alpha_1\lambda^2+\alpha_2\lambda+1$ both α_1 and α_2 are arbitrary in the $GF[p^n]$, we can choose

$$\alpha_1 \equiv \gamma+1/\delta, \quad -\alpha_2 \equiv \delta+\gamma/\delta,$$
so that
$$\Delta(\lambda) \equiv -(\lambda-1/\delta)(\lambda^2-\gamma\lambda+\delta),$$

where γ and δ are arbitrary in the $GF[p^n]$. In particular, G contains a substitution T whose characteristic determinant has an irreducible quadratic factor which vanishes for a primitive root μ of the $GF[p^{2n}]$. The canonical form of T is then B. The order of T is therefore the least integer t for which

$$\mu^t = \mu^{tp^n} = \mu^{-t(p^n+1)}$$

i. e., for which both $t(p^n-1)$ and $t(p^n+2)$ are divisible by $p^{2n}-1$. But $3t$ and $t(p^n-1)$ are both divisible by $p^{2n}-1$, for t a minimum, if and only if

$t=p^{2n}-1$, when $p^n=3^n$ or $3l-1$; $t=\frac{1}{3}(p^{2n}-1)$, when $p^n=3l+1$.

Hence the order of T is $\frac{1}{d}(p^{2n}-1)$.

OPERATORS AND CYCLIC SUBGROUPS etc. 247

By § 220, the most general substitution of G commutative with T has the canonical form
$$x' = \mu^r x, \quad y' = \mu^r \rho^n y, \quad z' = \mu^{-r(p^n+1)} z$$
and hence is T^r. Hence T is one of a set of $dN \div (p^{2n}-1)$ distinct conjugate substitutions. The only distinct powers of S which have the same multipliers as S are S and S^{p^n}. Hence G contains $\frac{1}{2} \frac{dN}{p^{2n}-1}$ distinct conjugate cyclic subgroups of order $\frac{1}{d}(p^{2n}-1)$.

The number of substitutions of G whose orders are factors of $\frac{1}{d}(p^{2n}-1)$ without being factors of $\frac{1}{d}(p^n-1)$, and hence not of p^n-1, is

237) $\qquad \frac{1}{2} N p^n/(p^n+1).$

In fact, such substitutions form in all
$$\frac{1}{2d}[(p^{2n}-1)-(p^n-1)] \equiv \frac{1}{2d}(p^n-1)p^n$$
different sets, those in each set having the same characteristic determinant. Each set contains $dN \div (p^{2n}-1)$ distinct conjugate substitutions. The product of the two numbers gives formula 237).

228. We can exhibit G as a permutation-group on $p^{2n}+p^n+1$ letters. Every linear function $A\xi_1 + B\xi_2 + C\xi_3$, in which A, B, C are marks not all zero of the $GF[p^n]$, can be put into one of the forms,
$$\mu(\xi_3 + \varrho\xi_2 + \sigma\xi_1), \quad \mu(\xi_2 + \varrho\xi_1), \quad \mu\xi_1,$$
where μ, ϱ, σ are marks of the $GF[p^n]$ and $\mu \neq 0$. Combining into one system $\{A\xi_1 + B\xi_2 + C\xi_3\}$ the p^n-1 linear functions
$$\mu(A\xi_1 + B\xi_2 + C\xi_3),$$
μ denoting in succession the p^n-1 marks $\neq 0$ of the field, we obtain $p^{2n}+p^n+1$ distinct systems,
$$\{\xi_3 + \varrho\xi_2 + \sigma\xi_1\}, \quad \{\xi_2 + \varrho\xi_1\}, \quad \{\xi_1\} \qquad [\varrho, \sigma \text{ arbitrary marks}].$$

Any ternary homogeneous linear substitution replaces the functions $\mu(A\xi_1 + B\xi_2 + C\xi_3)$, comprising one system, by linear functions
$$\mu(A\xi_1' + B\xi_2' + C\xi_3') \equiv \mu(\alpha\xi_1 + \beta\xi_2 + \gamma\xi_3)$$
all belonging to a single system. Hence it permutes the above $p^{2n}+p^n+1$ symbols amongst themselves. It follows that G is isomorphic with a permutation-group G' on these symbols. But a homogeneous substitution altering none of the symbols must have the form
$$\xi_1' = \alpha\xi_1, \quad \xi_2' = \alpha\xi_2, \quad \xi_3' = \alpha\xi_3.$$

248   CHAPTER XI.

If it have determinant unity, it corresponds in G to the identity. Hence G is *simply* isomorphic with G'.

The permutation-group G' is doubly-transitive. We need only prove that G' contains a permutation converting $\{\xi_1\}$, $\{\xi_2 + \xi_1\}$ into respectively
$$\{A\xi_1 + B\xi_2 + C\xi_3\}, \quad \{A'\xi_1 + B'\xi_2 + C'\xi_3\}$$
the latter being any two distinct symbols, viz.,
$$A:B:C \neq A':B':C'.$$

For the corresponding homogeneous substitution, we may take
$$\xi_1' = A\xi_1 + B\xi_2 + C\xi_3, \quad \xi_2' = (A'-A)\xi_1 + (B'-B)\xi_2 + (C'-C)\xi_3,$$
$$\xi_3' = \alpha\xi_1 + \beta\xi_2 + \gamma\xi_3,$$
where α, β, γ are chosen in any manner such that the determinant of the substitution is unity, viz.,
$$\alpha \begin{vmatrix} B & C \\ B' & C' \end{vmatrix} + \beta \begin{vmatrix} C & A \\ C' & A' \end{vmatrix} + \gamma \begin{vmatrix} A & B \\ A' & B' \end{vmatrix} = 1.$$

By hypothesis the determinants are not all zero, so that solutions α, β, γ in the $GF[p^n]$ certainly exist.

229. *Type D for $\alpha^3 \neq 1$.* Let α be a primitive root in the $GF[p^n]$, the cases $p^n = 2$ and $p^n = 2^2$ being necessarily excluded. For such an α, substitution D generates a cyclic group of order $\frac{1}{d}p(p^n - 1)$.

Considered as an operation of the isomorphic permutation-group, D belongs to a subgroup of G which leaves fixed the symbols $\{x\}$ and $\{y\}$. The general substitution of G possessing this property has the form
$$R: \quad x' = \gamma x, \quad y' = \beta y, \quad z' = \alpha z + \alpha' y + \alpha'' x \quad (\alpha\beta\gamma = 1).$$

In order that R shall have the order $\frac{1}{d}p(p^n - 1)$, it is necessary and sufficient that α be a primitive root in the $GF[p^n]$ and that either

(i) $\alpha' \neq 0, \quad \alpha = \beta \neq \gamma;$ or (ii) $\alpha'' \neq 0, \quad \alpha = \gamma \neq \beta.$

In fact, if both β and γ differ from α, R may be given the form
$$x' = \gamma x, \quad y' = \beta y, \quad Z' = \alpha Z,$$
whose $(p^n - 1)^{\text{st}}$ power is unity, by introducing in place of z the index
$$Z \equiv z + \frac{\alpha'}{\alpha - \beta} y + \frac{\alpha''}{\alpha - \gamma} x.$$
Hence, if $\alpha \neq \beta$, we may take $\alpha = \gamma$. Then $\alpha'' \neq 0$; for, if $\alpha'' = 0$, R multiplies $z + \frac{\alpha'}{\alpha - \beta} y$ by α, so that R would have as order a factor

OPERATORS AND CYCLIC SUBGROUPS etc. 249

of p^n-1. Similarly, if $\alpha \neq \gamma$, then must $\alpha = \beta$, $\alpha' \neq 0$. Finally, if $\alpha = \beta = \gamma$, each may be taken equal to unity. Then, by induction,

$$R^r: \quad x' = x, \quad y' = y, \quad z' = z + r\alpha'y + r\alpha''x,$$

so that R would have the period p. Hence either (i) or (ii) must be satisfied.

Suppose, inversely, that relations (i) are satisfied. Setting

$$Y \equiv \frac{\alpha'}{\alpha} y, \quad Z \equiv z + \frac{\alpha'' x}{\beta - \gamma},$$

R takes the form

$$x' = \alpha^{-2}x, \quad Y' = \alpha Y, \quad Z' = \alpha(Z+Y),$$

and is thus of period $\frac{1}{d}p(p^n-1)$ if, and only if, α be a primitive root of the $GF[p^n]$. Interchanging x with y, the proof follows for case (ii).

Using the theorem just proved, we proceed to determine the number and conjugacy of the cyclic subgroups of order $\frac{1}{d}p(p^n-1)$ which leave the symbols $\{x\}$ and $\{y\}$ fixed. For case (i),

$$R: \quad x' = \alpha^{-2}x, \quad y' = \alpha y, \quad z' = \alpha z + \alpha' y + \alpha'' x \quad (\alpha' \neq 0, \ \alpha^3 \neq 1),$$

where α is a primitive root of the $GF[p^n]$. By induction we find

$$R^t: \quad x' = \alpha^{-2t}x, \quad y' = \alpha^t y, \quad z' = \alpha^t z + t\alpha' \alpha^{t-1}y + \alpha'' \alpha^{t-1}\left(\frac{\alpha^{-3t}-1}{\alpha^{-3}-1}\right)x.$$

In order that $\Theta^r R^t$ shall be identical with the substitution

$$x' = \alpha^{-2}x, \quad y' = \alpha y, \quad z' = \alpha z + \varrho'y + \varrho''z,$$

it is necessary and sufficient that

$$\theta^r \alpha^{t-1} = 1, \quad t\alpha' = \varrho', \quad \alpha'' = \varrho''.$$

Let M_i denote any one of the $(p^n-1)/(p-1)$ distinct marks M_1, $M_2, \ldots$ such that no two have as their ratio an integral mark[1]). If α be a fixed mark $\neq 0$ and M an arbitrary mark, the $p^n(p^n-1)/(p-1)$ substitutions

238) $\quad x' = \alpha^{-2}x, \quad y' = \alpha y, \quad z' = \alpha z + M_i y + Mx$

have the property that no power of any one of them reduces to one of the set. We therefore obtain that number of cyclic subgroups of order $\frac{1}{d}p(p^n-1)$.

Furthermore, every substitution V of the subgroup leaving $\{x\}$ and $\{y\}$ fixed, and having $\alpha = \beta$, and of order a divisor of $\frac{1}{d}p(p^n-1)$

---

1) The marks $M_1, M_2, \ldots$ are evidently the multipliers in a rectangular array of the marks $\neq 0$ of the $GF[p^n]$, the first row being formed by the integral marks $1, 2, \ldots, p-1$.

250   CHAPTER XI.

without being a factor of p or p^n-1, is contained in one of the above cyclic subgroups. In fact, by the earlier argument, we may set

V: $x' = \alpha^{-2s}x, \quad y' = \alpha^s y, \quad z' = \alpha^s z + \alpha' y + \alpha'' x \quad (\alpha' \neq 0, \alpha^{3s} \neq 1).$

Let M_i be a mark $\neq 0$ such that its ratio to $\alpha'\alpha^{1-s}$ is an integral mark. The power $s + k(p^n-1)$ of 238) gives

$$x' = \alpha^{-2s}x, \quad y' = \alpha^s y,$$
$$z' = \alpha^s z + [s + k(p^n-1)]\alpha^{s-1}M_i y + M\alpha^{s-1}\left(\frac{\alpha^{-3s}-1}{\alpha^{-s}-1}\right)x.$$

By choice of k and M, we can make the coefficient of y in z' equal α' and that of x equal α''.

Hence there are $p^n(p^n-1)/(p-1)$ cyclic subgroups of G of order $\frac{1}{d}p(p^n-1)$ for which $\alpha = \beta$, and as many more for which $\alpha = \gamma$, each leaving the symbols $\{x\}$ and $\{y\}$ fixed, and together containing all substitutions having the last property and having an order not p nor a factor of p^n-1.

These cyclic subgroups are all conjugate within G and, indeed, within the subgroup which leaves fixed $\{x\}$ and $\{y\}$ or merely permutes them. First, the substitution

$$x' = x, \quad y' = y, \quad z' = z + \frac{M'-M}{\alpha^{-2}-\alpha}x$$

transforms 238) into a like substitution with M' in place of M. Also

$$x' = \lambda^{-1}\varrho^{-2}x, \quad y' = \varrho y, \quad z' = \lambda \varrho z$$

transforms 238) into the substitution

$$x' = \alpha^{-2}x, \quad y' = \alpha y, \quad z' = \alpha z + \lambda M_i y + \lambda^2 \varrho^3 M x.$$

Hence the cyclic subgroups given by $\alpha = \beta$ are all conjugate within the group leaving fixed $\{x\}$ and $\{y\}$. These symbols are interchanged by

$$x' = y, \quad y' = -x, \quad z' = z,$$

which transforms 238) into the substitution

$$x' = \alpha x, \quad y' = \alpha^{-2}y, \quad z' = \alpha z - My + M_i x.$$

Hence the set of cyclic subgroups given by $\alpha = \beta$ are conjugate to the set given by $\alpha = \gamma$ within the group leaving fixed the symbols $\{x\}$ and $\{y\}$ or permuting them. The latter group consequently contains $2p^n(p^n-1)/(p-1)$ conjugate cyclic groups of order $\frac{1}{d}p(p^n-1)$ and those substitutions of these groups whose orders are not divisors of p or p^n-1 are all distinct. Since the permutation-group isomorphic with G is doubly transitive, it contains

OPERATORS AND CYCLIC SUBGROUPS etc. 251

$$\frac{1}{2}(p^{2n}+p^n+1)(p^{2n}+p^n)$$

conjugate subgroups leaving fixed or permuting the two symbols. *Hence there are altogether*

$$2p^n\left(\frac{p^n-1}{p-1}\right)\cdot\frac{1}{2}(p^{2n}+p^n+1)(p^{2n}+p^n) \equiv \frac{dN}{p^n(p^n-1)(p-1)}$$

conjugate cyclic subgroups of order $\frac{1}{d}p(p^n-1)$. Each contains $p+\frac{1}{d}(p^n-1)-1$ substitutions of period p or a divisor of $\frac{1}{d}(p^n-1)$. There remain in each cyclic group $(p-1)\left[\frac{1}{d}(p^n-1)-1\right]$ substitutions. Hence G contains

239) $\qquad N(p^n-1-d) \div p^n(p^n-1)$

substitutions whose orders divide $\frac{1}{d}p(p^n-1)$ but not p or p^n-1.

For the cases $p^n=2$ and $p^n=2^2$ above excluded, formula 239) reduces to zero. Hence the result is always true.

230. Type D when $\alpha^3=1$. We are to consider substitutions of period p having the canonical form:

$$D':\qquad x'=x,\quad y'=y,\quad z'=z+y.$$

From the investigation at the beginning of § 229 it follows that the only substitutions of period p which leave fixed the symbols $\{x\}$ and $\{y\}$ have the form

240) $\quad x'=x,\quad y'=y,\quad z'=z+\alpha x+\beta y\quad$ (α and β not both zero).

There are $p^{2n}-1$ distinct substitutions of this form. They are all conjugate to D' within G. In fact, if $\beta\neq 0$, the substitution

$$x'=x,\quad y'=y+\varrho x,\quad z'=z$$

transforms 240) into

$$x'=x,\quad y'=y,\quad z'=z+(\alpha-\beta\varrho)x+\beta y.$$

By choice of ϱ, we can make $\alpha-\beta\varrho=0$. If $\beta=0$, we transform 240) by
$$x'=y,\quad y'=x,\quad z'=-z,$$
and get
$$x'=x,\quad y'=y,\quad z'=z-\alpha y.$$

In either case we reach a substitution of the form 230) but having $\alpha=0$, $\beta\neq 0$. It is transformed into D' by the substitution of G

$$x'=\beta^{-1}x,\quad y'=\beta y,\quad z'=z.$$

The $p^{2n}-1$ substitutions 230) determine $(p^{2n}-1)/(p-1)$ conjugate cyclic subgroups of order p contained in the subgroup of G which leaves fixed the symbols $\{x\}$ and $\{y\}$ and hence also $\{x+\varrho y\}$, ϱ being an arbitrary mark of the $GF[p^n]$.

Each such group therefore leaves fixed p^n+1 (and no more) symbols. But the $p^{2n}+p^n+1$ symbols furnish

$$\frac{\frac{1}{2}(p^{2n}+p^n+1)(p^{2n}+p^n)}{\frac{1}{2}(p^n+1)p^n} = p^{2n}+p^n+1$$

such sets of symbols. *Hence G contains*

$$(p^{2n}+p^n+1)\frac{(p^{2n}-1)}{(p-1)} = \frac{dN}{p^{3n}(p^n-1)(p-1)}$$

such conjugate cyclic subgroups, all of whose substitutions are conjugate under G. Each such subgroup is therefore contained self-conjugately within a subgroup of order $\frac{1}{d}p^{3n}(p^n-1)(p-1)$. The total number of distinct substitutions of G of order p of the type considered has thus been shown to be

241) $$\frac{dN}{p^{3n}(p^n-1)}.$$

231. Types E_i. By induction we find that

$$E_0^t: \quad x' = x, \quad y' = y + tx, \quad z' = z + ty + \frac{1}{2}t(t-1)x.$$

Hence E_0 is of period p or 4 according as $p > 2$ or $p = 2$. The most general substitution of G transforming E_0 into itself is

$$x' = ax, \quad y' = ay + bx, \quad z' = az + by + cx \qquad (a^3 = 1).$$

Exactly p^{2n}. of these substitutions are distinct in the group G.

Suppose first that $p > 2$. For any positive integer $t < p$, the substitution

242) $$x' = \frac{1}{t}x, \quad y' = y - \frac{t-1}{2t}x, \quad z' = tz$$

is of determinant unity and transforms E_0 into E_0^t. Taking

$$t = 1, 2, \ldots, p-1,$$

we see that G contains exactly $p^{2n}(p-1)$ distinct substitutions which transform into itself the cyclic group generated by E_0. *The cyclic group* $\{E_0\}$ *is, for* $p > 2$, *one of* $N/p^{2n}(p-1)$ *distinct conjugate subgroups of G*. In particular, G contains N/p^{2n} distinct conjugate substitutions of the type E_0.

Suppose next that $p = 2$. Then E_0 is of period 4. Since

$$E_0^2: \qquad x' = x, \quad y' = y, \quad z' = z + x$$

leaves fixed the 2^n+1 symbols $\{x\}$, $\{y+\lambda x\}$, λ any mark of the $GF[2^n]$, while E_0 leaves fixed but one symbol $\{x\}$, the two substitutions are not conjugate under G. But E_0 is transformed into E_0^3 by the substitution 242) for $t=3$, viz.,

$$x' = x, \quad y' = y + x, \quad z' = z.$$

The cyclic group generated by E_0 is therefore transformed into itself by exactly $2 \cdot 2^{2n}$ substitutions of G. For $p=2$, $\{E_0\}$ *is one of a complete set of* $N/2^{2n+1}$ *conjugate cyclic subgroups of* G. Just two of the four substitutions of every such cyclic group are of type E_0, while the remaining one not the identity is of type D with $\alpha^3=1$. Hence, for $p=2$, G contains $N/2^{2n}$ distinct substitutions conjugate with E_0.

Since E_1 and E_2 are conjugate to E_0 within the general ternary linear homogeneous group in the $GF[p^n]$, the number of substitutions of G conjugate to E_0 within G equals the number conjugate to E_1 or the number conjugate to E_2. Hence G contains altogether

243) $$3N/p^{2n}$$

distinct substitutions of the canonical forms E_i; they form three distinct sets of conjugate substitutions under G. Also, E_0, E_1, E_2 *each lead to the same number of conjugate cyclic subgroups of* G.

232. Type C. The substitutions of canonical form C are of order a divisor of p^n-1. Of the $(p^n-1)^2$ sets of solutions in the $GF[p^n]$ of $\alpha\beta\gamma = 1$, d sets have $\alpha = \beta = \gamma$ and hence each equal to θ^r $(r=0, 1, \text{ or } 2)$. If α be any mark different from $0, 1, \theta, \theta^2$, and if $\beta = \alpha$, then $\gamma = \alpha^{-2} \neq \alpha$. Hence there are $3(p^n - d - 1)$ sets of solutions in which two and only two of the quantities α, β, γ are equal. There remain

$$(p^n-1)^2 - 3(p^n-d-1) - d \equiv p^{2n} - 5p^n + 4 + 2d$$

sets of solutions in which α, β, γ are all distinct. Dividing this number by 6 to allow for permutations, we obtain the number of distinct sets of unequal multipliers of ternary homogeneous substitutions C.

If, for $d=3$, α, β, γ do not form a permutation of $1, \theta, \theta^2$, the three sets

$$\alpha, \beta, \gamma; \quad \theta\alpha, \theta\beta, \theta\gamma; \quad \theta^2\alpha, \theta^2\beta, \theta^2\gamma,$$

are not equivalent sets of multipliers in the homogeneous group, but are equivalent in the non-homogeneous group G. The number of sets of unequal multipliers in G is therefore

$$1 + \frac{1}{3}\left(\frac{p^{2n} - 5p^n + 4 + 2d}{6} - 1\right), \text{ for } d=3; \quad \frac{p^{2n} - 5p^n + 4 + 2d}{6}, \text{ for } d=1.$$

254 CHAPTER XI.

We proceed to prove that the total number of substitutions of G of canonical form C with α, β, γ distinct is, for $d=1$ or 3,

244) $$\frac{N}{(p^n-1)^2} \cdot \frac{p^{2n}-5p^n+4+2d}{6}.$$

By § 220, the only ternary homogeneous substitutions commutative with C with α, β, γ distinct are the $(p^n-1)^2$ substitutions

$$T: \qquad x'=ax, \quad y'=by, \quad z'=cz \qquad (abc=1).$$

For $d=1$, each set of unequal multipliers therefore leads to $N/(p^n-1)^2$ conjugate substitutions, so that we obtain the number 244). For $d=3$, the substitutions T give only $\frac{1}{d}(p^n-1)^2$ distinct substitutions in G. Furthermore, by § 102, C can be transformed into ΘC if, and only if, the multipliers α, β, γ form a permutation of 1, θ, θ^2. The special substitution C,

$$x'=x, \quad y'=\theta y, \quad z'=\theta^2 z$$

is transformed into C, ΘC or $\Theta^2 C$ by exactly the $3(p^n-1)^2$ products T, $(xyz)T$, $(xzy)T$. The corresponding substitution is therefore one of $N/(p^n-1)^2$ distinct conjugate substitutions under G. Each of the remaining substitutions C with unequal multipliers is one of a set of $N \div \frac{1}{3}(p^n-1)^2$ conjugate substitutions under G.

Corresponding to the p^n-d-1 sets of multipliers α, β, γ of which two are equal, there are $\frac{1}{d}(p^n-d-1)$ substitutions C' of G, no two of which are conjugate. Such a substitution

$$C': \qquad x'=\alpha x, \quad y'=\alpha y, \quad z'=\gamma z \qquad (\alpha^2\gamma=1, \; \gamma \neq \alpha)$$

cannot be transformed into $\Theta C'$. By § 218, the most general ternary linear homogeneous substitution which transforms C' into itself is

$$x'=ax+by, \quad y'=a'x+b'y, \quad z'=c''z.$$

The number of such substitutions in the $GF[p^n]$ of determinant unity is
$$(p^{2n}-1)(p^{2n}-p^n).$$

Hence the total number of substitutions in G of the canonical form C' is

245) $$\frac{1}{d}(p^n-d-1) \cdot \frac{N}{\frac{1}{d}(p^{2n}-1)(p^{2n}-p^n)}.$$

233. As a check upon the accuracy of our enumeration of the substitutions of G, we may verify that the numbers given by the formulae 236), 237), 239), 241), 243), 244) and 245), together with unity, to count the identical substitution, give as total sum the order N of the group G.

234. To complete the enumeration of the cyclic subgroups of G, it remains to determine those generated by substitutions of the canonical forms C. The method will be sufficiently illustrated if we confine the investigation to the case $d=1$.[1]) If α be a primitive root of the $GF[p^n]$, we may set

$$C: \quad x' = \alpha^r x, \quad y' = \alpha^s y, \quad z' = \alpha^{-r-s} z,$$

where r and s are integers chosen from the series $0, 1, \ldots, p^n - 2$. Let g denote the greatest common divisor of r and s. The period of C is the least positive integer l for which lr and ls, and therefore also lg, are multiples of $p^n - 1$. Hence C is of period $p^n - 1$ if, and only if, g be relatively prime to $p^n - 1$. In general, C is the gth power of a similar substitution with the multipliers $\alpha^{r/g}$, $\alpha^{s/g}$, $\alpha^{(-r-s)/g}$, the latter of period $p^n - 1$. Hence, for $d = 1$, the substitutions of type C are all included in the cyclic groups generated by those substitutions of type C which have the period $p^n - 1$. We may therefore confine our attention to these largest cyclic groups. The exponents r, s in the expression of any substitution C of period $p^n - 1$ must occur among the sets of two positive integers less than $p^n - 1$ and having their greatest common divisor prime to $p^n - 1$. Denote by $F(p^n - 1)$ the number of such sets. A similar remark holds for the couples s, r; $r, -r-s$; $-r-s, r$; $s, -r-s$; $-r-s, s$; provided $-r-s$ be replaced by its least positive residue modulo $p^n - 1$. If $r, s, -r-s$ be distinct, the above couples form six of the $F(p^n - 1)$ sets, but lead to the same set of three multipliers in C. If two of the exponents be equal and therefore different from the third, we may take them to be $r, r, -2r$. Then the couples r, r; $r, -2r$; $-2r, r$ form three of the $F(p^n - 1)$ sets, but lead to the same set of multipliers in C. Here r may be any one of the $\Phi(p^n - 1)$ integers less than and prime to $p^n - 1$. Hence there are $3\Phi(p^n - 1)$ sets leading to $\Phi(p^n - 1)$ distinct sets of multipliers two of which are equal, while the remaining sets lead to $\frac{1}{6}[F(p^n - 1) - 3\Phi(p^n - 1)]$ distinct sets of three unequal multipliers, together yielding all the substitutions C of period $p^n - 1$. The value of $F(p^n - 1)$ is given by the following theorem.[2])

The number of sets of two integers, not both zero, chosen from the series $0, 1, \ldots, k-1$ so that their greatest common divisor is prime to k is

$$F(k) = k^2 - \sum_{i=1}^{\varkappa} \frac{k^2}{q_i^2} + \sum_{i,j}^{1,\ldots,\varkappa} \frac{k^2}{q_i^2 q_j^2} - \cdots \equiv k^2 \left(1 - \frac{1}{q_1^2}\right)\left(1 - \frac{1}{q_2^2}\right) \cdots \left(1 - \frac{1}{q_\varkappa^2}\right)$$

where $q_1, q_2, \ldots, q_\varkappa$ are the distinct prime factors of k.

---

1) The case $d = 3$ is more intricate and the results quite complicated. The results are given in the *Amer. Journ.*, vol. XXII, p. 251; the proofs in vol. XXIV.
2) Jordan, Traité, p. 96.

Of the k^2 sets of two integers each $< k$, k^2/q_i^2 have their integers chosen from the k/q_i multiples of q_i and are to be excluded. We thereby exclude, in particular, the sets of integers each of which is one of the k/q_iq_j multiples of q_iq_j. Hence, in afterwards excluding the sets of integers each of which is a multiple of q_j, we subtract the number $k^2/q_j^2 - k^2/q_i^2q_j^2$. After the required exclusions have all been made, there evidently remains the number of sets indicated by $F(k)$. Among the latter sets, the couple 0, 0 does not occur since

$$1 - \varkappa + \frac{\varkappa(\varkappa-1)}{1 \cdot 2} - \frac{\varkappa(\varkappa-1)(\varkappa-2)}{1 \cdot 2 \cdot 3} + \cdots \equiv (1-1)^\varkappa = 0.$$

235. A cyclic group generated by a substitution C of period $p^n - 1$ will be called *special* if two of its substitutions C^a, C^b of period $p^n - 1$ are conjugate within G, i. e., have the same set of multipliers. Since a and b must be prime to $p^n - 1$, the condition requires that C and C^{ba_1} shall have the same set of multipliers, where a_1 is determined from $aa_1 \equiv 1 \pmod{p^n - 1}$. It thus suffices to investigate when C and C^m have the same multipliers, m being prime to $p^n - 1$ and $1 < m < p^n - 1$. The three distinct ways in which the two sets

$$\alpha^r, \alpha^s, \alpha^t; \quad \alpha^{mr}, \alpha^{ms}, \alpha^{mt} \qquad r+s+t \equiv 0 \pmod{p^n - 1}$$

may be identical in some order will be considered in turn.

i) If $\alpha^{mr} = \alpha^r$, $\alpha^{ms} = \alpha^s$, $\alpha^{mt} = \alpha^t$, then $r(m-1)$, $s(m-1)$, and therefore also $g(m-1)$, are divisible by $p^n - 1$. Since g is prime to $p^n - 1$, $m - 1$ must be divisible by $p^n - 1$, contrary to hypothesis.

ii) If $\alpha^{mr} = \alpha^s$, $\alpha^{ms} = \alpha^r$, $\alpha^{mt} = \alpha^t$, then must

$$mr \equiv s, \quad ms \equiv r, \quad m^2r \equiv r \pmod{p^n - 1}.$$

Then r must be prime to $p^n - 1$; for a common factor would divide s in virtue of the first congruence, whereas the greatest common divisor of r and s is prime to $p^n - 1$. Hence, by the last congruence,

246) $$m^2 \equiv 1 \pmod{p^n - 1}.$$

Inversely, if m be any solution of 246) and if r be any integer less than and prime to $p^n - 1$ and if s be determined by

$$s \equiv mr \pmod{p^n - 1},$$

then C and C^m have the same multipliers. Moreover, C is the rth power of a substitution with the multipliers α, α^m, α^{-m-1}, which may therefore be taken in place of C as generator of the special cyclic group.

If 2^k be the highest power of 2 contained in $p^n - 1$ and if $\varkappa = 0$ when $k = 0$ or 1, $\varkappa = 1$ when $k = 2$, $\varkappa = 2$ when $k \geq 3$, and if μ be

OPERATORS AND CYCLIC SUBGROUPS etc. 257

the number of distinct odd prime factors of p^n-1, then the congruence 246) has exactly $2^{\varkappa+\mu}$ solutions m.[1]) The solution $m \equiv 1$ is to be excluded. Consider the $2^{\varkappa+\mu}-1$ substitutions with the multipliers α, α^m, α^{-m-1}, $m > 1$. They generate as many cyclic groups. In fact, $(\alpha^m)^x = \alpha$ requires $x \equiv m \pmod{p^n-1}$; while $(\alpha^{-m-1})^y = \alpha$ is impossible since $m+1$ has a factor > 1 in common with p^n-1. Moreover, the sets of multipliers of the substitutions of period p^n-1 in each cyclic group are the same in pairs. Hence these special cyclic groups contain altogether $\frac{1}{2}\Phi(p^n-1)(2^{\varkappa+\mu}-1)$ distinct sets of unequal multipliers.

(iii) If $\alpha^{mr} = \alpha^s$, $\alpha^{ms} = \alpha^t$, $\alpha^{mt} = \alpha^r$, we find that

$$r(m^2+m+1) \equiv t+s+r \equiv 0, \quad s(m^2+m+1) \equiv r+t+s \equiv 0 \pmod{p^n-1}.$$

Hence $M \equiv m^2 + m + 1$ must be divisible by p^n-1. Since $m(m+1)$ is even, M is an odd number. Hence p^n-1 must be odd and therefore $p^n = 2^n$. Since $d = 1$, 3 is not a factor of p^n-1. Hence each prime factor q of p^n-1 is of one of the forms $6k+5$, $6k+1$. Now M and hence also m^3-1 must be divisible by q. If $q = 6k+5$, Fermat's theorem gives $m^{6k+4} \equiv 1 \pmod{q}$. Since $m^3 \equiv 1$, we have $m \equiv 1 \pmod{q}$ and therefore $M \equiv 3 \equiv 0 \pmod{q}$, which is impossible. Hence must $q = 6k+1$. Inversely, if $q = 6k+1$, $m^{6k}-1 \equiv 0 \pmod{q}$ has $6k$ distinct integral solutions. But the left member is divisible by m^3-1 and therefore by M. Hence $M \equiv 0 \pmod{q}$ has two distinct solutions. Each of these solutions leads to one, and but one, solution of $M \equiv 0 \pmod{q^\tau}$. To give a proof by induction from $\tau = e$ to $\tau = e+1$, let $m^3 - 1 = Qq^e$. Then

$$(m + xq^e)^3 - 1 \equiv Qq^e + 3m^2xq^e \pmod{q^{2e}}$$

and will therefore be divisible by q^{e+1} if, and only if,

$$Q + 3m^2x \equiv 0 \pmod{q}.$$

Since 3 and m are prime to q, x is uniquely determined mod q. Hence each m determines one solution $y \equiv m + xq^e$ of

$$y^3 - 1 \equiv 0 \pmod{q^{e+1}}.$$

Hence, if $m^2 + m + 1$ be divisible by q^e, $y - 1$ will be prime to q and hence $y^2 + y + 1$ will be divisible by q^{e+1}. Supposing that the prime factors of $2^n - 1$ are all of the form $6k + 1$ and that the number of distinct ones is γ, it follows that $M \equiv 0 \pmod{2^n - 1}$ has 2^γ solutions m. But, if m be a solution, then $-m - 1$ will be

---

[1]) Dirichlet, Zahlentheorie, § 37.

258   CHAPTER XI.

a second solution. Hence C is the rth power of one of the $2^{\gamma-1}$ substitutions with the multipliers α, α^m, α^{-m-1}. These generate distinct cyclic groups, since $(\alpha^m)^x = \alpha$ requires $x = -m - 1$. Hence there are $2^{\gamma-1}$ of these special cyclic groups and the substitutions of period $p^n - 1$ in each give just $\frac{1}{3}\Phi(p^n - 1)$ distinct sets of multipliers.

Excluding the special sets of multipliers of types (ii) and (iii), there remain

$$\frac{1}{6}[F(p^n-1) - 3\Phi(p^n-1)] - \frac{1}{2}\Phi(p^n-1)(2^{\varkappa+\mu}-1) - \frac{1}{3}\Phi(p^n-1)2^{\gamma-1}$$

sets of unequal multipliers, the last term occurring only for certain values of p^n. The corresponding substitutions C lie in sets of $\Phi(p^n-1)$ in cyclic subgroups not conjugate under G. Noting that $F(p^n-1)$ is divisible by $\Phi(p^n-1)$, giving the quotient

$$\Psi(p^n-1) \equiv (p^n-1)\left(1+\frac{1}{q_1}\right)\left(1+\frac{1}{q_2}\right)\cdots\left(1+\frac{1}{q_\gamma}\right),$$

where $q_1, q_2, \ldots, q_\gamma$ are the distinct prime factors of $p^n - 1$, we may combine our results in the theorem:

If $p^n - 1$ be not divisible by 3, the substitutions C generate the following types of cyclic groups of order $p^n - 1$ not conjugate under G:

a) one group generated by the substitution with multipliers α, α, α^{-2};

b) $2^{\varkappa+\mu} - 1$ generated by substitutions with multipliers α, α^m, α^{-m-1}, where $m^2 \equiv 1 \pmod{p^n - 1}$, $\varkappa$ and μ defined in (ii);

c) $2^{\gamma-1}$ generated by similar substitutions with

$$m^2 + m + 1 \equiv 0 \pmod{p^n - 1},$$

occurring only when $p^n - 1 \equiv 2^n - 1$ has only prime factors (γ distinct ones) of the form $6j + 1$;

d) $\frac{1}{6}[\Psi(p^n-1) - 3] - \frac{1}{2}(2^{\varkappa+\mu}-1) - \frac{1}{3}\cdot 2^{\gamma-1}$ further groups.

236. As a first example, let $p^n = 8$, so that $\mu = 1$, $\varkappa = 0$, $\gamma = 1$. There is just one cyclic group of each of the first three types. The generators have the sets of multipliers α, α, α^{-2}; α, α^{-1}, 1; α, α^2, α^{-3} respectively.

As second example, let $p^n = 17$, so that $\mu = 0$, $\varkappa = 2$, while the third type of group does not occur. There are three cyclic groups of the second type determined by the sets of multipliers α, α^{-1}, 1; α, α^7, α^8; α, α^9, α^6. The two cyclic groups of the fourth type may be determined by the sets of multipliers α, α^2, α^{13}; α, α^3, α^{12}.

OPERATORS AND CYCLIC SUBGROUPS etc.   259

237. It remains to determine the number of cyclic subgroups of G conjugate with each group of the types a), b), c), d). Type a) is generated by the substitution

$$x' = \alpha x, \quad y' = \alpha y, \quad z' = \alpha^{-2} z \qquad (\alpha^{-2} \neq \alpha)$$

and is commutative with exactly $(p^{2n} - 1)(p^{2n} - p^n)$ substitutions of G, viz.,
$$x' = ax + by, \quad y' = cx + dy, \quad z' = ez.$$

The cyclic group of order $p^n - 1$ generated by the substitution

$$x' = \alpha x, \quad y' = \alpha^m y, \quad z' = \alpha^{-m-1} z, \qquad m^2 \equiv 1 \pmod{p^n - 1}$$

is transformed into itself by $2(p^n - 1)^2$ substitutions, viz.,

$$S: \qquad x' = ax, \quad y' = by, \quad z' = cz \qquad (abc = 1)$$

and the products TS, where T replaces x by y and y by $-x$. When cyclic groups of the third type exist, each is transformed into itself by the $3(p^n - 1)^2$ substitutions S, $(xyz)S$, $(xzy)S$. Each cyclic group of the fourth type is transformed into itself by exactly the $(p^n - 1)^2$ substitutions S.

238. For $p^n = 2^2$, the simple group G has the order $N \equiv 20160$. There is, by 244), a single canonical form C, not the identity, its multipliers being 1, θ, θ^2. The $N/(p^n - 1)^2 \equiv 2240$ substitutions of G of period 3 are therefore all conjugate and generate a single set of conjugate cyclic groups. Applying the results of §§ 226—231 to the case $p^n = 2^2$, we see that G contains

960 conjugate cyclic groups of order 7 with 5760 substitutions of period 7
2016 „ „ „ „ „ 5 „ 8064 „ „ „ 5
630 „ „ „ „ „ 4 „ 1260 „ „ „ 4
630 „ „ „ „ „ 4 „ 1260 „ „ „ 4
630 „ „ „ „ „ 4 „ 1260 „ „ „ 4
1120 „ „ „ „ „ 3 „ 2240 „ „ „ 3
315 „ „ „ „ „ 2 „ 315 „ „ „ 2
                                  1 „ „ „ 1
                              20160

The substitutions of period 2 are all contained in the cyclic groups of order 4.

The group G differs in structure from the alternating group on 8 letters, likewise of order 20,160. Indeed, the latter contains 5760 substitutions of type (1234567), 3360 of type (123456)(78), 1344 of type (12345), 2688 of type (12345)(678), 2520 of type (1234)(56), 1260 of type (1234)(5678), 112 of type (123), 1120 of type (123)(456),

260   CHAPTER XII.

1680 of type (123)(45)(67), 210 of type (12)(34), 105 of type (12)(34)(56)(78), and the identity. The alternating group has substitutions of periods 6 and 15, while G does not. Both groups contain the same number of substitutions of period 7, the same number of period 4, the same number of period 2. But the distribution into sets of conjugates of the substitutions of period 2, or of period 3, or of period 4, differs in the two groups. In particular, G is not isomorphic with the alternating group on 8 letters, each group being simple and of order 20160.[1])

CHAPTER XII.

SUBGROUPS OF THE LINEAR FRACTIONAL GROUP $LF(2, p^n)$.[2])

239. In § 108 was defined the group of linear fractional substitutions

$$S: \qquad z' = \frac{\alpha z + \beta}{\gamma z + \delta} \qquad (\Delta \equiv \alpha\delta - \beta\gamma \neq 0)$$

on an arbitrary variable z with coefficients in the $GF[p^n]$. We proceed to represent it as a permutation-group on $p^n + 1$ letters. Suppose z runs through the series of marks of the $GF[p^n]$. For $\gamma = 0$, z' will also run through the series of marks. For $\gamma \neq 0$, the value $z = -\delta/\gamma$ gives $z' = \frac{-\Delta/\gamma}{0}$, so that z' can not be determined as a mark of the field. We may, however, obtain a set of elements which are merely permuted by S by adjoining to the series of marks a new element $\infty \equiv \frac{\mu}{0}$, necessarily the same for every mark $\mu \neq 0$, since $\frac{1}{0} \equiv \frac{\mu \cdot 1}{\mu \cdot 0} = \frac{\mu}{0}$, and assumed to combine with the marks $\lambda \neq 0$ of the field according to the laws

$$\infty + \lambda = \lambda + \infty = \infty, \quad \lambda\infty = \infty\lambda = \infty, \quad \lambda/\infty = 0, \quad \infty/\lambda = \infty,$$

while the indeterminate fraction $\frac{\alpha\infty + \beta}{\gamma\infty + \delta}$ is assumed to equal α/γ. Setting henceforth $s \equiv p^n$, the group $LF(2, s)$ of linear fractional substitutions of determinant unity in the $GF[s]$ may therefore be

---

[1]) Miss Schottenfels established this theorem by direct calculations, *Annals of Mathematics*, (2) vol. 1, pp. 147—152.

[2]) Moore, Mathematical Papers Chicago Congress of 1893, pp. 208—242, *Math. Ann.*, vol. 55 (56?); Wiman, Sweedish Acad., vol. 25 (1899), pp. 1—47; Burnside, *Proc. Lond. Math. Soc.*, vol. 25 (1894), p. 132. The work of Galois, Mathieu and Gierster is cited in the exposition for $n = 1$ in Klein-Fricke, Modulfunctionen I, p. 411 and pp. 419—491.

SUBGROUPS OF THE LINEAR FRACTIONAL GROUP $LF(2, p^n)$.

represented concretely as a permutation-group $G_{M(s)}^{s+1}$ on $s+1$ letters and having the order

247) $$M(s) \equiv \frac{s(s^2-1)}{2;1} \quad (2;1 \text{ according as } p > 2; p = 2).$$

The group of all substitutions S has the order $(2;1)M(s)$. For $p > 2$, it may be represented as a permutation-group $G_{2M(s)}^{s+1}$. For $p = 2$, it is the former group.

The group $G_{M(s)}^{s+1}$ is doubly transitive. It is only necessary to prove that a substitution T with coefficients in the field and of determinant unity may be found which will replace two arbitrary distinct elements ϱ, σ by the elements $0, \infty$. If both ϱ and σ are marks of the field, we may take as T

$$z' = \frac{\varkappa(z-\varrho)}{z-\sigma}, \quad \varkappa \equiv \frac{1}{\varrho - \sigma}.$$

If ϱ is a mark and $\sigma = \infty$, we may take T to be $z' = z - \varrho$.

The inverse of $S \equiv \begin{pmatrix} \alpha, \beta \\ \gamma, \delta \end{pmatrix}$ of determinant unity is $S^{-1} \equiv \begin{pmatrix} \delta, -\beta \\ -\gamma, \alpha \end{pmatrix}$, so that S is of period two if and only if $\alpha + \delta = 0$.

240. A substitution S, not the identity, of the group $G_{M(s)}^{s+1}$ leaves fixed at most two elements. The fixed elements are given by the equation

248) $$\gamma z^2 + (\delta - \alpha) z - \beta = 0.$$

By § 15, it has at most two roots in the field $GF[s]$ unless $\gamma = \beta = 0$, $\alpha = \delta$, when S is the identity. Now S leaves ∞ fixed only when $\infty = \alpha/\gamma$, whence $\gamma = 0$. The other fixed elements are given by $(\delta - \alpha) z - \beta = 0$, which, for $S \neq I$, is satisfied only by $z = \infty$ or $z = $ mark according as $\delta - \alpha = 0$ or $\neq 0$.

If S leaves fixed two distinct elements z_1 and z_2, it can be transformed by a suitably chosen substitution T of the group into a substitution with the fixed elements 0 and ∞, having therefore the form

$\Sigma:$ $$z' = \frac{az}{b} \qquad (ab = 1).$$

Its period is a divisor of $\frac{1}{2}(p^n - 1)$ or $p^n - 1$ according as $p > 2$ or $p = 2$.

If S leaves fixed a single element $z_1 \equiv z_2$, it can be transformed into

$$z' = z + \beta \qquad (\beta \text{ in field})$$

leaving fixed the single element ∞. Its period is therefore p. But the condition for a double root of 248) is $(\alpha + \delta)^2 = 4$.

If S leaves no element fixed, the quadratic 248) is irreducible in the $GF[p^n]$. By the corollary of § 31, its roots z_1 and z_2 are

262   CHAPTER XII.

marks of the $GF[p^{2n}]$ conjugate with respect to the $GF[p^n]$. Now S multiplies the function $(z - z_1) \div (z - z_2)$ by the constant a/b, where

$$a \equiv \alpha - \gamma z_1, \quad b \equiv \alpha - \gamma z_2.$$

The product ab reduces to $\alpha\delta - \beta\gamma = 1$. Also a and b are conjugate (§ 73). Hence

$$a^{-1} = b = a^{p^n}, \quad a^{p^n+1} = 1.$$

Hence S can be transformed into a substitution of the form Σ, whose period is a divisor of $\frac{1}{2}(p^n + 1)$ or $p^n + 1$ according as $p > 2$ or $p = 2$.

In particular, the substitutions of period p are characterized by the invariant $(\alpha + \delta)^2 = 4$.

241. *Commutative subgroups of order p^n.* The substitutions

$$S_\mu \equiv \begin{pmatrix} 1, & \mu \\ 0, & 1 \end{pmatrix}, \qquad \mu \text{ in the } GF[p^n],$$

form a commutative subgroup $G_s^{(\infty)}$ of order $s \equiv p^n$, containing all the substitutions of $G_{M(s)}$ leaving the single element ∞ fixed and containing no other substitutions. Each of its substitutions except the identity is of period p. Hence there are $(p^n - 1)/(p - 1)$ cyclic subgroups G_p of order p in the $G_s^{(\infty)}$. To determine the conjugacy of these substitutions and subgroups under $G_{M(s)}$, we transform S_μ ($\mu \neq 0$) by $V = \begin{pmatrix} \alpha, & \beta \\ \gamma, & \delta \end{pmatrix}$ and (see formula of composition at end of § 108) obtain the substitution[1])

$$V^{-1} S_\mu V = \begin{pmatrix} 1 - \alpha\gamma\mu, & \alpha^2\mu \\ -\gamma^2\mu, & 1 + \alpha\gamma\mu \end{pmatrix}.$$

This substitution belongs to $G_s^{(\infty)}$ if, and only if, $\gamma = 0$, when it becomes $S_{\alpha^2\mu}$. In particular, S_μ is transformed into itself only by the substitutions $\begin{pmatrix} 1, & \beta \\ 0, & 1 \end{pmatrix}$. Within $G_{M(s)}$ any substitution S_μ ($\mu \neq 0$) is self-conjugate in exactly the $G_s^{(\infty)}$, while the $G_s^{(\infty)}$ is self-conjugate in exactly the $G_{s(s-1)/2;1}^{(\infty)}$ composed of all the substitutions leaving the element ∞ invariant, viz., $\begin{pmatrix} \alpha, & \beta \\ 0, & \alpha^{-1} \end{pmatrix}$. As to the order of the latter group, β may be any mark of the $GF[p^n]$ and α any mark $\neq 0$; but $-\alpha, -\beta$ gives the same substitution as $+\alpha, +\beta$.

---

[1]) This order of the factors of a product is employed by Wiman, the reverse order by Moore.

Within $G_{M(s)}$, S_μ is conjugate only with the substitutions $S_{a^2\mu}$. Hence the $s-1$ substitutions, not the identity, of $G_s^{(\infty)}$ are all conjugate if $p=2$, but separate into two sets of $\frac{1}{2}(s-1)$ conjugate substitutions if $p>2$. The $p-1$ substitutions of a cyclic group G_p generated by S_μ belong half to one and half to the other set if $p>2$ and n be odd, but all belong to the same set if n be even (§ 62).

In place of ∞ the fixed element may be any one of the p^n marks of the $GF[p^n]$. Since $G_{M(s)}$ permutes the p^n+1 elements $\varkappa$ transitively, it contains p^n+1 conjugate commutative groups $G_s^{(\varkappa)}$. This result also follows from the numerical identity

$$\frac{s(s^2-1)}{2;1} \div \frac{s(s-1)}{2;1} = s+1 \equiv p^n+1.$$

Each $G_s^{(\varkappa)}$ is *defined* by any one of its substitutions not the identity as *the* group in which that substitution is self-conjugate. These p^n+1 groups have therefore no substitution in common except the identity and contain in all $p^{2n}-1$ distinct substitutions of period p.

242. *Cyclic subgroups of order* $\frac{s-1}{2;1}$. If ϱ be a primitive root of the $GF[p^n]$, the substitution

$$P = \begin{pmatrix} \varrho, & 0 \\ 0, & \varrho^{-1} \end{pmatrix}$$

generates a cyclic group of order $\frac{1}{2}(p^n-1)$ if $p>2$, but of order p^n-1 if $p=2$. It contains all the substitutions

$$\Sigma = \begin{pmatrix} a, & 0 \\ 0, & a^{-1} \end{pmatrix} \qquad (a \text{ in the } GF[p^n]).$$

Since it contains all the substitutions which leave fixed the elements ∞ and 0 and no other substitutions, it will be denoted by $G_{\frac{s-1}{2;1}}^{(\infty,0)}$. Any new substitution transforming this cyclic group into itself must interchange the elements ∞ and 0 and hence have the form

$$B = \begin{pmatrix} 0, & \beta \\ -\beta^{-1}, & 0 \end{pmatrix}.$$

Inversely, every B transforms Σ into its reciprocal Σ^{-1}. These $\frac{s-1}{2;1}$ substitutions B of period two together with the substitutions Σ form a dihedron-group[1]) $G_{2\frac{s-1}{2;1}}^{(\infty,0)}$, which is the largest subgroup of $G_{M(s)}$ within which the above cyclic group is self-conjugate.

---

1) See the definition given in § 245.

264   CHAPTER XII.

Since ∞, 0 form only one of the $\frac{1}{2}p^n(p^n+1)$ pairs of the p^n+1 elements, $G_{M(s)}$ contains exactly $\frac{1}{2}p^n(p^n+1)$ conjugate cyclic groups $G^{(\varkappa,\lambda)}_{\frac{s-1}{2;1}}$, each self-conjugate in exactly a dihedron $G^{(\varkappa,\lambda)}_{2\frac{s-1}{2;1}}$. Each of these cyclic groups is *defined* by any one of its substitutions not the identity as *the largest* cyclic group containing that substitution. These $\frac{1}{2}p^n(p^n+1)$ groups have therefore no substitution in common except the identity and contain in all $\frac{1}{4}s(s+1)(s-3)$ or $\frac{1}{2}s(s+1)(s-2)$ substitutions (not the identity) according as $p > 2$ or $p = 2$.

243. *Cyclic subgroups of order* $\frac{s+1}{2;1}$. By § 144, $LF(2, p^n)$ is holoedrically isomorphic with the group $H \equiv HO(2, p^{2n})$ of binary hyperorthogonal substitutions of determinant unity in the $GF[p^{2n}]$ when taken fractionally, viz.,

$$V = \begin{pmatrix} A, & B \\ -\overline{B}, & \overline{A} \end{pmatrix} \qquad (A\overline{A} + B\overline{B} = 1),$$

where $\overline{A} \equiv A^{p^n}$ is the conjugate of A with respect to the $GF[p^n]$. The reciprocal of V is, by § 142,

$$V^{-1} = \begin{pmatrix} \overline{A}, & -B \\ \overline{B}, & A \end{pmatrix}.$$

If J be a primitive root of $J^{p^n+1} = 1$, so that $\overline{J} = J^{-1}$, the following substitution of H,

$$Q = \begin{pmatrix} J, & 0 \\ 0, & \overline{J} \end{pmatrix}$$

generates a cyclic group $G_{\frac{s+1}{2;1}}$ composed of the substitutions

$$Q^g = \begin{pmatrix} J^g, & 0 \\ 0, & \overline{J}^g \end{pmatrix}.$$

Any substitution V of H transforms Q^g into

$$V^{-1}Q^g V = \begin{pmatrix} A\overline{A}J^g + B\overline{B}\overline{J}^g, & -AB(J^g - \overline{J}^g) \\ -\overline{A}\overline{B}(J^g - \overline{J}^g), & A\overline{A}\overline{J}^g + B\overline{B}J^g \end{pmatrix}.$$

This substitution belongs to the cyclic group generated by Q if and only if $AB = 0$. Two cases arise.

If $B = 0$, then $A\overline{A} = 1$ so that $V = \begin{pmatrix} A, & 0 \\ 0, & \overline{A} \end{pmatrix}$ belongs to the cyclic group and evidently transforms every Q^g into itself.

SUBGROUPS OF THE LINEAR FRACTIONAL GROUP $LF(2, p^n)$. 265

If $A = 0$, then $B\bar{B} = 1$, so that $V = \begin{pmatrix} 0, & B \\ -B, & 0 \end{pmatrix}$. The latter transforms Q^g into $\begin{pmatrix} \bar{J}^g, & 0 \\ 0, & J^g \end{pmatrix} \equiv Q^{-g}$, which is distinct from Q^g unless the latter be of period two.

The largest subgroup of H within which the cyclic group $G_{\frac{s+1}{2;1}}$ is self-conjugate is therefore a dihedron-group of order $2 \cdot \frac{s+1}{2;1}$. Hence H, and consequently also $G_{M(s)}$, contains

$$\frac{s(s^2-1)}{2;1} \div 2\frac{s+1}{2;1} = \frac{1}{2}s(s-1)$$

cyclic groups conjugate with $G_{\frac{s+1}{2;1}}$. Each of these is *defined* by any substitution lying in it (the identity excepted) as *the largest* cyclic group containing that substitution. The $\frac{1}{2}s(s-1)$ groups have therefore only the identity in common and contain in all $\frac{1}{4}s(s-1)^2$ or $\frac{1}{2}s^2(s-1)$ further substitutions according as $p > 2$ or $p = 2$.

244. To verify that we have now enumerated all the individual operators of $G_{M(s)}$ and consequently all the largest cyclic subgroups, we note that

$$1 + (s^2 - 1) + \frac{1}{4}s(s+1)(s-3) + \frac{1}{4}s(s-1)^2 = \frac{1}{2}s(s^2-1), \quad p > 2;$$

$$1 + (s^2 - 1) + \frac{1}{2}s(s+1)(s-2) + \frac{1}{2}s^2(s-1) = s(s^2-1), \quad p = 2.$$

It was shown that if any substitution S of a cyclic $G_{\frac{s \mp 1}{2;1}}$ be of period > 2 (viz., $S \neq S^{-1}$), then S is transformed into itself by no substitutions of $G_{M(s)}$ other than those of the cyclic $G_{\frac{s \mp 1}{2;1}}$. Hence the latter is the largest commutative subgroup of $G_{M(s)}$ which contains the substitution S. A commutative subgroup containing an operator of period > 2 and different from p is therefore a cyclic group. A commutative group containing an operator of period p contains only operators of period p (§ 241). Hence if a commutative subgroup of $G_{M(s)}$, $p > 2$, contains an operator of period > 2, it contains at most one operator of period 2.

245. *Cyclic and dihedron groups and their subgroups.* The abstract dihedron-group G_{2k} may be generated by operators A, B subject only to the generational relations

$$A^k = I, \quad B^2 = I, \quad AB = BA^{-1}.$$

From the latter two follow the relations (holding for any integer r)
$$B^{-1}A^r B = A^{-r}, \quad A^{-r}BA^r = BA^{2r}.$$

The cyclic subgroup G_k generated by A is therefore self-conjugate under G_{2k}. The latter is said to have the *cyclic base* G_k. The k operators
$$BA^i \qquad (i = 0, 1, \ldots, k-1)$$
are of period two. For k odd, they are all conjugate under G_{2k} since B transforms BA into $BA^{-1} \equiv BA^{k-1}$, which belongs to the series $B, BA^2, BA^4, \ldots$ For k even, they form two sets of conjugate operators

249) $\quad B, \quad BA^2, \quad BA^4, \ldots, BA^{k-2};$
$\quad\quad BA, \quad BA^3, \quad BA^5, \ldots, BA^{k-1}.$

According as k is odd or even, they generate cyclic groups G_2 forming one set or two sets of conjugate subgroups.

For every divisor d of k, G_k contains a single cyclic subgroup G_d, which is formed by the operators
$$A^\delta, \quad A^{2\delta}, \quad A^{3\delta}, \ldots, A^{d\delta} \equiv I \qquad (\delta \equiv k/d).$$
If μ be a given one of the integers $1, 2, \ldots, \delta$, the following d operators
$$BA^{\mu+\delta}, \quad BA^{\mu+2\delta}, \quad BA^{\mu+3\delta}, \ldots, BA^{\mu+d\delta} \equiv BA^\mu$$
extend the cyclic group G_d to the same dihedron G_{2d}. There are exactly δ such dihedron-groups. If k be odd, these G_{2d} are all conjugate under G_{2k}. If δ be odd, but k be even, the exponents μ, $\mu + \delta$, $\mu + 2\delta$, ... are alternately even and odd, so that each G_{2d} contains operators of both of the sets 249); the groups G_{2d} are therefore all conjugate under G_{2k}. If δ be even and hence k even, the exponents are all even or all odd, so that the operators all belong to a single one of the two sets 249); the groups G_{2d} thus belong to two distinct systems of conjugate subgroups of G_{2k}.

If $d > 2$, G_{2d} has a single cyclic G_d and G_{2k} a single cyclic G_k, so that the above process furnishes every dihedron subgroup G_{2d} of G_{2k}. The theorem stated below therefore follows if $d > 2$.

We consider next the case $d = 2$, k even and > 2. The only operators of period two in G_{2k} are then $A^{k/2}$ and
$$BA^i \qquad (i = 0, 1, \ldots, k-1).$$
Hence any dihedron G_4 must contain two operators BA^r, BA^s $(r \neq s)$ and therefore their product $BA^r BA^s \equiv A^{s-r}$. Hence every G_4 must contain $A^{k/2}$ and may therefore be based on the subgroup G_2 of G_k. The theorem then follows as before. The $k/2$ possible groups G_4 in G_{2k} are given by the formula
$$\{I, \quad A^{k/2}, \quad BA^r, \quad BA^{r+k/2}\} \quad \left(r = 0, 1, \ldots, \frac{k}{2}-1\right).$$

SUBGROUPS OF THE LINEAR FRACTIONAL GROUP $LF(2, p^n)$. 267

Theorem. — *For every divisor d of k the dihedron G_{2k} contains exactly k/d dihedrons G_{2d} forming one system or two systems of conjugate subgroups according as k/d is odd or even.*

246. *Cyclic and dihedron subgroups of $G_{M(s)}$ whose cyclic bases are subgroups of the cyclic $G_{\frac{s \mp 1}{2;1}}$.* By §§ 242—243, $G_{M(s)}$ contains $\frac{1}{2} s(s \pm 1)$ conjugate cyclic subgroups $G_{\frac{s \mp 1}{2;1}}$ each self-conjugate in a dihedron subgroup $G_{2\frac{s \mp 1}{2;1}}$, but self-conjugate in no larger subgroup of $G_{M(s)}$. Hence these dihedrons are all conjugate under the main group.[1]) Let $d_{\mp}$ be any divisor of $\frac{s \mp 1}{2;1}$ and denote the quotient by $\delta_{\mp}$. $G_{M(s)}$ contains $\frac{1}{2} s(s \pm 1)$ conjugate cyclic groups $G_{\delta_{\mp}}$, each of which is (§ 245) the cyclic base for $\delta_{\mp}$ dihedron subgroups $G_{2d_{\mp}}$. Under $G_{2\frac{s \mp 1}{2;1}}$ they form one system or two systems of conjugate subgroups according as $\delta_{\mp}$ is odd or even.

For $d_{\mp} > 2$, two subgroups $G_{2d_{\mp}}$ of $G_{2\frac{s \mp 1}{2;1}}$ are conjugate within the latter if conjugate within $G_{M(s)}$; indeed, the transforming substitution must be commutative with $G_{d_{\mp}}$, the only cyclic group of order $d_{\mp}$ in either $G_{2d_{\mp}}$, and therefore commutative with the cyclic $G_{\frac{s \mp 1}{2;1}}$ determined by it. *Hence if $d_{\mp}$ be any divisor > 2 of $\frac{s \mp 1}{2;1}$ and the quotient be $\delta_{\mp}$, $G_{M(s)}$ contains in all $M(s)/2d_{\mp}$ dihedron $G_{2d_{\mp}}$ forming one system or two systems of conjugate groups according as $\delta_{\mp}$ is odd or even. In the former case, a $G_{2d_{\mp}}$ is self-conjugate only under itself; in the latter case, self-conjugate under a dihedron $G_{2 \cdot 2d_{\mp}}$. These $G_{2d_{\mp}}$ are all conjugate within $G_{(2;1)M(s)}$.*

For $d_{\mp} = 2$, we have $p > 2$ since $s - 1$ is not divisible by 2 for $p = 2$. Then $s \equiv p^n$ is of the form $4h \pm 1$ according as the Jacobi-Legendre symbol $\left(\frac{-1}{s}\right)$ is ± 1; hence $\frac{1}{2}\left[s - \left(\frac{-1}{s}\right)\right]$ is even, say $= 2\sigma$. Then all the substitutions V_2 of period two of $G_{M(s)}$ belong to the conjugate cyclic $G_{2\sigma}$. It remains to study the *four-groups* G_4, each a dihedron $G_{2 \cdot 2}$ containing three cyclic G_2. Now $G_{M(s)}$ contains $\frac{1}{2} s\left[s + \left(\frac{-1}{s}\right)\right]$ conjugate cyclic G_2. Each G_2 lies in $\frac{1}{4}\left[s - \left(\frac{-1}{s}\right)\right]$ four-groups G_4. Hence, if $p > 2$, $G_{M(s)}$ contains in

---

[1]) For every operator commutative with a group G is transformed into an operator commutative with G' by the operator which transforms G into G'.

268   CHAPTER XII.

all $M(s)/12$ *four-groups*. Also the σ four-groups contained in a dihedron $G_{4\sigma}$ form (under the latter) one system or two systems of conjugate subgroups according as σ is odd or even, viz., according as $s \equiv p^n$ has the form $8h \pm 3$ or $8h \pm 1$. Since the $G_{4\sigma}$ are all conjugate within $G_{M(s)}$, it follows, for σ odd, that all the four-groups of $G_{M(s)}$ are conjugate; while, for σ even, they form *at most* two systems of conjugate subgroups under $G_{M(s)}$. For σ even, each G_4 is one of $\sigma/2$ conjugate subgroups of a certain $G_{4\sigma}$ and is therefore self-conjugate under a subgroup of order 8 of $G_{4\sigma}$. Suppose that, for σ even, the subgroups G_4 of $G_{M(s)}$ form a single system of conjugate subgroups. Then each G_4 would be one of $M(s)/12$ conjugate subgroups and consequently commutative with exactly the 12 operators of a subgroup G_{12}. By an earlier remark, the G_4 is commutative with a subgroup G_8. Since 8 is not a divisor of 12, our hypothesis is untenable. Hence, for σ even, the G_4 form exactly two systems of conjugate subgroups of $G_{M(s)}$. *For*[1]) $p > 2$, *the $M(s)/12$ four-groups G_4 contained in $G_{M(s)}$ form one system or two systems of conjugate subgroups according as $s \equiv p^n$ has the form $8h \pm 3$ or $8h \pm 1$. In the former case, a G_4 is self-conjugate under a G_{12}; in the latter case, under a G_{24}. In the $G_{2M(s)}$ the G_4 form a single system of conjugate subgroups and each is self-conjugate under a G_{24}.* Each G_{12} is not a commutative group by § 244 and so is of the tetrahedral type (§ 247). Likewise, each G_{24} contains a tetrahedral subgroup G_{12}. The latter is of index 2 and consequently self-conjugate under G_{24}. Since G_{12} contains a set of 4 conjugate G_3, the G_{24} will contain a complete system of 4 conjugate G_3. Each is self-conjugate under a G_6, which is a dihedron since it is not commutative (§ 244). Finally, no operator of period 2 is self-conjugate under G_{24}; for it is self-conjugate only under a dihedron $G_{s \mp 1}$ which contains no tetrahedral subgroup and hence none of the present G_{24}. Then by § 248 each G_{24} is an octahedral group.

247. *A non-commutative group of order 12 having a self-conjugate four-group is of the tetrahedral type.*

Let the operators of the four-group be I, V_2, V_2', V_2'', so that they are commutative and the product of any two V's gives the third V. The G_{12} contains at least one operator V_3 of period 3. The products
$$V_3^i, \quad V_2 V_3^i, \quad V_2' V_3^i, \quad V_2'' V_3^i \qquad (i = 0, 1, 2)$$
are all distinct and so give all the operators of G_{12}. The G_{12} would be a commutative group if V_3 were commutative with V_2, V_2', V_2''.

---

1) For $p = 2$, the four-groups are determined in § 249. There are $\frac{1}{6}(2^n - 1)(2^n - 2)$ sets.

SUBGROUPS OF THE LINEAR FRACTIONAL GROUP $LF(2, p^n)$. 269

Since therefore V_3 does not transform each V into itself and since it does not permute two of them, its period being $\neq 2$, it must permute them in a cycle. Fixing the notation, we thus have

$$V_3^{-1} V_2 V_3 = V_2', \quad V_3^{-1} V_2' V_3 = V_2'' \quad V_3^{-1} V_2'' V_3 = V_2,$$
$$(V_3 V_2)^3 \equiv V_3 V_2 V_3^{-1} \cdot V_3^{-1} V_2 V_3 \cdot V_2 = V_2'' V_2' V_2 = I.$$

Hence V_3, V_2 generate G_{12} and satisfy the generational relations

$$V_3^3 = I, \quad V_2^2 = I, \quad (V_3 V_2)^3 = I.$$

of the tetrahedral group, an abstract group of order 12 holoedrically isomorphic with the alternating group on 4 letters (§ 265).

248. *A group of order 24 having no self-conjugate operator of period 2 and having a set of 4 conjugate G_3 each self-conjugate in a dihedron G_6 is of the octahedral type.*

The 4 conjugate G_3 are transformed into each other by the operators of G_{24}. Hence G_{24} is isomorphic with a substitution-group on 4 letters. The isomorphism will be holoedric and consequently the latter the symmetric group $G_{24}^{(4)}$, if the identity be the only operator of G_{24} which transforms each G_3 into itself, i. e., if the four G_6 have only the identity in common. But if a substitution of period 3 were common to the dihedron G_6, it would be common to the G_3, and these would be identical contrary to hypothesis. If the G_6 contain in common two substitutions of period 2, they would contain in common the product of the two which is a substitution not the identity of the cyclic bases G_3 (§ 245). Finally, if the conjugate G_6 contain in common a single substitution of period 2, it would be self-conjugate under G_{24} contrary to hypothesis. Now the $G_{24}^{(4)}$ is of the octahedral type

249. *Subgroups of the $s+1$ commutative $G_s^{(\varkappa)}$.* Since these groups are all conjugate under $G_{M(s)}$, it suffices to determine the subgroups of $G_s^{(\infty)}$ formed of the commutative substitutions S_μ of period p. If a subgroup contain $S_{\mu_1}, S_{\mu_2}, \ldots, S_{\mu_t}$, it will contain S_μ, where $\mu = c_1 \mu_1 + c_2 \mu_2 + \cdots + c_t \mu_t$, the c_i running independently through the series $0, 1, \ldots, p-1$. Hence to every subgroup G_{p^m} of order $p^m \leq p^n$, there corresponds an additive-group in the $GF[p^n]$ of rank m with respect to the $GF[p]$ and inversely. Hence, by § 69, the number of distinct subgroups G_{p^m} of G_{p^n} is

$$\frac{(p^n-1)(p^n-p)(p^n-p^2)\ldots(p^n-p^{m-1})}{(p^m-1)(p^m-p)(p^m-p^2)\ldots(p^m-p^{m-1})}.$$

Let G_{p^m} be one such group composed of the substitutions S_λ, where λ ranges over an additive-group $[\lambda_1, \ldots, \lambda_m]$ of rank m with

270 CHAPTER XII.

respect to the $GF[p]$. By § 241, G_{p^m} is transformed into itself only by substitutions of the form $V = \begin{pmatrix} \alpha, & \beta \\ 0, & \alpha^{-1} \end{pmatrix}$. Since V transforms S_λ into $S_{\alpha^2 \lambda}$, a further condition is that $\alpha^2 \lambda$ and λ should run simultaneously through the series of marks of the $[\lambda_1, \ldots, \lambda_m]$. Suppose that there are in the $GF[p^n]$ exactly e marks $\varepsilon_1, \ldots, \varepsilon_e$ such that $[\lambda_1, \ldots, \lambda_m] = [\varepsilon_1^2 \lambda_1, \ldots, \varepsilon_i^2 \lambda_m]$. Then, according as $p > 2$; $p = 2$, the $\dfrac{se}{2;1}$ substitutions

$$V = \begin{pmatrix} \varepsilon_i, & \varepsilon_i \beta \\ 0, & \varepsilon_i^{-1} \end{pmatrix} \equiv \begin{pmatrix} 1, & \beta \\ 0, & 1 \end{pmatrix} \begin{pmatrix} \varepsilon_i, & 0 \\ 0, & \varepsilon_i^{-1} \end{pmatrix},$$

where β ranges over the $GF[s]$, constitute the largest subgroup H of $G_{M(s)}$ under which G_{p^m} is self-conjugate. But the multipliers $\varkappa$ of the additive-group $[\lambda_1, \ldots, \lambda_m]$ are (§ 70) the marks $\varkappa \neq 0$ of the multiplier $GF[p^k]$, k being a divisor of m and n. It remains to distinguish which of them are squares of marks ε_i of the $GF[p^n]$. For the respective cases

$$p > 2 \text{ with } n/k \text{ even}, \quad p > 2 \text{ with } n/k \text{ odd}; \quad p = 2,$$

there are (§ 62) exactly $e = (2, 1; 1)(p^k - 1)$ marks ε_i, so that H is of order $\dfrac{se}{2;1} = \dfrac{s(p^k-1)}{1,2;1}$. Hence G_{p^m} is one of a system of

$$(p^{2n} - 1) \div (2, 1; 1)(p^k - 1)$$

conjugate subgroups of $G_{M(s)}$. Here the value of k depends on the individual G_{p^m} chosen. Given k, the number of the corresponding sets of G_{p^m} follows from § 71.

250. *Non-commutative subgroups of the $s + 1$ conjugate $G^{(\varkappa)}_{\frac{s(s-1)}{2;1}}$.* It suffices to study the group G given by $\varkappa = \infty$. It is composed of the substitutions

$$\begin{pmatrix} \alpha, & \beta \\ 0, & \alpha^{-1} \end{pmatrix} = \begin{pmatrix} 1, & \mu \\ 0, & 1 \end{pmatrix} \begin{pmatrix} \alpha, & 0 \\ 0, & \alpha^{-1} \end{pmatrix}, \qquad \beta \equiv \alpha \mu.$$

For a given mark $\alpha \neq 0$, μ and β run simultaneously through the series of marks of the $GF[s]$. A rectangular array of the substitutions of G may be formed by taking as the first row the substitutions S_μ, which form the self-conjugate subgroup $G_s^{(\infty)}$, and as right-hand multipliers the substitutions $P_\alpha \equiv \begin{pmatrix} \alpha, & 0 \\ 0, & \alpha^{-1} \end{pmatrix}$ of the cyclic $G^{(\infty, 0)}_{\frac{s-1}{2;1}}$. In any subgroup G' of G the totality of substitutions of period p give rise to a commutative group G_{p^m} of substitutions S_λ, where λ ranges over an additive-group $[\lambda_1, \ldots, \lambda_m]$. Hence G_{p^m} is self-conjugate under G'. A rectangular array of the substitutions of G'

with those of G_{p^m} in the first row has the property that the substitutions in each row are all found in a corresponding row of the rectangular array for G. In fact, two operators A, B of G' lie in the same or in different rows of the array for G' according as AB^{-1} is or is not in G_{p^m}. But AB^{-1} belongs to G' and hence belongs to G_{p^m} if, and only if, it occurs among the substitutions in the first row of the array for G. Hence each row for G' lies wholly in a row for G. The quotient-group G'/G_{p^m} is therefore a subgroup G_{d-} of the quotient-group $G/G_s^{(\infty)}$, the latter being a cyclic $G_{\frac{s-1}{2;1}}$. Indeed, these quotient-groups may be obtained concretely as groups of the permutations of the rows of G induced by applying as right-hand multipliers the substitutions of G or G'. But all the substitutions in the same row of G (and, à fortiori, all in the same row of G') give rise to the same permutation. Hence G_{d-} is an abstract cyclic group. Now G contains s cyclic $G_{s-1}^{(\infty, \tau)}$, where τ runs through the series of marks of the $GF[s]$, all conjugate under the transformers S_μ. Leaving different elements τ fixed, they have no substitution other than the identity in common. Counting also the $s-1$ substitutions of period p, we have accounted for all the substitutions of G. Besides the cyclic subgroups of $G_s^{(\infty)}$, G therefore contains no cyclic subgroups other than the $G_d^{(\infty, \tau)}$, for the various divisors d of $\frac{s-1}{2;1}$. Among these cyclic groups occurs one whose substitutions may be chosen as the right-hand multipliers in forming the above array for G'. In fact, within the row of G' corresponding to the generator of the quotient cyclic G_d there must exist a substitution A such that A^d, and no lower power, belongs to the group G_{p^m} whose substitutions form the first row. The right-hand multipliers for the array may thus be chosen to be $I, A, A^2, \ldots, A^{d-1}$. Hence G' is given by the extension of the G_{p^m} by a certain $G_d^{(\infty, \tau)}$, within which G_{p^m} is self-conjugate. But the largest subgroup of $G_{M(s)}$ within which G_{p^m} is self-conjugate is (§ 249) the group H of order sK, $K = \frac{p^k-1}{1, 2; 1}$, given by the extension of $G_s^{(\infty)}$ by a cyclic $G_K^{(\infty, 0)}$. In particular, d must be a divisor of K, so that d depends upon the G_{p^m}. The cyclic $G_K^{(\infty, 0)}$ contains a single cyclic $G_d^{(\infty, 0)}$. Hence, by transforming G' by a suitably chosen S_μ, we obtain a group $G'_{p^m d}$ (conjugate with G' under G) given by the extension of G_{p^m} by the subgroup $G_d^{(\infty, 0)}$ of $G_K^{(\infty, 0)}$. The substitutions S_λ of G_{p^m} transform that subgroup into p^m conjugate cyclic $G_d^{(\infty, 0)}$, since S_λ replaces the fixed elements ∞, 0 by elements ∞, λ. These p^m groups together with G_{p^m} contain all the substitutions of $G'_{p^m d}$, as shown by simple

272   CHAPTER XII.

enumeration. The largest subgroup of $G_{M(s)}$ transforming $G'_{p^m d}$ into itself must therefore transform G_{p^m} into itself (and thus be a subgroup of H) and transform the groups of the single set of conjugate $G_d^{(\infty, \lambda)}$ amongst themselves. Of substitutions of period p, it must therefore contain only the S_λ. The required group is thus a subgroup of the group H' of order $p^m K$ given by the extension of G_{p^m} by $G_K^{(\infty, 0)}$. Moreover, it is H' itself since any substitution $\begin{pmatrix} \alpha, & 0 \\ 0, & \alpha^{-1} \end{pmatrix}$ of $G_K^{(\infty, 0)}$, α such that $\alpha^2 \lambda = \lambda'$ is of the $[\lambda_1, \ldots, \lambda_m]$, replaces the elements ∞, λ by elements ∞, λ' and consequently transforms $G_d^{(\infty, \lambda)}$ into $G_d^{(\infty, \lambda')}$. Hence the group $G'_{p^m d}$ is one of a system of

$$\frac{(p^n+1)(p^n-1)p^{n-m}}{(2, 1; 1)(p^k-1)}$$

conjugate groups. Finally, if the subgroup G' contains no substitution of period p, it is a cyclic subgroup $G_d^{(\infty, \varkappa)}$ of one of the cyclic $G_{\frac{s-1}{2;1}}^{(\infty, \varkappa)}$.

251. *Subgroups of $G_{M(s)}$ containing operators of period p.* — The substitutions of period p of a subgroup G_Ω of the $G_{M(s)}$ distribute themselves over certain $s+1$ subgroups $G_{p^{m_\mu}}^{(\mu)}$ of the $s+1$ conjugate $G_s^{(\mu)}$ (§ 241). By hypothesis at least one of the orders p^{m_μ} is > 1. By suitable transformation within $G_{M(s)}$, we arrange it so that $p^m > 1$, $m \equiv m_\infty > 0$. Under the p^m transformers S_β of the $G_{p^m}^{(\infty)}$, the remaining $G_{p^{m_\mu}}^{(\mu)}$ with $m_\mu > 0$ ($\mu \neq \infty$), if any, arrange themselves in sets each consisting of p^m conjugate groups. Under the G_Ω the $G_{p^m}^{(\infty)}$ is then one of a set of $1 + fp^m$ conjugate groups, f being a positive integer or zero. The G_Ω contains no group $G_{p^{m_\mu}}^{(\mu)}$ ($m_\mu > 0$) other than the $1 + fp^m$ groups of this set. For, any such group would be one of a set of n_μ conjugate groups, where n_μ would necessarily have at the same time the forms $1 + f_\mu p^{m_\mu}$ and $f'_\mu p^m$. Hence: *Every G_Ω which contains operators of period p contains these operators in $1 + fp^m$ groups $G_{p^m}^{(\mu)}$ conjugate under G_Ω, where for each G_Ω, f and m are properly determined integers $f \gtreqless 0$, $m > 0$.*

The groups G_Ω with $f = 0$ have been enumerated in §§ 249—250. Consider the group G_Ω with $f \gtreqless 1$, $m > 0$. It contains $1 + fp^m$ groups conjugate with a certain $G_{p^m}^{(\infty)}$ formed of the substitutions S_λ, where λ ranges over the an additive-group $[\lambda_1, \ldots, \lambda_m]$. The $G_{p^m}^{(\infty)}$ is (§ 250) self-conjugate within G_Ω under a certain largest subgroup $G'_{p^m d}$. Hence

250) $\Omega = (1 + fp^m)p^m d.$

As in § 250, we transform[1]) by a suitable S_μ and obtain a $G_{p^m d}^{(\infty)}$ given by the extension of the group G_{p^m} of the S_λ by the cyclic group $G_d^{(\infty,\,0)}$ of the substitutions P_η contained within the cyclic group $G_{e/(2;\,1)}^{(\infty,\,0)}$ of substitutions P_{ε_i}. The group $G_{p^m d}^{(\infty)}$ is thus composed of the substitutions[2])

$$V_{\eta,\,\lambda} \equiv \begin{pmatrix} \eta, & \eta\lambda \\ 0, & \eta^{-1} \end{pmatrix} \equiv \begin{pmatrix} 1, & \lambda \\ 0, & 1 \end{pmatrix} \begin{pmatrix} \eta, & 0 \\ 0, & \eta^{-1} \end{pmatrix} = S_\lambda P_\eta.$$

Since $-\eta$ and $+\eta$ lead to the same P_η, there are $(2;1)d$ marks η, the distinct powers of a primitive root of $\eta_0^d = -1$. Since each η is an ε_i, each η^2 is a multiplier of the additive-group $[\lambda_1, \ldots, \lambda_m]$.

To normalize G_Ω we transform by P_σ:

$$P_\sigma^{-1} \begin{pmatrix} \alpha, & \beta \\ \gamma, & \delta \end{pmatrix} P_\sigma = \begin{pmatrix} \alpha, & \sigma^2\beta \\ \sigma^{-2}\gamma, & \delta \end{pmatrix}, \quad P_\sigma^{-1} V_{\eta,\,\lambda} P_\sigma = V_{\eta,\,\sigma^2\lambda}.$$

Taking $\sigma = \sqrt{\lambda_0^{-1}}$, the transformer P_σ is a substitution $z' = \lambda_0^{-1} z$ of the $G_{(2;1)M(e)}$; $G_{M(s)}$ is transformed into itself and G_Ω into G'_Ω. The new additive-group $[\lambda'_1, \ldots, \lambda'_m]$ contains the mark $\sigma^2 \lambda_0 \equiv 1$ and hence all the marks $\neq 0$ of its multiplier $GF[p^k]$. We suppose this transformation to have been made and the primes dropped from G', λ'_i.

The G_Ω of order 250) is obtained by extending the $G_{p^m d}^{(\infty)}$, formed of the $V_{\eta,\,\lambda}$, by certain fp^m extenders V_j ($j = 1, \ldots, fp^m$),

$$V_j = \begin{pmatrix} \alpha_j, & \beta_j \\ \gamma_j, & \delta_j \end{pmatrix}, \quad V_{\eta,\,\lambda} V_j = \begin{pmatrix} \alpha_j\eta, & \alpha_j\eta\lambda + \beta_j\eta^{-1} \\ \gamma_j\eta, & \gamma_j\eta\lambda + \delta_j\eta^{-1} \end{pmatrix} \quad (\gamma_j \neq 0).$$

It was shown above that G_Ω contains $(1 + fp^m)(p^m - 1)$ substitutions of period p. Of these $p^m - 1$ are the S_λ lying in $G_{p^m}^{(\infty)}$. The remaining $fp^m(p^m - 1)$ are substitutions $V_{\eta,\,\lambda} V_j$ satisfying the necessary and sufficient conditions for period p (§ 240),

251) $$\alpha_j \eta + \delta_j \eta^{-1} + \gamma_j \eta \lambda = \pm 2.$$

Given V_j ($\gamma_j \neq 0$) and η ($\eta \neq 0$), there are at most $(2;1)$ values λ satisfying 251). For a given V_j ($\gamma_j \neq 0$) there are consequently at most $(2;1)d$ substitutions $V_{\eta,\,\lambda} \equiv V_{-\eta,\,\lambda}$ such that $V_{\eta,\,\lambda} V_j$ is of period p. Hence the various V_j lead to at most $fp^m(2;1)d$ such substitutions.

---

[1]) For $p^k = 3$, n/k odd, we have $d = 1$, so that this transformation is here unnecessary and is reserved for use in § 252.

[2]) The non-fractional substitutions (viz., with $\gamma = 0$) of G_Ω are all of the form $V_{\eta,\,\lambda}$. Indeed, they form a group G' leaving the element ∞ invariant. Its substitutions of period p form the subgroup $G_{p^m}^{(\infty)}$ which must be self-conjugate under G'. Hence $G' \equiv G_{p^m d}^{(\infty)}$.

274   CHAPTER XII.

Comparing this maximum with the actual number $fp^m(p^m-1)$, we have $p^m-1 \overline{\lessgtr} (2;1)d$. Since each of the corresponding $(2;1)d$ marks η must be one of the e marks ε_i' of § 249, then $(2;1)d \overline{\lessgtr} e$, where $e = (2,1;1)(p^k-1)$ Finally (§ 70), k is a divisor of m. Hence

252) $p^m - 1 \overline{\lessgtr} (2;1)d \overline{\lessgtr} (2,1;1)(p^k-1) \overline{\lessgtr} (2,1;1)(p^m-1)$.

Since the third number is always $\overline{\lessgtr} 2(p^k-1) < 2p^k-1$, we have $p^m < 2p^k$, so that $m = k$, m being a divisor of k. The additive-group $[\lambda_1, \ldots, \lambda_m]$ is therefore its own multiplier $GF[p^k]$ and every λ is zero or a multiplier $\varkappa$.

There are in all two cases:

[A] $m = k$, $p^k - 1 = (2;1)d$, $\Omega = (1+fp^k)p^k(p^k-1)/(2;1)$
[B] $m = k$, $p > 2$, n/k even, $p^k - 1 = d$, $\Omega = (1+fp^k)p^k(p^k-1)$,

where for $(2;1)$ we read 2 or 1 according as $p > 2$ with n/k odd or $p = 2$.

The following lemma finds repeated application below:

If V_j ($\gamma_j \neq 0$) be of period 2, the ratio α_j/γ_j differs from the α_i/γ_i of every other V_i and so is a characteristic invariant of the $V_{\eta,\lambda}V_j$.

For $i \neq j$, $V_i V_j$ is not of the form $V_{\eta,\lambda}$, since otherwise

$$V_i = V_{\eta,\lambda}V_j,$$

contrary to the choice of the extenders V_i. Hence in $V_i V_j$ the term corresponding to γ is $\neq 0$, viz., $\alpha_i\gamma_j + \gamma_i\delta_j \neq 0$. Dividing by $\gamma_i\gamma_j$ and applying $\delta_j = -\alpha_j$ (V_j being of period 2) we find that

$$\alpha_i/\gamma_i - \alpha_j/\gamma_j \neq 0.$$

252. *For case* [A] *with* $p^k > 2$, *the group* G_Ω *is the group* $G_{M(p^k)}$ *of all linear fractional substitutions of determinant unity in the* $GF[p^k]$. *For* $p^k = 2$, G_Ω *is a dihedron* $G_{2(1+2f)}$, *which for* $f = 1$ *is the* $G_{M(2)}$.

For $p^k > 2$, it is shown that every V_j may be chosen so that $\alpha_j, \beta_j, \gamma_j, \delta_j$ all belong to the $GF[p^k]$. Hence G_Ω is a subgroup of $G_{M(p^k)}$. But, if $f > 1$, $\Omega > M(p^k)$. Hence must $f = 1$, $\Omega = M(p^k)$, so that $G_\Omega \equiv G_{M(p^k)}$.

For case [A], relations 252) become equalities, so that the earlier argument shows that, for V_j and η given ($\gamma_j \neq 0$, $\eta \neq 0$), there exist *exactly* $(2;1)$ marks λ of the $[\lambda_1, \ldots, \lambda_m]$ which satisfy 251). The given η may be any one of the multipliers $\varkappa$, since the number $(2;1)d$ of η's equals the number p^k-1 of $\varkappa$'s.

The extender V_j may be replaced by any one of the products $V_{\eta,\lambda}V_j$ and in particular by one of period p, having therefore $\alpha_j + \delta_j = \pm 2$. Changing if necessary the signs of all four coefficients of V_j, we may take $\alpha_j + \delta_j = 2$. With this normalized V_j, the condition 251) becomes (upon setting $\eta = \varkappa$)

253) $\qquad \alpha_j(\varkappa - \varkappa^{-1}) + \gamma_j \varkappa \lambda = \pm 2 - 2\varkappa^{-1}.$

For any given j and any given mark $\varkappa \neq 0$ of the $GF[p^k]$ and for each sign $\pm$, this equation must determine a mark $\lambda \equiv \lambda_{j,\varkappa,\pm}$ of the $GF[p^k]$. If $p > 2$, 253) for $\varkappa = 1$ gives

$$\gamma_j \lambda_{j,1,-} = -4 \neq 0,$$

so that γ_j belongs to the $GF[p^k]$. For $p^k > 3$, $\varkappa$ has a value different from ± 1 and from zero; for such a $\varkappa$ 253) requires that α_j belong to the $GF[p^k]$. Then $\delta_j = 2 - \alpha_j$ belongs to the field. The determinant being unity, β_j also belongs to the field.

For $p^k = 3$, the non-vanishing marks γ_j, η may be restricted to the value $+1$. Since $\alpha_j + \delta_j = 2$ in V_j, the $\alpha + \delta$ of $V_{1,1}V_j \equiv V_j'$ has the value $\alpha_j + \delta_j + 1 = 0$ in the field. Hence V_j' takes the form

$$\begin{pmatrix} \alpha, & -1-\alpha^2 \\ 1, & -\alpha \end{pmatrix} \equiv W_\alpha.$$

The W_α may be taken as extenders in place of the V_j. The subgroup $G_{p^m d}$ is here composed of three substitutions $V_{1,\lambda}$, $\lambda = 0, \pm 1$. Hence every substitution of G_Ω has as its γ and $\alpha + \delta$ marks of the $GF[p^k]$. Transforming the group by $S_{-\alpha_0}$, where α_0 is a particular α, each $V_{\eta,\lambda} \equiv S_\lambda$ is transformed into itself and each W_α into $W_{\alpha - \alpha_0}$. Hence, in the transformed group each γ and $\alpha + \delta$ belong to the $GF[p^k]$. Among the new extenders $W_a \equiv W_{\alpha - \alpha_0}$ occurs W_0. Hence G_Ω contains

$$W_0 \equiv \begin{pmatrix} 0, & -1 \\ 1, & 0 \end{pmatrix}, \quad W_a W_0 = \begin{pmatrix} -1, & a \\ a, & -1-a^2 \end{pmatrix},$$

so that the mark a, being in the position of a γ, belongs to the $GF[p^k]$.

For $p = 2$, $k > 1$, there exist marks $\varkappa$ different from 0 and 1 $(+1 = -1)$; for such a $\varkappa$, 253) shows that α_j/γ_j is a mark λ_j of the $GF[2^k]$. Since $p = 2$, $\alpha_j + \delta_j = 2$ gives $\alpha_j = \delta_j$, and $\delta_j/\gamma_j = \lambda_j$. There are fp^k substitutions V_j and $fp^k > 2$. The product $V_i V_j$ $(i \neq j)$ belongs to G_Ω and is not of the form $V_{\varkappa,\lambda}$ since $V_i \neq V_j V_{\varkappa,\lambda}$ and since V_j is of period 2. Hence we may set $V_i V_j = V_{\varkappa,\lambda} V_l$. Since $i \neq j$, $\lambda_i + \lambda_j \neq 0$ (end of § 251). We find that

$$\lambda_l = \frac{\alpha_i \alpha_j + \gamma_i \beta_j}{\alpha_i \gamma_j + \gamma_i \alpha_j} = \frac{\lambda_i \lambda_j}{\lambda_i + \lambda_j} + \frac{1}{\lambda_i + \lambda_j} \frac{\beta_j}{\gamma_j}.$$

Hence every β_j/γ_j belongs to the $GF[p^k]$. Then $\alpha_j \delta_j - \beta_j \gamma_j = 1$ requires that γ_j^2 belong to that field and hence also γ_j, p being 2. Then α_j, β_j, δ_j belong to the field since their ratios to γ_j do.

276 CHAPTER XII.

For $p=2$, $k=1$, the group G_Ω of order $2(1+2f)$ is given by the extension of $G_2^{(\infty)}$, formed of the substitutions $S_0 \equiv I$ and S_1, by certain $2f$ extenders V_j $(j=1,\ldots,2f)$ each of period 2. By § 251, all the substitutions of period 2 in G_Ω form one set of $1+2f$ conjugate substitutions. Setting $V_0 \equiv S_1$, the substitutions of period 2 in G_Ω are V_j $(j=0,1,\ldots,2f)$ and the remaining substitutions $V_0 V_j \equiv U_j$ are of period $\neq 2$. Hence no U is conjugate with a V. The product $V_{j'} U_j$ cannot be a U; for the substitution of G_Ω which transforms $V_{j'}$ into V_0 transforms the product into $V_0 U_i \equiv V_i$, but transforms the U into some U. Hence $V_{j'} U_j$ is of the form $V_{j''}$ so that $V_0 U_{j'} U_j = V_0 U_{j''}$. Hence every product $U_{j'} U_j$ is a $U_{j''}$. *The substitutions U form a group $G_{\frac{1}{2}\Omega}$.* Since $U_j = V_0 V_j$, we have, for every j,

254) $$V_0^{-1} U_j V_0 = U_j^{-1}$$

For U_j and $U_{j'}$ arbitrary, there exists in the $G_{\frac{1}{2}\Omega}$ a U_i such that

$$U_j U_{j'} = (U_{j'}^{-1} U_j^{-1})^{-1} = (U_i)^{-1} = V_0^{-1} U_i V_0 = V_0^{-1} (U_{j'}^{-1} U_j^{-1}) V_0 = U_{j'} U_j.$$

The group G_{1+2f} of the U's is therefore commutative and contains substitutions of period >2. By § 244, it is a cyclic subgroup of $G_{\ast \mp 1}$. In view of 254) the group G_Ω is a dihedron $G_{2(1+2f)}$ based on the cyclic G_{1+2f} (§ 245). These groups G_Ω have therefore been enumerated in § 246 and may be dropped from further consideration.

253. For case [B], $p>2$, n/k is even and $p^k-1=d$. The $2d$ marks η are the square roots of the p^k-1 marks $\varkappa$ and hence are the distinct powers of $\eta_0 = \sqrt{\varkappa_0}$, where $\varkappa_0$ is a primitive root of the $GF[p^k]$. In particular, there is a mark $\eta = \sqrt{-1}$.

Within G_Ω there are exactly $1+fp^k$ groups conjugate with the $G_{p^k(p^k-1)}^{(\infty)}$. The latter contains p^k conjugate cyclic $G_{p^k-1}^{(\infty,\lambda)}$ and hence in all p^k substitutions T of period 2, each conjugate with

$$T_0 = \begin{pmatrix} \sqrt{-1}, & 0 \\ 0, & -\sqrt{-1} \end{pmatrix}.$$

Under G_Ω of order $\Omega = (1+fp^k)p^k(p^k-1)$, this T_0 is one of a system of $(1+fp^k)p^k$ or $\frac{1}{2}(1+fp^k)p^k$ conjugate substitutions T according as T_0 is within G_Ω self-conjugate under the cyclic $G_{p^k-1}^{(\infty,0)}$ or under a dihedron obtained by extending the former by a substitution T_0' which interchanges the elements ∞, 0 (§§ 242, 246). In the respective cases there would be at least fp^{2k} or $\frac{1}{2}(fp^k-1)p^k$ substitutions $V_{\eta,\lambda} V_j$ $(j>0)$ of period 2, necessarily satisfying the relation 251),

$$\alpha_j \eta + \delta_j \eta^{-1} + \gamma_j \eta \lambda = 0.$$

For each of the fp^k extenders V_j ($j > 0$, $\gamma_j \neq 0$), each value of η gives a single value of λ, which may or may not belong to the $GF[p^k]$. Hence there are at most $fp^k(p^k-1)$ substitutions $V_{\eta,\lambda}V_j$ of period 2. The second alternative therefore holds, so that G_Ω contains a substitution of the form

$$T_0' = \begin{pmatrix} 0, & -\tau^{-1} \\ \tau, & 0 \end{pmatrix}.$$

Also $\frac{1}{2}(1 + fp^k)$ is an integer so that f is odd.

In case a V_j ($j > 0$) gives rise to one or more substitutions $T = V_{\eta,\lambda}V_j$, we replace V_j by one such T, so that the new V_j has $\alpha_j + \delta_j = 0$. Let N denote the number of these V_j for which there exists a product $V_{\eta,\lambda}V_j$ distinct from V_j and of period 2. For such a V_j the equation

$$\alpha_j(\eta - \eta^{-1}) + \gamma_j\eta\lambda = 0 \qquad (\eta \neq 0)$$

will be satisfied by a pair η, $\lambda \neq \pm 1, 0$, such that η^2 and λ belong to the $GF[p^k]$. Hence will

$$\alpha_j/\gamma_j = \eta^2\lambda/(1 - \eta^2)$$

belong to that field. Inversely, if α_j/γ_j belong to that field, and η^2 be an arbitrary mark $\neq 0$ of that field, there exists an unique solution λ in the field, so that there will be $p^k - 1$ substitutions $V_{\eta,\lambda}V_j$ of period 2. By the lemma at the end of § 251, the N substitutions V_j have distinct values for α_j/γ_j, here shown to belong to the $GF[p^k]$. Hence $N \leqq p^k$. Let M denote the number of the V_j leading to a single $V_{\eta,\lambda}V_j$ of period 2. Then $M \leqq fp^k - N$. The total number of the $V_{\eta,\lambda}V_j$ ($j > 0$) of period 2 is therefore

$$N(p^k - 1) + M \leqq N(p^k - 1) + fp^k - N.$$

The second member is greatest when N has its maximum value p^k. By comparing the minimum and maximum numbers for the

$$V_{\eta,\lambda}V_j \qquad (j > 0)$$

of period 2 in G_Ω, we have

255) $\qquad \frac{1}{2}(fp^k - 1)p^k \leqq p^k(p^k - 1) + (f - 1)p^k.$

Hence must $f = 1$ or 3, leading to the two cases:

$(f = 1)\quad p > 2, \quad n/k$ even, $\quad \Omega = (p^k + 1)p^k(p^k - 1) \equiv 2M(p^k)$
$(f = 3)\quad p = 3, \quad k = 1, \quad n$ even, $\quad \Omega = 60.$

Consider first the case $f = 1$. G_Ω contains the transformed of T_0' by S_λ,

$$T_\lambda' = \begin{pmatrix} \lambda\tau, & -\lambda^2\tau - \tau^{-1} \\ \tau, & -\lambda\tau \end{pmatrix}.$$

278   CHAPTER XII.

Letting λ run through the series of marks of the $GF[p^k]$, the ratio $\alpha/\gamma \equiv \lambda\tau/\tau \equiv \lambda$ takes p^n distinct values. By the lemma at the end of § 251, the T_λ' may be chosen as the $p^k \equiv fp^k$ extenders V_j. For each V_j the ratios α_j/γ_j, δ_j/γ_j are marks λ_j, λ_j'' of the $GF[p^k]$. As in case [A] for $p=2$, $k>1$, the ratio β_j/γ_j is a mark λ_j' of the field. The determinant being unity, γ_j^2 belongs to the field, so that γ_j is some η_j. Hence

$$V_j = \begin{pmatrix} \lambda_j\eta_j, & \lambda_j'\eta_j \\ \eta_j, & \lambda_j''\eta_j \end{pmatrix}.$$

According as η_j is an even or an odd power of $\eta_0 \equiv \sqrt{\varkappa_0}$, V_j or $V_{\eta_0,0} V_j$ has its coefficients in the $GF[p^k]$. The one having this property is denoted by V_j'. These p^k substitutions V_j' serve to extend the group $G_{\frac{1}{2}p^k(p^k-1)}^{(\infty)}$ of the $V_{\varkappa,\lambda}$ to the group $G_{M(p^k)}$ of all linear fractional substitutions of determinant unity in the $GF[p^k]$. It is transformed into itself by

$$P_{\eta_0} = \begin{pmatrix} \eta_0, & 0 \\ 0, & \eta_0^{-1} \end{pmatrix} \equiv \begin{pmatrix} \varkappa_0, & 0 \\ 0, & 1 \end{pmatrix}$$

whose square $P_{\eta_0^2} \equiv P_{\varkappa_0}$ belongs to $G_{M(p^k)}$. Hence P_{η_0} extends the latter to the group $G_{2M(p^k)}$ of all linear fractional substitutions in the $GF[p^k]$. The latter is a subgroup of G_Ω and is of order Ω. G_Ω is therefore identical with the linear fractional group $G_{2M(p^k)}$.

For the case $f=3$, $p^k=3$, the relation 255) becomes an equality, so that there are exactly $12 + 3 = 15$ substitutions T of period 2 in G_{60}. At the beginning of the section, each T was shown to be self-conjugate within G_{60} under exactly a dihedron G_4. The 15 substitutions T are therefore all conjugate under G_{60} and form 5 conjugate four-groups G_4. By § 251, G_{60} contains one set of $1 + fp^k = 10$ conjugate G_3. Hence, if the G_{60} exists, it is of the icosahedral type (§ 254). For n even, $5 \equiv \frac{1}{2}(3^2+1)$ divides $\frac{1}{2}(3^{2n}-1)$, so that the existence of icosahedral subgroups of $G_{M(3^n)}$ follows from § 259. The question of the conjugacy of the icosahedral subgroups is answered in that section.

254. *A group of order* 60 *is of the icosahedral type if it contains exactly ten conjugate G_3 and exactly* 15 *operators of period* 2 *lying in* 5 *conjugate four-groups.*

Since there is a complete set of 5 conjugate G_4 within the G_{60}, each G_4 is self-conjugate under exactly a subgroup G_{12}. The latter is of the tetrahedral type by § 247; for if commutative it would contain a self-conjugate G_3 which would be one of a set of at most

5 conjugate subgroups of G_{60}. Hence G_{60} contains a set of 5 conjugate tetrahedral G_{12}. No two of them are identical since each contains a single four-group. They have only the identity in common. Indeed, their common operators form a self-conjugate subgroup of G_{60} and hence a self-conjugate subgroup of each G_{12}. Aside from the identity and G_{12} itself (cases requiring no further discussion), the only self-conjugate subgroup of a tetrahedral G_{12} is its four-group. But the 5 four-groups are all distinct. Hence the identity is the only operator of G_{60} which transforms each G_4 into itself. Applied as transformers, the operators of G_{60} permute the 5 conjugate G_4, so that G_{60} is holoedrically isomorphic with a substitution-group on 5 letters. Being of order 60, the latter is necessarily the alternating group on 5 letters.[1]) Hence the G_{60} is of the icosahedral type (§ 267).

255. It remains to study the conjugacy of the linear fractional subgroups $G_{M(p^k)}$ and $G_{2M(p^k)}$ of $G_{M(s)}$. Within $G_{M(s)}$ the $G_{M(p^k)}$ is self-conjugate exactly under $G_{2M(p^k)}$, $G_{M(p^k)}$; $G_{M(p^k)}$ according as $p > 2$ with n/k even, $p > 2$ with n/k odd; or $p = 2$, and hence is one of a system of $M(s)/(2,1;1)M(p^k)$ conjugate groups. In proof, we note that a substitution $V = \begin{pmatrix} \alpha, \beta \\ \gamma, \delta \end{pmatrix}$ of $G_{M(s)}$ transforms (§ 240) the substitutions $\begin{pmatrix} 1, \sigma \\ 0, 1 \end{pmatrix}$ and $\begin{pmatrix} 1, 0 \\ \sigma, 1 \end{pmatrix}$ into respectively

$$\begin{pmatrix} 1-\alpha\gamma\sigma, & \alpha^2\sigma \\ -\gamma^2\sigma, & 1+\alpha\gamma\sigma \end{pmatrix}, \quad \begin{pmatrix} 1+\beta\delta\sigma, & -\beta^2\varrho \\ \delta^2\sigma, & 1-\beta\delta\sigma \end{pmatrix}.$$

If σ belongs to the $GF[p^n]$, these substitutions belong to that field if, and only if, α and γ are each marks μ of the $GF[p^k]$ or are each of the form $\mu\sqrt{\nu}$, where ν is a not-square in the $GF[p^k]$, and β, δ are each marks μ or are each of the form $\mu\sqrt{\nu}$. Since $\alpha\delta - \beta\gamma = 1$, α, β, γ, δ are all of the form μ or all of the form $\mu\sqrt{\nu}$. Hence V is either a substitution S of $G_{M(p^k)}$ or else a product $SP_{\sqrt{\nu}}$. The latter alternative does not occur if $p = 2$. Also, if $p > 2$, $\sqrt{\nu}$ belongs to the $GF[p^n]$ if, and only if, n/k is even. Hence $G_{M(p^k)}$ is self-conjugate within $G_{M(s)}$ in a larger group, viz., $G_{2M(p^k)}$, if and only if $p > 2$ with n/k even.

Within $G_{M(s)}$ the $G_{2M(p^k)}$, when existent, is self-conjugate only under itself. For any substitution of the former which transforms the latter into itself must transform its self-conjugate subgroup $G_{M(p^k)}$

---

1) If a $G_{60}^{(5)}$ contained odd substitutions, it would have a subgroup $G_{30}^{(5)}$ of even substitutions. The latter would be of index two under the alternating group $G_{60}^{(5)}$ and hence self-conjugate under it, whereas it is simple.

280   CHAPTER XII.

into itself and hence belong to $G_{2M(p^k)}$. The latter thus forms one of a system of $M(s)/2M(p^k)$ conjugate subgroups.

It remains to determine the number of systems of conjugate subgroups of these two types; indeed, in § 251, there entered the transformer $P_{\sqrt{\lambda_0-1}}$ which belongs to $G_{M(s)}$ if and only if λ_0 is a square in the $GF[p^n]$. For $p=2$, λ_0 is necessarily a square; for $p>2$, n/k odd, λ_0 may be chosen as a square, since every additive-group $[\lambda_1, \ldots, \lambda_k]$ with the multiplier $GF[p^k]$ has half of its non-vanishing marks squares in the $GF[p^n]$. In these two cases there is evidently but one system of conjugate subgroups $G_{M(p^k)}$ of $G_{M(s)}$. For $p>2$, n/k even, all the marks of $[\lambda_1, \ldots, \lambda_k]$ are squares or all are not-squares in the $GF[p^n]$; indeed, they are all obtained from a single one by multiplication by the p^n marks of the multiplier $GF[p^k]$ and the latter are all squares in the $GF[p^n]$. In this case there are consequently two systems of conjugate subgroups $G_{M(p^k)}$ and two systems of conjugate $G_{2M(p^k)}$, the systems of each type being interchanged upon transformation by $P_{\sqrt{\nu}}$, belonging to $G_{2M(s)}$, where ν is any not-square in the $GF[p^n]$. *Hence there are $(2,1;1)$ systems of conjugate $G_{M(p^k)}$ and $(2,0;0)$ systems of conjugate $G_{2M(p^k)}$ within $G_{M(s)}$.*

256. *Subgroups of $G_{M(s)}$ containing no operators of period p.* Every substitution of such a subgroup G_Ω lies in and determines a largest cyclic subgroup G_d of G_Ω (§§ 242—243). Two such groups G_d have only the identity in common. According as G_d is self-conjugate within G_Ω only under itself or under a dihedron G_{2d} based on G_d, it is one of a system of Ω/d or $\Omega/2d$ conjugate subgroups of G_Ω. Let r denote the number of such systems. The enumeration of the substitutions of G_Ω leads to the relations

256) $\qquad \Omega = 1 + \sum_{i=1}^{r}(d_i-1)\dfrac{\Omega}{f_i d_i} \qquad (f_i = 1 \text{ or } 2)$

257) $\qquad \Omega \gtreqless f_i d_i \qquad\qquad\qquad (i=1, 2, \ldots, r).$

If two non-conjugate cyclic G_{d_i}, G_{d_j} of *odd* order are present in G_Ω, there are at least d_j groups in the system determined by G_{d_i}, viz., the transformed of the latter by the operators of G_{d_j}, and vice versa, so that

258) $\qquad\qquad \Omega \gtreqless d_i(d_j-1) + d_j(d_i-1) + 1.$

Solving 256) for $1/\Omega$, we get

259) $\qquad\qquad 1 - \sum_{i=1}^{r}\dfrac{(d_i-1)}{f_i d_i} = \dfrac{1}{\Omega}.$

Since $f_i = 1$ or 2, the least value of $(d_i-1)/f_i d_i$ is $1/4$. Since 259) must be positive, there can be at most three terms in the sum, whence $r \lessgtr 3$.

For $r = 1$, the reciprocal of 259) is not an integer if $f_1 = 2$. For $f_1 = 1$, $\Omega = d_1$, and the G_Ω is a cyclic group considered in §§ 242—243.

For $r = 2$, we have
$$1 - \frac{1}{\Omega} = \frac{1}{f_1}\left(1 - \frac{1}{d_1}\right) + \frac{1}{f_2}\left(1 - \frac{1}{d_2}\right).$$
If $f_1 = f_2 = 1$, the left member is < 1 and the right member is $\gtreqless 1$.
If $f_1 = f_2 = 2$,
$$\frac{2}{\Omega} = \frac{1}{d_1} + \frac{1}{d_2}, \quad \frac{2}{\Omega} \lessgtr \frac{1}{d_1} \qquad \text{[by 257)]}.$$
Hence these two cases are to be excluded. The case $f_1 = 2$, $f_1 = 1$ differs only in notation from the case $f_1 = 1$, $f_2 = 2$. In the latter case,
$$\frac{1}{\Omega} = \frac{1}{d_1} + \frac{1}{2d_2} - \frac{1}{2} \gtreqless \frac{1}{d_1} - \frac{1}{4},$$
so that $d_1 < 4$. For $d_1 = 2$, $\Omega = 2d_2$, so that G_Ω is a dihedron G_{2d_2} with d_2 odd (§ 245) yielding a group considered in § 246. For $d_1 = 3$, d_2 must be 2, whence $\Omega = 12$. The operator of period $d_2 = 2$ is self-conjugate within G_{12} under exactly a dihedron G_4, so that G_{12} is not a commutative group. Since the operators of period 2 fall into a single set of 3 conjugate operators, there is a single subgroup G_4, so that it is self-conjugate under G_{12}. By § 247, the G_{12} is a tetrahedral group.

For $r = 3$, then $f_1 = f_2 = f_3 = 2$. For if $f_1 = 1$, for example, 259) becomes
$$\frac{1}{d_1} - \frac{(d_2-1)}{f_2 d_2} - \frac{(d_3-1)}{f_3 d_3} \gtreqless \frac{1}{d_1} - \frac{1}{4} - \frac{1}{4} \lessgtr 0.$$
Setting each $f_i = 2$, equation 259) may be written
$$1 + \frac{2}{\Omega} = \frac{1}{d_1} + \frac{1}{d_2} + \frac{1}{d_3}.$$
If every $d_i \gtreqless 3$, the right member would be $\lesseqgtr 1$. Setting $d_3 = 2$,
$$\frac{1}{2} + \frac{2}{\Omega} = \frac{1}{d_1} + \frac{1}{d_2}.$$
If either d_1 or d_2 is 2, we may take $d_2 = 2$, whence $\Omega = 2d_1$ and G_Ω is a dihedron G_{2d_1} with d_1 even (§ 245) yielding a group considered in § 246. In the contrary case, $d_1 > 2$, $d_2 > 2$. Then both do not exceed 3, since otherwise the right member would be at most $\frac{1}{4} + \frac{1}{4} = \frac{1}{2}$. Taking $d_2 = 3$, we have
$$\frac{1}{6} + \frac{2}{\Omega} = \frac{1}{d_1}.$$

282   CHAPTER XII.

Hence $d_1 < 6$. For $d_1 = 3, 4, 5$ we find $\Omega = 12, 24, 60$ respectively. But $d_1 = 3$, $d_2 = 3$, $d_3 = 2$, $\Omega = 12$ is excluded by 258). For $d_1 = 4$, $d_2 = 3$, $d_3 = 2$, the G_{24} is of the octahedral type (§ 248). For $d_1 = 5$, $d_2 = 3$, $d_3 = 2$, the G_{60} is of the icosahedral type (§ 254).

257. *The tetrahedral and octahedral subgroups of the $G_{M(s)}$.* A group of either type must contain a self-conjugate four-group. For $p > 2$, the desired groups are therefore given by the theorem at the end of § 246. For $p = 2$, they contain operators of period 2 and are therefore to be sought among the subgroups determined in §§ 250—253. But for $p = 2$, the dihedron $G_{2(1+2f)}$ and the $G_{M(p^k)}$ are neither of the tetrahedral and neither of the octahedral type. There remain for consideration only the subgroups of the $G_{s(s-1)}^{(\varkappa)}$ of § 250. There is no octahedral subgroup of $G_{s(s-1)}^{(\infty)}$ since the substitutions of period $p = 2$ in the latter are all commutative. In a tetrahedral group the three substitutions of period 2 are all commutative. Hence if there be a tetrahedral subgroup of $G_{s(s-1)}^{(\infty)}$, $p = 2$, then must $2^m = 4$, $d = 3$ and n even (since 3 must divide $2^n - 1$). Inversely, if $m = 2$, $p = 2$, n even, there exists a subgroup $G_{2^m d} \equiv G_{12}$ of $G_{s(s-1)}^{(\infty)}$. The G_{12} is not commutative, since it would then contain only operators of period $p = 2$ (§ 241), and therefore G_{12} has the tetrahedral type (§ 247). We may state the complete theorems:

For $s \equiv p^n = 8h \pm 1$, the $G_{M(s)}$ contains two systems each of $M(s)/24$ conjugate octahedral groups G_{24} and two systems each of $M(s)/24$ conjugate tetrahedral groups G_{12}. Every G_{12} is self-conjugate under a G_{24}. The two systems are conjugate under $G_{2M(s)}$.

For $s = 8h \pm 3$ or $s = 2^n$, n even, the $G_{M(s)}$ contains no octahedral G_{24} but contains one system of $M(s)/12$ conjugate tetrahedral G_{12}. For $p > 2$, the $G_{2M(s)}$ contains one system of $M(s)/12$ conjugate octahedral G_{24} each containing one G_{12}. For $s = 2^n$, n odd, $G_{M(s)}$ contains no octahedral and no tetrahedral groups.

258. *Icosahedral subgroups of $G_{M(s)}$ for $p = 5$.* An icosahedral G_{60} is generated by two operators V_5, V_2 different from the identity and subject to the generational relations (§ 267)

$$V_5{}^5 = V_2{}^2 = I, \quad (V_5 V_2)^3 = I.$$

$G_{M(5)}$ contains $4(5+1) = 24$ substitutions of period 5 and each is conjugate within $G_{M(5)}$ with one of the substitutions (§ 241)

$$V_5 = \begin{pmatrix} 1, & \mu \\ 0, & 1 \end{pmatrix} \qquad \mu \not\equiv 0 \pmod{5}.$$

SUBGROUPS OF THE LINEAR FRACTIONAL GROUP $LF(2, p^n)$. 283

The only substitutions V_2 of period 2 of $G_{M(5)}$ which satisfy the condition[1]) $(V_5 V_2)^3 = I$ are seen to be the following five

$$V_2 = \begin{pmatrix} \alpha, & -\mu(1+\alpha^2) \\ \mu^{-1}, & -\alpha \end{pmatrix} \qquad (\alpha \equiv 0, 1, 2, 3, 4).$$

Hence $G_{M(5)}$ is an icosahedral group[2]) and contains $24 \cdot 5 = 120$ pairs of generators V_5, V_2. By § 255, $G_{M(5^n)}$ contains $M(5^n)$ 60 *icosahedral subgroups forming two systems or one system of conjugate groups according as n is even or odd.*

259. *Icosahedral subgroups of $G_{M(s)}$ for $p \neq 5$.* The order $p^n(p^{2n}-1)/(2; 1)$ of $G_{M(s)}$ is divisible by 60 if, and only if, $p^{2n}-1$ be divisible by 5 and hence either $p^n + 1$ or $p^n - 1$ divisible by 5. In either case $G_{M(s)}$ contains cyclic subgroups G_5 all of which are conjugate (§§ 242, 243).

(i) Let $p^n - 1$ be divisible by 5 and set $\lambda \equiv (p^n - 1)/5$. Let ϱ be a primitive root of the $GF[p^n]$, so that ϱ^λ is of period 5. Setting

260)
$$V_5 = \begin{pmatrix} \varrho^\lambda, & 0 \\ 0, & \varrho^{-\lambda} \end{pmatrix}, \quad V_2 = \begin{pmatrix} \alpha, & \beta \\ \gamma, & -\alpha \end{pmatrix}$$
$$-\alpha^2 - \beta\gamma = 1,$$

we seek the conditions under which the product

$$V_5 V_2 = \begin{pmatrix} \alpha\varrho^\lambda, & \beta\varrho^\lambda \\ \gamma\varrho^{-\lambda}, & -\alpha\varrho^{-\lambda} \end{pmatrix}$$

shall be of period 3. The necessary and sufficient condition is

$$\alpha(\varrho^\lambda - \varrho^{-\lambda}) = \pm 1.$$

The upper sign may be chosen, changing if necessary the signs of α, β, γ in V_2. Hence α is determined uniquely. Combining with 260),

$$\beta\gamma = -1 - (\varrho^\lambda - \varrho^{-\lambda})^{-2} \neq 0.$$

Indeed, if the second member vanish, $\varrho^{4\lambda} - \varrho^{2\lambda} + 1 = 0$, so that $\varrho^{6\lambda} + 1 = 0$ and therefore $\varrho^{2\lambda} = +1$, whereas ϱ^λ is of period 5. Hence to each of the $p^n - 1$ values $\neq 0$ of β corresponds a single value of γ. But $G_{M(s)}$ contains (§ 242) exactly $\frac{1}{2} p^n(p^n + 1)$ distinct cyclic G_5. Hence there are $2p^n(p^{2n}-1)$ *pairs of generators V_5, V_2 of icosahedral subgroups.*

---

1) It is readily verified that a substitution $\begin{pmatrix} \alpha, & \beta \\ \gamma, & \delta \end{pmatrix}$ of determinant unity is of period 3 if, and only if, $\alpha + \delta = \pm 1$.
2) Cf. § 280.

284   CHAPTER XII.

(ii) For p^n+1 divisible by 5, let $g=(p^n+1)/5$ and set (§ 243)

$$V_5 = \begin{pmatrix} J^g, & 0 \\ 0, & \overline{J}^g \end{pmatrix}, \quad V_2 = \begin{pmatrix} A, & B \\ -\overline{B}, & -\overline{A} \end{pmatrix},$$

$$J\overline{J} = 1, \quad -A^2 + B\overline{B} = 1, \quad \overline{\overline{A}} = -A.$$

The condition $(V_5 V_2)^3 = I$ is satisfied if, and only if,

$$A(J^g - \overline{J}^g) = 1.$$

The A thus determined satisfies the condition $\overline{A} = -\acute{A}$. Then must

$$B\overline{B} \equiv B^{p^n+1} = 1 + A^2 = 1 + (J^g - \overline{J}^g)^{-2}.$$

The last term is a mark $\mu \neq 0$ of the $GF[p^n]$. Hence $B^{p^n+1} = \mu$ has a solution B_0 in the $GF[p^{2n}]$ and consequently p^n+1 distinct solutions $B_0, B_0J, B_0J^2, \ldots, B_0J^{p^n}$. But $G_{M(s)}$ contains exactly $\frac{1}{2}p^n(p^n-1)$ conjugate cyclic G_5 (§ 243). Hence there are $2p^n(p^{2n}-1)$ pairs of generators V_5, V_2 of icosahedral subgroups.

Since each icosahedral group contains (§ 258) exactly 120 pairs of generators V_5, V_2, it follows that, for $p^{2n}-1$ divisible by 5, $G_{M(p^n)}$ contains in all $p^n(p^{2n}-1)/60$ icosahedral subgroups.

For $p=2$, $2^{2n}-1$ is divisible by $5 \equiv 2^2+1$ if and only if n be even. If n be even, $G_{M(2^n)}$ contains a single system of $M(s)/60$ subgroups $G_{M(2^2)}$ (§ 255), the latter being icosahedral by case (ii). *Hence $G_{M(2^n)}$ contains no icosahedral groups if n be odd, but, for n even, contains $2^n(2^{2n}-1)/60$ icosahedral groups forming a single system of conjugate groups.*

To determine, for $p > 2$, the distribution of the icosahedral subgroups into sets of conjugates within $G_{M(s)}$ and within $G_{2M(s)}$, consider first the case (i) and set $\varepsilon^2 = \varrho$, so that only the even powers of ε belong to the $GF[p^n]$. Then will

$$z' = \varepsilon^e z / \varepsilon^{-e} \equiv \varrho^e z$$

transform V_5 into itself, but transforms V_2 into

$$\begin{pmatrix} \alpha, & \beta\varrho^e \\ \gamma\varrho^{-e}, & -\alpha \end{pmatrix}.$$

Hence the groups G_{60} are all conjugate under $G_{2M(s)}$ and form *at most* two systems of conjugate subgroups under $G_{M(s)}$. But if there were a single system, their number would be at most $M(s)/60$, whereas it is $M(s)/30$. *Hence there are two systems each of $M(s)/60$ conjugate icosahedral groups within $G_{M(s)}$ and each is self-conjugate only under itself.*

SUBGROUPS OF THE LINEAR FRACTIONAL GROUP $LF(2, p^n)$. 285

For case[1]) (ii), let $E^2 = J$, so that $E\overline{E} = J^{(p^n+1)/2} = -1$. Then
$$z' = E^e z / E^{-e}$$
transforms V_5 into itself and transforms V_2 into
$$\begin{pmatrix} A, & BJ^e \\ -\overline{B}J^{-e}, & -A \end{pmatrix}.$$

Taking $e = 0, 1, \ldots, p^n$, we reach the various $p^n + 1$ substitutions V_2. If e be even, the transformer belongs to the hyperorthogonal group since $\overline{J} = J^{-1}$. For e odd, it may be given the hyperorthogonal form with determinant a not-square. In fact, there exist in the $GF[p^{2n}]$ solutions of $X^{p^n-1} = -1$, so that $\overline{X} = -X$. Then
$$\begin{pmatrix} E, & 0 \\ 0, & E^{-1} \end{pmatrix} = \begin{pmatrix} E, & 0 \\ 0, & -\overline{E} \end{pmatrix} = \begin{pmatrix} XE, & 0 \\ 0, & -XE \end{pmatrix} = \begin{pmatrix} XE, & 0 \\ 0, & \overline{XE} \end{pmatrix}$$
of determinant X^2. Hence the groups G_{60} are all conjugate within $G_{2M(s)}$ but form two systems of conjugates within $G_{M(s)}$.

260. Summary of the subgroups of $G_{M(s)}$, $s \equiv p^n$:

$s + 1$ conjugate commutative groups of order s;

$\frac{1}{2} s(s \pm 1)$ conjugate cyclic groups of order $\frac{s \mp 1}{2;1}$, 2; 1 according as $p > 2$; $p = 2$;

$\frac{1}{2} s(s \pm 1)$ conjugate cyclic $G_{d_{\mp}}$ for every divisor $d_{\mp}$ of $\frac{s \mp 1}{2;1}$;

$M(s)/2 d_{\mp}$ conjugate dihedral $G_{2d_{\mp}}$, for $d_{\mp}$ odd;

two systems each of $M(s)/4d_{\mp}$ conjugate dihedral $G_{2d_{\mp}}$, for $d_{\mp}$ even and > 2;

for $p^n = 8h \pm 3$, one set of $M(s)/12$ conjugate four-groups;

for $p^n = 8h \pm 1$, two sets each of $M(s)/24$ conjugate four-groups[2]);

$\frac{(p^n-1)(p^n-p)\ldots(p^n-p^{m-1})}{(p^m-1)(p^m-p)\ldots(p^m-p^{m-1})}$ sets each of $\frac{p^{2n}-1}{(2, 1; 1)(p^k-1)}$ conjugate commutative groups of order p^m, where $(2, 1; 1)$ is read 2, 1, or 1 according as $p > 2$ with n/k an even integer, $p > 2$ with n/k an odd integer, or $p = 2$ with n/k an integer, and where k is a divisor of m depending on the particular G_{p^m};

---

1) This case may be made to depend on (i) since 5 divides $p^{2n}-1$. Hence each G_{60} is self-conjugate only under itself within the group $G_{M(s^2)}$ and so within its subgroup $G_{2M(s)}$. Hence each G_{60} is one of a system of $2M(s)/60$ conjugate groups within $G_{2M(s)}$, so that the icosahedral subgroups all form a single system of conjugates within $G_{2M(s)}$. They fall into two systems in $G_{M(s)}$.

2) For $p = 2$, the four-groups occur among the groups of order $p^m = 2^2$ given later.

286 CHAPTER XII. SUBGROUPS OF THE LINEAR FRACTIONAL etc.

certain sets of $\frac{(p^{2n}-1)p^{n-m}}{(2,1;1)(p^k-1)}$ conjugate $G_{p^m d_-}$, where k and d_- depend on m;

$(2,1;1)$ sets each of $M(s)/(2,1;1)M(p^k)$ conjugate $G_{M(p^k)}$, k a divisor of n, each group being isomorphic with the group of linear fractional substitutions of determinant unity in the $GF[p^k]$;

two systems each of $M(s)/2M(p^k)$ conjugate $G_{2M(p^k)}$, $p>2$, n/k an even integer, each group isomorphic with the linear fractional group in the $GF[p^k]$;

for $s = 8h \pm 1$, two sets each of $M(s)/24$ conjugate octahedral G_{24};
for $s = 8h \pm 1$, two sets each of $M(s)/24$ conjugate tetrahedral G_{12};
for $s = 8h \pm 3$ or $s = 2^n$, n even, $M(s)/12$ conjugate tetrahedral G_{12};
for $s = 10l \pm 1$, two sets each of $M(s)/60$ conjugate icosahedral G_{60}.[1])

261. Theorem. — *If $p^n > 3$, the linear fractional group $G_{M(s)}$ is simple.*

Indeed, the only cases in which the number of groups in a set of conjugate subgroups is unity are the following two:

$p^n = 2$, $d_+ = 3$, $M(s)/2d_+ = 1$, when the G_6 has a self-conjugate G_3;
$p^n = 3$, $M(s)/12 = 1$, when the G_{12} has a self-conjugate four-group.

262. Theorem.[2]) — *The group $G_{M(s)}$ always has subgroups of index $s + 1$, but has subgroups of lower index only when*

$$s = 2, 3, 5, 7, 3^2, 11.$$

Every subgroup of $G_{M(s)}$ is contained in one of the following: $G_{s(s-1)\atop 2;1}$, dihedral $G_{s\mp 1}$ ($p > 2$), $G_{M(p^k)}$ (n/k an odd integer if $p > 2$), $G_{2M(p^k)}$ ($p > 2$, n/k an even integer), G_{12} ($s = 8h \pm 3$), G_{24} ($s = 8h \pm 1$), G_{60} ($s = 10l \pm 1$). The first group is always of order greater than the $G_{M(p^k)}$ and $G_{2M(p^k)}$; indeed, since $k \leq n/2$,

$$p^k(p^{2k}-1) < p^k(p^n-1) < \frac{s(s-1)}{2;1}.$$

Also $s(s-1)/(2;1) > s+1 > s-1$ if $s > 3$ and $s(s-1)/(2;1) > 60$ if $s > 11$. Hence $G_{s(s-1)\atop 2;1}$ of index $s+1$ has the maximum order if $s > 11$. The same result holds for $s = 2^3$ since the $G_{M(2^3)} \equiv G_{60}$ is then not a subgroup; likewise for $s = 2^2$ since it is (§ 257) then

---

1) For $p = 2$ or $p = 5$ the icosahedral subgroups are of the type $G_{M(2^2)}$ or $G_{M(5)}$ given earlier.

2) For $n = 1$, this is the celebrated theorem stated without proof by Galois in the letter to his friend Auguste Chevalier written before the fatal duel. For references to the proofs by Betti, Gierster, etc., see Klein, *Math. Ann.*, vol. 14.

CHAPTER XIII. AUXILIARY THEOREMS ON ABSTRACT GROUPS. etc. 287

the tetrahedral G_{12}. For $s = 11$, 3^2, 7, 5, the subgroups of maximum order are G_{60}, G_{60}, G_{24}, G_{12} respectively, the index under $G_{M(s)}$ being 11, 6, 7, 5 and hence $< s + 1$. For $s = 2, 3$ the $G_{M(s)}$ is a dihedron G_6, a tetrahedron G_{12}, respectively, and has a subgroup of maximum order G_3, G_4 respectively.

263. A simple group can be represented as a transitive substitution-group on N letters if, and only if, it contains a complete system of N conjugate subgroups.[1]) For $s > 3$, $G_{M(s)}$ is simple (§ 261). Hence $G_{M(s)}$ can be represented as a transitive group on $< s + 1$ letters only when $s = 5, 7, 3^2, 11$. For $s = 2, 3$ it can be represented as a transitive group on 3, 4 letters respectively, but on no fewer, being of order G_6, G_{12}. If a simple group be represented as an intransitive substitution-group on D letters, D must equal the sum of the degrees of two or more transitive representations; for $G_{M(s)}$ we have always $D > s + 1$. Hence *the linear fractional group $G_{M(s)}$ may be represented as a substitution-group on $s + 1$ letters but on no fewer number except when $s = 5, 7, 9, 11$, for which the minimum number of letters is 5, 7, 6, 11 respectively.*

CHAPTER XIII.

AUXILIARY THEOREMS ON ABSTRACT GROUPS. ABSTRACT FORMS OF VARIOUS LINEAR GROUPS.[2])

264. Theorem. — *The symmetric substitution-group on k letters is holoedrically isomorphic with the abstract group $G(k)$ generated by the operators $B_1, B_2, \ldots, B_{k-1}$ with the generational relations*

261) $B_1^2 = B_2^2 = \cdots = B_{k-1}^2 = I$,
262) $B_i B_j = B_j B_i$ $(i = 1, 2, \ldots, k-3; j = i+2, i+3, \ldots, k-1)$,
263) $B_j B_{j+1} B_j = B_{j+1} B_j B_{j+1}$ $(j = 1, 2, \ldots, k-2)$.

The symmetric group $G_{k!}^{(k)}$ on the letters $l_1, l_2, \ldots, l_k$ may be generated by the transpositions

$$S_d \equiv (l_d l_{d+1}) \qquad (d = 1, 2, \ldots, k-1),$$

which satisfy the relations 261), 262), 263) prescribed for the generators B_d of the abstract group $G(k)$ and conceivably also other

---

1) For a proof of this theorem due to Dyck see Burnside, The Theory of Groups, § 123.
2) The theorems of §§ 264, 265 are due to Professor Moore, *Proceed. Lond. Math. Soc.*, vol. XXVIII, pp. 357—366. The proofs given in §§ 264, 266 are due to the author; for that in § 264 see *Proceed. Lond. Math. Soc.*, vol. XXXI, 351—353; for that in § 266 see *Math. Ann.*, vol. 54, pp. 564—569.

288   CHAPTER XIII.

relations not derivable therefrom. The order $O(k)$ of $G(k)$ is therefore $\geq k!$.

Denote by G the subgroup $G(k-1)$ generated by $B_1, B_2, \ldots, B_{k-2}$ and consider the following sets of operators[1]) of $G(k)$:

$$O_k \equiv G, \quad O_{k-1} \equiv GB_{k-1}, \quad O_{k-2} \equiv GB_{k-1}B_{k-2}, \ldots, O_1 \equiv GB_{k-1}B_{k-2}\ldots B_1.$$

It will be shown that these sets of operators are merely permuted amongst themselves upon applying as right-hand multipliers the generators B_r $(r = 1, \ldots, k-1)$. Since $B_r^2 = I$, we have

$$O_{r+1}B_r \equiv GB_{k-1}\ldots B_{r+1}B_r \equiv O_r,$$
$$O_rB_r \equiv GB_{k-1}\ldots B_rB_r \equiv O_{r+1}.$$

If $i > r+1$, we find, on applying 262) to move B_r to the left of $B_i, \ldots, B_{k-1}$,

$$O_iB_r \equiv GB_{k-1}\ldots B_iB_r = GB_rB_{k-1}\ldots B_i = GB_{k-1}\ldots B_i \equiv O_i.$$

If $i < r$, we find, on moving B_r to the left of $B_i, B_{i+1}, \ldots, B_{r-2}$,

$$O_iB_r \equiv GB_{k-1}\ldots B_iB_r = GB_{k-1}\ldots B_{r+1}B_rB_{r-1}B_rB_{r-2}\ldots B_i.$$

By 263), we may replace $B_rB_{r-1}B_r$ by $B_{r-1}B_rB_{r-1}$. We then move the first B_{r-1} to the left of $B_{r+1}, \ldots, B_{k-1}$ and merge it into G and get

$$O_iB_r = GB_{k-1}\ldots B_{r+1}B_rB_{r-1}B_{r-2}\ldots B_i \equiv O_i.$$

Hence the right-hand multiplier B_r gives rise to the transposition (O_rO_{r+1}) on the k sets $O_1, \ldots, O_k$. It follows that the product of any operator of these k sets by an arbitrary operator of $G(k)$ is an operator belonging to these sets. Taking for the former operator the identity, we see that these k sets include all the operators of the group $G(k)$. The number of operators in $G(k)$ is therefore at most k times the number in $G(k-1)$. Hence

$$O(k) \leq k \cdot O(k-1) \leq \cdots \leq k!$$

Combining this result with the earlier one, we have $O(k) = k!$. The proof of the holoedric isomorphism of $G(k)$ and $G_{k!}^{(k)}$ is therefore complete.

The relations 261), 262), 263) may be combined into the formulae

264) $I = B_i^2 = (B_iB_{i+1})^3 = (B_iB_j)^2 \quad (i,j = 1, \ldots, k-1; j > i+1).$

---

[1]) It turns out that these sets form a rectangular table for $G(k)$ with the operators of G in the first line.

265. Theorem. — *The alternating group on k letters is holoedrically isomorphic with the abstract group $G\{k\}$ generated by the operators $E_1, E_2, \ldots, E_{k-2}$ subject to the generational relations*

265) $\qquad I = E_1^3 = E_{i+1}^2 = (E_i E_{i+1})^3 = (E_i E_j)^2 \quad (i,j=1,\ldots,k-2;\, j>i+1).$

The abstract symmetric group $G(k)$ may be generated by B_1 and

266) $\qquad\qquad E_d \equiv B_{d+1} B_1 \qquad (d = 1, 2, \ldots, k-2).$

From the relations 264) we readily derive 265) together with

267) $\qquad\qquad B_1^2 = I, \quad E_d B_1 = B_1 E_d^{-1} \qquad (d=1,2,\ldots,k-2).$

Inversely, from 265) and 267), we can easily get relations 264). Hence $B_1, E_1, E_2, \ldots, E_{k-2}$, subject only to the relations 265) and 267), generate the abstract group $G(k)$. Upon extending $G\{k\}$ by the operator B_1 subject to the relations 267), we obtain a group whose operators are of the form E or EB_1, E being derived from $E_1, E_2, \ldots, E_{k-2}$, and hence of order $2O\{k\}$. But the extended group was shown to be $G(k)$. Hence $G\{k\}$ is a subgroup of $G(k)$ of order $\frac{1}{2} k!$. It is readily shown to be the abstract alternating group $G_{\frac{1}{2}k!}$. Since the generational relations 264) involve the generators B_i evenly, the various expressions for an operator of $G(k)$ in terms of its generators involve all an even or all an odd number of the generators, so that its operators may be classed into even and odd operators. By 266), the operators of the subgroup $G\{k\}$ are all even, so that it is a subgroup of $G_{\frac{1}{2}k!}$. Since its order is $\frac{1}{2}k!$, it is identical with the latter.

266. The last theorem may be readily proved by the direct method of § 264. The generational relations 265) are seen to be satisfied by the substitutions

$$A_d \equiv (l_{d+1} l_{d+2})(l_1 l_2) \equiv S_{d+1} S_1 \qquad (d = 1, \ldots, k-2)$$

which generate the alternating group on $l_1, l_2, \ldots, l_k$. Hence

$$O\{k\} \geq \frac{1}{2} k!$$

The theorem being evident if $k = 3$, we take $k \geq 4$. Denote by Γ the subgroup $G\{k-1\}$ generated by $E_1, E_2, \ldots, E_{k-3}$ and consider the following sets of operators of $G\{k\}$:

$R_{k-1} \equiv \Gamma,\ R_{k-2} \equiv \Gamma E_{k-2},\ R_{k-3} \equiv \Gamma E_{k-2} E_{k-3}, \ldots, R_2 \equiv \Gamma E_{k-2} E_{k-3} \ldots E_2,$
$\qquad\qquad R_1 \equiv \Gamma E_{k-2} \ldots E_2 E_1, \quad R_k \equiv \Gamma E_{k-2} \ldots E_2 E_1^2.$

The reader may readily verify, as in § 264, that E_1 and E_r $(r > 1)$, when applied as right-hand multipliers to the above sets, give rise

290   CHAPTER XIII.

to the permutations $(R_1 R_k R_2)$ and $(R_r R_{r+1})(R_1 R_k)$ respectively. The sets $R_1, \ldots, R_k$ therefore include all the operators of $G\{k\}$, so that

$$O\{k\} \gtreqqless k \cdot O\{k-1\} \gtreqqless \cdots \gtreqqless k(k-1) \cdots 4 \cdot O\{3\} \equiv \frac{1}{2} k!$$

Combining this result with the earlier one, $O\{k\} = \frac{1}{2} k!$

267. Theorem. — *The abstract alternating group $G_{\frac{1}{2}5!}$ may be generated by two operators V and W subject to the generational relations*

268) $\qquad V^5 = W^2 = I, \quad (VW)^3 = I.$

For $k = 5$, the relations 265) defining $G_{\frac{1}{2}5!}$ may be written

269) $\quad E_1^3 = E_2^2 = E_3^2 = (E_1 E_2)^3 = (E_2 E_3)^3 = (E_1 E_3)^2 = I.$

The group contains two operators $V \equiv E_1 E_2 E_3$, $W \equiv E_3$ such that $W^2 = I$, $(VW)^3 \equiv (E_1 E_2)^3 = I$. To prove that $V^5 = I$, we apply 269) and find that

$$V^2 = E_1 E_2 E_1^2 E_3 E_2 E_3 = E_1 E_2 E_1^2 E_2 E_3 E_2 = E_1^2 E_2 E_3 E_1^2 E_2,$$
$$V^4 = E_1^2 E_2 E_3 E_2 E_1 E_3 E_1^2 E_2 = E_1^2 E_3 E_2 E_1 E_2 = E_3 E_2 E_1 = V^{-1}.$$

Inversely, if V, W satisfy 268) and we set[1])

$$E_3 = W, \quad E_2 = V^2 W V^{-2}, \quad E_1 = V W E_2 = W V^{-2} W V^2,$$

the relations 269) will follow. We have at once $E_3^2 = I$, $E_2^2 = I$, $(E_1 E_3)^2 = I$, $(E_1 E_2)^3 = I$. Also $(E_2 E_3)^3 = I$ and $E_1^3 = I$. In fact

$$(E_2 E_3)^2 = (V^2 W V^{-2} W)^2 = V^2 W V^2 W V^{-2} W V^2 W$$
$$= V W V^{-2} W V^{-1} \cdot V^2 W V^{-2} W V^{-1} = V W V^2 W V^2 W V^{-1}$$
$$= W V^{-2} W V^{-1} \cdot V^2 W V^{-1} = W V^2 W V^{-2} = (E_2 E_3)^{-1}.$$

$$E_1^3 = V W V^2 W V^3 \cdot W V^{-2} W V^2 = V W V^2 \cdot V W V^2 W V \cdot V^{-2} W V^2$$
$$= V \cdot V W V^2 W V \cdot V^2 W V^{-1} W V^2 = V^2 W V^2 W V^3 \cdot V W V \cdot V^2$$
$$= V^2 W V^2 \cdot V W V \cdot V^3 = V^2 W V^{-2} W V^{-1} = V^{-2} W V^2 W = E_1^{-1}.$$

268. Theorem.[2]) — *The general linear homogeneous group $GLH(4,2)$ is holoedrically isomorphic with the alternating group on 8 letters.*

---

1) The later reductions depend upon the formulae

$VWV = WV^{-1}W, \quad WV^2 W = V^{-1} W V^{-2} W V^{-1}, \quad W V^{-2} W = V W V^2 W V.$

2) Jordan, Traité des substitutions, No. 516; Moore, *Math. Annalen*, vol. 51, pp. 417—444; Dickson, *ibid* vol. 54, pp. 564—569.

AUXILIARY THEOREMS ON ABSTRACT GROUPS. etc. 291

The following substitutions of $GLH(4,2)$

$$E_1 = \begin{pmatrix} 1 & 1 & 1 & 1 \\ 0 & 0 & 0 & 1 \\ 1 & 1 & 0 & 0 \\ 0 & 1 & 0 & 1 \end{pmatrix}, \quad E_2 = \begin{pmatrix} 0 & 1 & 0 & 1 \\ 0 & 0 & 1 & 0 \\ 0 & 1 & 0 & 0 \\ 1 & 0 & 1 & 0 \end{pmatrix}, \quad E_3 = \begin{pmatrix} 0 & 1 & 1 & 1 \\ 0 & 1 & 0 & 1 \\ 1 & 1 & 0 & 0 \\ 0 & 0 & 0 & 1 \end{pmatrix},$$

$$E_4 = \begin{pmatrix} 1 & 0 & 1 & 0 \\ 0 & 1 & 0 & 0 \\ 0 & 0 & 1 & 0 \\ 0 & 1 & 0 & 1 \end{pmatrix}, \quad E_5 = \begin{pmatrix} 0 & 0 & 1 & 0 \\ 0 & 1 & 0 & 1 \\ 1 & 0 & 0 & 0 \\ 0 & 0 & 0 & 1 \end{pmatrix}, \quad E_6 = \begin{pmatrix} 0 & 1 & 1 & 1 \\ 0 & 0 & 1 & 0 \\ 0 & 1 & 0 & 0 \\ 1 & 1 & 1 & 0 \end{pmatrix}$$

satisfy the relations 265) for $k = 8$ and therefore generate a subgroup L which is isomorphic with the alternating group on the letters $1, 2, \ldots, 8$. The latter group being simple, the isomorphism is holoedric. Since the order of $GLH(4,2)$ equals $\frac{1}{2}8!$ by § 99, it coincides with its subgroup L. The correspondence of generators of $L \equiv GLH(4,2)$ and $G_{\frac{1}{2}8!}$ is as follows:

270) $\quad E_1 \sim (23)(12), \quad E_2 \sim (34)(12), \quad E_3 \sim (45)(12),$
$\quad E_4 \sim (56)(12), \quad E_5 \sim (67)(12), \quad E_6 \sim (78)(12).$

269. To effect the inversion of 270), so that we shall be able to pass readily from an arbitrary substitution of L to the corresponding substitution of $G_{\frac{1}{2}8!}$, we begin with the simple identities,

$(\xi_1 \xi_3) B_{24} = E_5, \quad (\xi_2 \xi_4) B_{31} = E_5 E_4 E_5, \quad (\xi_1 \xi_4)(\xi_2 \xi_3) = E_2 E_4,$
$B_{12} B_{43} = E_2 E_6, \quad (\xi_3 \xi_4) B_{21} = E_1 E_6, \quad (\xi_1 \xi_2)(\xi_3 \xi_4) B_{32} = E_2 E_3 E_1 E_5.$

Since these relations can be solved for $E_5, E_4, E_2, E_6, E_1, E_3$ in order, their left members may be chosen as generators of L. By 270), we have

$(\xi_1 \xi_3) B_{24} \sim (67)(12), \quad (\xi_2 \xi_4) B_{31} \sim (57)(12), \quad (\xi_1 \xi_4)(\xi_2 \xi_3) \sim (34)(56),$
$B_{12} B_{43} \sim (34)(78), \quad (\xi_3 \xi_4) B_{21} \sim (23)(78),$
$\alpha \equiv (\xi_1 \xi_2)(\xi_3 \xi_4) B_{32} \sim (67)(2354).$

From these generators of L, we obtain in succession the substitutions

$(\xi_2 \xi_3) B_{32} \equiv B_{32} B_{23} = \alpha^{-1}(\xi_1 \xi_4)(\xi_2 \xi_3) \alpha (\xi_1 \xi_4)(\xi_2 \xi_3),$
$(\xi_1 \xi_2 \xi_4 \xi_3) = \alpha [(\xi_2 \xi_3) B_{32}]^{-1},$
$(\xi_3 \xi_4) B_{12} = (\xi_1 \xi_2 \xi_4 \xi_3) \cdot (\xi_2 \xi_4) B_{31} \cdot (\xi_1 \xi_2 \xi_4 \xi_3)^{-1},$
$(\xi_1 \xi_2) B_{12} \equiv B_{12} B_{21} = (\xi_3 \xi_4) B_{12} \cdot (\xi_3 \xi_4) B_{21},$
$(\xi_1 \xi_2)(\xi_3 \xi_4) = (\xi_1 \xi_2) B_{12} \cdot (\xi_3 \xi_4) B_{12},$
$B_{34}(\xi_3 \xi_4) \equiv (\xi_3 \xi_4) B_{43} = (\xi_3 \xi_4) B_{12} \cdot B_{12} B_{43},$
$B_{42} = [B_{34}(\xi_3 \xi_4)]^{-1} \cdot B_{32} \cdot B_{34}(\xi_3 \xi_4).$

292 CHAPTER XIII.

These results lead at once to the following correspondences:
$(\xi_2\xi_3)B_{32} \sim (265)(347)$, $(\xi_1\xi_2\xi_4\xi_3) \sim (27)(3645)$, $(\xi_3\xi_4)B_{12} \sim (24)(17)$,
$(\xi_1\xi_2)B_{12} \sim (187)(243)$, $(\xi_1\xi_2)(\xi_3\xi_4) \sim (18)(34)$, $(\xi_3\xi_4)B_{43} \sim (187)(234)$,
$B_{42} \sim (16)(25)(34)(78)$, $B_{32} \sim (23)(45)(67)(18)$, $(\xi_2\xi_3) \sim (18)(27)(35)(46)$.
By simple transformations, we complete the proof of the Theorem. — *The correspondences* 270) *give reciprocally*

$(\xi_1\xi_2) \sim (13)(27)(48)(56)$, $(\xi_1\xi_3) \sim (16)(27)(34)(58)$,
$(\xi_1\xi_4) \sim (18)(27)(36)(45)$,
$(\xi_2\xi_3) \sim (18)(27)(35)(46)$, $(\xi_2\xi_4) \sim (15)(27)(34)(68)$,
$(\xi_3\xi_4) \sim (14)(27)(38)(56)$,
$B_{12} \sim (12)(38)(47)(56)$, $B_{31} \sim (17)(25)(34)(68)$,
$B_{32} \sim (18)(23)(45)(67)$,
$B_{14} \sim (18)(23)(46)(57)$, $B_{24} \sim (17)(26)(34)(58)$,
$B_{43} \sim (12)(37)(48)(56)$.

By § 100, these relations enable us to pass from an arbitrary substitution of the linear group on 4 indices modulo 2 to the corresponding even substitution on 8 letters.

Abstract form of the simple group $FO(5, 3)$[1]), §§ 270—274.

270. By the notation of § 194, $FO(5,3)$ denotes the group $O'_1(5,3)$. By §§ 189 and 181, it is of order 25920 and is generated by the substitutions[2])
$$C_iC_j, \quad (\xi_i\xi_j)(\xi_k\xi_l), \quad w \equiv W_{1234} \quad (i,j,k,l = 1, \ldots, 5).$$
It has a commutative subgroup L_{16} composed of the substitutions I, C_1C_2, C_1C_3, C_1C_4, C_1C_5, C_2C_3, C_2C_4, C_2C_5, C_3C_4, C_3C_5, C_4C_5, $C_1C_2C_3C_4$, $C_1C_2C_3C_5$, $C_1C_3C_4C_5$ and $C_2C_3C_4C_5$. The $(\xi_i\xi_j)(\xi_k\xi_l)$ generate a subgroup L_{60} of the even linear substitutions on $\xi_1, \ldots, \xi_5$. The groups L_{16} and L_{60} are commutative with each other and have only the identity in common; hence they generate a subgroup A_{960} of $FO(5,3)$. We readily determine the abstract forms of these subgroups. By § 265, we have the theorem:

---

[1]) Taken from the author's papers, *Comptes Rendus*, vol. 128, pp. 873—875; *Proceed. Lond. Math. Soc.*, vol. 32, pp. 3—10. In the earlier paper, *Proceed. Lond. Math. Soc.*, vol. 31, pp. 30—68, another set of generators was determined by a more complicated analysis.

[2]) For $p^n = 3$, $O^{\alpha,\beta}_{i,j}$ is either the identity, C_iC_j, $(\xi_i\xi_j)C_i$ or $(\xi_i\xi_j)C_j$, the first two alone being of the form Q_{ij}. Here $(\xi_i\xi_j)$ denotes the *linear* substitution $\xi'_i = \xi_j$, $\xi'_j = \xi_i$. They are to be compounded as linear substitutions; for example, $(\xi_1\xi_3)(\xi_1\xi_2) = (\xi_1\xi_2\xi_3)$. Also C_i denotes the substitution changing the sign of the index ξ_i.

AUXILIARY THEOREMS ON ABSTRACT GROUPS. etc. 293

The abstract group G_{60} generated by E_1, E_2, E_3 subject to the relations
$$271)\quad E_1^3 = E_2^2 = E_3^2 = I, \quad (E_1 E_2)^3 = (E_2 E_3)^3 = (E_1 E_3)^2 = I$$
is put into holoedric isomorphism with L_{60} by the correspondences
$$272)\quad E_1 \sim (\xi_1 \xi_2 \xi_3), \quad E_2 \sim (\xi_3 \xi_4)(\xi_1 \xi_2), \quad E_3 \sim (\xi_4 \xi_5)(\xi_1 \xi_2).$$

The following theorem is quite evident:

The abstract group G_{16} generated by B_1, B_2, B_3, B_4 subject to the relations
$$273)\qquad B_i^2 = I, \quad B_i B_j = B_j B_i \qquad (i, j = 1, 2, 3, 4)$$
is put into holoedric isomorphism with L_{16} by the correspondences
$$274)\quad B_1 \sim C_1 C_2, \quad B_2 \sim C_2 C_3, \quad B_3 \sim C_3 C_4, \quad B_4 \sim C_4 C_5.$$

If we impose the relations 275) below, the two groups G_{60} and G_{16} will be permutable. Writing the analogous relations between the corresponding orthogonal substitutions 272), 274), we readily see that they are satisfied. We have therefore the theorem:

The abstract group generated by $E_1, E_2, E_3, B_1, B_2, B_3, B_4$ subject to the generational relations 271), 273), and

$$275)\quad\begin{cases} E_1^{-1} B_1 E_1 = B_1 B_2, & E_1^{-1} B_2 E_1 = B_1, & E_1^{-1} B_3 E_1 = B_2 B_3, \\ & E_1^{-1} B_4 E_1 = B_4, & \\ E_2^{-1} B_1 E_2 = B_1, & E_2^{-1} B_2 E_2 = B_1 B_2 B_3, & E_2^{-1} B_3 E_2 = B_3, \\ & E_2^{-1} B_4 E_2 = B_3 B_4, & \\ E_3^{-1} B_1 E_3 = B_1, & E_3^{-1} B_2 E_3 = B_1 B_2, & E_3^{-1} B_3 E_3 = B_3 B_4, \\ & E_3^{-1} B_4 E_3 = B_4, & \end{cases}$$

is of order 960 and is holoedrically isomorphic with the linear group A_{960}.

271. Theorem. — *The abstract group G_{960} of § 270 may be generated by the operators E_1, E_2, E_3, B_1 subject to the generational relations*
$$276)\quad\begin{array}{c} E_1^3 = E_2^2 = E_3^2 = B_1^2 = I, \quad (E_1 E_2)^3 = (E_2 E_3)^3 = (B_1 E_1)^3 = I, \\ (E_1 E_3)^2 = (B_1 E_2)^2 = (B_1 E_3)^2 = I. \end{array}$$

These relations follow immediately from 271), 273), 275), with the exception of $(B_1 E_1)^3 = I$, which is derived from the first two of 275):
$$E_1^{-1} B_1 E_1 B_1 = B_2 = E_1 B_1 E_1^{-1},$$
together with $E_1^3 = B_1^2 = I$. Furthermore, we have by 275),
$$277)\quad B_2 = E_1 B_1 E_1^2, \quad B_3 = E_1 E_2 E_1^2 B_1 E_1 E_2 E_1^2, \quad B_4 = E_2 E_3 B_3 E_3 E_2.$$

Inversely, if B_2, B_3, B_4 be defined by 277), the relations 271), 273), 275) all follow from 276). Since $B_1 E_1$ is of period 3,

$$B_1 B_2 = B_1 E_1 B_1 E_1^2 = E_1^2 B_1 E_1,$$
$$B_1 B_2 B_3 = E_1^2 B_1 E_1 \cdot B_3 = E_1^2 B_1 E_1^2 E_2 E_1^2 B_1 E_1 E_2 E_1^2$$
$$= E_1^2 B_1 \cdot E_2 E_1 E_2 \cdot B_1 E_1 E_2 E_1^2$$
$$= E_1^2 E_2 B_1 E_1 B_1 E_2 E_1 E_2 E_1^2 \text{ (interchanging } B_1 E_2 \text{ with } E_2 B_1 \text{)}$$
$$= E_1^2 E_2 \cdot E_1^2 B_1 E_1^2 \cdot E_1^2 E_2 E_1^2 \cdot E_1^2 = E_2 E_1 E_2 B_1 E_1 E_2 E_1$$
$$= E_2 E_1 E_2 B_1 \cdot E_2 E_1^2 E_2 = E_2 E_1 B_1 E_1^2 E_2 \equiv E_2^{-1} B_2 E_2.$$

$$E_3^{-1} B_2 E_3 = E_3 E_1 B_1 E_1^2 E_3 = E_1^2 E_3 B_1 E_3 E_1 = E_1^2 B_1 E_1 = B_1 B_2.$$

Upon setting $B_1 E_2 = E_2 B_1$, $B_1 E_1^2 B_1 E_1 = E_1 B_1 E_1^2$, we find that
$$B_2 B_3 = E_1 B_1 E_2 E_1^2 B_1 E_1 E_2 E_1^2 = E_1 E_2 \cdot E_1 B_1 E_1^2 \cdot E_2 E_1^2$$
$$= E_2 \, E_1^2 E_2 B_1 E_2 E_1 E_2 = E_2 E_1^2 B_1 E_1 E_2 \equiv E_1^{-1} B_3 E_1.$$

Since $E_2 E_1 E_2 E_1^2 = E_1 E_2 E_1^2 E_2$, we get
$$E_2^{-1} B_3 E_2 = E_2 E_1 E_2 E_1^2 B_1 E_1 E_2 E_1^2 E_2$$
$$= E_1 E_2 E_1^2 (E_2 B_1 E_2) E_1 E_2 E_1^2 = B_3.$$

Since $E_1 E_2 E_1^2 E_2 \cdot E_3 E_1 E_2 E_1^2 = E_2 E_1 E_2 E_1^2 \cdot E_3 E_1 E_2 E_1^2$
$$= E_2 E_1 E_2 E_1 E_3 E_2 E_1^2 = E_1^2 E_2 E_3 E_2 E_1^2 = E_1^2 E_3 E_2 E_3 E_1^2$$
$$= E_3 E_1 E_2 E_1 E_3 = E_3 E_2 E_1^2 E_2 E_3,$$

we find by 277) that
$$B_3 B_4 = E_1 E_2 E_1^2 B_1 \cdot E_3 E_2 E_1^2 E_2 E_3 \cdot B_1 E_1 E_2 E_1^2 E_3 E_2$$
$$= E_1 E_2 E_1^2 E_3 E_2 B_1 E_1^2 B_1 E_2 E_3 E_1 E_2 E_3 E_1 E_2$$
$$= E_1 E_2 E_3 E_1 E_2 \cdot E_1 B_1 E_1 \cdot E_2 E_1^2 E_3 E_2 E_3 E_1 E_2$$
$$= E_1 E_2 E_3 E_2 E_1^2 (E_2 B_1 E_2) E_1 E_2 E_1^2 E_3 E_1^2 E_2 E_1^2$$
$$= E_1 E_3 E_2 E_1 (E_3 B_1 E_3) E_1^2 E_2 E_3 E_1^2$$
$$= E_3 \cdot E_1 E_2 E_1^2 B_1 E_1 E_2 E_1^2 \cdot E_3 \equiv E_3^{-1} B_3 E_3,$$

upon setting $E_3 B_1 E_3 = B_1 = E_2 B_1 E_2$, $E_1^2 E_2 E_1 E_2 = E_1 E_2 E_1^2$ and applying also the equation given by taking the reciprocals of the last substitutions. Using 277) and the last result,
$$E_2^{-1} B_4 E_2 = E_3 B_3 E_3 = B_3 B_4.$$

In order to prove that $E_1^{-1} B_4 E_1 = B_4$, we note that
$$E_1^{-1} (E_3 B_3 E_3) E_1 = E_1^{-1} (B_3 B_4) E_1,$$
or
$$E_3 E_1 B_3 E_1^2 E_3 = B_2 B_3 \cdot E^{-1} B_4 E_1.$$

But the left member equals $B_2 B_3 B_4$. Indeed, by the earlier results,
$$E_1^{-1} B_1 B_2 B_3 E_1 = B_1 B_2 \cdot B_1 \cdot B_2 B_3 = (E_1^2 B_1 E_1)^2 B_3 = B_3.$$
Hence
$$E_1 B_3 E_1^2 = B_1 B_2 B_3, \quad E_3 E_1 B_2 B_3 E_3 = B_1 \cdot B_1 B_2 \cdot B_3 B_4.$$

Finally,
$$E_3^{-1}B_4E_3 = E_3E_2E_3B_3E_3E_2E_3 = E_2E_3(E_2B_3E_2)E_3E_2 = B_4.$$

We have now derived from 276) all of the relations 275). It remains to derive 273). Since B_2, B_3, B_4 are conjugate with B_1 by 277), they are of period 2. By 275), B_1B_2 is conjugate with B_1, B_2B_3 with B_3, B_3B_4 with B_3. Hence they are of period 2 and therefore B_3 is commutative with B_2 and B_4, B_1 with B_2. Since $E_2^{-1}B_2E_2$ is its own reciprocal, we have
$$B_1B_3B_2 = B_1B_2B_3 = (B_1B_2B_3)^{-1} = B_3B_2B_1 = B_3B_1B_2,$$
so that $B_1B_3 = B_3B_1$. Since $B_2B_3B_4$ was shown to be the transformed of B_3 by $E_1^2E_3$, we have
$$B_3B_2B_4 = B_2B_3B_4 = (B_2B_3B_4)^{-1} = B_4B_3B_2 = B_3B_4B_2.$$
Hence B_2 is commutative with B_4. Since B_1 is commutative with B_3, E_2 and E_3, it is commutative with B_4 by 277).

272. Theorem. — *Every substitution of $FO(5, 3)$ is given once and but once by the following 27 sets, in which A denotes the subgroup A_{960}:*

$$R_t \equiv Aw^t \qquad (t = 0, 1, 2)$$
$$R_{sit} \equiv Aw^s(\xi_1\xi_2)(\xi_i\xi_5)w^t \qquad \begin{pmatrix} s = 1, 2;\ t = 0, 1, 2 \\ i = 1, 2, 3, 4 \end{pmatrix}.$$

Since w is not in A, a substitution of R_t belongs to R_τ if and only if $t = \tau$. If a substitution of R_t belong to $R_{si\tau}$, the product
$$w^s(\xi_1\xi_2)(\xi_i\xi_5)w^{\tau-t}$$
must belong to A, whereas it replaces ξ_5 by a linear function of $\xi_1, \xi_2, \xi_3, \xi_4$, every coefficient being ± 1.

If a substitution of R_{sit} belong to $R_{\sigma j \tau}$, the product
$$S \equiv w^s(\xi_1\xi_2)(\xi_i\xi_5)w^{t-\tau}(\xi_j\xi_5)(\xi_1\xi_2)w^{-\sigma}$$
must belong to A. Supposing first that $t - \tau \neq 0$, we show that S replaces ξ_5 by a function involving more than one index and therefore does not belong to A. In fact, $w^{-s}S$ replaces ξ_5 by a function of the form
$$f \equiv \pm \xi_a \pm \xi_b \pm \xi_c \pm \xi_5,$$
where a, b, c are three of the integers 1, 2, 3, 4. Then w^s replaces f by $f_1 \pm \xi_5$, where f_1 is a linear function of $\xi_1, \xi_2, \xi_3, \xi_4$ with coefficients not all $\equiv 0$ (mod 3). Hence S replaces ξ_5 by $f_1 \pm \xi_5$, involving two or more indices. Suppose, however, that $t = \tau$. If then $i \neq j$, S replaces ξ_5 by a linear function of $\xi_1, \xi_2, \xi_3, \xi_4$ with coefficients ± 1. If $i = j$, $S \equiv w^{s-\sigma}$, which belongs to A only if $s = \sigma$. But in the latter case, the two sets R_{sit} and $R_{\sigma j \tau}$ are themselves identical.

273. Theorem. — *The abstract group O generated by the operators*[1]) E_1, E_2, E_3, B_1, W *subject to the generational relations* 271) *and*

278) $\quad W^3 = I, \quad W^{-1}E_1W = B_1E_1, \quad W^{-1}E_2W = B_1E_2,$
$\qquad W^{-1}B_1W = B_2E_2B_2,$
279) $\qquad\qquad WB_4W = B_4E_1B_1E_2E_1^2,$
280) $\quad (WE_3E_2E_1W)E_3 = E_1^2E_2E_3E_2E_1(WE_3E_2E_1W),$

B_2 *and* B_4 *being defined by* 277), *is holoedrically isomorphic with* $FO(5,3)$.

Writing these relations for the corresponding orthogonal substitutions as defined by 272), 274) and $W \sim w$, we obtain relations which reduce to identities modulo 3. The order Ω of O is therefore $\geqq 25920$. The holoedric isomorphism will be established when it is shown that $\Omega \leqq 25920$. To prove this statement, consider the following 27 sets[2]) of operators of O, those of the first set being the operators of $G \equiv G_{960}$:

$R_t \equiv GW^t, \quad R_{s4t} \equiv GW^sE_3W^t, \quad R_{s3t} \equiv GW^sE_3E_2W^t \quad \begin{pmatrix} t=0,1,2 \\ s=1,2 \end{pmatrix}.$
$R_{s2t} \equiv GW^sE_3E_2E_1W^t, \quad R_{s1t} \equiv GW^sE_3E_2E_1^2W^t$

It is shown in the next section that the generators E_1, E_2, E_3, W, and therefore an arbitrary operator α of the group O, gives rise to a mere interchange of the above 27 sets when applied as a right-hand multipliers. Since the first set G contains the identity I, the product $I\alpha \equiv \alpha$ lies in one of the 27 sets. Hence O contains at most $27 \cdot 960 \equiv 25920$ operators. In particular, it follows that the 27 sets form a rectangular table for O with the operators G_{960} in the first row.

We make use of the formulae derived from 271), 278), 279), 280), 277):

$\qquad E_3E_2E_1^2B_3 = B_2E_3E_2E_1^2, \quad E_3E_2E_1B_3 = B_1B_2E_3E_2E_1,$
$\qquad WE_2 = B_3W, \quad W^2E_2 = B_1E_2W^2, \quad WE_1 = B_3E_2E_1W,$
$\qquad W^2E_1 = B_1E_1W^2,$
281) $\quad WB_1 = B_3E_2W, \quad W^2B_1 = B_3B_1E_2W^2, \quad WB_3 = E_2B_1W,$
$\qquad E_1W = WB_1E_1,$
$\qquad E_1^2W = WE_1^2B_1, \quad E_2W = WE_2B_1, \quad E_2W^2 = W^2B_3,$
$\qquad E_2E_1W = WE_2E_1.$

---

1) For simplicity B_1 is retained. It may be dropped since
$$B_1 = W^{-1}E_1WE_1^{-1} = W^{-1}E_2WE_2^{-1}.$$

2) They correspond in $FO(5,3)$ with the 27 rows of the rectangular table.

AUXILIARY THEOREMS ON ABSTRACT GROUPS. etc. 297

274. Theorem. — *When applied as right-hand multipliers to the above 27 sets, the generators W, E_1, E_2, E_3 give rise to the respective permutations:*

$[W]$: $(R_0 R_1 R_2)(R_{si0} R_{si1} R_{si2})$,
$[E_1]$: $(R_{s10} R_{s30} R_{s20})(R_{s21} R_{s31} R_{2s41})(R_{s22} R_{2s12} R_{s32})$,
$[E_2]$: $(R_{s10} R_{s20})(R_{s30} R_{s40})(R_{s22} R_{2s12})(R_{s32} R_{s42})(R_{131} R_{231})(R_{141} R_{241})$,
$[E_3]$: $(R_s R_{s40})(R_{s10} R_{s20})(R_{211} R_{242})(R_{221} R_{132})(R_{112} R_{222})(R_{122} R_{141})$
$\qquad(R_{212} R_{231})(R_{131} R_{241})$,

where $i = 1, 2, 3, 4$ and $s = 1, 2$, while the first subscript $2s$ is to be reduced modulo 3.

The form of $[W]$ is evident. Consider the multiplier E_2.

$R_0 E_2 = R_0$, $\quad R_1 E_2 = GWE_2 = GB_3 W = GW = R_1$.
$\qquad R_2 E_2 = GW^2 E_2 = GB_1 E_2 W^2 = GW^2 = R_2$.
$R_{s11} E_2 = GW^s E_3 E_2 E_1^2 W E_2 = GW^s E_3 E_2 E_1^2 B_3 W = R_{s11}$
$\qquad\qquad\qquad\qquad\qquad\qquad\qquad\qquad\qquad$ [by 281)].
$R_{s21} E_2 = GW^s E_3 E_2 E_1 W E_2 = GW^s E_3 E_2 E_1 B_3 W = R_{s21}$
$\qquad\qquad\qquad\qquad\qquad\qquad\qquad\qquad\qquad$ [by 281)].
$R_{s10} E_2 = GW^s E_3 E_2 E_1^2 E_2 = GW^s E_3 E_1 E_2 E_1$
$\qquad = GW^s E_1^2 E_3 E_2 E_1 = R_{s20}$.
$R_{s30} E_2 = GW^s E_3 E_2 \cdot E_2 = R_{s40}$.
$R_{s22} E_2 = GW^s E_3 E_2 E_1 \cdot B_1 E_2 W^2 = GW^s B_4 E_1 E_3 E_2 E_1^2 W^2$
$\qquad = R_{2s12}$,

by 279), since $E_3 E_2 E_1 B_1 E_2 = B_2 B_3 B_4 E_1 E_3 E_2 E_1^2$.

$R_{s32} E_2 = GW^s E_3 E_2 W^2 E_2 = GW^s E_3 E_2 \cdot B_1 E_2 W^2$
$\qquad = GW^s B_1 E_3 W^2 = R_{s42}$.
$R_{131} E_2 = GW E_3 E_2 \cdot B_3 W = GW B_3 B_4 E_3 E_2 W$
$\qquad = GW^2 E_3 E_2 W = R_{231}$.
$R_{141} E_2 = GW E_3 \cdot B_3 W = GW^2 E_3 W = R_{241}$.

Next, $\quad R_{s21} E_1 = GW^s E_3 E_2 E_1 \cdot B_3 E_2 E_1 W = GW^s E_3 (E_2 E_1)^2 W$
$\qquad = GW^s E_1 E_3 E_2 W = R_{s31}$, upon applying 281).
$R_{s22} E_1 = GW^s E_3 E_2 E_1 \cdot B_1 E_1 W^2 = GW^s B_2 B_3 B_4 E_3 E_2 E_1^2 W^2$
$\qquad = R_{2s12}$.
$R_{2s12} E_1 = GW^{2s} E_3 E_2 E_1^2 \cdot B_1 E_1 W^2$
$\qquad = GW^{2s} B_1 B_2 B_3 B_4 E_3 E_2 W^2 = R_{s32}$.
$R_{s11} E_1 = GW^s E_3 E_2 E_1^2 \cdot B_3 E_2 E_1 W = GW^s E_3 E_2 E_1^2 \cdot E_2 E_1 W$
$\qquad\qquad\qquad\qquad\qquad\qquad\qquad\qquad\qquad$ [by 281)]
$\qquad = GW^s E_3 E_1 E_2 E_1^2 W = GW^s E_1^2 E_3 E_2 E_1^2 W = R_{s11}$.

The remaining cases follow immediately.

298   CHAPTER XIII.

For the right-hand multiplier E_3, the calculations are not so simple.

$R_{121}E_3 = GWE_3E_2E_1WE_3 = GWE_3E_2E_1W = R_{121}$ [by 280)].
$R_{221}E_3 = GW(E_1^2E_2E_3E_2E_1WE_3E_2E_1W)$
$= GWE_3E_2E_1WE_3 \cdot E_2E_1W.$
$= GWE_3E_2E_1W \cdot E_2E_1W = GWE_3E_2E_1 \cdot E_2E_1W^2 = R_{132}.$
$R_{211}E_3 = GW^2E_3E_2E_1 \cdot WB_1E_1 \cdot E_3 = GW^2E_3E_2E_1WE_3B_1E_1^2$
$= R_{221}E_3B_1E_1^2 = R_{132}B_1E_1^2 = GWE_3E_2 \cdot B_1B_2E_2E_1^2W^2$
$= GWB_1B_2B_3B_4E_3E_1^2W^2 = GW^2E_3W^2 = R_{242}.$
$R_{231}E_3 = GW^2E_3E_2 \cdot E_1WE_1^2B_1 \cdot E_3 = R_{221}E_3E_1B_1 = R_{132}E_1B_1$
$= GWE_3E_2 \cdot B_1E_1B_3B_1E_2W^2 = GWB_3B_4E_1 \cdot E_3E_2E_1^2W^2$
$= R_{212}.$
$R_{131}E_3 = GWE_3E_2 \cdot E_1WE_1^2B_1 \cdot E_3 = GWE_3E_2E_1WE_1B_1$ [by 280)]
$= GWE_3E_2E_1 \cdot B_3E_2E_1B_3E_2W = GW^2E_3W = R_{241}$
[by 281)].
$R_{232}E_3 = GW^2E_3 \cdot W^2B_3 \cdot E_3 = GW^2E_3W^2E_3B_3B_4$
$= R_{242}E_3B_3B_4 = R_{211}B_3B_4$
$= GW^2E_3E_2E_1^2 \cdot E_2B_1 \cdot B_4E_1B_1E_2E_1^2W^2$
$= GW^2B_1B_3E_3E_1^2E_2W^2 = R_{232}.$
$R_{122}E_3 = GWE_3E_2 \cdot B_1W^2E_1 \cdot E_3 = GWE_3E_2W^2E_3E_1^2$
$= R_{132}E_3E_1^2 = R_{221}E_1^2 = R_{141}.$
$R_{222}E_3 = GW^2E_3E_2 \cdot B_1W^2E_1 \cdot E_3 = GW^2E_3E_2W^2E_3E_1^2$
$= R_{232}E_3E_1^2 = R_{232}E_1^2 = R_{112}.$
$R_{111}E_3 = GWE_3E_2 \cdot WE_1^2B_1 \cdot E_3 = R_{131}E_3E_1B_1 = R_{241}E_1B_1 = R_{121}B_1$
$= GWE_3E_2E_1 \cdot B_3E_2W = GWE_3E_2E_1E_2W = R_{111}$ [by 281)].
$R_{142}E_3 = GWE_3W^2E_3 = GWE_3E_2W^2B_3E_3$
$= GWE_3E_2W^2E_3B_3B_4 = R_{132}E_3B_3B_4$
$= R_{221}B_3B_4 = GW^2E_3E_2E_1 \cdot E_2B_1 \cdot B_4E_1B_1E_2E_1^2W^2$
$= GW^2B_2B_4E_3E_1^2W^2 = GWE_3W^2 = R_{142}.$

275. Theorem. — *The simple group $HA(4, 2^2)$ is put into holoedric isomorphism with the abstract group O by the correspondences of generators*

$$W \sim \begin{Bmatrix} 0 & 1 & 1 & 0 \\ 1 & 1 & 1 & 0 \\ 0 & 0 & 1 & 0 \\ 1 & 0 & 1 & 1 \end{Bmatrix}, \quad E_1 \sim \begin{Bmatrix} 1 & 0 & I^2 & 0 \\ 0 & 0 & I^2 & I^2 \\ I & 0 & 0 & 0 \\ I & I & 1 & 1 \end{Bmatrix}, \quad E_2 \sim \begin{Bmatrix} 0 & 0 & 1 & 0 \\ 0 & 0 & 1 & 1 \\ 1 & 0 & 0 & 0 \\ 1 & 1 & 0 & 0 \end{Bmatrix},$$

$$E_3 \sim \begin{Bmatrix} 0 & 1 & I^2 & 0 \\ 0 & 0 & I^2 & I^2 \\ I & 0 & 1 & 1 \\ I & I & 1 & 1 \end{Bmatrix}, \quad B_1 \sim \begin{Bmatrix} 1 & 1 & 0 & 0 \\ 0 & 1 & 0 & 0 \\ 0 & 0 & 0 & 1 \\ 0 & 0 & 1 & 0 \end{Bmatrix}, \quad B_2 \sim \begin{Bmatrix} 1 & 1 & I^2 & I^2 \\ 0 & 1 & 0 & 0 \\ 0 & I & 1 & 0 \\ 0 & I & 0 & 1 \end{Bmatrix},$$

$$B_3 \sim \begin{Bmatrix} 1 & 0 & 1 & 1 \\ 0 & 1 & 0 & 0 \\ 0 & 1 & 1 & 0 \\ 0 & 1 & 0 & 1 \end{Bmatrix}, \quad B_4 \sim \begin{Bmatrix} 1 & 0 & I^2 & I^2 \\ 0 & 1 & 0 & 0 \\ 0 & I & 1 & 0 \\ 0 & I & 0 & 1 \end{Bmatrix},$$

where I is a root of the irreducible congruence $x^2 \equiv x + 1 \pmod{2}$.
Indeed, it may be verified that these correspondences preserve the generational relations (§ 273) prescribed for the generators of O. Furthermore, by § 132 the order of $HA(4, 2^2)$ is 25920, so that the isomorphism is holoedric.

276. The correspondences established in the last section enable us to pass readily from any orthogonal substitution S to the corresponding substitution of $HA(4, 2^2)$. In fact, we have only to express S in terms of the simple generators w, $(\xi_1 \xi_2 \xi_3)$, $(\xi_3 \xi_4)(\xi_1 \xi_2)$, $(\xi_4 \xi_5)(\xi_1 \xi_2)$, $C_1 C_2$, $C_2 C_3$, $C_3 C_4$, $C_4 C_5$ of $FO(5, 3)$.

It is not difficult to invert these correspondences and obtain the orthogonal substitutions which correspond to the simplest set of generators of $HA(4, 2^2)$, viz.: —

$$L_{1,1} \sim C_2 C_3 C_4 C_5, \quad M_2 \sim C_1 C_3 C_4 C_5, \quad J \sim (\xi_3 \xi_5 \xi_4) C_3 C_5,$$

$$M_1 \sim \begin{Bmatrix} 1 & 0 & 1 & 1 & 2 \\ 0 & 2 & 0 & 0 & 0 \\ 1 & 0 & 1 & 2 & 1 \\ 1 & 0 & 2 & 1 & 1 \\ 2 & 0 & 1 & 1 & 1 \end{Bmatrix}, \quad L_{2,1} \sim \begin{Bmatrix} 2 & 0 & 0 & 0 & 0 \\ 0 & 1 & 1 & 1 & 2 \\ 0 & 1 & 1 & 2 & 1 \\ 0 & 1 & 2 & 1 & 1 \\ 0 & 2 & 1 & 1 & 1 \end{Bmatrix}, \quad R_{2,1,1} \sim \begin{Bmatrix} 0 & 2 & 1 & 1 & 1 \\ 2 & 0 & 1 & 1 & 1 \\ 1 & 1 & 2 & 0 & 2 \\ 1 & 1 & 0 & 2 & 2 \\ 1 & 1 & 2 & 2 & 0 \end{Bmatrix}.$$

Here J denotes the hyperabelian substitution of period 3:
$$\xi_1' = I\xi_1, \quad \eta_1' = I\eta_1, \quad \xi_2' = I^2 \xi_2, \quad \eta_2' = I^2 \eta_2.$$

277. By § 189, the orthogonal group $FO(5, 3)$ is holoedrically isomorphic with the Abelian group $A(4, 3)$. Given an arbitrary Abelian substitution, the process of forming the second compound and a subsequent transformation of indices (§ 189) enables us to find quite readily the corresponding orthogonal substitution. The inverse problem is solved by employing the set[1] of Abelian substitutions which correspond to the simplest orthogonal generators w, $(\xi_1 \xi_2 \xi_3)$, etc.

---
[1] *Transact. Amer. Math. Soc.*, July, 1900, p. 366.

300   CHAPTER XIII.

278. Theorem.[1]) — *The special linear homogeneous group $SLH(2, p^n)$ of binary linear substitutions of determinant unity in the $GF[p^n]$ is holoedrically isomorphic with the abstract group L generated by the operators T and S_λ, where λ runs through the series of p^n marks of the field, subject to the generational relations*

a) $S_0 = I$, $S_\lambda S_\mu = S_{\lambda+\mu}$ $\qquad$ (λ, μ any marks)
b) $T^4 = I$, $S_\lambda T^2 = T^2 S_\lambda$,
c) $S_\lambda T S_\mu T S_{\frac{\lambda-1}{\lambda\mu-1}} T S_{-(\lambda\mu-1)} T S_{\frac{\mu-1}{\lambda\mu-1}} T = I$ $\quad$ (λ, μ any marks, $\lambda\mu \neq 1$).

Since the relations a), b), c) are satisfied by the substitutions

$$T = \begin{pmatrix} 0, & -1 \\ 1, & 0 \end{pmatrix}, \quad S_\lambda = \begin{pmatrix} 1, & \lambda \\ 0, & 1 \end{pmatrix}$$

which (§ 100, Cor. II) serve to generate $SLH(2, p^n)$, the order l of the abstract group is at least $p^n(p^{2n} - 1)$. We proceed to prove that l is at most $p^n(p^{2n} - 1)$. Then will $SLH(2, p^n)$ and L be of equal order and so holoedrically isomorphic.

Consider the following sets of operators of L

$$S_\sigma T S_\alpha T S_\alpha^{-1}, \quad S_\sigma T S_\alpha T S_\tau T \qquad (\sigma, \alpha, \tau \text{ arbitrary, } \alpha \neq 0).$$

At most $p^n(p^n - 1) + p^{2n}(p^n - 1) \equiv (p^n - 1)p^n(p^n + 1)$ of them are distinct. If it be shown that every operator of L occurs in these sets, it will follow that $l \leq p^n(p^{2n} - 1)$. The proof consists in showing that the product of any operator of the sets by T or by any S_λ equals an operator of the sets. Since an arbitrary operator Σ of L is derived from T and S_λ, it will follow that $I\Sigma \equiv \Sigma$ belongs to the sets.

In view of a) the reciprocal of S_λ is $S_{-\lambda}$. For $\lambda = 1$, $\mu \neq 1$, c) gives

d) $\qquad S_1 T^3 S_1 T S_1 T \equiv (S_1 T^3)^3 = I$.

Applying T as a right-hand multiplier, the product of any operator of the first set by T gives one of the second set. We next show that

$$S_\sigma T S_\alpha T S_{\alpha^{-1}} T \cdot T = S_{\sigma - 2\alpha^{-1}} T S_{-\alpha} T S_{-\alpha^{-1}}.$$

Applying a) and b) the condition for this identity is seen to be

e) $\qquad T S_\alpha T S_{2\alpha^{-1}} T S_\alpha T S_{2\alpha^{-1}} T^2 = I$.

For $p = 2$, it reduces to an identity. For $p > 2$, we have by c)

$$T S_\alpha T S_{2\alpha^{-1}} T S_{\alpha-1} T S_{-1} T S_{2\alpha^{-1}-1} = I.$$

From this e) follows upon replacing $S_{-1} T S_{-1} T S_{-1}$ by T^3 as allowed by d).

---

1) Due to Professor Moore, who gave a different proof.

For operators of the second set with $\alpha \neq 0$, $\tau \neq \alpha^{-1}$, we prove that
$$S_\sigma T S_\alpha T S_\tau T \cdot T = S_{\sigma_1} T S_{\alpha_1} T S_{\tau_1} T,$$
where σ_1, α_1, τ_1 are suitably chosen marks, $\alpha_1 \neq 0$. The equivalent condition
$$S_\alpha T S_\tau T S_{-\tau_1} T S_{-\alpha_1} T S_{-\sigma_1+\sigma} T = I$$
may be satisfied by c) by proper choice of τ_1, α_1, σ_1, with
$$\alpha_1 \equiv \alpha\tau - 1 \neq 0.$$

We next apply S_ϱ as a right-hand multiplier. $S_\sigma T S_\alpha T S_{\alpha^{-1}} S_\varrho$ will be of the form $S_{\sigma_1} T S_{\alpha_1} T S_{\tau_1} T^3$, and consequently belong to the sets by the previous proof, if we have
$$S_\alpha T S_{\alpha^{-1}+\varrho} T S_{-\tau_1} T S_{-\alpha_1} T S_{-\sigma_1+\sigma} T = I.$$
Since $\alpha(\alpha^{-1}+\varrho) \equiv 1 + \alpha\varrho \neq 1$, this condition is of the form c) if α_1, σ_1, τ_1 be suitably chosen. If $\varrho = \alpha/(\alpha\tau - 1)$, so that $\alpha\tau \neq 1$, we have, by c),
$$S_\sigma T S_\alpha T S_\tau T \cdot S_\varrho = S_{\sigma + \frac{1-\tau}{\alpha\tau-1}} T S_{\alpha\tau - 1} T S_{\frac{1}{\alpha\tau-1}}.$$
For the case $A \equiv \alpha - \varrho(\alpha\tau - 1) \neq 0$, we prove that
f) $\qquad S_\sigma T S_\alpha T S_\tau T \cdot S_\varrho = S_{\sigma+\tau\varrho A^{-1}} T S_A T S_{\alpha\tau A^{-1}} T.$
If $\alpha\tau \neq 1$, we replace $T S_\alpha T S_\tau T$ by its equivalent derived from c) and find that condition f) becomes
$$S_A T S_{\alpha\tau A^{-1}} T S_{\frac{A-1}{\alpha\tau-1}} T S_{-(\alpha\tau-1)} T S_{\frac{\alpha\tau A^{-1}-1}{\alpha\tau-1}} T = I,$$
and hence is satisfied from c). If, however, $\alpha\tau = 1$, so that $A = \alpha$, then f) takes the simpler form
f') $\qquad T S_\alpha T S_{\alpha^{-1}} T S_\varrho = S_{\alpha^{-2}\varrho} T S_\alpha T S_{\alpha^{-1}} T.$
If also $\varrho \neq \alpha$, we replace $T S_{\alpha^{-1}} T S_\varrho$ by its equivalent derived from c) and find the condition, where $\nu \equiv \alpha^{-1}\varrho - 1$,
$$T S_{(1-\alpha)/\nu} T^{-1} S_\nu T S_{(1-\alpha^{-1})/\nu} T = S_{\alpha^{-2}\varrho} T S_\alpha T S_{\alpha^{-1}} T.$$
This reduces to the identity c) for $\lambda = \alpha^{-2}\varrho$, $\mu = \alpha$, whence $\lambda\mu \neq 1$. In particular, f') is true if $\varrho = \alpha + \varkappa$, $\varkappa \neq 0$, so that
$$(T S_\alpha T S_{\alpha^{-1}} T S_\alpha) S_\varkappa = S_{\alpha^{-2}\varkappa} (S_{\alpha^{-1}} T S_\alpha T S_{\alpha^{-1}} T).$$
The products in the parentheses are identical and so f') is true for $\varrho = \alpha$, if the following condition be true for any particular mark $\varkappa \neq 0$,
$$S_{-\alpha^{-2}\varkappa}(T S_\alpha T S_{\alpha^{-1}} T S_\alpha) S_\varkappa = (T S_\alpha T S_{\alpha^{-1}} T S_\alpha).$$
The latter is of the form f') for $\varrho \equiv -\varkappa$ and hence is true if $-\varkappa \neq \alpha$. But marks $\varkappa \neq 0$, $-\alpha$ exist if $p^n > 2$. For $p^n = 2$, $\alpha = 1$, so that f') is true for any ϱ by d).

Corollary. — *The quotient-group $LF(2, p^n)$ is holoedrically isomorphic with the abstract group F generated by the operators T and S_λ subject to the relations $T^2 = I$ together with* a) *and* c).

279. For $\lambda = 0$ or 1 or for $\mu = 0$ or 1, relations c) always reduce to d) upon applying a) and b). For the group $LF(2, p^n)$, d) becomes

D) $\qquad (S_1 T)^3 = I$.

If neither λ nor μ is 0 or 1, the product of any two consecutive subscripts in c) is not unity, the first subscript λ being regarded as consecutive with the last subscript $(\mu - 1)/(\lambda \mu - 1)$. Using any two consecutive subscripts as the initial λ, μ, the resulting identity c) is seen to be an immediate consequence of the given identity c). Taking for λ any one of the $p^n - 2$ marks $\neq 0, 1$ and for μ any of the $p^n - 3$ marks $\neq 0, 1, \lambda^{-1}$, the remaining subscripts in c) are different from 0 and 1. Hence those identities c) which do not reduce to D) are equivalent in sets of five, an exception being those with all subscripts equal to λ, where $\lambda^2 + \lambda = 1$. If the latter has σ solutions in the $GF[p^n]$, it follows that there are exactly

$$N \equiv \sigma + \frac{1}{5}\{(p^n - 2)(p^n - 3) - \sigma\}$$

distinct identities c) not immediately reducible to D). For $p = 2$, $\sigma = 0$ or 2 according as n is odd or even; for $p = 5$, $\sigma = 1$; for $p \neq 2, \neq 5$, $\sigma = 0$ or 2 according as $p^n = 5k \pm 2$ or $p^n = 5k \pm 1$.

280. For the group $LF(2, 5)$ of order 60, the $N = 2$ relations c) are

$$(S_2 T)^5 = I, \quad S_2 TS_4 TS_3 TS_3 TS_4 T = I.$$

These may both be derived from a), D) and $T^2 = I$, so that $LF(2, 5)$ is generated by $A \equiv S_1$, $B \equiv T$ subject to the relations

282) $\qquad A^5 = I, \quad B^2 = I, \quad (AB)^3 = I$.

In proof, we apply D) repeatedly and find that

$$(S_2 T)^5 = (S_1 T S_{-1} T S_1 T)^2 S_2 T = S_1 T S_{-3} T S_1 T S_2 T$$
$$= S_1 T S_1 \cdot S_1 T S_1 T S_1 \cdot S_1 T = S_1 T S_1 \cdot T \cdot S_1 T = I.$$

Hence also $(TS_3)^5 = I$, so that the second relation becomes

$$S_2 T S_1 (S_3 T)^4 T S_1 T = S_2 T S_1 T S_2 T S_1 T = S_1 (S_1 T)^2 S_1 (S_1 T)^2 = I.$$

281. The group $LF(2, 2^2)$ of order 60 may be generated by $A = TS_i$ and $B = S_{i^2}$ subject to the relations 282), where i and i^2 are the roots of $x^2 + x \equiv 1 \pmod{2}$. Indeed, the $N = \sigma = 2$ relations c) to be considered in addition to D) are

$$(S_i T)^5 = I, \quad (S_{i^2} T)^5 = I.$$

The latter only serves to define the operator T in terms of A and B:
$$T = BABA^{-1}BABA^{-1}B.$$
The resulting expressions for S_i, S_1, S_{i^2} are seen to be commutative and of period 2, so that relations a) follow from 282).

282. The group $LF(2, 7)$ of order 168 is defined by relations a), D), $T^2 = I$, together with the following $N = 4$ relations
$$S_2 TS_2 TS_5 TS_4 TS_5 T = I, \quad S_3 TS_3 TS_2 TS_6 TS_2 T = I,$$
$$S_4 TS_4 TS_3 TS_6 TS_3 T = I, \quad S_5 TS_5 TS_6 TS_4 TS_6 T = I.$$
Applying a), D) and $T^2 = I$, the second and third relations become
$$S_3 TS_3 TS_3 \cdot S_{-1} TS_6 TS_{-1} \cdot S_3 T = (S_3 T)^4 = I,$$
$$S_4 TS_4 T = TS_4 TS_1 TS_4 = TS_3 \cdot S_1 TS_1 TS_1 \cdot S_3 = TS_3 TS_3.$$
The first relation may be written $S_2 TS_2 TS_1 \cdot S_4 TS_4 TS_4 \cdot S_1 T = I$ or
$$S_4 TS_4 TS_4 = S_{-1} TS_{-2} TS_{-2} TS_{-1} = (S_{-1} TS_{-1})^3 = (TS_1 T)^3 = TS_3 T.$$
The fourth relation becomes an identity if we replace $S_5 TS_5 T$ by $TS_5 TS_4 TS_5$ as derived from the first relation. Hence the G_{168} may be generated by S_1 and T subject only to the generational relations[1])

283) $\quad T^2 = I, \quad S_1^7 = I, \quad (S_1 T)^3 = I, \quad (S_1^4 T)^4 = I.$

Corollary. — *The group $LF(3, 2)$ of order 168 is isomorphic with $LF(2, 7)$.* In fact, the relations 283) are satisfied by the substitutions
$$T = \begin{pmatrix} 1 & 0 & 0 \\ 0 & 1 & 0 \\ 0 & 1 & 1 \end{pmatrix}, \quad S_1 = \begin{pmatrix} 1 & 1 & 1 \\ 1 & 0 & 1 \\ 1 & 0 & 0 \end{pmatrix}.$$

CHAPTER XIV.

GROUP OF THE EQUATION FOR THE 27 STRAIGHT LINES ON A GENERAL SURFACE OF THE THIRD ORDER.[2])

283. A general cubic surface contains 27 straight lines such that[3])

1^0. Any one of the lines A meets ten others which intersect two by two, forming with A five triangles. The total number of such triangles on the cubic surface is $5 \cdot 27/3 = 45$.

---

1) Dyck, *Math. Ann.*, vol. 20, p. 41; Burnside, The Theory of Groups, p. 305.
2) Compare Jordan, Traité, pp. 316—329, 365—369; Dickson, *Comptes Rendus*, vol. 128, pp. 873—875.
3) Steiner, *Crelle*, vol. 53.

304   CHAPTER XIV.

2^0. Any two triangles ABC and $A'B'C'$ having no side in common determine uniquely a third triangle $A''B''C''$ such that the corresponding sides of the three triangles intersect and form three new triangles $AA'A''$, $BB'B''$, $CC'C''$. The former set of three triangles is said to constitute a *trieder*, which will be designated $[ABC, A'B'C', A''B''C'']$.

These two properties completely define the configuration of the 45 triangles formed by the 27 lines on the cubic surface.

Denoting the lines by R_t, R_{sit} ($s = 1, 2$; $i = 1, 2, 3, 4$; $t = 0, 1, 2$), it will be shown that the 45 triangles are given by the notation[1])

$$R_0 R_1 R_2, \quad R_{s10} R_{s11} R_{s12} \qquad [s = 1, 2]$$
$$R_t R_{1it} R_{2it} \qquad [t = 0, 1, 2;\ i = 1, 2, 3, 4]$$
$$R_{s2t} R_{s3\,t\pm 1} R_{s4\,t\mp 1} \qquad [s = 1, 2;\ t \equiv 0, 1, 2 \ (\text{mod } 3)]$$
$$R_{s1t} R_{sj\,t-1} R_{2sj\,t+1} \qquad [s = 1, 2;\ j = 2, 3, 4;\ t \equiv 0, 1, 2]$$

where the subscript $2s$ is to be replaced by 1 when $s = 2$.

Each element R lies in exactly five of these sets. Thus R_t lies in the sets $R_0 R_1 R_2$, $R_t R_{1it} R_{2it}$ ($i = 1, 2, 3, 4$); R_{s1t} lies in the 5 sets

$$R_{s10} R_{s11} R_{s12}, \quad R_t R_{11t} R_{21t}, \quad R_{s1t} R_{sj\,t-1} R_{2sj\,t+1} \quad (j = 2, 3, 4);$$

finally, R_{sjt} lies in the following 5 sets, in the last two of which τ is to be suitably chosen modulo 3:

$$R_t R_{1jt} R_{2jt}, \quad R_{s1\,t+1} R_{sjt} R_{2sj\,t+2}, \quad R_{2s1\,t-1} R_{2sj\,t-2} R_{sjt},$$
$$R_{s2\tau} R_{s3\,\tau\pm 1} R_{s4\,\tau\mp 1}.$$

Hence each element can be associated with exactly ten other elements to determine a set. Property 1^0 thus holds for the 45 sets.

The set $R_0 R_1 R_2$ lies in exactly the following sixteen trieders:

$$[R_0 R_1 R_2, \quad R_{110} R_{111} R_{112}, \quad R_{210} R_{211} R_{212}],$$
$$[R_t R_{t+1} R_{t-1}, \quad R_{12t} R_{13\,t\pm 1} R_{14\,t\mp 1}, \quad R_{22t} R_{23\,t\pm 1} R_{24\,t\mp 1}],$$
$$[R_t R_{t-1} R_{t+1}, \quad R_{11t} R_{1j\,t-1} R_{2j\,t+1}, \quad R_{21t} R_{2j\,t-1} R_{1j\,t+1}],$$

where $j = 2, 3$ or 4, $t \equiv 0, 1$ or 2 (mod 3). Property 2^0 therefore holds for the set $R_0 R_1 R_2$ in conjunction with any set no one of whose elements is R_0, R_1 or R_2. It is next shown that the property holds for an arbitrary pair of sets ABC, $A'B'C'$ which have no element in common. By the next section the 45 sets are merely permuted by the substitutions $[W]$, $[E_1]$, $[E_2]$, $[E_3]$ given in § 274. The latter generate a substitution-group $[O]$ holoedrically isomorphic

---

[1]) The connection with the 27 sets of orthogonal substitutions exhibited in § 272 will be shown in the sequel.

GROUP OF THE EQUATION FOR THE 27 STRAIGHT LINES etc. 305

with the abstract simple group O of § 273. From its origin $[O]$ is transitive and hence contains a substitution S which replaces R_0 by an arbitrary element A. We proceed to prove that $[O]$ contains a substitution S_1 which leaves R_0 fixed and replaces R_1 by an arbitrary one of the ten elements R_1, R_2, R_{1i0}, R_{2i0} ($i = 1, 2, 3, 4$) which lie in sets with R_0. The substitutions $[E_3]$, $[E_3][E_2]$, $[E_3][E_2][E_1]$, $[E_3][E_2][E_1]^2$ replace R_1 by R_{140}, R_{130}, R_{120}, R_{110} respectively, without altering R_0. The transformed of $[E_1]$ by $[W]$ gives the substitution

$$(R_{s11}R_{s31}R_{s21})(R_{s22}R_{s32}R_{2s42})(R_{s20}R_{2s10}R_{s30})$$

which replaces R_{120} by R_{210}, R_{110} by R_{230}. Then $[E_1]$ and $[E_2]$ replace R_{230} by R_{220} and R_{240} respectively. Finally, $[E_3]$ replaces R_{240} by R_2. It follows that $[O]$ contains a substitution $S_1 S$ which replaces the set $R_0 R_1 R_2$ by a set ABC in which A is any one of the 27 elements and B any of the 10 elements which lie in sets with A. Hence $[O]$ contains a substitution Σ replacing the set $R_0 R_1 R_2$ by an arbitrary one of the 45 sets. Then Σ^{-1} replaces the given pair ABC, $A'B'C'$ by a pair $R_0 R_1 R_2$, $A_1 B_1 C_1$ having no elements in common. The latter sets determine a trieder by the earlier proof. Applying to it the substitution Σ, which was derived from $[W]$ and $[E_i]$ and therefore replaces sets by sets, we obtain a trieder containing ABC, $A'B'C'$ and determined by them. Hence the above distribution of the 27 elements R into 45 sets is a suitable notation for the configuration of the 45 triangles formed by the 27 lines on a general cubic surface.

284. The next step is to verify that the substitutions $[W]$, $[E_1]$, $[E_2]$ and $[E_3]$ of § 274 permute amongst themselves the 45 triangles. $[W]$ gives rise to the following *even* substitution:

$$(R_0 R_{1i0} R_{2i0}, \quad R_1 R_{1i1} R_{2i1}, \quad R_2 R_{1i2} R_{2i2})$$
$$(R_{s20} R_{s31} R_{s42}, \quad R_{s21} R_{s32} R_{s40}, \quad R_{s22} R_{s30} R_{s41})$$
$$(R_{s20} R_{s32} R_{s41}, \quad R_{s21} R_{s30} R_{s42}, \quad R_{s22} R_{s31} R_{s40})$$
$$(R_{s10} R_{sj2} R_{2sj1}, \quad R_{s11} R_{sj0} R_{2sj2}, \quad R_{s12} R_{sj1} R_{2sj0})$$

where $i = 1, 2, 3, 4$; $j = 2, 3, 4$; $s = 1, 2$.

$[E_2]$ gives rise to the even substitution on the 45 triangles:

$$(R_0 R_{110} R_{210}, \; R_0 R_{120} R_{220})(R_{s10} R_{s11} R_{s12}, \; R_{s20} R_{s11} R_{2s22})$$
$$(R_0 R_{130} R_{230}, \; R_0 R_{140} R_{240})(R_{s10} R_{s22} R_{2s21}, \; R_{s20} R_{2s12} R_{2s21})$$
$$(R_2 R_{112} R_{212}, \; R_2 R_{122} R_{222})(R_{s20} R_{s31} R_{s42}, \; R_{s10} R_{2s31} R_{s32})$$
$$(R_2 R_{132} R_{232}, \; R_2 R_{142} R_{242})(R_{s20} R_{s32} R_{s41}, \; R_{s10} R_{s42} R_{2s41})$$
$$(R_{s11} R_{s30} R_{2s32}, \; R_{s11} R_{s40} R_{2s42})(R_{s21} R_{s32} R_{s40}, \; R_{s21} R_{s42} R_{s30})$$
$$(R_{s22} R_{s30} R_{s41}, \; R_{2s12} R_{s40} R_{2s41})(R_{s22} R_{s31} R_{s40}, \; R_{2s12} R_{2s31} R_{s30}).$$

Similarly $[E_1]$ and $[E_3]$ give rise to even permutations of the 45 triangles.

306 CHAPTER XIV. GROUP OF THE EQUATION FOR THE 27 STR. LINES etc.

285. Theorem. — *The group G of the equation for the 27 lines on a general cubic surface is of order 51840 and has a subgroup of index 2 holoedrically isomorphic with the abstract group O.*

The group G is formed of the substitutions on the 27 elements R which permute the 45 triangles. These substitutions can replace R_0 by at most 27 elements. Those leaving R_0 fixed can replace R_1 by no element other than the ten lying with R_0 in some triangle; namely, R_1, R_2, R_{1i0}, R_{2i0} ($i = 1, 2, 3, 4$). The substitutions leaving R_0 and R_1 fixed and consequently the triangle $R_0 R_1 R_2$ cannot alter R_2 and must replace R_{130} by one of the 8 elements

$$R_{si0} \qquad (s = 1, 2;\ i = 1, 2, 3, 4)$$

which enter the four remaining triangles containing R_0. The substitutions leaving R_0, R_1, R_{130} fixed cannot alter R_2 or R_{230}, and must permute amongst themselves the triangles which contain R_1 and likewise the triangles which contain R_{130}. Hence they must permute the pairs R_{111}, R_{211}; R_{121}, R_{221}; R_{131}, R_{232}; R_{141}, R_{241}; and likewise permute the pairs R_{111}, R_{232}; R_{212}, R_{231}; R_{122}, R_{141}; R_{121}, R_{142}. Hence the elements R_{111}, R_{121}, R_{231}, R_{141} common to the two sets must be permuted amongst themselves, which can be done in at most 24 ways. Finally, a substitution of G which leaves fixed R_0, R_1, R_2, R_{130}, R_{230}, R_{111}, R_{121}, R_{231} and R_{141} must not alter R_{211}, R_{221}, R_{131}, R_{241}, R_{232}, R_{212}, R_{122} and R_{142} and therefore must leave fixed the third element in each of the triangles $R_{211} R_{212} R_{210}$, $R_{241} R_{232} R_{220}$, $R_{221} R_{122} R_{110}$, $R_{210} R_{121} R_{222}$, $R_{230} R_{131} R_{112}$, $R_{230} R_{221} R_{242}$, $R_{131} R_{142} R_{120}$, $R_{s22} R_{s31} R_{s40}$, and $R_{121} R_{140} R_{132}$. Such a substitution therefore leaves fixed every element and is therefore the identity. The order of G is therefore at most $27 \cdot 10 \cdot 8 \cdot 24 = 51840$.

But G contains the subgroup $[O]$ of order 25920 whose substitutions permute the 45 triangles *evenly*. Also G contains

$$T \equiv \Pi\,(R_{s2t} R_{s4t}) \qquad (s = 1, 2;\ t = 0, 1, 2)$$

which gives rise to the following odd substitution on the triangles:

$$(R_t R_{12t} R_{22t},\ R_t R_{14t} R_{24t})$$
$$(R_{12t} R_{11\,t+1} R_{22\,t-1},\ R_{14t} R_{11\,t+1} R_{24\,t-1})$$
$$(R_{22t} R_{21\,t+1} R_{12\,t-1},\ R_{24t} R_{21\,t+1} R_{14\,t-1})$$
$$(R_{s2t} R_{s3\,t+1} R_{s4\,t-1},\ R_{s2\,t-1} R_{s4t} R_{s3\,t+1})$$

containing $3 + 3 + 3 + 6 = 15$ transpositions. The order of G is therefore at least $2 \cdot 25920$. The order is consequently 51840.

286. Certain subgroups of the abstract group O of order 25920 appear at once by considering the various isomorphic linear groups.

CHAPT. XV. SUMMARY OF THE KNOWN SYSTEMS OF SIMPLE GROUPS. 307

By §§ 118 and 133, the simple group $HA(4, 2^2)$, which is isomorphic with O by § 275, has a complete set of 36 conjugate subgroups $A(4, 2)$ holoedrically isomorphic with the symmetric group on 6 letters. By § 136, $HA(4, 2^2)$ has a complete set of 216 conjugate subgroups $LF(2, 2^2)$, holoedrically isomorphic with the alternating group on 5 letters. By §§ 270—274, O has a subgroup G_{960} of index 27. The quotient-group $A(4, 3)$ of the special Abelian group $SA(4, 3)$ is (§ 189) holoedrically isomorphic with $FO(5, 3)$ and therefore with the abstract group O. By § 114, $SA(4, 3)$ contains $3^3(3^2-1)3$ substitutions which leave ξ_1 fixed, so that $A(4, 3)$ contains a subgroup of index $25920 \div 8 \cdot 3^4 \equiv 40$. By § 121, $SA(4, 3)$ contains exactly $(3^2+1)3^2$ substitutions conjugate with $T_{1,-1}$. But the latter is conjugate with $T_{2,-1}$, the two being identical in the quotient-group $A(4, 3)$. Hence $A(4, 3)$ has a subgroup of index 45. *Hence the simple group O has subgroups of indices* 27, 36, 40, 45, 216. By a lengthy analysis[1]), it has been shown that O contains no subgroup of index < 27. The problem of the determination of the 27 straight lines on a general cubic surface has therefore resolvent equations of degrees 27, 36, 40, 45 but none of degree < 27.

Since O is isomorphic with $A(4, 3)$, our problem is identical with the problem of the trisection of the periods of hyperelliptic functions with four periods.[2])

CHAPTER XV.

SUMMARY OF THE KNOWN SYSTEMS OF SIMPLE GROUPS.

287. In the preceding chapters were derived the following systems of simple groups, with the specified restrictions upon the prime number p and the positive integers m and n[3]):

$$LF(m, p^n): \frac{1}{d}(p^{nm}-1)p^{n(m-1)}(p^{n(m-1)}-1)p^{n(m-2)} \ldots (p^{2n}-1)p^n$$

where $p^n > 3$ if $m = 2$, and d is the greatest common divisor of m and $p^n - 1$.

$$HO(m, p^{2n}): \frac{1}{g}[p^{nm}-(-1)^m]p^{n(m-1)}[p^{n(m-1)}-(-1)^{m-1}]p^{n(m-2)} \ldots$$
$$[p^{2n}-1]p^n$$

where $p^n > 3$ if $m = 2$, $p^n > 2$ if $m = 3$, and g is the greatest common divisor of m and $p^n + 1$.

---
1) Jordan, Traité, pp. 319—329.
2) Jordan, pp. 354—369.
3) The notations were introduced in §§ 108, 119, 148, 194 and end of 209.

308   CHAPTER XV.

$$A(2m, p^n): \frac{1}{a}(p^{n(2m)}-1)p^{n(2m-1)}(p^{n(2m-2)}-1)p^{n(2m-3)}\ldots(p^{2n}-1)p^n$$

where $p^n > 3$ if $m = 1$, $p^n > 2$ if $m = 2$, and $a = 1$ if $p = 2$, $a = 2$ if $p > 2$.

$$FO(2m+1, p^n): \frac{1}{2}(p^{n(2m)}-1)p^{n(2m-1)}(p^{n(2m-2)}-1)p^{n(2m-3)}\ldots\\(p^{2n}-1)p^n$$

where $p > 2$ and, for $m = 1$, $p^n > 3$.

$$FO(2m, p^n): \frac{1}{4}[p^{n(2m-1)}-\varepsilon^m p^{n(m-1)}](p^{n(2m-2)}-1)p^{n(2m-3)}\ldots(p^{2n}-1)p^n$$

where $p > 2$ and $m > 2$, while $\varepsilon = \pm 1$ according as $p^n = 4l \pm 1$.

$$SO(2m, p^n): \frac{1}{2}[p^{n(2m-1)}+\varepsilon^m p^{n(m-1)}](p^{n(2m-2)}-1)p^{n(2m-3)}\ldots(p^{2n}-1)p^n$$

where $p > 2$ and $m > 1$, $\varepsilon = \pm 1$ according as $p^n = 4l \pm 1$.

$$FH(2m, 2^n): (2^{nm}-1)(2^{2n(m-1)}-1)2^{2n(m-1)}(2^{2n(m-2)}-1)2^{2n(m-2)}\ldots\\(2^{2n}-1)2^{2n}$$

where $m > 2$.

$$SH(2m, 2^n): (2^{nm}+1)(2^{2n(m-1)}-1)2^{2n(m-1)}\ldots(2^{2n}-1)2^{2n}, \; m > 1.$$

In addition to these systems may be added the cyclic groups of prime order and the alternating group on $n > 4$ letters.

288. Between certain of the above groups there exists holoedric isomorphism, a relation indicated by the symbol $\sim$. For $p > 2$, the following isomorphisms were established in § 178, §§ 187—190, 197—198:

$FO(3, p^n) \sim LF(2, p^n)$; $FO(6, p^n) \sim LF(4, p^n)$, for $p^n = 4l+1$;
$FO(5, p^n) \sim A(4, p^n)$; $SO(6, p^n) \sim LF(4, p^n)$, for $p^n = 4l+3$;
$FO(6, p^n) \sim HO(4, p^{2n}), p^n = 4l+3$; $SO(6, p^n) \sim HO(4, p^{2n}), p^n = 4l+1$;
$SO(4, p^n) \sim LF(2, p^{2n})$

the latter holding also for $p^n = 3$, a case not treated in §§ 197—198. For any p,
$$LF(2, p^n) \sim A(2, p^n) \sim HO(2, p^{2n}).$$

For $p = 2$, it was shown in §§ 198, 206, 207 that

$$SH(4, 2^n) \sim LF(2, 2^{2n}), \; FH(6, 2^n) \sim LF(4, 2^n), \; SH(6, 2^n) \sim HO(4, 2^{2n}).$$

By chapter XIII,

$$FO(5, 3) \sim HO(4, 2^2), \quad LF(4, 2) \sim G^{(8)}_{\frac{1}{2}81}, \quad LF(3, 2) \sim LF(2, 7).$$

SUMMARY OF THE KNOWN SYSTEMS OF SIMPLE GROUPS. 309

289. Theorem.[1]) — *The simple groups* $A(2m, p^n)$ *and*
$$FO(2m+1, p^n), \quad p > 2,$$
of equal order are not isomorphic if $m > 2$.

The proof consists in showing that the orthogonal group contains a greater number of sets of conjugate operators of period two than the Abelian group. By § 122, $A(2m, p^n)$, $p > 2$, has exactly $\frac{1}{2}(m+2)$ or $\frac{1}{2}(m+1)$ distinct sets of conjugate operators of period two according as m is even or odd. But $FO(2m+1, p^n)$ contains the following m distinct substitutions of period two,

$$C_1 C_2, \quad C_1 C_2 C_3 C_4, \ldots, C_1 C_2 C_3 C_4 \ldots C_{2m-1} C_{2m},$$

having the respective characteristic determinants,

$$(1+K)^2(1-K)^{2m-1}, \quad (1+K)^4(1-K)^{2m-3}, \ldots, (1+K)^{2m}(1-K).$$

By § 102, no two of these m substitutions are conjugate under linear transformation.

For $m = 1$ or for $m = 2$, the corresponding groups are isomorphic (§ 288).

290. The following table gives the 53 known simple groups of composite order less than one million. The alternating group on n letters is designated by its order $\frac{1}{2}n!$. The isomorphisms indicated in § 288 are not given in the table.

60	$LF(2,5) \sim LF(2,2^2) \sim \frac{1}{2}5!$	6072	$LF(2, 23)$
168	$LF(2,7) \sim LF(3,2)$	7800	$LF(2, 5^2)$
360	$LF(2, 3^2) \sim \frac{1}{2}6!$	7920	Group on 9 letters[2])
504	$LF(2, 2^3)$	9828	$LF(2, 3^3)$
660	$LF(2, 11)$	12180	$LF(2, 29)$
1092	$LF(2, 13)$	14880	$LF(2, 31)$
2448	$LF(2, 17)$	20160	$LF(4, 2) \sim \frac{1}{2}8!$
2520	$\frac{1}{2}7!$	20160	$LF(3, 2^2)$
3420	$LF(2, 19)$	25308	$LF(2, 37)$
4080	$LF(2, 2^4)$	25920	$A(4, 3) \sim HO(4, 2^2)$
5616	$LF(3, 3)$	32736	$LF(2, 2^5)$
6048	$HO(3, 3^2)$	34440	$LF(2, 41)$
		39732	$LF(2, 43)$

---

1) The existence of two non-isomorphic groups of order $\frac{1}{2}8!$ was noted in § 238.
2) Cole, *Quart. Journ. of Math.*, vol. 27, p. 48, foot-note.

310 CHAPT. XV. SUMMARY OF THE KNOWN SYSTEMS OF SIMPLE GROUPS.

51 888	$LF(2, 47)$	265 680	$LF(2, 3^4)$
58 800	$LF(2, 7^2)$	285 852	$LF(2, 83)$
62 400	$HO(3, 2^4)$	352 440	$LF(2, 89)$
74 412	$LF(2, 53)$	372 000	$LF(3, 5)$
95 040	Group on 12 letters[1])	443 520	Group on 22 letters[2])
102 660	$LF(2, 59)$	456 288	$LF(2, 97)$
113 460	$LF(2, 61)$	515 100	$LF(2, 101)$
126 000	$HO(3, 5^2)$	546 312	$LF(2, 103)$
150 348	$LF(2, 67)$	612 468	$LF(2, 107)$
178 920	$LF(2, 71)$	647 460	$LF(2, 109)$
181 440	$\frac{1}{2} 9!$	721 392	$LF(2, 113)$
194 472	$LF(2, 73)$	885 720	$LF(2, 11^2)$
246 480	$LF(2, 79)$	976 500	$LF(2, 5^3)$
262 080	$LF(2, 2^6)$	979 200	$A(4, 2^2)$

Aside from the simple groups $LF(2, p^n)$, the known simple groups of composite orders between one million and one billion are the following:

1 451 520	$A(6, 2)$	42 456 960	$LF(3, 3^2)$
1 814 400	$\frac{1}{2} 10!$	42 573 600	$HO(3, 3^4)$
1 876 896	$LF(3, 7)$	70 915 680	$HO(3, 11^2)$
3 265 920	$HO(4, 3^2)$	138 297 600	$A(4, 7)$
4 680 000	$A(4, 5)$	174 182 400	$FH(8, 2)$
5 515 776	$HO(3, 2^6)$	197 406 720	$SH(8, 2)$
5 663 616	$HO(3, 7^2)$	212 427 600	$LF(3, 11)$
6 065 280	$LF(4, 3)$	239 500 800	$\frac{1}{2} 12!$
9 999 360	$LF(5, 2)$	244 823 040	Group on 24 letters[2])
10 200 960	Group on 23 letters[2])	270 178 272	$LF(3, 13)$
13 685 760	$HO(5, 2^2)$	811 273 008	$HO(3, 13^2)$
16 482 816	$LF(3, 2^3)$	987 033 600	$LF(4, 2^2)$
19 958 400	$\frac{1}{2} 11!$		

---

1) Mathieu, *Journal de Mathématiques*, 1861, p. 270; proof of simplicity by Miller.
2) Miller, *Bull. Soc. Math. de France*, vol. 28, p. 266 (1900).

INDEX OF SUBJECTS.

(The numbers refer to pages; g or G denotes group.)

Abelian g, 89, 110, 115, 117, 151, 179, 200, 201, 299, 309.
abstract field, 9, 13.
abstract g, 287, 289, 292, 300.
additive-field, 5.
additive-g, 49, 269.
alternating g, 4 letters, 269.
—— 5 letters, 279, 290.
—— 8 letters, 259, 290.
—— k letters, 289.
basis-system, 49.
Betti-Mathieu g, 64, 67, 69.
canonical form, 221, 237, 244.
characteristic determinant, 80.
—— equation, 222.
class of quantic, 29.
—— residue, 3, 6, 7.
commutative g, 262, 265.
—— substitution, 193, 229.
compound of g, 145.
configuration 27 lines, 303.
congruent, 3.
conjugate, 52, 100, 236.
cubic surface, 303, 306.
cyclic base, 266.
dihedron g, 265.
doubly-transitive, 248, 261.
exercises, 19, 42, 70, 216.
existence of Galois F, 14, 19.
exponent of mark, 11.
—— of function, 19.
factors of composition, 81, 91, 94, 191, 192.
Fermat's theorem, 4, 11.
field, 5.
first hypoabelian, 201, 208.
first orthogonal, 131, 159, 191, 292, 299, 309.
four-group, 267.
Galois Field, 6, 14.
general linear homogeneous g, 69, 75, 77, 124, 146, 147, 236, 290.

group, 65; G_{168}, 303; G_{20160}, 259; G_{25920}, 293, 296; G_{51840}, 306;
see alternating, icosahedral, dihedron, tetrahedral, octahedral, symmetric, linear, general, special, simple.
Hermite's theorem, 59.
homogeneous, see general, special.
hyperabelian g, 115, 183, 209, 298.
hyperelliptic, 307.
hyperorthogonal g, 131, 264.
hypoabelian, see first, second.
icosahedral, 278, 283, 302.
index of subgroup, 286, 307.
infinity (mark), 260.
invariant, quadratic, 144, 153, 156, 191, 194, 197, 206.
—— of degree 2, 126, 218.
irreducible, 10, 15, 44.
isomorphic, 99, 164, 174, 183, 194, 208, 209, 287, 298, 308.
linear independence, 10, 52.
linear fractional g, 87, 126, 132, 164, 174, 179, 193, 194, 208, 242, 259, 260, 286, 302, 303.
mark, 9.
modulus, 3, 6.
multiplier Galois F, 51, 270.
Newton's identities, 53.
non-isomorphic, 260, 309.
not-square, 44, 48.
octahedral g, 269, 282.
order of field, 5, 10.
orthogonal, see first, second.
period of mark, 11.
Pfaffian, 147, 172.
primitive root, 13, 36.
—— irreducible quantic, 21, 35, 44.
quadratic equation, 46.
—, see invariant.
rank, 49.
reduced quantic, 63.

312 INDEX OF SUBJECTS. — ERRATA.

representation of substitutions, 55.
residue, 3, 6.
self-conjugate, 82, 117, 279.
second hypoabelian, 201, 209.
—— orthogonal, 159, 191, 194.
simple g, 87, 97, 100, 120, 138, 152, 191, 212, 260, 286, 307, 309.
special linear homogeneous g, 82, 125, 147, 151, 153, 300.
squares, 44, 48.

substitution-quantic, 55, 63.
surface third order, 303.
symmetric g, 6 letters, 99.
—— k letters, 287.
tetrahedral g, 268, 282.
transformation of indices, 80.
transformed subst., 81, 288.
transitive, 248, 261.
trieder, 304.

ERRATA.

Page 14, line 12, read $GF[p^m]$ for $GF[p^n]$.

„ 17, „ 31, read $y = x^{p^{nm_1}-1}$ for $y = x^{p^{nm_1}}-1$.
„ 20, „ 21, read q_i for qi.
„ 48, „ 2 of § 67, read number of squares.
„ 71, „ 5 of Ex. 6, read $\lambda^{-1}\lambda_1$ for $\lambda^{-1}\lambda$.
„ 78, „ 15, read $B_{r,s,\lambda}$ for $B_{r,s\lambda}$.
„ 93, „ 6, read $\gamma'_{12}, \alpha'_{1m}$; line 2, read M_1 for M_j.
„ 95, „ 30, read α_{21} for a_{21}.
„ 102, „ 17, read $T^{-1}_{m-1,\delta_{1m-1}}$; line 16, read $T_{m,\alpha_{1m}}$.
„ 113, „ 3, read j_1 for j.
„ 132, „ 28 and line 33, read —— for —— ——.
„ 139, „ 8, read $\bar{\alpha}^{p}_{11}{}^{s}{+1} \neq 1$.
„ 152, „ 5, read 139) for 139,.
„ 172, „ 16, p. 175, l. 14, read $G'_{4,2}$ for $G_{4,2}$.
„ 189, „ 3 of § 192, delete comma before "are".
„ 209, „ 1, for hyperabelian read hypoabelian.
„ 221, „ 14, for $\{L\}$ read $\{B\}$.
„ 227, read $Y_{is} \equiv y_s + y'_s K_i + y''_s K_i^2 + \cdots +$
„ 267, line 10, for $G_{\delta_{\mp}}$ read $G_{d_{\mp}}$.
„ 272, „ 3 from bottom, delete "an".
„ 300, „ 16, for S_α^{-1} read $S_{\alpha-1}$.

DOVER PHOENIX EDITIONS

A series of hardcover reprints of major works in mathematics, science and engineering

Mathematics

A Treatise on the Calculus of Finite Differences, George Boole. Reprint of the 2nd and last revised ed. 352pp. 5⅜ x 8½. 49523-X $47.50

A History of Geometrical Methods, Julian Lowell Coolidge. Reprint of the 1940 1st ed. 480pp. 5⅜ x 8½. 49524-8 $55.00

Continuous Groups of Transformations, Luther Pfahler Eisenhart. Reprint of the 1933 1st ed. 320pp. 5⅜ x 8½. 49525-6 $45.00

Transcendental and Algebraic Numbers, A. O. Gelfond. Reprint of the 1960 ed. 208pp. 5⅜ x 8½. 49526-4 $37.50

Mathematical Methods and Theory in Games, Programming, and Economics: Two Volumes Bound as One, Samuel Karlin. Reprint of the 1959 ed. 848pp. 5⅜ x 8½.
49527-2 $90.00

Lectures on the Icosahedron, Felix Klein. Reprint of the 1913 2nd revised ed. 304pp. 5⅜ x 8½. 49528-0 $42.50

The Variational Theory of Geodesics, M. M. Postnikov. Reprint of the 1967 ed. 208pp. 6½ x 9¼. 49529-9 $37.50

Elements of Number Theory, I. M. Vinogradov. Reprint of the 1954 1st ed. 240pp. 5⅜ x 8½. 49530-2 $40.00

Asymptotic Expansions for Ordinary Differential Equations, Wolfgang Wasow. Reprint of the 1976 ed. 384pp. 5⅜ x 8½. 49518-3 $40.00

Lectures on Ordinary Differential Equations, Witold Hurewicz. Reprint of the 1958 ed. 144pp. 5⅜ x 8½. 49510-8 $27.50

The Theory of Branching Processes, Theodore E. Harris. Corrected reprint of the 1963 ed. 256pp. 5⅜ x 8½. 49508-6 $32.50

A Survey of Minimal Surfaces, Robert Osserman. Reprint of the corrected and enlarged 1986 ed. 224pp. 5⅜ x 8½. 49514-0 $32.50

Physics

The Analytical Theory of Heat, Joseph Fourier. Reprint of the 1878 ed. 496pp. 5⅜ x 8½. 49531-0 $57.50

Theoretical Physics, A. S. Kompaneyets. Reprint of the 1961 ed. 592pp. 5⅜ x 8½.
49532-9 $65.00

Quantum Mechanics, H. A. Kramers. Reprint of the 1957 ed. 512pp. 5⅜ x 8½.
49533-7 $60.00

Semiconductor Statistics, J.S. Blakemore. Corrected reprint of the 1962 ed. 400pp. 5⅜ x 8½. 49502-7 $42.50

Treatise on Irreversible and Statistical Thermophysics: An Introduction to Nonclassical Thermodynamics, Wolfgang Yourgrau, Alwyn van der Merwe and Gough Raw. Corrected reprint of the 1966 ed. 288pp. 5⅜ x 8½. 49519-1 $35.00

Thermodynamics of Small Systems, Terrell L. Hill. Corrected reprint of the 1966 ed. 408pp. 6½ x 9¼. 49509-4 $45.00

The Conceptual Foundations of the Statistical Approach in Mechanics, Paul and Tatiana Ehrenfest. Reprint of the 1959 ed. 128pp. 5⅜ x 8½. 49504-3 $25.00